Lecture Notes in Artificial Intelli

Edited by R. Goebel, J. Siekmann, and W. W

Subseries of Lecture Notes in Computer Science

T0238577

Zhi-Hua Zhou   Takashi Washio (Eds.)

Advances in Machine Learning

First Asian Conference
on Machine Learning, ACML 2009
Nanjing, China, November 2-4, 2009
Proceedings

 Springer

Series Editors

Randy Goebel, University of Alberta, Edmonton, Canada
Jörg Siekmann, University of Saarland, Saarbrücken, Germany
Wolfgang Wahlster, DFKI and University of Saarland, Saarbrücken, Germany

Volume Editors

Zhi-Hua Zhou
Nanjing University
National Key Laboratory for Novel Software Technology
Nanjing, China
E-mail: zhouzh@nju.edu.cn

Takashi Washio
Osaka University
The Institute of Scientific and Industrial Research
Osaka, Japan
E-mail: washio@ar.sanken.osaka-u.ac.jp

Library of Congress Control Number: 2009936922

CR Subject Classification (1998): I.2, I.2.6, K.3.1, H.3.3, I.5.3, I.2.11

LNCS Sublibrary: SL 7 – Artificial Intelligence

ISSN   0302-9743
ISBN-10   3-642-05223-1 Springer Berlin Heidelberg New York
ISBN-13   978-3-642-05223-1 Springer Berlin Heidelberg New York

springer.com

© Springer-Verlag Berlin Heidelberg 2009
Printed in Germany

Typesetting: Camera-ready by author, data conversion by Scientific Publishing Services, Chennai, India
Printed on acid-free paper   SPIN: 12774401   06/3180   5 4 3 2 1 0

Preface

The First Asian Conference on Machine Learning (ACML 2009) was held at Nanjing, China during November 2–4, 2009. This was the first edition of a series of annual conferences which aim to provide a leading international forum for researchers in machine learning and related fields to share their new ideas and research findings.

This year we received 113 submissions from 18 countries and regions in Asia, Australasia, Europe and North America. The submissions went through a rigorous double-blind reviewing process. Most submissions received four reviews, a few submissions received five reviews, while only several submissions received three reviews. Each submission was handled by an Area Chair who coordinated discussions among reviewers and made recommendation on the submission. The Program Committee Chairs examined the reviews and meta-reviews to further guarantee the reliability and integrity of the reviewing process. Twenty-nine papers were selected after this process.

To ensure that important revisions required by reviewers were incorporated into the final accepted papers, and to allow submissions which would have potential after a careful revision, this year we launched a *"revision double-check"* process. In short, the above-mentioned 29 papers were *conditionally accepted*, and the authors were requested to incorporate the *"important-and-must"* revisions summarized by area chairs based on reviewers' comments. The revised final version and the revision list of each conditionally accepted paper was examined by the Area Chair and Program Committee Chairs. Papers that failed to pass the examination were finally rejected.

After the reviewing process, 27 papers were accepted for oral presentation at the conference. ACML 2009 was a single-track conference which enabled all the accepted papers to have good visibility for all the audience. These 27 papers are included in this conference proceedings. In addition to the accepted submissions, the conference proceedings also include articles or extended abstracts of the keynote speech given by Thomas Dietterich, and invited talks given by Masashi Sugiyama and Qiang Yang.

ACML 2009 would not have been a success without the support of many people and organizations. We wish to sincerely thank the Area Chairs, Program Committee members and external reviewers for their efforts and engagements in providing a rich and rigorous scientific program for ACML 2009. We wish to express our gratitude to our General Chairs, Hiroshi Motoda and Jian Lu, for their invaluable support and concerns, to the Tutorial Chair, Tu Bao Ho, for selecting and coordinating the fruitful tutorials, to the keynote speaker, Thomas Dietterich, and invited speakers, Masashi Sugiyama and Qiang Yang, for their insightful talks, and to the tutorial speakers, Wray Buntine, Nitesh Chawla and Hang Li. We are also grateful to the Local Arrangements Chairs, Yang Gao and

Ming Li, as well as the Local Organizing Committee, whose great effort ensured the success of the conference.

We greatly appreciate the support from various institutions. The conference was organized by the LAMDA Group of Nanjing University, Nanjing, China. We are also indebted to the National Science Foundation of China (NSFC) and IEEE Computer Society Nanjing Chapter for sponsoring the conference.

Last but not least, we also want to thank all authors and all conference participants for their contribution and support. We hope all participants took this opportunity to share and exchange ideas with one another and enjoyed ACML 2009.

August 2009   Zhi-Hua Zhou
                                                               Takashi Washio

Organization

ACML 2009 Conference Committee

General Chairs

Hiroshi Motoda   AFOSR/AOARD and Osaka University, Japan
Jian Lu   Nanjing University, China

Program Committee Chairs

Zhi-Hua Zhou   Nanjing University, China
Takashi Washio   Osaka University, Japan

Tutorial Chair

Tu Bao Ho   JAIST, Japan

Local Arrangements Chairs

Yang Gao   Nanjing University, China
Ming Li   Nanjing University, China

Publicity Chair

Yuhong Guo   Temple University, USA

Publication Chair

Min-Ling Zhang   Hohai University, China

Registration Chair

Yang Yu   Nanjing University, China

Web Chair

Sheng-Jun Huang   Nanjing University, China

ACML 2009 Program Committee

Chairs

Zhi-Hua Zhou   Nanjing University, China
Takashi Washio   Osaka University, Japan

Area Chairs

Sung-Bae Cho   Yonsei University, Korea
Eibe Frank   University of Waikato, New Zealand
João Gama   University of Porto, Portugal
Rong Jin   Michigan State University, USA
James Kwok   HKUST, Hong Kong, China
Hang Li   Microsoft Research Asia, China
Charles X. Ling   University of Western Ontario, Canada
Huan Liu   Arizona State University, USA
Yuji Matsumoto   NAIST, Japan
Katharina Morik   University of Dortmund, Germany
Bernhard Pfahringer   University of Waikato, New Zealand
Michèle Sebag   CNRS, France
Masashi Sugiyama   Tokyo Institute of Technology, Japan
Kai Ming Ting   Monash University, Australia
Koji Tsuda   AIST, Japan
Limsoon Wong   National University of Singapore, Singapore
Qiang Yang   HKUST, Hong Kong, China
Changshui Zhang   Tsinghua University, China

Members

Shin Ando   Peter Andreae
Annalisa Appice   Marta Arias
Antonio Bahamonde   Kristin Bennett
Gilles Bisson   Hendrik Blockeel
Christian Bockermann   Remco Bouckaert
Wray Buntine   Tru Hoang Cao
Rouveirol Céline   Jonathan Chan
Nitesh Chawla   Sanjay Chawla
Phoebe Chen   Songcan Chen
Xilin Chen   Zheng Chen
Hong Cheng   Jian Cheng
Li Cheng   Xueqi Cheng
Kwok Wai, William Cheung   Mingmin Chi

Masashi Shimbo
Hyunjung (Helen) Shin
Eduardo Spinosa
Einoshin Suzuki
Pang-Ning Tan
Qing Tao
Ryota Tomioka
Vincent S. Tseng
Christel Vrain
Hansheng Wang
Lei Wang
Shijun Wang
Fei Wu
Dong Xu
Takehisa Yairi
Jun Yan
Jieping Ye
Jian Yu
Lei Yu
Filip Zelezny
Bo Zhang
Dell Zhang
Jian Zhang
Kai Zhang
Liqing Zhang
Min-Ling Zhang
Wei Zhong
Xingquan Zhu

Nobuyuki Shimizu
Le Song
Jian-Tao Sun
Ah-Hwee Tan
Dacheng Tao
Alexandre Termier
Ivor Tsang
Takeaki Uno
Fei Wang
Jun Wang
Liwei Wang
Xizhao Wang
Mingrui Wu
Gui-Rong Xue
Kenji Yamanishi
Shuicheng Yan
Jie (Jessie) Yin
Kai Yu
Shipeng Yu
Bernard Zenko
Daoqiang Zhang
Harry Zhang
Junping Zhang
Kun Zhang
Mengjie Zhang
Zhongfei (Mark) Zhang
Yan Zhou
Jean-Daniel Zucker

ACML 2009 External Reviewers

Anna Ciampi
Chuan Sheng Foo
Robert Jäschke
Debora Medeiros
Luis Peña
Hiroshi Sakamoto
Yukihiro Tagami
Zhe Wang
Jing Zhou

Nuno Escudeiro
Yann Guermeur
Yoshitaka Kameya
Claudia Milaré
Lishan Qiao
Kazutaka Shimada
Celine Vens
Jin Yuan
Jinfeng Zhuang

Organized by

Nanjing University

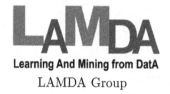

LAMDA Group

Sponsored by

National Science
Foundation of China

IEEE Computer Society
Nanjing Chapter

Table of Contents

Keynote and Invited Talks

Regular Papers

Machine Learning and Ecosystem Informatics: Challenges and Opportunities

Thomas G. Dietterich

Oregon State University, Corvallis OR 97331, USA
tgd@cs.orst.edu
http://web.engr.oregonstate.edu/~tgd

Abstract. Ecosystem Informatics is the study of computational methods for advancing the ecosystem sciences and environmental policy. This talk will discuss the ways in which machine learning—in combination with novel sensors—can help transform the ecosystem sciences from small-scale hypothesis-driven science to global-scale data-driven science. Example challenge problems include optimal sensor placement, modeling errors and biases in data collection, automated recognition of species from acoustic and image data, automated data cleaning, fitting models to data (species distribution models and dynamical system models), and robust optimization of environmental policies. The talk will also discuss the recent development of The Evidence Tree Methodology for complex machine learning applications.

1 Introduction

There are many different ways of conducting scientific research. At one extreme—which we might call "science-in-the-small"—individual scientists formulate hypotheses, perform experiments, gather data, and analyze that data to test and refine their hypotheses. This approach provides profound scientific understanding, but it tends to yield slow progress, because each individual scientist can only study a small collection of phenomena limited in time and space. At the other extreme—which we might call "science-in-the-large"—automated instruments collect massive amounts of observational data, which are then analyzed via machine learning and data mining algorithms to formulate and refine hypotheses. This approach to science can lead to rapid progress, but because it is driven by data rather than hypotheses, it tends not to result in deep causal understanding. Progress can be both fast and deep if we can combine these two approaches to scientific research.

Until the early 1990s, molecular biology was conducted exclusively as science-in-the-small. But the development of automated sequencing methods and algorithms for proteins, DNA sequences, and ultimately whole genomes has permitted the rapid development of science-in-the-large. We are now witnessing a strong and healthy interaction between these two forms of research that is producing rapid progress.

Z.-H. Zhou and T. Washio (Eds.): ACML 2009, LNAI 5828, pp. 1–5, 2009.

Our planet is facing numerous challenges, including global climate change, species extinctions, disease epidemics, and invasive species, that require the development of robust, effective policies based on sound scientific understanding. However, in most cases, the scientific understanding is lacking, because the ecosystem sciences are still in their infancy. Ecology is still dominated by science-in-the-small. Individual investigators collect observational data in the field or conduct controlled experiments in order to refine and test hypotheses. But there is relatively little science-in-the-large. The goal of our research at Oregon State University is to develop novel computer science methods to help promote science-in-the-large in the ecosystem sciences to address critical policy questions.

Fig. 1. The Ecosystem Informatics Pipeline

Figure 1 shows a conceptual model of the information processing pipeline for ecological science-in-the-large. Each box corresponds to a computational problem that requires novel computer science (and often machine learning) research to solve. Let us consider each problem in turn.

- **Sensor Placement.** Many novel sensors are being developed including wireless sensor nodes and fiber-optic-based distributed sensors. The first decision that must be made is where to place these sensors in order to most effectively collect data. This problem can be formulated statically, or it can be considered dynamically, as a form of active learning in which the sensors are moved around (or new sensors are added) based on information collected from previous sensors. Among some of the objective functions that must be considered are (a) maximizing the probability of detecting the phenomenon (e.g., detecting a rare or endangered species), (b) improving model accuracy (as in active learning), (c) improving causal understanding, and (d) improving policy effectiveness and robustness (which is related to exploration in reinforcement learning).

- **Data Collection.** The process of data collection can introduce biases and errors into the data. For example, the Cornell Lab of Ornithology runs a large citizen-science project called ebird (`www.ebird.org`) in which amateur bird watchers can fill out bird watching checklists and upload them to a web site. In this case, humans are the "sensors", and they may introduce several kinds of noise. First, they introduce sampling bias because they tend to go bird watching at locations near their homes. Second, even if a bird is present at a location, they may not detect it because it is not singing and it is hidden in dense foliage. Third, the humans may misidentify the species and report the wrong species to ebird.org. Machine learning algorithms need to have ways of dealing with these problems.

- **Feature Extraction.** It is almost always necessary to transform the raw data to extract higher-level information. For example, image data collected from cameras must be processed to recognize animals and classify them according to species. In some applications, such as counting the number of endangered wolves or bears, it is important to recognize individual animals so that they are not counted multiple times. A related problem is to track individuals as they are detected by multiple instruments over time. At Oregon State, we have been working on identifying the species of arthropods.

- **Data Cleaning.** As large numbers of inexpensive sensors are placed in the environment, the quantity of data increases greatly, but the quality of that data decreases due to sensor failures, networking failures, and other sources of error (e.g., recognition failures in image or acoustic data). This gives rise to the need for automatic methods for data cleaning. This is an important area for machine learning research.

- **Model Fitting.** The computational task of fitting models to data is a core problem in machine learning with many existing algorithms available. However, ecological problems pose several novel challenges. One problem is the simultaneous prediction of the spatio-temporal distribution of thousands of species. A simplified view of species distribution modeling is that it is simple supervised learning, where the goal is to learn $P(y|x)$, where x is a description of a site (elevation, rainfall, temperature, soil, etc.) and y is a boolean variable indicating whether a particular species is present or absent there. However, we are often interested in the presence/absence of thousands of species, and these species are not independent. Species can be positively correlated (e.g., because they have similar environmental requirements or because one of them eats the other) or negatively correlated (e.g., because they compete for the same limited resources). While each species could be treated as a separate boolean classification problem, it should be possible to exploit these correlations to make more accurate predictions. This is a form of very-large-scale multi-task learning. Existing methods are unlikely to scale to handle thousands of species.

Another machine learning challenge is to predict the behavior of species. For example, many bird species are migratory. Ecologists need models that can predict when the birds will migrate (north or south), what paths they will take, and where they will stop. It may be possible to formulate this problem as a form of structured prediction problem.

A third novel machine learning challenge is to fit dynamical systems models to populations of single or multiple species. This involves fitting nonlinear differential equations to observations, which is a problem that has received very little attention in the machine learning community. Such models can exhibit exponential increases and decreases as well as chaotic behavior, so this presents formidable statistical challenges.

Figure 1 shows an arrow from model fitting back to sensor placement. This is intended to capture the opportunity to apply active learning methods to improve sensor placement based on fitted models.

- **Policy Optimization.** In many cases, the models that are fit to data become the inputs to subsequent optimization steps. Consider, for example, the problem of designing ecological reserves to protect the habitat of endangered species. The goal is to spend limited funds to purchase land that will ensure the survival (and even the recovery) of endangered species. An important challenge here is to develop solutions that are robust to the errors that may exist in fitted models. Can we develop ways of coupling optimization with model fitting so that the solutions are robust to the uncertainties introduced throughout the data pipeline?

The arrow leading from policy optimization back to sensor placement suggests one possible solution to this challenge—position more sensors to collect more data to reduce the uncertainties in the fitted models.

2   Summary Remarks

Machine Learning has the potential to transform the ecosystem sciences by enabling science-in-the-large. Although many problems in ecology are superficially similar to previously-studied problems (e.g., active learning, density estimation, model fitting, optimization), existing methods are not directly applicable or do not completely solve the ecological problems.

The keynote talk will describe three instances of this. First, standard methods for generic object recognition do not provide sufficient accuracy for recognizing arthropods. The talk will describe two novel machine learning methods that can solve this problem and that also work well for generic object recognition tasks. Second, standard methods for multi-task learning do not suffice to jointly predict thousands of species. The talk will provide evidence that joint prediction is important, but no known method can currently solve this problem. Third, the talk will describe a spatio-temporal Markov decision problem for managing forests to reduce catastrophic forest fires. In principle, this can be solved by existing dynamic programming algorithms. But in practice, the size of the state and action spaces makes existing algorithms completely infeasible.

I urge everyone to work on these interesting research problems. Given the ecological challenges facing our planet, there is an urgent need to develop the underlying science that can guide policy making and implementation. Ecology is poised to make rapid advances through science-in-the-large. But ecologists can't do this alone. They need computer scientists to accept the challenge and develop the novel computational tools that can make these advances a reality.

Density Ratio Estimation:
A New Versatile Tool for Machine Learning

Masashi Sugiyama

Department of Computer Science, Tokyo Institute of Technology

A new general framework of statistical data processing based on the ratio of probability densities has been proposed recently and gathers a great deal of attention in the machine learning and data mining communities [1,2,3,4,5,6,7,8,9,10,11,12,13,14,15,16,17]. This density ratio framework includes various statistical data processing tasks such as non-stationarity adaptation [18,1,2,4,13], outlier detection [19,20,21,6], and conditional density estimation [22,23,24,15]. Furthermore, mutual information— which plays a central role in information theory [25]—can also be estimated via density ratio estimation. Since mutual information is a measure of statistical independence between random variables [26,27,28], density ratio estimation can be used also for variable selection [29,7,11], dimensionality reduction [30,16], and independent component analysis [31,12].

Accurately estimating the density ratio is a key issue in the density ratio framework. A naive approach is to estimate each density separately and then take the ratio of the estimated densities. However, density estimation is known to be a hard problem and thus this two-step approach may not be accurate in practice. A promising approach would be to estimate the density ratio directly without going through density estimation. Several direct density ratio estimation methods have been proposed so far, including kernel mean matching [3], logistic regression [32,33,5], the Kullback-Leibler importance estimation procedure [8,9], least-squares importance fitting [10,17], and unconstrained least-squares importance fitting [10,17]. Note that the importance refers to the density ratio, derived from *importance sampling* [34]. Furthermore, a density ratio estimation method incorporating dimensionality reduction has also be developed [35], which is shown to be useful in high-dimensional cases. A review of the density ratio framework is available in [36].

The density ratio approach was shown to be useful in various real-world applications including brain-computer interface [4,37], robot control [38,39,40,41,15], speaker identification [42,43], natural language processing [14], bioinformatics [11], visual inspection of precision instruments [44], spam filtering [5], and HIV therapy screening [45].

The author thanks MEXT Grant-in-Aid for Young Scientists (A) 20680007, SCAT, and AOARD for their support.

References

1. Zadrozny, B.: Learning and evaluating classifiers under sample selection bias. In: Proceedings of the Twenty-First International Conference on Machine Learning, pp. 903–910. ACM Press, New York (2004)

Z.-H. Zhou and T. Washio (Eds.): ACML 2009, LNAI 5828, pp. 6–9, 2009.
© Springer-Verlag Berlin Heidelberg 2009

2. Sugiyama, M., Müller, K.R.: Input-dependent estimation of generalization error under covariate shift. Statistics & Decisions 23(4), 249–279 (2005)
3. Huang, J., Smola, A., Gretton, A., Borgwardt, K.M., Schölkopf, B.: Correcting sample selection bias by unlabeled data. In: Schölkopf, B., Platt, J., Hoffman, T. (eds.) Advances in Neural Information Processing Systems 19, pp. 601–608. MIT Press, Cambridge (2007)
4. Sugiyama, M., Krauledat, M., Müller, K.R.: Covariate shift adaptation by importance weighted cross validation. Journal of Machine Learning Research 8, 985–1005 (2007)
5. Bickel, S., Brückner, M., Scheffer, T.: Discriminative learning for differing training and test distributions. In: Proceedings of the 24th International Conference on Machine Learning, pp. 81–88 (2007)
6. Hido, S., Tsuboi, Y., Kashima, H., Sugiyama, M., Kanamori, T.: Inlier-based outlier detection via direct density ratio estimation. In: Giannotti, F., Gunopulos, D., Turini, F., Zaniolo, C., Ramakrishnan, N., Wu, X. (eds.) Proceedings of IEEE International Conference on Data Mining (ICDM 2008), Pisa, Italy, December 15–19, pp. 223–232 (2008)
7. Suzuki, T., Sugiyama, M., Sese, J., Kanamori, T.: Approximating mutual information by maximum likelihood density ratio estimation. In: Saeys, Y., Liu, H., Inza, I., Wehenkel, L., de Peer, Y.V. (eds.) JMLR Workshop and Conference Proceedings. New Challenges for Feature Selection in Data Mining and Knowledge Discovery, vol. 4, pp. 5–20 (2008)
8. Sugiyama, M., Nakajima, S., Kashima, H., von Bünau, P., Kawanabe, M.: Direct importance estimation with model selection and its application to covariate shift adaptation. In: Platt, J.C., Koller, D., Singer, Y., Roweis, S. (eds.) Advances in Neural Information Processing Systems 20, pp. 1433–1440. MIT Press, Cambridge (2008)
9. Sugiyama, M., Suzuki, T., Nakajima, S., Kashima, H., von Bünau, P., Kawanabe, M.: Direct importance estimation for covariate shift adaptation. Annals of the Institute of Statistical Mathematics 60(4), 699–746 (2008)
10. Kanamori, T., Hido, S., Sugiyama, M.: Efficient direct density ratio estimation for non-stationarity adaptation and outlier detection. In: Koller, D., Schuurmans, D., Bengio, Y., Botton, L. (eds.) Advances in Neural Information Processing Systems 21, pp. 809–816. MIT Press, Cambridge (2009)
11. Suzuki, T., Sugiyama, M., Kanamori, T., Sese, J.: Mutual information estimation reveals global associations between stimuli and biological processes. BMC Bioinformatics 10(1), S52 (2009)
12. Suzuki, T., Sugiyama, M.: Estimating squared-loss mutual information for independent component analysis. In: Adali, T., Jutten, C., Romano, J.M.T., Barros, A.K. (eds.) Independeqnt Component Analysis and Signal Separation. LNCS, vol. 5441, pp. 130–137. Springer, Berlin (2009)
13. Quiñonero-Candela, J., Sugiyama, M., Schwaighofer, A., Lawrence, N. (eds.): Dataset Shift in Machine Learning. MIT Press, Cambridge (2009)
14. Tsuboi, Y., Kashima, H., Hido, S., Bickel, S., Sugiyama, M.: Direct density ratio estimation for large-scale covariate shift adaptation. Journal of Information Processing 17, 138–155 (2009)
15. Sugiyama, M., Takeuchi, I., Suzuki, T., Kanamori, T., Hachiya, H.: Least-squares conditional density estimation. Technical Report TR09-0004, Department of Computer Science, Tokyo Institute of Technology (February 2009)
16. Suzuki, T., Sugiyama, M.: Sufficient dimension reduction via squared-loss mutual information estimation. Technical Report TR09-0005, Department of Computer Science, Tokyo Institute of Technology (February 2009)
17. Kanamori, T., Hido, S., Sugiyama, M.: A least-squares approach to direct importance estimation. Journal of Machine Learning Research (to appear, 2009)
18. Shimodaira, H.: Improving predictive inference under covariate shift by weighting the log-likelihood function. Journal of Statistical Planning and Inference 90(2), 227–244 (2000)

19. Breunig, M.M., Kriegel, H.P., Ng, R.T., Sander, J.: LOF: Identifying density-based local out-liers. In: Chen, W., Naughton, J.F., Bernstein, P.A. (eds.) Proceedings of the ACM SIGMOD International Conference on Management of Data (2000)
20. Schölkopf, B., Platt, J.C., Shawe-Taylor, J., Smola, A.J., Williamson, R.C.: Estimating the support of a high-dimensional distribution. Neural Computation 13(7), 1443–1471 (2001)
21. Hodge, V., Austin, J.: A survey of outlier detection methodologies. Artificial Intelligence Review 22(2), 85–126 (2004)
22. Bishop, C.M.: Pattern Recognition and Machine Learning. Springer, New York (2006)
23. Takeuchi, I., Le, Q.V., Sears, T.D., Smola, A.J.: Nonparametric quantile estimation. Journal of Machine Learning Research 7, 1231–1264 (2006)
24. Takeuchi, I., Nomura, K., Kanamori, T.: Nonparametric conditional density estimation using piecewise-linear solution path of kernel quantile regression. Neural Computation 21(2), 533–559 (2009)
25. Cover, T.M., Thomas, J.A.: Elements of Information Theory. John Wiley & Sons, Inc., New York (1991)
26. Kraskov, A., Stögbauer, H., Grassberger, P.: Estimating mutual information. Physical Review E 69(6), 066138 (2004)
27. Hulle, M.M.V.: Edgeworth approximation of multivariate differential entropy. Neural Computation 17(9), 1903–1910 (2005)
28. Suzuki, T., Sugiyama, M., Tanaka, T.: Mutual information approximation via maximum like-lihood estimation of density ratio. In: Proceedings of 2009 IEEE International Symposium on Information Theory (ISIT 2009), Seoul, Korea, June 28–July 3, pp. 463–467 (2009)
29. Guyon, I., Elisseeff, A.: An introduction to variable and feature selection. Journal of Machine Learning Research 3, 1157–1182 (2003)
30. Song, L., Smola, A., Gretton, A., Borgwardt, K.M., Bedo, J.: Supervised feature selection via dependence estimation. In: Proceedings of the 24th International Conference on Machine learning, pp. 823–830. ACM, New York (2007)
31. Hyvärinen, A., Karhunen, J., Oja, E.: Independent Component Analysis. Wiley, New York (2001)
32. Qin, J.: Inferences for case-control and semiparametric two-sample density ratio models. Biometrika 85(3), 619–639 (1998)
33. Cheng, K.F., Chu, C.K.: Semiparametric density estimation under a two-sample density ratio model. Bernoulli 10(4), 583–604 (2004)
34. Fishman, G.S.: Monte Carlo: Concepts, Algorithms, and Applications. Springer, Berlin (1996)
35. Sugiyama, M., Kawanabe, M., Chui, P.L.: Dimensionality reduction for density ratio estimation in high-dimensional spaces. Neural Networks (to appear)
36. Sugiyama, M., Kanamori, T., Suzuki, T., Hido, S., Sese, J., Takeuchi, I., Wang, L.: A density-ratio framework for statistical data processing. IPSJ Transactions on Computer Vision and Applications (to appear, 2009)
37. Li, Y., Koike, Y., Sugiyama, M.: A framework of adaptive brain computer interfaces. In: Proceedings of the 2nd International Conference on BioMedical Engineering and Informatics (BMEI 2009), Tianjin, China, October 17–19 (to appear, 2009)
38. Hachiya, H., Akiyama, T., Sugiyama, M., Peters, J.: Adaptive importance sampling with automatic model selection in value function approximation. In: Proceedings of the Twenty-Third AAAI Conference on Artificial Intelligence (AAAI 2008), Chicago, Illinois, USA, pp. 1351–1356. The AAAI Press, Menlo Park (2008)
39. Hachiya, H., Akiyama, T., Sugiyama, M., Peters, J.: Adaptive importance sampling for value function approximation in off-policy reinforcement learning. Neural Networks (to appear, 2009)

40. Akiyama, T., Hachiya, H., Sugiyama, M.: Active policy iteration: Efficient exploration through active learning for value function approximation in reinforcement learning. In: Proceedings of the Twenty-first International Joint Conference on Artificial Intelligence (IJCAI 2009), Pasadena, California, USA, July 11–17 (to appear, 2009)
41. Hachiya, H., Peters, J., Sugiyama, M.: Efficient sample reuse in EM-based policy search. In: Machine Learning and Knowledge Discovery in Databases. LNCS. Springer, Berlin (to appear, 2009)
42. Yamada, M., Sugiyama, M., Matsui, T.: Covariate shift adaptation for semi-supervised speaker identification. In: Proceedings of 2009 IEEE International Conference on Acoustics, Speech, and Signal Processing (ICASSP 2009), Taipei, Taiwan, April 19–24, pp. 1661–1664 (2009)
43. Yamada, M., Sugiyama, M., Matsui, T.: Semi-supervised speaker identification under covariate shift. Signal Processing (to appear, 2009)
44. Takimoto, M., Matsugu, M., Sugiyama, M.: Visual inspection of precision instruments by least-squares outlier detection. In: Proceedings of The Fourth International Workshop on Data-Mining and Statistical Science (DMSS 2009), Kyoto, Japan, July 7–8, pp. 22–26 (2009)
45. Bickel, S., Bogojeska, J., Lengauer, T., Scheffer, T.: Multi-task learning for HIV therapy screening. In: McCallum, A., Roweis, S. (eds.) Proceedings of 25th Annual International Conference on Machine Learning (ICML2008), Helsinki, Finland, July 5–9, pp. 56–63. Omnipress (2008)

Transfer Learning beyond Text Classification

Qiang Yang

Department of Computer Science and Engineering
Hong Kong University of Science and Technology, Hong Kong
qyang@cse.ust.hk
http://www.cse.ust.hk/~qyang

Abstract. Transfer learning is a new machine learning and data mining framework that allows the training and test data to come from different distributions or feature spaces. We can find many novel applications of machine learning and data mining where transfer learning is necessary. While much has been done in transfer learning in text classification and reinforcement learning, there has been a lack of documented success stories of novel applications of transfer learning in other areas. In this invited article, I will argue that transfer learning is in fact quite ubiquitous in many real world applications. In this article, I will illustrate this point through an overview of a broad spectrum of applications of transfer learning that range from collaborative filtering to sensor based location estimation and logical action model learning for AI planning. I will also discuss some potential future directions of transfer learning.

1 Introduction

Supervised machine learning and data mining techniques have been widely studied and applied. However, a major assumption in applying these techniques is that the training and test data must come from the same distribution and feature space. When this assumption is violated, we may often find the performance of learning to reduce dramatically. In the worst case, learning would become impossible, as in the case when the training data come in the form of English text, and the test data are the Chinese documents. Relabeling the data, however, is time consuming and expensive, because it incurs much human effort.

In response to this problem, there is a recent effort on the study of transfer learning, to allow useful knowledge to be transferred from one or more domains into a domain of interest, or a target domain, where the training and test domains can be different either in data distributions, or in feature representations. With this motivation, transfer learning has been studied in various contexts [1,2,3]. For example, [4] provided a theoretical justification for transfer learning through multi-task learning. Daumé and Marcu [5] investigated how to train a general model with data from both a source domain (known as in-domain data) and a target domain (out-of-domain data), to train a statistical classifier for a natural language mention-type classification task. [6] applied redundant copies of features to facilitate the transfer of knowledge.

Despite much effort in transfer learning, many of the applications have been mainly confined to the area of text classification [7,8,9,10,11], or in the area of reinforcement

Z.-H. Zhou and T. Washio (Eds.): ACML 2009, LNAI 5828, pp. 10–22, 2009.

learning [12]. In this invited article, I will present an overview of transfer learning for work beyond document classification and reinforcement learning. I will present three applications in different application domains, such as object tracking in wireless sensor networks, collaborative filtering, and logical modeling of actions in AI planning. I will also explore some potential future works of transfer learning.

2   An Overview of Transfer Learning

Transfer learning aims at transferring knowledge from source tasks to a target task, where the training data from source domains and the test data from a target domain may follow different distributions or are represented by different features [3]. In [13], we provide a comprehensive survey of transfer learning. Informally, transfer learning can be defined based on the notion of a domain and a task. A domain consists of a feature space in which to describe the attributes of the problem, and a marginal probability distribution of the data. In this definition, a domain does not involve the label space. Instead, the label space is part of a task; that is, a learning task consists of a label space and a mapping function to be learned that maps from the problem features to the labels. Using these concepts, a transfer learning problem can be defined as follows. Given a source domain and a source learning task T_S, a target domain and a target learning task T_T, *transfer learning* aims to improve the performance of learning of a target function in the target domain using the knowledge in the source domain and the source task T_S. What distinguishes transfer learning from other traditional learning is that either the source and target domains, or the target and source tasks, or both, are different.

Many applications of transfer learning have been explored recently. Raina *et al.* [14] and Dai *et al.* [7,8] proposed to use transfer learning techniques to mine text data across domains, respectively. Blitzer *et al.* [9] used structural-correspondence learning (SCL) for solving natural language processing tasks. An extension of SCL was proposed in [10] for solving sentiment classification problems. Wu and Dietterich [15] proposed to use both inadequate target domain data and plenty of low quality source domain data for image classification problems. Arnold *et al.* [16] proposed to use *transductive transfer learning* methods to solve name entity recognition problems. In [17], a novel Bayesian multiple instance learning algorithm is proposed, which can automatically identify the relevant feature subset and use inductive transfer for learning multiple (conceptually related) classifiers in a computer aided design (CAD) domain. In [18], Ling *et al.* proposed an information-theoretic approach for transfer learning, to address the cross-language classification problem from English to Chinese.

It should be emphasized that one misconception for transfer learning is that it is the same as concept drift in database and data mining research. In fact, transfer learning is very different from *Concept Drift*. As defined in [19], concept drift assumes that the target concept as well as its data distribution change over time [20,21]. Approaches for solving the concept drift problem include the so-called *aging techniques* and *windowing techniques*, whereby a fixed or variable-sized window is used to scan the data. Works on instance selection [22], instance weighting [20] or ensemble learning [23,24,25] have focused on how to make the best selection of the source instances for learning a model that can be applied to target data. Instance weighting approaches assume that a different instance has a different impact on a task. Thus, these approaches process higher weighted instances first, and use these weighted instances to build a classifier for prediction. Ensemble learning approaches train several classifiers based on different sub-sets of stream data, and use voting strategies to make a final decision. A major difference between transfer learning and concept drift research is that in transfer learning, data can come from different domains, and the difference may not be associated with time. For example, the transfer of knowledge can be between English and Chinese Web pages, or between playing chess and business negotiation. However, concept drift is tied to time-varying domains. Another major difference is that in transductive transfer learning, the learner has the entire training and test set to learn from, whereas in concept drift only a small number of current data is available.

For more information about transfer learning, the reader can refer to the survey, which will be updated online regularly.[1]

3   Selected Applications of Transfer Learning

In this section, we present a number of interesting applications of transfer learning in various application scenarios. For each application, we describe the main idea, but refer the readers to their respective detailed publications. Specifically, in this section, we will give an overview of the following applications:

WiFi-based Localization [26,27]: transfer learning is used to reduce the calibration effort of learning a model for telling where a client device is in a wireless network.

Cross-language classification [28]: transfer learning is help classify a collection of documents in one language using the labeled documents in another language.

Knowledge engineering for AI planning [29]: transfer learning is exploited for learning action models for planning from a source domain and from target-domain plan traces.

Transfer learning for collaborative filtering [30]: transfer learning is applied to help solve the sparsity and cold-start problems for rating prediction in collaborative filtering.

3.1   Transfer Learning in WiFi Localization Applications

In this section, we present a transfer learning application in wireless sensor networks. Accurately locating a mobile device in an indoor or outdoor environment is an important task in many AI and ubiquitous computing applications. While GPS is widely used, in many places outdoor where high buildings can block signals, and in most indoor places, GPS cannot work well.

To understand the nature of the problem, let us review machine-learning-based WiFi localization models, which work in two phases. In an *offline* phase, a mobile device (or multiple ones) moving around the wireless environment is used to collect wireless signals from various access points (APs). Then, the received signal strength (RSS) values (*e.g.* the signal vector of $(-30dBm, -50dBm, -70dBm)$ as shown in the figure), together with the physical location information (*i.e.* the location coordinates where the

---

mobile device is), are used as the training data to learn a statistical localization model. In an *online* phase, the learned localization model is used to estimate the locations of a mobile device according to the real-time RSS values.

A machine-learning-based localization model assumes that the learning happens in two phases. In the offline phase, a lot of labeled training data are available. In the online phase, the localization model learned in the offline phase can be used to accurately locate mobile device online without any adaptation. However, these assumptions may not hold in many real-world WiFi-based localization problem for several reasons. First, the data distribution may be a function of time and space, making it expensive to collect the training data at all locations in a large building. Also, the data distribution may be a function of the client devices, making the model trained for one type of device (say Cisco) to be invalid when applied to another device (say Intel).

To illustrate how transfer learning helps alleviate the re-calibration problem, we will consider transfer learning across *devices* for a two-dimensional WiFi indoor localization problem. We denote the WiFi signal data collected by a *device* **A** as D^a and denote WiFi signal data collected by another device **B** as D^b. We assume D^a to be fully labeled whereas D^b to have only a few labeled examples and some unlabeled ones that can be easily obtained by quickly walking through the environment. We collected a lot of labeled data by the device **A** while only collected a few labeled data by the device **B**. Note that, although the devices may be different from each other, the learning tasks on these devices are related since they all try to learn a mapping function from a signal space to a *same* location space. This motivates us to model the multi-device localization problem as a multi-task learning problem [3].

Many existing multi-task learning methods assume that the hypotheses learned from the original feature space for related tasks can be similar. This potentially requires the data distributions for related tasks to be similar in the high-dimensional feature space. However, in our multi-device localization problem, the data distributions are expected to be quite different from each other. Therefore, we extend the multi-task learning for multi-device localization problem by only requiring that the hypotheses learned from a *latent* feature space are similar. In other words, we look for appropriate feature mappings, by which we can map different devices' data to a well-defined low-dimensional feature space. In this latent space, new devices can benefit from integrating the data collected before on other devices to train a localization model. Basically, our algorithm also combines both *feature representation-transfer* and *parameter-transfer* for transfer learning.

Given a *source* device, by which we have collected a large amount of labeled signal data $D_{src} = \{(\mathbf{x}_{src}^{(i)}, y_{src,j}^{(i)})\}$, $i = 1, ..., n_s, j = 1, 2$ for 2-D coordinates. Here $\mathbf{x}_{src}^{(i)}$ is a signal vector and $y_{src,j}^{(i)}$ is a corresponding location coordinate. Our objective is to use D_{src} to help calibrate a *target* device, by which we will only collect a small amount of labeled signal data $D_{tar}^l = \{\mathbf{x}_{tar}^{(i)}, y_{tar,j}^{(i)}\}$, $i = 1, ..., n_t^l, j = 1, 2$. We can then formulate our latent multi-task learning solution under a soft-margin support vector regression (SVR) framework. A difficulty is that the overall optimization problem is not convex. However, by separately considering the parameters in the objective function, we can reduce the objective function to a convex optimization problem. For details, please consult [27].

To test the above algorithms, we used the benchmark data from the first IEEE ICDM Data Mining Contest (IEEE ICDM DMC'07)[31], which consists of collecting the data sets in an academic building in the Hong Kong University of Science and Technology. Empirical results show that for transfer learning across both space and devices, our algorithms give superior performance as compared to traditional learning algorithms [26,27].

3.2 Transfer Learning between Different Natural Languages

An important problem in Web mining is Web document classification. In this area, it is widely known that the classification quality depends on the availability of the labeled data; the more labeled data one has, the better the performance. In the English language, the work on labeling documents have started much earlier. For example, the Open Directory Project and Wikipedia provide thousands of labeled documents with relatively high quality. However, in some other languages, such as Chinese, there is a shortage of labeled data, especially in some newer topics. In such cases, it is helpful to be able to make use of the labeled data in English to help classify the documents in Chinese.

To address this problem, a naive idea is to first translate the Chinese pages to English, or English pages to Chinese, and then build a classifier based on the training data in the same feature space as the test data. However, when taken into account the language and topical differences between cultures, we notice that the naive translation based method introduces much noise, so that the performance of learning is quite low.

A better approach, which is described in [28] is to filter out the noisy data using an *information bottleneck* idea [32]. In this approach, training and *roughly translated* target texts (from Chinese to English, for example) are encoded together, allowing all the information to be put through a *bottleneck*. The bottleneck is represented by a limited number of *codewords* that are the category labels in the classification problem (we assume that the label spaces are the same in the source and target tasks). This information bottleneck based approach can maintain most of the common information and disregard the irrelevant information. Thus we can approximate the ideal situation where similar training and translated test pages, which is the common part, are encoded into the same codewords.

This approach, as described in detail in [28], is an instance of *cross-language learning* as well, besides the instance-based transfer learning mentioned in last subsection. In the cross-language learning area, [33] investigated two scenarios in cross-language classification between English and Spanish. [34] considered the English-Chinese topic-aspect classification problem by translating Chinese data into English and applying a k-NN classifier. [35] used an EM algorithm for a English-Italian cross-language classification task.

Let X_e be a set of the Web pages in English with class labels and X_c be the set of the Web pages in Chinese without labels. It is assumed that the English Web pages X_e and Chinese Web pages X_c share some common information. Our objective is to classify the pages in X_c into C, the predefined class label set, which is the same for the pages in X_e.

In our approach, we first *unify* the data in X_e and X_c into a single feature space, which we denote as Y. The most common way is to translate the pages in one language into the other one; in [28], the test Chinese Web page set is translated to English. Let the translated text is denoted as X_c^T and let $X = X_e \cup X_c^T$. We use $\tilde{X} = \{\tilde{x}\}$ to denote a categorization of X for the hypothesis h, where $\tilde{x} = \{x'|h(x') = h(x)\}$. Clearly, $|\tilde{X}|$ is equal to $|C|$, since h maps the instances in X to the class-labels in C. The joint probability distribution of X and Y under the categorization $\tilde{X}$ is denoted by $\tilde{p}(X, Y)$, where

$$\tilde{p}(x, y) = p(\tilde{x}, y)p(x|\tilde{x}) = p(\tilde{x}, y)\frac{p(x)}{p(\tilde{x})} \tag{1}$$

where $x \in \tilde{x}$, and $p(x|\tilde{x}) = \frac{p(x)}{p(\tilde{x})}$ since x totally depends on $\tilde{x}$.

Based on the above formalism, we can define the objective function as

$$\min I(X; Y) - I(\tilde{X}; Y), \tag{2}$$

which can be derived from the objective function as described in [32]. Minimization of the function guarantees the effectiveness of the encoding.

In [28], we show how to transform the objective function in Equation (2) into another representation by KL-divergence [36], as shown below:

$$I(X; Y) - I(\tilde{X}; Y) = D(p(X, Y) \parallel \tilde{p}(X, Y)) \tag{3}$$

Our objective becomes one of optimizing the KL-divergence between $p(X, Y)$ and $\tilde{p}(X, Y)$. In [28], we describe an algorithm to iteratively optimize the difference term $D(p(Y|x) \parallel \tilde{p}(Y|\tilde{x}))$ for each instance x, and show that the objective function will decrease monotonically.

To test our algorithm, we have conducted our evaluation on 5 binary, one 3-way, one 4-way and one 5-way classification tasks constructed from 14 top level categories in the Open Directory Project (ODP). Detailed information about these data sets can be found in [28]. Three baseline methods, Naive Bayes classifiers (NBC), support vector machines (SVM) and transductive support vector machines (TSVM), were introduced for comparison purpose. Note that the supervised learning baseline methods, NBC and SVM, are trained on X_e and tested on X_c^T. The semi-supervised learning method TSVM takes X_e and unlabeled X_c^T for training and X_c^T for testing. To evaluate the performance of IB and other baseline methods, we adapted the precision, recall and F_1-measure as the evaluation criterions.

Table 1 and Table 2 present the performance on each data set given by NBC, SVM, TSVM and our algorithm IB. From the table, we can see that IB significantly outperforms the other three methods. Although SVM and TSVM is slightly better than IB on the **Arts vs Computers data set**, IB is still comparable. But, on some of the other data sets, e.g. **Computers vs Sports and Reference vs Shopping**, both SVM and TSVM fail, while IB is much better than the two discriminative methods. In addition, NBC is always worse than IB, but never fails a lot. On average, IB gives the best performance in all the three evaluation metrics.

3.3   Transfer Learning for Acquiring Action Models in Planning

In this section, we consider another example for illustrating applications of transfer learning is in the domain of knowledge engineering in AI planning. In this area, an

Table 1. The precision, recall and F_1-measure for each classifier on each binary classification data set. These results are quoted from [28].

Data Set	Precision				Recall				F_1-Measure			
	NBC	SVM	TSVM	IB	NBC	SVM	TSVM	IB	NBC	SVM	TSVM	IB
Games vs News	0.749	0.737	0.747	0.798	0.747	0.739	0.779	0.817	0.748	0.738	0.762	0.807
Arts vs Computers	0.731	0.783	0.768	0.767	0.728	0.801	0.785	0.782	0.730	0.792	0.776	0.774
Recreation vs Science	0.836	0.874	0.883	0.903	0.832	0.877	0.891	0.906	0.834	0.876	0.887	0.905

Table 2. The precision, recall and F_1-measure for each classifier on each multiple classification data set. Note that, SVM and TSVM are not included here, because they are designed for binary classification. These results are quoted from [28].

Data Set	Precision		Recall		F_1-Measure	
	NBC	IB	NBC	IB	NBC	IB
3Categories	0.647	0.661	0.630	0.665	0.638	0.663
4Categories	0.445	0.592	0.570	0.648	0.500	0.619
5Categories	0.550	0.608	0.488	0.582	0.517	0.627
Average	0.547	0.620	0.563	0.632	0.552	0.636

important problem is encode user knowledge about actions in the form of precondition-postcondition rules (details of the method can be found in [29]). These rules, or actions, can be used by automated reasoning systems such as theorem provers or planning systems to generate new plans. Traditionally, this is done by humans. However, in our research, we try to alleviate the effort of human experts, and explore how actions can be learned automatically.

Currently, algorithms to learn action models in AI planning are mostly based on the assumption that, the training data in the target domain is largely sufficient in order to learn high-quality action models. In real world applications, however, it is difficult, or very expensive, to attain such large training data. However, we observe that many parts of different planning domains are in fact similar, providing an opportunity for us to reuse previously encoded action models in our new domain. Thus, it is interesting to see how *transfer knowledge* can be done from one or more existing planning domains to help learn the action models in a new target domain.

We have developed a series of transfer learning algorithms to help learn action models in a new domain. In this section, we summarize our action-model learning algorithm that transfers from a source domain to a target domain. We call the algorithm *t*-LAMP, which stands for *transfer Learning Action Models from Plan traces*. As an example, we may have a domain *elevator*[2] where action models are already encoded. In this domain, an action is the action 'up(?f1, ?f2)' which means the elevator can go up from a floor 'f1' to a floor 'f2'. This action has a precondition 'lift-at(?f1)' and an effect 'lift-at(?f2)'. Now, suppose we wish to learn the model of an action 'move(?l1,?2)' in a new *briefcase*[1] domain, where a case is moved from location 'l1' to location 'l2'. In this new domain we have a precondition 'is-at(?l1)' and an effect 'is-at(?l2)'. Because of the similarity between the two domains, we may be able to discover that these two actions share the common knowledge on the actions that can cause changes in locations. Thus,

---

[2] http://www.cs.toronto.edu/aips2000/

to learn one action model, transferring the knowledge from the other action model is likely to be helpful.

The basic idea of t-LAMP is to use the source domain actions as a bias on the probability prior for learning the models in the target domain. The influence from similar actions in the source domain present themselves in the form of larger weighted formulae in the learning process. Eventually, when learning stops, the higher-weighted target domain formulas are selected to be the action models.

The t-LAMP algorithm can be described in four steps. In the first step, we encode the plan traces into propositional formulae. Then, in the second step, we generate a collection of logical formulae according to typical constraints that define a correct plan for the plan traces. These generated formulae are taken as the input of a Markov Logic Network (MLN) [37], which denoted as M. A Markov logic network is a collection of first order logic formulas that are associated with weights. Together they form a Markov network, where each formula is a clique with a potential function taking on binary value. With the MLN, we can incorporate the training plan traces and set the learning component to run to complete. This will give a set of final weights, which are indicative of the set of formulas that should be used for the target action model. Thus, in the third step, we merge the action models from the source domain into another MLN, denoted as M^*, that include the traces in the target domain. In the learning process, weights are then used to help transfer the knowledge from M^* to M. Finally, we can learn the most likely subset of candidate formulae in M, which can then be converted to the final formulae for the target domain action models.

Details of the t-LAMP algorithm can be found in [29], and here we will summarize our experimental results. We collect plan traces from *briefcase* and *elevator* domains . Traces are generated by generating plans from the given initial and goal states in these planning domains using the human encoded action models and a planning algorithm, FF planner [3]. These domains have the characteristics we need to evaluate our t-LAMP algorithm: they have enough similarities and hence we have the intuition that one can borrow knowledge from the other while learning the action models (as shown in the examples of earlier sections). Error rates are defined as the differences between our learned action models and the hand-written action models that are considered as the "ground truth" from IPC-2. The error rate of an action model is defined as $R(a) = \frac{1}{2}(E(pre)/T(pre) + E(eff)/T(eff))$, where $E(.)$ denotes the number of errors, and $T(.)$ denote the number of true preconditions or effects. In this way, we can measure the performance of our learning algorithms clearly.

The evaluation results of t-LAMP in the two domains 'briefcase' and 'elevator' are described in detail in [29]. The results show that generally the threshold can be neither too large nor too small, but the performance is not very sensitive to the value.

In general, we have found that learning from one source domain, that is similar to the target domain, would improve the learning performance. However, how to pick a similar domain is a difficult matter. In [29], we consider how to measure the similarity between two planning domains. Once we have this measure, we can then transfer from multiple domains, where a certain amount of knowledge on action models can be transferred and

---

[3] http://members.deri.at/ joergh/ff.html

integrated to result in the final action models in a target domain. We will consider this issue in our future work.

3.4 Transfer Learning for Collaborative Filtering

Collaborative filtering (CF) is widely used in many real-world recommender systems. A particularly serious problem is that users of CF system can only rate a very limited number of items. As a result, the rating matrices are usually too sparse to model well. This is particularly true for a new rating matrix in a new domain. As an example, when we open a new online service, the rating matrices are often very sparse, which gives rise to the so-called *cold-start* problem.

To alleviate such sparsity and cold-start problems, we proposed a new approach of transfer learning for CF, in order to pool together the rating knowledge from multiple rating matrices in related but different domains. Here we give an overview of our method, which was presented in detail in [30]. In the real world, many web sites for recommending similar items, e.g., movies, books, and music, are closely related. For example, users who visit a online bookstore can be partitioned into similar groups as users who visit an online movie store. Similarly, books can be partitioned into groups just as movies can. In other words, items share some common properties (e.g., genre and style) and users share some of the same population-wide properties as well. If we can uncover their intrinsic relationship at the group, or cluster, levels, we can then transfer the rating knowledge from a dense rating matrix to a sparse rating matrix by discovering what is common among multiple rating matrices in related domains to share useful knowledge.

More specifically, suppose that we have Z rating matrices in related domains. In the z-th rating matrix, a set of users, $U_z = \{u_1^{(z)}, \ldots, u_{n_z}^{(z)}\} \subset \mathcal{U}$, make ratings on a set of items, $V_z = \{v_1^{(z)}, \ldots, v_{m_z}^{(z)}\} \subset \mathcal{V}$. We assume $\bigcap_z U_z = \emptyset$ and $\bigcap_z V_z = \emptyset$. The rating data in the z-th rating matrix is a set of triplets $D_z = \{(u_1^{(z)}, v_1^{(z)}, r_1^{(z)}), \ldots, (u_{s_z}^{(z)}, v_{s_z}^{(z)}, r_{s_z}^{(z)})\}$. The ratings in $\{D_1, \ldots, D_Z\}$ should be in the same rating scales R (e.g., $1-5$).

We wish to learn a rating-matrix generative model (RMGM) for all related tasks on $\bigcup_z D_z$. According to the model, the z-th rating matrix can be viewed as drawing U_z and V_z from the learned RMGM. The missing values in the z-th rating matrix can be generated by the RMGM. To uncover the implicit correspondence in users and items, we can construct a *cluster-level rating matrix* as a "bridge" to connect all the related rating matrices. By permuting (co-clustering) the rows and columns in each rating matrix, we can obtain three block-level rating matrices. We can further reduce the block matrices to the cluster-level rating matrices, in which each row corresponds to a *user cluster* and each column corresponds to an *item cluster*. The entries in the cluster-level rating matrices can be the average ratings of the corresponding co-clusters.

The user and item marginal distributions can be expressed in mixture models, $P_{\mathcal{U}}(u) = \sum_k P(c_{\mathcal{U}}^{(k)}) P(u|c_{\mathcal{U}}^{(k)})$ and $P_{\mathcal{V}}(v) = \sum_l P(c_{\mathcal{V}}^{(l)}) P(v|c_{\mathcal{V}}^{(l)})$, in which $P(.)$ denotes the probability, and each component corresponds to a latent user/item cluster. Then, the users and items can be drawn from the user and item mixture models, respectively.

$$\left(u_i^{(z)}, v_i^{(z)}\right) \sim \sum_{k,l} P(c_{\mathcal{U}}^{(k)}) P(c_{\mathcal{V}}^{(l)}) P(u|c_{\mathcal{U}}^{(k)}) P(v|c_{\mathcal{V}}^{(l)}). \tag{4}$$

Eq. (4) defines the *user-item joint mixture model*. Besides, the ratings can be drawn from the conditional distributions given the latent cluster variables

$$r_i^{(z)} \sim P(r|c_{\mathcal{U}}^{(k)}, c_{\mathcal{V}}^{(l)}). \tag{5}$$

Eq. (5) defines the *cluster-level rating model*.

Combining (4) and (5), we can obtain the *rating-matrix generative model* (RMGM) [30], which can generate rating matrices. Each rating matrix can thus be viewed as drawing a set of users U_z and items V_z from the user-item joint mixture model as well as drawing the corresponding ratings for (U_z, V_z) from the cluster-level rating model.

To enable the learning to happen, we can adopt the Expectation Maximization (EM) algorithm for RMGM training on $\bigcup_z D_z$. We need to learn five sets of model parameters in (4) and (5), i.e., $P(c_{\mathcal{U}}^{(k)})$, $P(c_{\mathcal{V}}^{(l)})$, $P(u|c_{\mathcal{U}}^{(k)})$, $P(v|c_{\mathcal{V}}^{(l)})$, and $P(r|c_{\mathcal{U}}^{(k)}, c_{\mathcal{V}}^{(l)})$, In the E-step, the joint posterior probability $P(c_{\mathcal{U}}^{(k)}, c_{\mathcal{V}}^{(l)}|u_i^{(z)}, v_i^{(z)}, r_i^{(z)})$ can be computed using the above five sets of model parameters. In the M-step, the five sets of model parameters for Z given tasks are updated. By alternating E-step and M-step, an RMGM can be obtained.

After training the RMGM, the missing values in the K given rating matrices can then be generated by

$$f_R(u_i^{(z)}, v_i^{(z)}) = \sum_r r \sum_{k,l} P(r|c_{\mathcal{U}}^{(k)}, c_{\mathcal{V}}^{(l)}) P(c_{\mathcal{U}}^{(k)}|u_i^{(z)}) P(c_{\mathcal{V}}^{(l)}|v_i^{(z)}). \tag{6}$$

To evaluate the transfer learning framework, we compare our RMGM-based multi-task method to two baseline single-task methods: Pearson correlation coefficients (PCC) and flexible mixture model (FMM). FMM can be viewed as a single-task version of RMGM. In this experiment, we randomly select 500 users and 1000 items from three real-world data sets (MovieLens, EachMovie, and Book-Crossing), respectively. Our method will learn a shared RMGM on the union of the three 500×1000 rating matrices. The evaluation metric we adopt is mean absolute error (MAE).

The comparison results are reported in Table 3[30]. One can see that our method clearly outperforms the two baseline methods on all the three data sets, which has

Table 3. MAE Comparison. Given# means the number of observed ratings for each user. These results are quoted from [30].

TRAIN	METHOD	GIVEN5	GIVEN10	GIVEN15
	PCC	0.935	0.896	0.888
MOVIELENS	FMM	0.885	0.868	0.846
	RMGM	**0.857**	**0.820**	**0.804**
	PCC	0.976	0.937	0.933
EACHMOVIE	FMM	0.952	0.930	0.924
	RMGM	**0.934**	**0.906**	**0.890**
	PCC	0.621	0.619	0.630
BOOK-CROSSING	FMM	0.615	0.604	0.596
	RMGM	**0.612**	**0.590**	**0.581**

validated that RMGM indeed can gain additional useful knowledge by pooling the rating data from multiple related domains to make these tasks benefit from one another.

4   Conclusion and Future Work

Transfer learning has been proposed as a new learning problem in machine learning, but in data mining community it is still considered a new problem. One of the reasons is that data mining emphasizes scaled up applications to real world problems, and such diverse applications have been rare in transfer learning. This situation is beginning to change with our recent works on transfer learning. In this article, I have illustrated several novel applications of transfer learning that cover diverse areas: WiFi-based location estimation, cross-language classification, action model learning in AI planning and collaborative filtering.

In the future, we will see many other applications of transfer learning to emerge, and to see more scaled up applications of transfer learning as well.

Acknowledgement

We thank the support of NEC China Lab. We thank Wenyuan Dai, Guirong Xue, Yuqiang Chen, Xiao Ling, Bin Li, Hankz Hankui Zhuo, Vincent Wenchen Zheng, Sinno Jialin Pan for their great help with this article.

References

1. Thrun, S., Mitchell, T.M.: Learning one more thing. In: Proceedings of the Fourteenth International Joint Conference on Artificial Intelligence, pp. 825–830. Morgan Kaufmann, San Francisco (1995)
2. Schmidhuber, J.: On learning how to learn learning strategies. Technical Report FKI-198-94, Fakultat fur Informatik, Palo Alto, CA (1994)
3. Caruana, R.: Multitask learning. Machine Learning 28(1), 41–75 (1997)
4. Ben-David, S., Schuller, R.: Exploiting task relatedness for multiple task learning. In: Proceedings of the Sixteenth Annual Conference on Learning Theory, pp. 825–830. Morgan Kaufmann, San Francisco (2003)
5. DauméIII, H., Marcu, D.: Domain adaptation for statistical classifiers. Journal of Artificial Intelligence Research 26, 101–126 (2006)
6. Daumé III, H.: Frustratingly easy domain adaptation. In: Proceedings of the 45th Annual Meeting of the Association of Computational Linguistics, Prague, Czech Republic, June 2007, pp. 256–263 (2007)
7. Dai, W., Xue, G., Yang, Q., Yu, Y.: Co-clustering based classification for out-of-domain documents. In: Proceedings of the 13th ACM SIGKDD International Conference on Knowledge Discovery and Data Mining, San Jose, California, USA (August 2007)
8. Dai, W., Xue, G., Yang, Q., Yu, Y.: Transferring naive bayes classifiers for text classification. In: Proceedings of the 22nd AAAI Conference on Artificial Intelligence (July 2007)
9. Blitzer, J., McDonald, R., Pereira, F.: Domain adaptation with structural correspondence learning. In: Proceedings of the Conference on Empirical Methods in Natural Language, Sydney, Australia, pp. 120–128 (2006)

10. Blitzer, J., Dredze, M., Pereira, F.: Biographies, bollywood, boom-boxes and blenders: Domain adaptation for sentiment classification. In: Proceedings of the 45th Annual Meeting of the Association of Computational Linguistics, Prague, Czech Republic, pp. 432–439 (2007)

11. Raina, R., Battle, A., Lee, H., Packer, B., Ng, A.Y.: Self-taught learning: Transfer learning from unlabeled data. In: Proceedings of the 24th International Conference on Machine Learning, Corvalis, Oregon, USA, June 2007, pp. 759–766 (2007)

12. Konidaris, G., Barto, A.: Autonomous shaping: Knowledge transfer in reinforcement learning. In: Proceedings of Twenty-Third International Conference on Machine Learning (2006)

13. Pan, S.J., Yang, Q.: A survey on transfer learning. Technical Report HKUST-CS08-08, Department of Computer Science and Engineering, Hong Kong University of Science and Technology, Hong Kong, China (November 2008)

14. Raina, R., Ng, A.Y., Koller, D.: Constructing informative priors using transfer learning. In: Proceedings of the 23rd International Conference on Machine Learning, Pittsburgh, Pennsylvania, USA, June 2006, pp. 713–720 (2006)

15. Wu, P., Dietterich, T.G.: Improving svm accuracy by training on auxiliary data sources. In: Proceedings of the 21st International Conference on Machine Learning, Banff, Alberta, Canada. ACM, New York (2004)

16. Arnold, A., Nallapati, R., Cohen, W.W.: A comparative study of methods for transductive transfer learning. In: Proceedings of the 7th IEEE International Conference on Data Mining Workshops, Washington, DC, USA, pp. 77–82. IEEE Computer Society, Los Alamitos (2007)

17. Raykar, V.C., Krishnapuram, B., Bi, J., Dundar, M., Rao, R.B.: Bayesian multiple instance learning: automatic feature selection and inductive transfer. In: Proceedings of the 25th International Conference on Machine learning, Helsinki, Finland, pp. 808–815. ACM, New York (2008)

18. Ling, X., Xue, G.R., Dai, W., Jiang, Y., Yang, Q., Yu, Y.: Can chinese web pages be classified with english data source? In: Proceedings of the 17th International Conference on World Wide Web, Beijing, China, pp. 969–978. ACM, New York (2008)

19. Widmer, G., Kubat, M.: Learning in the presence of concept drift and hidden contexts. Mach. Learn. 23(1), 69–101 (1996)

20. Klinkenberg, R.: Learning drifting concepts: Example selection vs. example weighting. Intell. Data Anal. 8(3), 281–300 (2004)

21. Tsymbal, A.: The problem of concept drift: Definitions and related work

22. Hulten, G., Spencer, L., Domingos, P.: Mining time-changing data streams. In: KDD 2001: Proceedings of the seventh ACM SIGKDD international conference on Knowledge discovery and data mining, pp. 97–106. ACM Press, New York (2001)

23. Kolter, J., Maloof, M.: Dynamic weighted majority: A new ensemble method for tracking concept drift. In: Proceedings of the Third IEEE International Conference on Data Mining, pp. 123–130. IEEE Press, Los Alamitos (2003)

24. Wang, H., Fan, W., Yu, P.S., Han, J.: Mining concept-drifting data streams using ensemble classifiers. In: KDD 2003: Proceedings of the ninth ACM SIGKDD international conference on Knowledge discovery and data mining, pp. 226–235. ACM Press, New York (2003)

25. Gao, J., Fan, W., Han, J., Yu, P.S.: A general framework for mining concept-drifting data streams with skewed distributions. In: Proceedings of the Seventh SIAM International Conference on Data Mining, Minneapolis, Minnesota, USA (2007)

26. Pan, S.J., Shen, D., Yang, Q., Kwok, J.T.: Transferring localization models across space. In: Proceedings of the 23rd AAAI Conference on Artificial Intelligence, pp. 1383–1388 (2008)

27. Zheng, V.W., Pan, S.J., Yang, Q., Pan, J.J.: Transferring multi-device localization models using latent multi-task learning. In: Proceedings of the 23rd AAAI Conference on Artificial Intelligence, Chicago, Illinois, USA, July 2008, pp. 1427–1432 (2008)

28. Ling, X., Xue, G.R., Dai, W., Jiang, Y., Yang, Q., Yu, Y.: Can chinese web pages be classified with english data source? In: WWW 2008: Proceeding of the 17th International conference on World Wide Web, pp. 969–978. ACM, New York (2008)

29. Zhuo, H., Yang, Q., Hu, D.H., Li, L.: Transferring knowledge from another domain for learning action models. In: Ho, T.-B., Zhou, Z.-H. (eds.) PRICAI 2008. LNCS (LNAI), vol. 5351, pp. 1110–1115. Springer, Heidelberg (2008)

30. Li, B., Yang, Q., Xue, X.: Transfer learning for collaborative filtering via a rating-matrix generative model. In: ICML, pp. 617–624 (2009)

31. Yang, Q., Pan, S.J., Zheng, V.W.: Estimating location using Wi-Fi. IEEE Intelligent Systems 23(1), 8–13 (2008), http://www.cse.ust.hk/~qyang/ICDMDMC07/

32. Tishby, N., Pereira, F.C., Bialek, W.: The information bottleneck method. In: Proc. of the 37th Annual Allerton Conference on Communication, Control and Computing, pp. 368–377

33. Bel, N., Koster, C.H.A., Villegas, M.: Cross-lingual text categorization. In: Koch, T., Sølvberg, I.T. (eds.) ECDL 2003. LNCS, vol. 2769, pp. 126–139. Springer, Heidelberg (2003)

34. Wu, Y., Oard, D.W.: Bilingual topic aspect classification with a few training examples. In: SIGIR 2008: Proceedings of the 31st annual international ACM SIGIR conference on Research and development in information retrieval, pp. 203–210. ACM, New York (2008)

35. Rigutini, L., Maggini, M., Liu, B.: An em based training algorithm for cross-language text categorization. In: WI 2005: Proceedings of the 2005 IEEE/WIC/ACM International Conference on Web Intelligence, Washington, DC, USA, pp. 529–535. IEEE Computer Society, Los Alamitos (2005)

36. Kullback, S., Leibler, R.A.: On information and sufficiency. Annals of Mathematical Statistics 22(1), 79–86 (1951)

37. Domingos, P., Kok, S., Lowd, D., Poon, H., Richardson, M., Singla, P., Sumner, M., Wang, J.: Markov logic: A unifying language for structural and statistical pattern recognition. In: da Vitoria Lobo, N., Kasparis, T., Roli, F., Kwok, J.T., Georgiopoulos, M., Anagnostopoulos, G.C., Loog, M. (eds.) S+SSPR 2008. LNCS, vol. 5342, pp. 3–3. Springer, Heidelberg (2008)

Improving Adaptive Bagging Methods for Evolving Data Streams

Albert Bifet[1], Geoff Holmes[1], Bernhard Pfahringer[1], and Ricard Gavaldà[2]

[1] University of Waikato, Hamilton, New Zealand
{abifet,geoff,bernhard}@cs.waikato.ac.nz
[2] Universitat Politècnica de Catalunya, Barcelona, Spain
{gavalda}@lsi.upc.edu

Abstract. We propose two new improvements for bagging methods on evolving data streams. Recently, two new variants of Bagging were proposed: ADWIN Bagging and Adaptive-Size Hoeffding Tree (ASHT) Bagging. ASHT Bagging uses trees of different sizes, and ADWIN Bagging uses ADWIN as a change detector to decide when to discard underperforming ensemble members. We improve ADWIN Bagging using Hoeffding Adaptive Trees, trees that can adaptively learn from data streams that change over time. To speed up the time for adapting to change of Adaptive-Size Hoeffding Tree (ASHT) Bagging, we add an error change detector for each classifier. We test our improvements by performing an evaluation study on synthetic and real-world datasets comprising up to ten million examples.

1 Introduction

Data streams pose several challenges on data mining algorithm design. First, algorithms must make use of limited resources (time and memory). Second, by necessity they must deal with data whose nature or distribution changes over time. In turn, dealing with time-changing data requires strategies for detecting and quantifying change, forgetting stale examples, and for model revision. Fairly generic strategies exist for detecting change and deciding when examples are no longer relevant. Model revision strategies, on the other hand, are in most cases method-specific.

The following constraints apply in the Data Stream model:

1. Data arrives as a potentially infinite sequence. Thus, it is impossible to store it all. Therefore, only a small summary can be computed and stored.
2. The speed of arrival of data is fast, so each particular element has to be processed essentially in real time, and then discarded.
3. The distribution generating the items may change over time. Thus, data from the past may become irrelevant (or even harmful) for the current prediction.

Under these constraints the main properties of an ideal classification method are the following: high accuracy and fast adaption to change, low computational

---

Z.-H. Zhou and T. Washio (Eds.): ACML 2009, LNAI 5828, pp. 23–37, 2009.

cost in both space and time, theoretical performance guarantees, and a minimal number of parameters.

Ensemble methods are combinations of several models whose individual predictions are combined in some manner (for example, by averaging or voting) to form a final prediction. Often, ensemble learning classifiers provide superior predictive performance and they are easier to scale and parallelize than single classifier methods.

In [6] two new state-of-the-art bagging methods were presented: ASHT Bagging using trees of different sizes, and ADWIN Bagging using a change detector to decide when to discard underperforming ensemble members. This paper improves on ASHT Bagging by speeding up the time taken to adapt to changes in the distribution generating the stream. It improves on ADWIN Bagging by employing Hoeffding Adaptive Trees, trees that can adaptively learn from evolving data streams. The paper is structured as follows: the state-of-the-art Bagging methods are presented in Section 2. Improvements to these methods are presented in Section 3. An experimental evaluation is conducted in Section 4. Finally, conclusions and suggested items for future work are presented in Section 5.

2   Previous Work

2.1   Bagging Using Trees of Different Size (ASHT Bagging)

In [6], a new method of bagging was presented using Hoeffding Trees of different sizes. A *Hoeffding tree* [10] is an incremental, anytime decision tree induction algorithm that is capable of learning from massive data streams, assuming that the distribution generating examples does not change over time. Hoeffding trees exploit the fact that a small sample can often be enough to choose an optimal splitting attribute. This idea is supported mathematically by the Hoeffding bound, which quantifies the number of observations (in our case, examples)

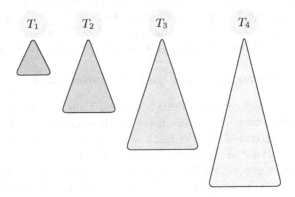

Fig. 1. An ensemble of trees of different size

needed to estimate some statistics within a prescribed precision (in our case, the goodness of an attribute). More precisely, the Hoeffding bound states that with probability $1 - \delta$, the true mean of a random variable of range R will not differ from the estimated mean after n independent observations by more than:

$$\epsilon = \sqrt{\frac{R^2 \ln(1/\delta)}{2n}}.$$

A theoretically appealing feature of Hoeffding Trees not shared by many other incremental decision tree learners is that it has sound theoretical guarantees of performance. IADEM-2 [9] uses Chernoff and Hoeffding bounds to give simi-lar guarantees. Using the Hoeffding bound one can show that the output of a Hoeffding tree is asymptotically nearly identical to that of a non-incremental learner using infinitely many examples. See [10] for details.

The Adaptive-Size Hoeffding Tree (ASHT) is derived from the Hoeffding Tree algorithm with the following differences:

- it has a value for the maximum number of split nodes, or *size*
- after one node splits, if the number of split nodes of the ASHT tree is higher than the maximum value, then it deletes some nodes to reduce its size

When the tree size exceeds the maximun size value, there are two different delete options:

- delete the oldest node, the root, and all of its children except the one where the split has been made. After that, the root of the child not deleted becomes the new root.
- delete all the nodes of the tree, that is, reset the tree to the empty tree

The intuition behind this method is as follows: smaller trees adapt more quickly to changes, and larger trees perform better during periods with little or no change, simply because they were built on more data. Trees limited to size s will be reset about twice as often as trees with a size limit of $2s$. This creates a set of different reset-speeds for an ensemble of such trees, and therefore a subset of trees that are a good approximation for the current rate of change. It is important to note that resets will happen all the time, even for stationary datasets, but this behaviour should not have a negative impact on the ensemble's predictive performance.

In [6] a new bagging method was presented that uses these Adaptive-Size Hoeffding Trees and that sets the size for each tree. The maximum allowed size for the n-th ASHT tree is twice the maximum allowed size for the $(n-1)$-th tree. Moreover, each tree has a weight proportional to the inverse of the square of its error, and it monitors its error with an exponential weighted moving average (EWMA) with $\alpha = .01$. The size of the first tree is 2.

With this new method, the authors attempted to improve bagging perfor-mance by increasing tree diversity. It has been observed [18] that boosting tends to produce a more diverse set of classifiers than bagging, and this has been cited as a factor in increased performance.

2.2 Bagging Using ADWIN (ADWIN Bagging)

ADWIN [4] is a change detector and estimator that solves in a well-specified way the problem of tracking the average of a stream of bits or real-valued numbers. ADWIN keeps a variable-length window of recently seen items, with the property that the window has the maximal length statistically consistent with the hypothesis "there has been no change in the average value inside the window".

More precisely, an older fragment of the window is dropped if and only if there is enough evidence that its average value differs from that of the rest of the window. This has two consequences: first, that change is reliably declared whenever the window shrinks; and second, that at any time the average over the existing window can be reliably taken as an estimate of the current average in the stream (barring a very small or very recent change that is still not statistically visible). A formal and quantitative statement of these two points (in the form of a theorem) appears in [4].

ADWIN is parameter- and assumption-free in the sense that it automatically detects and adapts to the current rate of change. Its only parameter is a confidence bound δ, indicating how confident we want to be in the algorithm's output, a property inherent to all algorithms dealing with random processes.

Also important for our purposes, ADWIN does not maintain the window explicitly, but compresses it using a variant of the exponential histogram technique. This means that it keeps a window of length W using only $O(\log W)$ memory and $O(\log W)$ processing time per item.

Bagging using ADWIN is implemented as ADWIN Bagging where the Bagging method is the online bagging method of Oza and Rusell [20] with the addition of the ADWIN algorithm as a change detector. The base classifiers used are Hoeffding Trees. When a change is detected, the worst classifier of the ensemble of classifiers is removed and a new classifier is added to the ensemble.

3 Improvements for Adaptive Bagging Methods

In this section we propose two new improvements for the adaptive bagging methods explained in the previous section.

3.1 ADWIN Bagging Using Hoeffding Adaptive Trees

The basic idea is to use Hoeffding trees, that are able to adapt to distribution changes instead of non-adaptive Hoeffding trees as the base classifier for the bagging ensemble method. We use the Hoeffding Adaptive Trees proposed in [5], where a new method for managing alternate trees is proposed. The general idea is simple: we place ADWIN instances at every node that will raise an alert whenever something worth attention happens at the node.

We use the variant of the *Hoeffding Adaptive Tree* algorithm (HAT for short) that uses ADWIN as a change detector. It uses one instance of ADWIN in each node, as a change detector, to monitor the classification error rate at that node. A significant increase in that rate indicates that the data is changing with respect to the time at which the subtree was created. This was the approach used by Gama *et al.* in [12], using another change detector.

When any instance of `ADWIN` at a node detects change, we create a new alternate tree without splitting any attribute. Using two `ADWIN` instances at every node, we monitor the average error of the subtree rooted at this node and the average error of the new alternate subtree. When there is enough evidence (as witnessed by `ADWIN`) that the new alternate tree is doing better than the original decision subtree, we replace the original decision subtree by the new alternate subtree.

3.2 DDM Bagging Using Trees of Different Size

We improve bagging using trees of different size, by adding a change detector for each tree in the ensemble to speed up the adaption to the evolving stream.

A *change detector* or *drift detection* method is an algorithm that takes as input a sequence of numbers and emits a signal when the distribution of its input changes. There are many such methods such as the CUSUM Test, the Geometric Moving Average test, etc. Change detection is not an easy task, since a fundamental limitation exists: the design of a change detector is a compromise between detecting true changes and avoiding false alarms. See [14] and [3] for a more detailed survey of change detection methods.

We use two drift detection methods (DDM and EDDM) proposed by Gama et al. [11] and Baena-García et al. [2]. These methods control the number of errors produced by the learning model during prediction. They compare the statistics of two windows: the first contains all of the data, and the second contains only the data from the beginning until the number of errors increases. Their methods do not store these windows in memory. They keep only statistics and a window of recent errors data.

The drift detection method (DDM) uses the number of errors in a sample of n examples, modelled by a binomial distribution. For each point i in the sequence that is being sampled, the error rate is the probability of misclassifying (p_i), with standard deviation given by $s_i = \sqrt{p_i(1 - p_i)/i}$. They assume (as stated in the PAC learning model [19]) that the error rate of the learning algorithm (p_i) will decrease while the number of examples increases if the distribution of the examples is stationary. A significant increase in the error of the algorithm, suggests that the class distribution is changing and, hence, the actual decision model is supposed to be inappropriate. Thus, they store the values of p_i and s_i when $p_i + s_i$ reaches its minimum value during the process (obtaining p_{pmin} and s_{min}). It then checks when the following conditions are triggered:

- $p_i + s_i \geq p_{min} + 2 \cdot s_{min}$ for the warning level. Beyond this level, the examples are stored in anticipation of a possible change of context.
- $p_i + s_i \geq p_{min} + 3 \cdot s_{min}$ for the drift level. Beyond this level the concept drift is supposed to be true, the model induced by the learning method is reset and a new model is learnt using the examples stored since the warning level triggered. The values for p_{min} and s_{min} are reset too.

This approach is good at detecting abrupt changes and gradual changes when the gradual change is not very slow, but has difficulties when the change is gradual and slow. In that case, the examples will be stored for a long time, the drift level then takes too long to trigger and the examples in memory can be exceeded.

Baena-García et al. proposed a new method EDDM [2] in order to improve DDM. It is based on the estimated distribution of the distances between classification errors. The window resize procedure is governed by the same heuristics.

4 Comparative Experimental Evaluation

Massive Online Analysis (MOA) [16] is a software environment for implementing algorithms and running experiments for online learning from data streams. The data stream evaluation framework and all algorithms evaluated in this paper were implemented in the Java programming language extending the MOA software. MOA includes a collection of offline and online methods as well as tools for evaluation. In particular, it implements boosting, bagging, and Hoeffding Trees, all with and without Naïve Bayes classifiers at the leaves.

One of the key data structures used in MOA is the description of an example from a data stream. This structure borrows from WEKA, where an example is represented by an array of double precision floating point values. This provides freedom to store all necessary types of value – numeric attribute values can be stored directly, and discrete attribute values and class labels are represented by integer index values that are stored as floating point values in the array. Double precision floating point values require storage space of 64 bits, or 8 bytes. This detail can have implications for memory usage.

We use the new experimental framework for concept drift presented in [6]. Considering data streams as data generated from pure distributions, we can model a concept drift event as a weighted combination of two pure distributions that characterizes the target concepts before and after the drift. This framework defines the probability that every new instance of the stream belongs to the new concept after the drift. It uses the sigmoid function, as an elegant and practical solution.

We see from Figure 2 that the sigmoid function

$$f(t) = 1/(1 + e^{-s(t-t_0)})$$

has a derivative at the point t_0 equal to $f'(t_0) = s/4$. The tangent of angle α is equal to this derivative, $\tan \alpha = s/4$. We observe that $\tan \alpha = 1/W$, and as $s = 4 \tan \alpha$ then $s = 4/W$. So the parameter s in the sigmoid gives the length of W and the angle α. In this sigmoid model we only need to specify two parameters : t_0 the point of change, and W the length of change.

Definition 1. *Given two data streams* a, b, *we define* $c = a \oplus_{t_0}^W b$ *as the data stream built joining the two data streams* a *and* b, *where* t_0 *is the point of change,* W *is the length of change and*

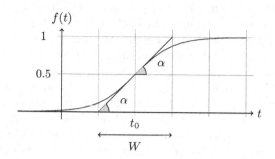

Fig. 2. A sigmoid function $f(t) = 1/(1 + e^{-s(t-t_0)})$

- $\Pr[c(t) = a(t)] = e^{-4(t-t_0)/W}/(1 + e^{-4(t-t_0)/W})$
- $\Pr[c(t) = b(t)] = 1/(1 + e^{-4(t-t_0)/W})$.

In order to create a data stream with multiple concept changes, we can build new data streams joining different concept drifts:

$$(((a \oplus_{t_0}^{W_0} b) \oplus_{t_1}^{W_1} c) \oplus_{t_2}^{W_2} d)\ldots$$

4.1 Datasets for Concept Drift

Synthetic data has several advantages – it is easier to reproduce and there is little cost in terms of storage and transmission. For this paper we use the data generators most commonly found in the literature.

SEA Concepts Generator. This artificial dataset contains abrupt concept drift, first introduced in [23]. It is generated using three attributes, where only the two first attributes are relevant. All the attributes have values between 0 and 10. The points of the dataset are divided into 4 blocks with different concepts. In each block, the classification is done using $f_1 + f_2 \leq \theta$, where f_1 and f_2 represent the first two attributes and θ is a threshold value. The most frequent values are 9, 8, 7 and 9.5 for the data blocks. In our framework, SEA concepts are defined as follows:

$$(((SEA_9 \oplus_{t_0}^{W} SEA_8) \oplus_{2t_0}^{W} SEA_7) \oplus_{3t_0}^{W} SEA_{9.5})$$

Rotating Hyperplane. It was used as testbed for CVFDT versus VFDT in [17]. A hyperplane in d-dimensional space is the set of points x that satisfy

$$\sum_{i=1}^{d} w_i x_i = w_0 = \sum_{i=1}^{d} w_i$$

where x_i, is the ith coordinate of x. Examples for which $\sum_{i=1}^{d} w_i x_i \geq w_0$ are labeled positive, and examples for which $\sum_{i=1}^{d} w_i x_i < w_0$ are labeled negative. Hyperplanes are useful for simulating time-changing concepts, because we can change the orientation and position of the hyperplane in a smooth manner by changing the relative size of the weights. We introduce change to this dataset adding drift to each weight attribute $w_i = w_i + d\sigma$, where σ is the probability that the direction of change is reversed and d is the change applied to every example.

Random RBF Generator. This generator was devised to offer an alternate complex concept type that is not straightforward to approximate with a decision tree model. The RBF (Radial Basis Function) generator works as follows: A fixed number of random centroids are generated. Each center has a random position, a single standard deviation, class label and weight. New examples are generated by selecting a center at random, taking weights into consideration so that centers with higher weight are more likely to be chosen. A random direction is chosen to offset the attribute values from the central point. The length of the displacement is randomly drawn from a Gaussian distribution with standard deviation determined by the chosen centroid. The chosen centroid also determines the class label of the example. This effectively creates a normally distributed hypersphere of examples surrounding each central point with varying densities. Only numeric attributes are generated. Drift is introduced by moving the centroids with constant speed. This speed is initialized by a drift parameter.

LED Generator. This data source originates from the CART book [7]. An implementation in C was donated to the UCI [1] machine learning repository by David Aha. The goal is to predict the digit displayed on a seven-segment LED display, where each attribute has a 10% chance of being inverted. It has an optimal Bayes classification rate of 74%. The particular configuration of the generator used for experiments (led) produces 24 binary attributes, 17 of which are irrelevant.

Data streams may be considered infinite sequences of (x, y) where x is the feature vector and y the class label. Zhang et al. [24] observe that $p(x, y) = p(x|t) \cdot p(y|x)$ and categorize concept drift in two types:

- *Loose Concept Drifting (LCD)* when concept drift is caused only by the change of the class prior probability $p(y|x)$,
- *Rigorous Concept Drifting (RCD)* when concept drift is caused by the change of the class prior probability $p(y|x)$ and the conditional probability $p(x|t)$

Note that the Random RBF Generator has RCD drift, and the rest of the dataset generators have LCD drift.

4.2 Real-World Data

It is not easy to find large real-world datasets for public benchmarking, especially with substantial concept change. The UCI machine learning repository [1] contains some real-world benchmark data for evaluating machine learning techniques. We consider three of the largest: Forest Covertype, Poker-Hand, and Electricity.

Table 1. Comparison of algorithms. Accuracy is measured as the final percentage of examples correctly classified over the 1 or 10 million test/train interleaved evaluation. Time is measured in seconds, and memory in MB. The best individual accuracies are indicated in boldface.

	Hyperplane Drift .0001			Hyperplane Drift .001		
	Time	Acc.	Mem.	Time	Acc.	Mem.
BagADWIN 10 HAT	3025.87	91.36 ± 0.13	9.91	2834.85	91.12 ± 0.15	1.23
DDM Bag10 ASHT W	1321.64	91.60 ± 0.09	0.85	1351.16	**91.43 ± 0.07**	2.10
EDDM Bag10 ASHT W	1362.31	91.43 ± 0.13	3.15	1371.46	91.17 ± 0.11	2.77
NaiveBayes	86.97	83.40 ± 2.11	0.01	86.87	77.54 ± 2.93	0.01
NBADWIN	308.85	91.30 ± 0.90	0.06	295.19	90.58 ± 0.59	0.06
HT	157.71	86.09 ± 1.60	9.57	159.43	82.99 ± 1.88	10.41
HT DDM	174.10	89.36 ± 0.24	0.04	180.51	89.07 ± 0.26	0.01
HT EDDM	207.47	88.73 ± 0.38	13.23	193.07	87.97 ± 0.51	2.52
HAT	500.81	89.75 ± 0.25	1.72	431.6	89.40 ± 0.30	0.15
BagADWIN 10 HT	1306.22	91.18 ± 0.12	11.40	1308.08	90.86 ± 0.15	5.52
Bag10 HT	1236.92	87.07 ± 1.64	108.75	1253.07	84.13 ± 1.79	114.14
Bag10 ASHT	1060.37	90.85 ± 0.53	2.68	1070.44	90.39 ± 0.33	2.69
Bag10 ASHT W	1055.87	91.38 ± 0.26	2.68	1073.96	91.01 ± 0.18	2.69
Bag10 ASHT R	995.06	91.44 ± 0.09	2.95	1016.48	91.17 ± 0.12	2.14
Bag10 ASHT W+R	996.52	**91.62 ± 0.08**	2.95	1024.02	91.40 ± 0.10	2.14
Bag5 ASHT W+R	551.53	90.83 ± 0.06	0.08	562.09	90.75 ± 0.06	0.09
OzaBoost	974.69	86.68 ± 1.19	130.00	959.14	84.47 ± 1.49	123.75

Forest Covertype Contains the forest cover type for 30 x 30 meter cells obtained from US Forest Service (USFS) Region 2 Resource Information System (RIS) data. It contains $581,012$ instances and 54 attributes, and it has been used in several papers on data stream classification [13, 21].

Poker-Hand Consists of $1,000,000$ instances and 11 attributes. Each record of the Poker-Hand dataset is an example of a hand consisting of five playing cards drawn from a standard deck of 52. Each card is described using two attributes (suit and rank), for a total of 10 predictive attributes. There is one Class attribute that describes the "Poker Hand". The order of cards is important, which is why there are 480 possible Royal Flush hands instead of 4.

Electricity Is another widely used dataset described by M. Harries [15] and analysed by Gama [11]. This data was collected from the Australian New South Wales Electricity Market. In this market, the prices are not fixed and are affected by demand and supply of the market. The prices in this market are set every five minutes. The ELEC dataset contains $45,312$ instances. Each example of the dataset refers to a period of 30 minutes, i.e. there are 48 instances for each time period of one day. The class label identifies the change of the price related to a moving average of the last 24 hours. The class level only reflect deviations of the price on a one day average and removes the impact of longer term price trends.

Table 2. Comparison of algorithms. Accuracy is measured as the final percentage of examples correctly classified over the 1 or 10 million test/train interleaved evaluation. Time is measured in seconds, and memory in MB.

	SEA W= 50000			RandomRBF No Drift 50 centers		
	Time	Acc.	Mem.	Time	Acc.	Mem.
BagADWIN 10 HAT	154.91	**88.88** ± 0.05	2.35	5378.75	95.34 ± 0.06	119.17
DDM Bag10 ASHT W	44.02	88.72 ± 0.05	0.65	1230.43	93.09 ± 0.05	3.21
EDDM Bag10 ASHT W	48.95	88.60 ± 0.05	0.90	1317.58	93.28 ± 0.03	3.76
NaiveBayes	5.52	84.60 ± 0.03	0.00	111.12	72.04 ± 0.02	0.01
NBADWIN	12.40	87.83 ± 0.07	0.02	396.01	72.04 ± 0.02	0.08
HT	7.20	85.02 ± 0.11	0.33	154.67	93.80 ± 0.10	6.86
HT DDM	7.88	88.17 ± 0.18	0.16	185.15	93.77 ± 0.12	13.72
HT EDDM	8.52	87.87 ± 0.22	0.06	185.89	93.75 ± 0.14	13.81
HAT	20.96	88.40 ± 0.07	0.18	794.48	93.80 ± 0.10	9.28
BagADWIN 10 HT	53.15	88.58 ± 0.10	0.88	1238.50	95.34 ± 0.06	67.79
Bag10 HT	30.88	85.38 ± 0.06	3.36	995.46	**95.35** ± 0.05	71.26
Bag10 ASHT	35.30	88.02 ± 0.08	0.91	1009.62	80.08 ± 0.14	3.73
Bag10 ASHT W	35.69	88.27 ± 0.08	0.91	986.90	92.84 ± 0.05	3.73
Bag10 ASHT R	33.74	88.38 ± 0.06	0.84	913.74	90.03 ± 0.03	2.65
Bag10 ASHT W+R	33.56	88.51 ± 0.06	0.84	925.65	92.85 ± 0.05	2.65
Bag5 ASHT W+R	20.00	87.80 ± 0.06	0.05	536.61	80.82 ± 0.07	0.06
OzaBoost	39.97	86.64 ± 0.06	4.00	964.75	94.85 ± 0.03	206.60

The size of these datasets is small, compared to tens of millions of training examples of synthetic datasets: $45,312$ for ELEC dataset, $581,012$ for CoverType, and $1,000,000$ for Poker-Hand. Another important fact is that we do not know when drift occurs or indeed if there is any drift. We may simulate RCD concept drift, joining the three datasets, merging attributes, and supposing that each dataset corresponds to a different concept.

$$\text{CovPokElec} = (\text{CoverType} \oplus_{581,012}^{5,000} \text{Poker}) \oplus_{1,000,000}^{5,000} \text{ELEC}$$

As all examples need to have the same number of attributes, we simple concatenate all the attributes, and set the number of classes to the maximum number of classes of all the datasets. The attribute values for a type not currently selected is a constant one, for example, zero.

4.3   Results

We use the datasets for evaluation explained in the previous sections. The experiments were performed on a 2.0 GHz Intel Core Duo PC machine with 2 Gigabyte main memory, running Ubuntu 8.10. The evaluation methodology used was Interleaved Test-Then-Train on 10 runs: every example was used for testing the

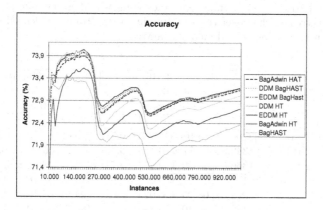

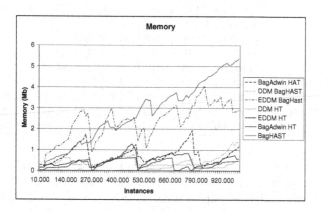

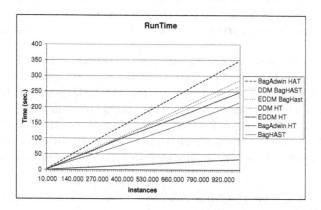

Fig. 3. Accuracy, runtime and memory on dataset LED with three concept drifts

Table 3. Comparison of algorithms. Accuracy is measured as the final percentage of examples correctly classified over the 1 or 10 million test/train interleaved evaluation. Time is measured in seconds, and memory in MB.

	RandomRBF Drift .0001 50 centers			RandomRBF Drift .001 50 centers		
	Time	Acc.	Mem.	Time	Acc.	Mem.
BagADWIN 10 HAT	2706.91	**85.98** ± 0.04	0.51	1976.62	67.20 ± 0.03	0.10
DDM Bag10 ASHT W	1349.65	83.77 ± 0.15	0.53	1441.22	**70.01** ± 0.31	3.09
EDDM Bag10 ASHT W	1366.65	84.20 ± 0.46	0.71	1422.31	69.99 ± 0.35	0.71
NaiveBayes	111.47	53.21 ± 0.02	0.01	113.37	53.18 ± 0.02	0.01
NBADWIN	272.58	68.05 ± 0.02	0.05	249.1	62.19 ± 0.02	0.04
HT	189.25	63.40 ± 0.10	9.86	186.47	55.40 ± 0.02	8.90
HT DDM	199.95	76.40 ± 0.08	0.02	206.41	62.40 ± 1.91	0.03
HT EDDM	214.55	76.39 ± 0.57	0.09	203.41	63.23 ± 0.71	0.02
HAT	413.53	79.12 ± 0.08	0.09	294.94	65.32 ± 0.03	0.01
BagADWIN 10 HT	1326.12	85.28 ± 0.03	0.26	1354.03	67.20 ± 0.03	0.03
Bag10 HT	1362.66	71.01 ± 0.06	106.20	1240.89	58.17 ± 0.03	88.52
Bag10 ASHT	1124.40	74.76 ± 0.04	3.05	1133.51	66.25 ± 0.02	3.10
Bag10 ASHT W	1104.03	75.55 ± 0.05	3.05	1106.26	66.81 ± 0.02	3.10
Bag10 ASHT R	1069.76	83.48 ± 0.03	3.74	1085.99	67.78 ± 0.02	2.35
Bag10 ASHT W+R	1068.59	84.20 ± 0.03	3.74	1101.10	69.14 ± 0.02	2.35
Bag5 ASHT W+R	557.20	79.45 ± 0.06	0.09	587.46	67.53 ± 0.02	0.10
OzaBoost	1312.00	71.64 ± 0.07	105.94	1266.75	58.21 ± 0.05	88.36

model before using it to train. This interleaved test followed by train procedure was carried out on 10 million examples from the hyperplane and RandomRBF datasets, and one million examples from the SEA dataset. Tables 1, 2 and 3 reports the final accuracy, and speed of the classification models induced on synthetic data. Table 4 shows the results for real datasets: Forest CoverType, Poker Hand, and CovPokElec. The results for the Electricity dataset were structurally similar to those for the Forest CoverType dataset and therefore not reported. Additionally, the learning curves and model growth curves for LED dataset are plotted (Figure 3).

The first, and baseline, algorithm (HT) is a single Hoeffding tree, enhanced with adaptive Naive Bayes leaf predictions. Parameter settings are $n_{min} = 1000$, $\delta = 10^{-8}$ and $\tau = 0.05$, as used in [10]. The HT DDM and HT EDDM are Hoeffding Trees with drift detection methods as explained in Section 3.2.

Bag10 is Oza and Russell online bagging using ten classifiers and Bag5 only five. BagADWIN is the online bagging version using ADWIN explained in Section 2.2. As described earlier, we implemented the following new variants of bagging:

- ADWIN Bagging using Hoeffding Adaptive Trees.
- Bagging ASHT using the DDM drift detection method
- Bagging ASHT using the EDDM drift detection method

Table 4. Comparison of algorithms on real data sets. Time is measured in seconds, and memory in MB. The results are based on a single run. Unlike synthetic datasets, it is not straightforward to add randomization to real datasets such that meaningful datasets and analysis of performance variances are obtained.

	Cover Type			Poker			CovPokElec		
	Time	Acc.	Mem.	Time	Acc.	Mem.	Time	Acc.	Mem.
BagADWIN 10 HAT	317.75	85.48	0.2	267.41	**88.63**	13.74	1403.40	**87.16**	0.62
DDM Bag10 ASHT W	249.92	**88.39**	6.09	128.17	75.85	1.51	876.18	84.54	19.74
EDDM Bag10 ASHT W	207.10	86.72	0.37	141.96	85.93	12.84	828.63	84.98	48.54
NaiveBayes	31.66	60.52	0.05	13.58	50.01	0.02	91.50	23.52	0.08
NBADWIN	127.34	72.53	5.61	64.52	50.12	1.97	667.52	53.32	14.51
HT	31.52	77.77	1.31	18.98	72.14	1.15	95.22	74.00	7.42
HT DDM	40.26	84.35	0.33	21.58	61.65	0.21	114.72	71.26	0.42
HT EDDM	34.49	86.02	0.02	22.86	72.20	2.30	114.57	76.66	11.15
HAT	55.00	81.43	0.01	31.68	72.14	1.24	188.65	75.75	0.01
BagADWIN 10 HT	247.50	84.71	0.23	165.01	84.84	8.79	911.57	85.95	0.41
Bag10 HT	138.41	83.62	16.80	121.03	87.36	12.29	624.27	81.62	82.75
Bag10 ASHT	213.75	83.34	5.23	124.76	86.80	7.19	638.37	78.87	29.30
Bag10 ASHT W	212.17	85.37	5.23	123.72	87.13	7.19	636.42	80.51	29.30
Bag10 ASHT R	229.06	84.20	4.09	122.92	86.21	6.47	776.61	80.01	29.94
Bag10 ASHT W+R	198.04	86.43	4.09	123.25	86.76	6.47	757.00	81.05	29.94
Bag5 ASHT W+R	116.83	83.79	0.23	57.09	75.87	0.44	363.09	77.65	0.95
OzaBoost	170.73	85.05	21.22	151.03	87.85	14.50	779.99	84.69	105.63

In general terms, we observe that ensemble methods perform better than single classifier methods, and that explicit drift detection is better. However, these improvements come at a cost of runtime and memory. In fact, the results indicate that memory is not as big an issue as the runtime accuracy tradeoff. We observe that the variance in accuracy goes up when the drift increases, and vice versa. All classifiers have a small accuracy variance for the dataset RandomRBF with no drift, as shown in Table 2.

RCD drift produces much greater differences than LCD - for example, the best result in Table 3 goes down 16% when increasing the drift from 0.0001 to 0.001 - in Hyperplane the same change in drift elicits only a fractional change in accuracy. For RCD drift all methods drop significantly. The bagging methods have the most to lose and go down between 14-17% (top three) and 10-18% (bottom of table). The base methods have less to lose going down (0-14%).

For all datasets one of the new methods always wins in terms of accuracy. Specifically, on the nine analysed datasets: BagADWIN 10 HAT wins four times out of nine, DDM Bag10 ASHT W wins three times, DDM Bag10 ASHT W wins once; this relationship is consistent in the following way: whenever the single HAT beats Bag10 ASHT W, then BagADWIN 10 HAT beats DDM Bag10 ASHT W, and vice versa. Note that the bad result for DDM Bag10 ASHT W on the Poker dataset must be due to too many false positive drift predictions that wrongly keep the model too small (this can be verified by observing the memory usage in this case), which is mirrored by the behavior of HT DDM on the Poker dataset.

The new improved ensemble methods presented in this paper are slower than the old ones, with `BagADWIN 10 HAT` being worst, because it pays twice: through the addition of `ADWIN` and `HAT`, the latter being slower than `HT` by a factor between two and five. Change detection can be time-consuming, the extreme case being `Naive Bayes` vs. `NBAdwin`, where `Naive Bayes` can be up to six times faster. However, change detection helps accuracy, with improvements of up to 20 percentage points (see, for example, CovPokElec).

Non-drift-aware non-adaptive ensembles like `Bag10 HT` and `OzaBoost` usually need the most memory, sometimes by a very large margin. `Bag10 ASHT W+R` needs a lot more memory than `Bag5 ASHT W+R`, because the last trees in a `Bag N ASHT W+R` needs as much space as the full `Bag N-1 ASHT W+R` ensemble.

Recall that data stream evaluation is fundamentally three-dimensional. When adding adaptability to evolving data streams using change detector methods, we increase the run-time, obtaining more accurate methods. For example, adding a change detector DDM to HT, or to `Bag10 ASHT W`, in Table 1, we observe a higher cost in runtime, but also an improvement in accuracy.

5   Conclusions and Future Work

We have presented two new improvements for bagging methods, using Hoeffding Adaptive Trees and change detection methods. In general terms, we observe that using explicit drift detection methods we improve accuracy. However, these improvements come at a cost of runtime and memory. It seems that the cost of improving accuracy in bagging methods for data streams, is large in runtime, but small in memory.

As future work, we would like to build new ensemble methods that perform with an accuracy similar to the methods presented in this paper, but using less runtime. These new methods could be a boosting ensemble method, or a bagging method using new change detection strategies. We think that a boosting method could improve bagging performance by increasing tree diversity, as it is shown by the increased performance of boosting for the traditional batch learning setting. This could be a challenging topic since in [6], the authors didn't find any boosting method in the literature [20, 22, 8] that outperformed bagging in the streaming setting.

References

[1] Asuncion, A., Newman, D.: UCI machine learning repository (2007)
[2] Baena-García, M., del Campo-Ávila, J., Fidalgo, R., Bifet, A., Gavaldà, R., Morales-Bueno, R.: Early drift detection method. In: Fourth International Workshop on Knowledge Discovery from Data Streams (2006)
[3] Basseville, M., Nikiforov, I.V.: Detection of abrupt changes: theory and application. Prentice-Hall, Inc., Upper Saddle River (1993)
[4] Bifet, A., Gavaldà, R.: Learning from time-changing data with adaptive windowing. In: SIAM International Conference on Data Mining, pp. 443–448 (2007)

[5] Bifet, A., Gavaldà, R.: Adaptive learning from evolving data streams. In: IDA (2009)

[6] Bifet, A., Holmes, G., Pfahringer, B., Kirkby, R., Gavaldà, R.: New ensemble methods for evolving data streams. In: KDD 2009. ACM, New York (2009)

[7] Breiman, L., et al.: Classification and Regression Trees. Chapman & Hall, New York (1984)

[8] Chu, F., Zaniolo, C.: Fast and light boosting for adaptive mining of data streams. In: Dai, H., Srikant, R., Zhang, C. (eds.) PAKDD 2004. LNCS (LNAI), vol. 3056, pp. 282–292. Springer, Heidelberg (2004)

[9] del Campo-Ávila, J., Ramos-Jiménez, G., Gama, J., Bueno, R.M.: Improving the performance of an incremental algorithm driven by error margins. Intell. Data Anal. 12(3), 305–318 (2008)

[10] Domingos, P., Hulten, G.: Mining high-speed data streams. In: Knowledge Discovery and Data Mining, pp. 71–80 (2000)

[11] Gama, J., Medas, P., Castillo, G., Rodrigues, P.: Learning with drift detection. In: SBIA Brazilian Symposium on Artificial Intelligence, pp. 286–295 (2004)

[12] Gama, J., Medas, P., Rocha, R.: Forest trees for on-line data. In: SAC 2004: Proceedings of the 2004 ACM symposium on Applied computing, pp. 632–636. ACM Press, New York (2004)

[13] Gama, J., Rocha, R., Medas, P.: Accurate decision trees for mining high-speed data streams. In: KDD 2003, August 2003, pp. 523–528 (2003)

[14] Gustafsson, F.: Adaptive Filtering and Change Detection. Wiley, Chichester (2000)

[15] Harries, M.: Splice-2 comparative evaluation: Electricity pricing. Technical report, The University of South Wales (1999)

[16] Holmes, G., Kirkby, R., Pfahringer, B.: MOA: Massive Online Analysis (2007), http://sourceforge.net/projects/moa-datastream

[17] Hulten, G., Spencer, L., Domingos, P.: Mining time-changing data streams. In: KDD 2001, San Francisco, CA, pp. 97–106. ACM Press, New York (2001)

[18] Margineantu, D.D., Dietterich, T.G.: Pruning adaptive boosting. In: ICML 1997, pp. 211–218 (1997)

[19] Mitchell, T.: Machine Learning. McGraw-Hill Education (ISE Editions), New York (1997)

[20] Oza, N., Russell, S.: Online bagging and boosting. In: Artificial Intelligence and Statistics 2001, pp. 105–112. Morgan Kaufmann, San Francisco (2001)

[21] Oza, N.C., Russell, S.: Experimental comparisons of online and batch versions of bagging and boosting. In: KDD 2001, August 2001, pp. 359–364 (2001)

[22] Pelossof, R., Jones, M., Vovsha, I., Rudin, C.: Online coordinate boosting (2008), http://arxiv.org/abs/0810.4553

[23] Street, W.N., Kim, Y.: A streaming ensemble algorithm (SEA) for large-scale classification. In: KDD 2001, pp. 377–382. ACM Press, New York (2001)

[24] Zhang, P., Zhu, X., Shi, Y.: Categorizing and mining concept drifting data streams. In: KDD 2008, pp. 812–820. ACM, New York (2008)

A Hierarchical Face Recognition Algorithm

Remco R. Bouckaert

Computer Science Department, University of Waikato, New Zealand
rrb@xm.co.nz, remco@cs.waikato.ac.nz

Abstract. In this paper, we propose a hierarchical method for face recognition where base classifiers are defined to make predictions based on various different principles and classifications are combined into a single prediction. Some features are more relevant to particular face recognition tasks than others. The hierarchical algorithm is flexible in selecting features relevant for the face recognition task at hand. In this paper, we explore various features based on outline recognition, PCA classifiers applied to part of the face and exploitation of symmetry in faces. By combining the predictions of these features we obtain superior performance on benchmark datasets (99.25% accuracy on the ATT dataset) at reduced computation cost compared to full PCA.

1 Introduction

Many face recognition algorithms rely on general purpose techniques like principle component analysis (PCA) [10], linear discriminant analysis (LDA) [9] or independent component analysis (ICA) [4,8][1]. These algorithms are general purpose image recognition algorithms that do not use background knowledge of the problem at hand, but transform the space represented by an image into another space where each of the dimensions spanning the space represent otherwise unknown informative features of the pictures of faces.

In this article, we exploit some properties related to face recognition and propose the following alternative approach; extract features from the picture that represent items known to contribute to the recognition of faces. Each of these features can be used to recognize faces, or at least make a ranking or distribution of the people one wants to recognize. These results can then be taken as attributes in a classification problem and a general purpose classifier trained with these attributes to give a combined classifier. Alternatively, we can develop a custom weighting scheme of the features that results in highly accurate classification.

To support the claim that this is indeed a fruitful approach, we consider as a feature the shape of the outline of the head. By calculating the difference between outlines, a person can be classified with the closest matching one. Even this simple feature gives already impressive accuracy results on a benchmark

---

[1] See http://www.face-rec.org/algorithms/ for an overview of various algorithms for face recognition.

Z.-H. Zhou and T. Washio (Eds.): ACML 2009, LNAI 5828, pp. 38–50, 2009.
© Springer-Verlag Berlin Heidelberg 2009

face recognition test set at little computational cost. Another type of feature is created by applying PCA on part of the image. By selecting only the left, right or top half of a picture, the number of pixels is halved and therefore the computational and memory requirements of PCA reduced by an eight and a quarter respectively.

The following section explains the technical details of our approach and the method for defining features. Section 3 discusses empirical results of outline features and Section 4 the results for PCA features and Section 5 considers both sets. We conclude with computational consideration (Section 6) some final remarks and directions for further research.

2   Outline Recognition Algorithm

The outline recognition algorithm gives us the feature that represents the outline of a persons' face as taken in a picture. Consider the two images shown in Figure 1 from the Yale face database [1]. First, we automatically cropped the images based leaving only the top part to be processed. Then, we applied a sharpening filter, changing all pixels above average intensity to white, and all below to black.

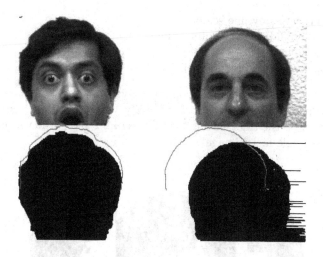

Fig. 1. Two examples, their transformations and outline detection

The next stage consists of finding the boundaries of the subjects head by scanning lines starting from the outside of the image towards the center. Lines are scanned parallel to the x-axis from left to right and from right to left. Also, lines are scanned parallel to the y-axis axis from the top to the bottom. A pixel change from white to black fairly reliably indicates that the outline is found. Exceptions occur due to background noise and shading of the head on the background (see Section 3 for more details). Figure 1 bottom left shows that scans on the x-axis from the right sometimes do not reach the head due to shading in the background surface. This line scan algorithm works under the assumption that the outline of someones head is fairly concave, which in general seems a reasonable assumption. However, unkempt hair styles can be cause for concern for this algorithm.

The outline consists of the pixels found by the three scans (from the right, from the left and from the top). The faces typically are not nicely centered in the image e.g. as in the Yale data set. In order for the outlines to match up, these outlines are translated to the center of the image, first by aligning the highest pixel to the top of the image, and in the x-direction by translating toward the average of the pixels found from the scan from the left.

Since we are interested in the top of the head only, we crop the outline to the pixels appearing within 100 pixels from the top of the outline. Finally, the same pixel can be found by scans from the top and scans from the side, therefore, we de-duplicate the outline as a last step.

In summary, these are the steps followed by the outline recognition algorithm;

1. Crop image
2. Sharpen image
3. Find boundaries by line scanning
4. Translate boundaries to center of image
5. Crop boundary pixels
6. De-duplicate image boundary pixels

We distinguish three methods of cropping, giving us three different outlines. Firstly, we crop around the whole face, secondly, we crop the left side of the face and thirdly we crop the right side of the face. Figure 2 shows the various outlines obtained for a single facial image.

Fig. 2. Different methods of cropping give different outlines

2.1   Distance Calculation

To recognize a face, we calculate the distance between the outline of an image to be classified and those in the training database. The distance is calculated as follows; let $\mathbf{x} = \{x_1, \ldots, x_n\}$ be the set of n distinct pixels in the outline of an image and $\mathbf{y} = \{y_1, \ldots, y_k\}$ the set of k pixels in an image from the database. Note that the number of pixels in the two outlines differ in general (i.e. n does not need to be equal to k). The distance between a point x_i ($1 \leq i \leq n$) and the outline $\mathbf{y}$ is the minimum Euclidean distance of x_i and any of the points in $\mathbf{y}$. The total distance between $\mathbf{x}$ and $\mathbf{y}$ is then the sum of the distances of the points in $\mathbf{x}$. To cater for the different number of pixels, we divide the total distance by the number of points in $\mathbf{x}$, which gives us an average distance for each of the pixels, making it easier to compare distances among different pictures. In summary, the distance $d(\mathbf{x}, \mathbf{y})$ is defined as

$$d(\mathbf{x}, \mathbf{y}) = \frac{\sum_{i=1}^{n} min_{j=1,\ldots,k} ||x_i - y_j||}{n}$$

where $||.||$ is the length of the vector. The face returned from the database containing m outlines $\mathbf{y}_1, \ldots, \mathbf{y}_m$ is

$$argmin_{i=1,\ldots,m} d(\mathbf{x}, \mathbf{y}_i)$$

2.2   PCA on Partial Image

One way of looking at image recognition is trying to find the closest point in a space where each pixel spans one dimension in that space. Since the number of pixels tends to be quite high, the dimensionality of the space can become too large for practical classification. Principle component analysis (PCA) is a method to reduce the space spanned by the pixels into a reduced dimension space.

To apply PCA to face recognition, an image with n pixels is considered to represent an n dimensional vector $\mathbf{x} = \{x_1, \ldots, x_n\}$ where the grey-scale value of a ith-pixel is taken as the entry x_i of that pixel. In order to compensate for different lighting conditions, the data is preprocessed by calculating the average over the m images in the training set $\mathbf{x_m} = \frac{1}{m} \sum_{i=1}^{m} \mathbf{x}_i$ and subtracting this from each image. So, instead of using the vector representation $\mathbf{x}$ we use $\mathbf{x} - \mathbf{x_m}$. From these vectors, we calculate the covariance matrix and determine the eigenvalues and eigenvectors. The eigenvectors with the largest values are used as principal components (PC). Each of the images in the training set is projected onto these PCs. To classify a new image, the mean image x_m is subtracted, the resulting vector projected on the PCs and the training image with PC of closest Euclidean distance is taken as the predicted class. This method has been applied to face recognition problems with good success.

Instead of taking the outlines of the heads, we took a 50% fraction of the picture being the left half of the picture, the right half, and the top half. Application of PCA on each of these half images gives us a classification of the picture. We used 50 eigenvalues in our experiments.

PCA is very sensitive to the position of the face with respect to the boundaries of the picture. In order to alleviate this problem to some extend the image we want to classify is cropped so that an image of size $l \times h$ is reduced to an image of size $l - c \times h - c$. There are $c \times c$ such images, and instead of classifying the original image, we classify the $c \times c$ images and select the class with the closest distance to any of these images. In our experiments, we used $c = 4$, so 16 images were used to classify each test image.

2.3   Exploiting Symmetry

Due to the symmetry found in face, it can be expected that a left half image can find some resemblance in the mirrored right half of a face image. Figure 3 illustrates the process. So, left halves can be matched against right half mirrored images (likewise for right halves). As we shall see in the empirical sections, this result in classifiers that are better than base line predictors, but much worse than classifications for left, right and top half images. Another approach is to match left halves to a set containing both left halves and right halve mirrored images.

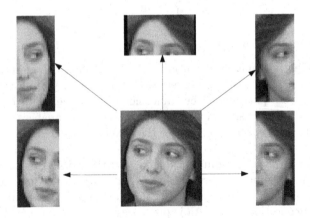

Fig. 3. Splitting of image into different sub-images

2.4   Combining Features

Sometimes, when using a single feature, the mostly likely person given a picture of a face will be incorrectly classified. In such cases, the second most likely person if often the one we are interested in. We can exploit the fact that features and classes are highly correlated by not only considering the most likely person indicated by a feature, but also the second most likely person. This way, a single feature gives rise to two attributes used for classification. Furthermore, the features we considered are ranked using a distance measure. This distance measure can be used as an attribute as well. So, a single feature can give rise to up to four attributes.

A straightforward way to combine the attributes based on the features of outlines is to train a new classifier on these attributes. We considered a range of standard classification schemes including naive Bayes, C4.5, nearest neighbor, SVM, AODE after discretizing data, bagging trees and others, but none of the off the shelve classifiers turned out to give satisfactory results (see Section 3.2).

We propose a simple method for combining features, namely by majority voting; each feature gets a vote for the closest and second closest person and the person selected is the one with the most votes. This results in a lot draws, and furthermore the second closest person for a feature clearly is not equally favored with the closest person. A simple refinement that addresses the second issue is to vote with the closest person getting a single vote and the second closest person gets only half a vote. However, this still results in a lot of draws in the voting. The mechanism to break ties that we used in our experiments is to select the lowest numbered person in the dataset.

This voting scheme can be refined by taking the distance information in account as follows. The closer the distance the larger the vote, so we divide the vote by the distance. Formally, the classifier returns

$$argmax_i \sum_j I(f_j = i)w_j/d_j$$

where f_j is the jth feature, $I(.)$ the indicator function, w_j is 1 for the closest person, $\frac{1}{2}$ for the second closest and d_j is the distance for feature j.

3   Empirical Results Outline Features

We use the Yale face database [1] containing images of 15 subjects with 11 different expressions. Original images were 320 by 243 pixels in 255 grey scale. The following configuration/expressions were used; (a) center-light, (b) with glasses, (c) happy, (d) without glasses, (e) normal, (f) sad, (g) sleepy, (h) surprised, (i) wink, (j) left lighted and (k) right lighted. For three expressions, namely right lighted, left lighted and front lighted images our simple outline recognition algorithm could not reliably detect outlines due to interference of shadows (see Figure 4). Outlines for these expressions showed no resemblance to a head outline whatsoever and consequently classifications based for these expressions turned out pour. Therefore, we will distinguish in our analysis between the non-lighted images (expressions b through i), non-side lighted images (expressions a through i) and all images (expressions a through k). First, we study the behavior of classification based on a single outline, then we study the combined effect of the outlines.

3.1   Single Outline Behavior

All experiments were done using leave one out cross validation. The accuracy for the left, right and top outlines is summarized in Table 1. Given the simplicity of

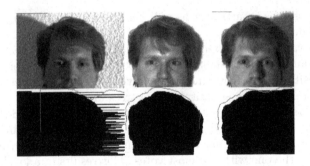

Fig. 4. Lighted images causing problems for detecting outlines

the features, these performs remarkably well with accuracies up to 95%. Closer inspection of the results reveals that most errors are made in the left and right lighted images and to a lesser extend in the front lighted images.

Table 1. Yale faces accuracy for single outlines (using leave one out cross validation)

	All 165 images	With frontal light (135)	No lighted images(120)	All images	With frontal light	No lighted images
	Closest person			Contained in closest two people		
Left outlines	79.4%	89.6%	93.3%	84.8%	93.3%	94.2%
Right outlines	77.6%	90.4%	93.3%	87.3%	97.0%	99.2%
Top outlines	75.8%	91.9%	95.0%	78.8%	94.1%	96.7%

Table 1 lists the percentage of images where the person was the closest, or second closest person. Clearly it frequently happens that if the closest matching outline is of a different person then the second closest matching outline (of a different person from the first one) is of the person that we are after. This information is useful when combining features, as we shall see shortly.

3.2   Combined Outline Behavior

Table 2 shows the results for various standard algorithms with different sets of features. Since the second closest person seems informative, we can take these as features as well. Nearest neighbor shows some promise, especially when closest person with distance features are used where it is consistently better than any of the single feature classifiers (Table 1). However, it cannot consistently exploit information on the second closest person. We looked at a few other standard methods (SVM, AODE after discretizing data, bagging trees) but none produced better results. Instead of using an off the shelf method, we turned to majority voting of the features. The first rows of Table 3 shows the accuracy for one vote per feature, and the second assigns a full vote to the closest person and half a vote

Table 2. Yale faces accuracy for single outlines (using leave one out cross validation)

	All images	With frontal light	No lighted images	All images	With frontal light	No lighted images
	Only closest person attributes					
	closest person only			second closest person as well		
Naive Bayes	80.6%	91.9%	95.0%	79.4%	90.4%	95.8%
Nearest Neighbor	77.6%	91.9%	96.7%	75.7%	89.6%	94.2%
C4.5	79.4%	91.9%	95.0%	79.4%	91.9%	95.0%
	With distance attributes as well					
Naive Bayes	76.4%	86.7%	87.5%	76.4%	86.7%	91.7%
Nearest Neighbor	80.0%	92.6%	96.7%	76.4%	89.6%	94.2%
C4.5	78.8%	90.4%	93.3%	79.4%	90.4%	93.3%

to the second closest. Accuracy raises considerably compared to single feature classifiers. However, there are quite a few instances where there are multiple persons with the same number of votes. The lowest numbered person is chosen to break those ties which result in some instances being correctly classified by luck, (indicates by numbers in braces in Table 3).

Instead of using full votes, the distance information can be used. Table 3 shows results in the last two rows. Clearly, this solves the problem with ties. Furthermore, accuracy is consistently higher than for single feature classifiers and off the shelf methods listed in Table 2. Using second closest person information increases accuracy consistently over using closest person only.

As comparison, we applied eigenfaces (also known as PCA) on the full image with the same resolution since it is the closest to our technique. Furthermore, we looked at a number of other classifiers, but the one we found to perform best over all was support vector machines, which are known to perform well on face recognition [5]. A linear kernel turned out to perform best. Table 3 shows the leave-one-out cross validation accuracy for these algorithms. Clearly, the simple outline features perform almost always better than these general purpose techniques. The differences become even larger when PCA features are taken in account, as we will see in Section 5.

Table 3. Yale faces majority voting algorithms (using leave one out cross validation). Accuracy in percentage, followed by number of ties with lucky ones in braces.

	All images	With frontal light	All lighted images
1st person	87.9% 27(7)	97.8% 10(4)	99.2% 5(2)
+2nd person	87.3% 15(5)	97.8% 5(3)	99.2% 2(1)
dist 1st person	82.4% 0	93.3% 0	96.7% 0
dist +2nd person	84.2% 0	95.6% 0	98.3% 0
Eigenfaces	71.5%	83.0%	92.5%
SVM	87.3%	95.5%	95.8%

4   Empirical Results PCA Features

The ATT face database (formerly Olivetti) [11] contains 400 images of 92 by 112 pixels at 8-bit grey levels. There are 40 different subjects with ten different poses each taken at different times and with different poses. The images are fairly well cropped around the face of the subjects, so very few outlines of heads are fully in the pictures. Therefore, the outline technique described in Section 2 are not applicable to this data set. However, PCA is applicable so we consider PCA applied to half images of the data set.

Table 4. Results on ATT faces database (using leave one out cross validation)

	accuracy (out of 400)	Ties (lucky)
Eigenfaces	78.25%	
SVM	98.25%	
PCA Left part of image:	95.5%	
PCA Right part of image:	92.75%	
PCA Top part of image:	95.25%	
PCA Left mirrored	42.5%	
PCA Right mirrored	44%	
PCA left+mirrored	96.25%	
PCA right+mirrored	95.25%	
Various combinations	96%-98.5%	2-7(1-3)
Left(+.m)+Right(+.m)+Top (using 2nd person)	99.25%	3(3)
Distance weighted left(+.m)+right(+.m)+top (using 2nd person)	99.25%	0

Table 4 lists classification accuracy based on just these half images. These results are for leave one out cross validation (leaving training data in the test set results in 100% accuracy, which is not very informative). To keep experimental time in check, we reduced the image resolution by combining every four pixels into a single pixel.

Classifying left halves using PCA on the right half images (taking care to leave out the right half for the image from the set of right half images) gives 42.5% accuracy and right halves on left half images gives 44% accuracy. This is significantly over the 2.5% accuracy of a default class classifier, but far from the high nineties accuracies we are aiming at as shown in Table 4. By adding the left mirrored images to the right half images we get a new dataset with a larger accuracy. Left half images get classified with 96.25% accuracy, which is 0.75% higher than when classified on just left side images. Right half images get classified at 95.25% accuracy which is an improvement of 2.5% over classification on right half images alone. Table 4 summarizes these results.

We considered various combinations of PCA features on half images and most combinations improved on the accuracy of the worst base classifier (see Table 4).

Ties were broken by selecting the lowest numbered person, which caused some instances to be classified correct by luck. Like for outlines, the second closest person gives extra information and fairy consistently improved classification accuracy. The most successful combination is the one where left, right and top images are taken together with the left+mirrored and right+mirrored features. Plain voting (i.e. 1 vote for the closest and $\frac{1}{2}$ for second closest) resulted in 99.25% accuracy. However, 3 images were classified correctly out of the 3 ties contributing $\frac{3}{4}$% to the accuracy.

In order to decrease the probability of ties occurring, we use the distance that the PCA algorithm produces for ranking the persons. The closest person is weighted 1/distance of closets person, and the second closest person is weighted 0.5/distance of second closest person. Table 4 has its entry labelled as *distance weighted*. The number of ties reduces to zero while accuracy increases to an unprecedented 99.25%. To our knowledge, this is the highest accuracy obtained on this dataset so far.

Reported accuracy for this dataset tends to be significantly lower, for example Eigenfaces [11] less than 90% HMM [11] less than 95% and Kernel PCA [12,7] between 97.5% and 98%. However, these results may not be directly comparable due to the different experimental setting. In order to gain a better comparison, we applied Eigenfaces and SVMs to the same resolution image and Table 4 shows the leave-one-out cross validation results. The combined PCA voting algorithm performs more favorably than these algorithms from the literature and performs similar with Kernel PCA applied on the whole image (i.e. not reduced in resolution like done in this article).

5   Results for PCA + Outline Features

For the Yale faces dataset, both PCA on (cropped) half images and outline features are available. First, we performed the same experiments as for the ATT database on the Yale faces dataset with only PCA features and the same trends as for the ATT database were observed. However, since we reduced resolution by combining blocks of 5×5 pixels into a single pixels for the Yale faces dataset, classification accuracy was not very high (in the 60s and 70s percent range for various individual features and combination of features).

However, since we now have outline features available as well, it makes sense to combine these with the PCA features. Just weighting the three outline features with their distance and giving the second closest person half weight, we get 84.6% with leave one out cross validation on all Yale faces. Leaving out the front lighted and side lighted increases accuracy to 95.6% and 98.3% respectively. Similarly weighing PCA features for left, right and top halves and left+mirrored and right+mirrored gives 76.4% accuracy for the full set 90.4% when side lighted images are ignored and 96.7% when front lighted faces are ignored as well. By combining both sets of features we get a new classifier. The average distance for outlines is calculated different than that for PCA images, and is 41.33 as large so to compensate the weights for PCA images are multiplied by this factor. The

Table 5. Yale faces results combined outlines and PCA

	All images	With frontal light	No lighted images
Distance weighted outlines	84.2%	95.6%	98.3%
Distance weighted PCA faces	76.4%	90.4%	96.7%
Distance weighted both	83.6%	97.0%	100.0%

combined classifier is never worse than any the individual ones, and reaches 100% accuracy on all images that are not front or side lighted. Table 5 summarizes the results.

6   Computational Considerations

For a single outline feature it took approximately 7 seconds for a full cross validation over the 165 images, including image reading from file on a 5600+ AMD Dual Core machine with implementation in Java. A single classification of a half image for Yale dataset took approximately 30 second, which makes about an hour and a quarter for a complete cross validation. So, the outline features' computational requirements can be considered negligible compared to that of the PCA features.

The benefit of combined PCA is that it only needs to deal with half the number of pixels for each of the images. PCA's computational cost is $O(mn^2 + n^3)$ and it requires $O(nm + n^2)$ memory where n is the number of pixels and m is the number of images. Since $n >> m$ the complexity is dominated by the $O(n^3)$ term and memory by the $O(n^2)$ term. Reducing the number of pixels by half thus reduces computational cost to 12.5% and memory requirements to a quarter.

7   Discussion and Relation to Other Work

Ensemble methods like bagging [3], boosting, stacking [14] are based on a philosophy that come closest to our method; use many base classifiers that each on their own are weak and combine them into a single classifier which tends to outperform each of the base classifiers and often performs a lot better. In our experiments outlined in Section 3 we found that stacking does not perform well on this task.

The original bagging method uses classifiers on different samples of the dataset and each give them a vote. It was found that in the situation where the base classifier produces a probability distribution over the classes, a better way to combine the base classifiers is to have them vote proportional to the probabilities instead of assigning a single vote to the most likely class [2]. The features that apply to face recognition do not produce probabilities, but do provide a ranking of the faces through a distance measure. As shown in our experiments, weighing votes by a function of the distance provided by the features results in a better classifier.

There are a large number of general face recognition methods [6], such as HMM, PCA, ICA, LDA, support vectors, case based methods, etc. Most methods are general purpose classification methods applied to the task of face recognition. Our method differs in that it is specific to the face recognition task and combines a number of methods into a single classifier. In our method, we used PCA as a base classifier. However, other methods such as kernel PCA, ICA or LDA may be more suited as base classifier. As demonstrated in our experiments, in particular support vector machines offer a good candidate as alternative base classifier to PCA.

Benefits of our method is that with simple features a well performing classifier can be constructed. There are a number of obvious limitations to our method, all having to do with the circumstances under which the method is applied.

The features outlined for the Yale dataset suffer from sensitivity to hair style, which is an easily changed feature. Likewise, changes of the appearance using facial hair features, sun glasses, ear rings and other ornaments can impact on the performance. Dealing with such occlusions are part of ongoing current research. However, when one of the features does not perform well (e.g. hair style) but there is sufficient information in the other features, the methods still performs reasonable, as shown in the experiments on the Yale dataset under different lighting conditions.

As many other face recognition techniques, out method is sensitive to lighting conditions, as shown in the performance results on the Yale dataset. Techniques exist to make the method less susceptible to changing lighting in an environment, for instance, using infrared cameras to obtain a face image [6].

An assumption made by our method is that the position of the face with respect to the camera is the same in train and test examples. This is a fair assumption when the environment is fully controlled. However, this does not apply to less restricted tasks such as face recognition in a crowds.

8   Conclusions

We proposed a general method for face recognition based on extraction of various features from an image, using a classifier on each of these feature and combining the classifications into a single classifier. The final prediction in this hierarchical classifier can be performed by any general purpose classifier. In this paper, we developed such a classifier based on weighing the votes of individual features. The voting method takes best and second best matches in account as well as the distance measure used for each of the features. A simple set of feature representing the outline of a head was taken from images cropped to the left, to the right and centering the head in an image. Another set of features was developed based on PCA applied to parts of the image. Empirical results on benchmark datasets give accuracies that are favorable compared to single feature classifiers and to results presented in the literature all at smaller computational cost.

In future research we plan to define more specific face recognition features. Furthermore, we like to combine the predictions of general purpose face recognition algorithms such as kernel PCA, ICA and LDA based systems with those based on simple features.

References

1. Belhumeur, P.N., Hespanha, J.P., Kriegman, D.J.: Eigenfaces vs. Fisherfaces: Recognition using class specific linear projection. IEEE Transactions on Pattern Analysis and Machine Intelligence 19(7), 711–720 (1997); Yale face database, http://cvc.yale.edu/
2. Bouckaert, R.R., Goebel, M., Riddle, P.J.: Generalized Unified Decomposition of Ensemble Loss. In: Australian Conference on Artificial Intelligence 2006, pp. 1133–1139 (2006)
3. Breiman, L.: Bagging predictors. Machine Learning 24(2), 123–140 (1996)
4. Delac, K., Grgic, M., Grgic, S.: Independent Comparative Study of PCA, ICA, and LDA on the FERET Data Set. International Journal of Imaging Systems and Technology 15(5), 252–260
5. Guo, G., Li, S.Z., Chan, K.L.: Support vector machines for face recognition Image and Vision Computing 19(9-10), 631–638 (2001)
6. Kong, S.G., Heo, J., Abidi, B.R., Paik, J., Abidi, M.A.: Recent advances in visual and infrared face recognition a review. Computer Vision and Image Understanding. Elsevier, Amsterdam (2005)
7. Kim, K.I., Jung, K., Kim, H.J.: Face recognition using kernel principal component analysis Signal Processing Letters. IEEE 9(2), 40–42 (2002)
8. Liu, C., Wechsler, H.: Comparative Assessment of Independent Component Analysis (ICA) for Face Recognition. In: Proc. of the Second International Conference on Audio- and Video-based Biometric Person Authentication, AVBPA 1999, Washington D.C., USA, March 22-24, pp. 211–216 (1999)
9. Lu, J., Plataniotis, K.N., Venetsanopoulos, A.N.: Face Recognition Using LDA-Based Algorithms. IEEE Trans. on Neural Networks 14(1), 195–200 (2003)
10. Moon, H., Phillips, P.J.: Computational and Performance aspects of PCA-based Face Recognition Algorithms. Perception 30, 303–321 (2001)
11. Samaria, F., Harter, A.: Parameterisation of a Stochastic Model for Human Face Identification. In: Proceedings of 2nd IEEE Workshop on Applications of Computer Vision, Sarasota FL (December 1994)
12. Yang, M.-H., Ahuja, N., Kriegman, D.: Face recognition using kernel eigenfaces Proceedings. In: 2000 International Conference on Image Processing, vol. 1, pp. 37–40 (2000)
13. Witten, I.H., Frank, E.: Data mining: Practical machine learning tools and techniques with Java implementations. Morgan Kaufmann, San Francisco (2000)
14. Wolpert, D.H.: Stacked generalization. Neural Networks 5, 241–259 (1992)
15. Zheng, W., Zou, C., Zhao, L.: Face recognition using two novel nearest neighbor classifiers. In: IEEE International Conference on Acoustics, Speech, and Signal Processing, vol. 5, pp. 725–728 (2004)

Estimating Likelihoods for Topic Models

Wray Buntine*

NICTA and Australian National University
Locked Bag 8001, Canberra, 2601, ACT Australia
wray.buntine@nicta.com.au

Abstract. Topic models are a discrete analogue to principle component analysis and independent component analysis that model *topic* at the word level within a document. They have many variants such as NMF, PLSI and LDA, and are used in many fields such as genetics, text and the web, image analysis and recommender systems. However, only recently have reasonable methods for estimating the likelihood of unseen documents, for instance to perform testing or model comparison, become available. This paper explores a number of recent methods, and improves their theory, performance, and testing.

1 Introduction

Topic models are a discrete analogue to principle component analysis (PCA) and independent component analysis (ICA) that model *topic* at the word level within a document. They have many variants such as NMF [LS] PLSI [Hof] and LDA [BNJ], and are used in many fields such as genetics [PSD], text and the web, image analysis and recommender systems. A unifying treatment of these models and their relationship to PCA and ICA is given by Buntine and Jakulin [BJ2]. The first Bayesian treatment was due to Pritchard, Stephens and Donnelly [PSD] and the broadest model is the Gamma-Poisson model of Canny [Can].

A variety of extensions exist to the basic models incorporating various forms of bierarchies [BGJT, MLM], combining topical and syntactic information [GSBT], jointly modelling text and citations/links [NAXC], models of information retrieval [AGvR]. The literature is extensive, especially in genetics following [PSD] and using NMF, and we cannot hope to cover the breadth here.

A continuing problem with these methods is how to do unbiased evaluation of different models, for instance using different topic dimensions or different variants and hierarchical models. The basic problem is that the likelihood of a single document (or datum/image) incorporates a large number of latent variables and thus its exact calculation is intractable. In this paper we present the problem according to the theory in its Dirichlet formulation such as LDA, but following the theory of [BJ2] our results apply more broadly to the wider class of topic models.

---

* Also a fellow at Helsinki Institute of IT.

Z.-H. Zhou and T. Washio (Eds.): ACML 2009, LNAI 5828, pp. 51–64, 2009.
© Springer-Verlag Berlin Heidelberg 2009

Innovative methods for estimating the likelihood given a model and a test set of documents have been tried. The first approach suggested was the harmonic mean [GS, BJ1], but has proven to be considerably biased. Li and McCallum noted the unstable performance of this and proposed Empirical Likelihood [LM], a non-probabilistic method that is not able to give individual document estimates but rather broad scores. The approach is not widely used. Another approach suggested is to hold out a set of words [RZGSS], rather than just a set of documents. Since training is not optimising for the scores of the held out words, unbiased perplexity scores can be computed for these words. While this approach is reasonable, it still does not address the whole problem, the quality of individual documents.

Recently, an advance has been made with a number of algorithms presented and tested in the groundbreaking breaking paper of [WMSM]. Testing of Wallach's left-to-right algorithm [Wal, Algorithm 3.3] under small controlled tests (reported in Section 4.1) indicates that it is still biased. In this paper we present one new method for likelihood evaluation, based on the left-to-right algorithm, and revisit an importance sampling one. For both, improved theory is presented. The algorithms are tested rigorously in a number of controlled situations to demonstrate that significant improvements are made over previous methods.

2   Notation and Problem

This section introduces the notation used and then the problem being considered.

2.1   Documents and Topics

In our data reduction approach, one normally has a collection of documents and are estimating the model parameters for the topic model from the documents. We will, however, consider only a single document, one whose likelihood we wish to estimate. The document has L terms (or words or tokens), and these are indexed $l = 0, \ldots, L - 1$. The l-th term in the document has dictionary value j_l. Assume the dictionary has J entries numbered $0, \ldots, J - 1$. The values of terms can be stored as a sequence in vector j, or "bagged" into a vector of sparse counts.

Now we will associate the terms with K topics, aspects or components, and give each document a vector of propensities for seeing the topics, represented as a K-dimensional probability vector q. This propensity vector is sampled per document.

We will also assign to each term indexed by (l) a hidden topic (also called aspect or component) denoted $k_l \in \{0, \ldots, K - 1\}$. This is modelled as a latent variable.

2.2   Model

The full probability for a document is given by a product of generative probabilities for each term. For the vectors of latent variables q, k and the data vector j.

$$q \sim \mathrm{Dirichlet}_K(\boldsymbol{\alpha}) \,,$$
$$k_l \sim \mathrm{Discrete}_K(\boldsymbol{q}) \qquad \text{for } l = 0, \ldots, L-1 \,,$$
$$j_l \sim \mathrm{Discrete}_J(\boldsymbol{\theta}_{k_l}) \qquad \text{for } l = 0, \ldots, L-1 \,.$$

Here the subscripts K, J indicate the dimensions of the distributions. Note the component indicators $\boldsymbol{k}$ can be aggregated in total counts per class, to become a K-dimensional counts vector $\boldsymbol{C}$ with entries

$$C_k = \sum_{l=0,\ldots,L-1} 1_{k_l = k} \,.$$

This aggregate $\boldsymbol{C}$ corresponds to entries in the *score matrix* in conventional PCA. The parameter matrix $\boldsymbol{\Theta}$ corresponds to the *loading matrix* in conventional PCA. Here the model parameters are

$\boldsymbol{\alpha}$: A K dimensional vector of Dirichlet parameters generating probabilities for the topic.

$\boldsymbol{\Theta}$: A $K \times J$ dimensional matrix defines term probabilities, with column vectors $\boldsymbol{\theta}_k$ giving them for topic k. This is the loading matrix in conventional PCA.

One can extend topic models in all sorts of ways, for instance placing a hierarchical prior on $\boldsymbol{\alpha}$ or $\boldsymbol{\Theta}$ or splitting the terms up into separate semantic parts, such as citations and words, and modelling them with separate processes. The literature here is extensive.

2.3   The Problem

The full likelihood for a document, after marginalising $\boldsymbol{q}$ takes the form

$$p(\boldsymbol{k}, \boldsymbol{j} \,|\, \boldsymbol{\alpha}, \boldsymbol{\Theta}) = \frac{Z_K(\boldsymbol{C} + \boldsymbol{\alpha})}{Z_K(\boldsymbol{\alpha})} \prod_{l=0,\ldots,L} \theta_{k_l, j_l} \,, \qquad (1)$$

where $Z_K()$ is the normalising constant for the Dirichlet distribution. Assume the model parameters are given, then the task we are considering is how to evaluate the marginal of this, $p(\boldsymbol{j} \,|\, \boldsymbol{\alpha}, \boldsymbol{\Theta})$, which means summing out over all K^L values for $\boldsymbol{k}$ in Equation (1). For L taking values in the 100's and K taking values in the 10's, the exact calculation is clearly impractical, and no alternative exact algorithms to brute force are known. Also note, that following [BJ2], the methods should be readily adapted to related models such as NMF.

3   Sampling Methods

We now consider several algorithms for the task of estimating document likelihoods for these basic topic models. Note that these should extend to more sophisticated topic models based on the same framework.

3.1 Importance Sampling

The method of importance sampling works as follows. Suppose one wishes to estimate $\mathcal{E}_{p(v)}[f(v)]$, and use a sampler $q(v)$ instead of $p(v)$, then importance sampling uses the N samples $\{v_n : n = 1, ..., N\}$ to make the unbiased estimate $\sum_n f(v_n)\frac{p(v_n)}{q(v_n)}$. Now if $q(v)$ is approximated via Gibbs, so its normalising constant is not known, then one uses

$$\frac{\sum_n f(v_n)\frac{p(v_n)}{q(v_n)}}{\sum_n \frac{p(v_n)}{q(v_n)}},$$

where the denominator estimates the inverse of the normaliser of $q(v)$.

3.2 The Harmonic Mean

The first method suggested for estimating the likelihood uses the second form of importance sampling with $v \to k$. $f(v) \to p(j \mid k, \alpha, \Theta)$, $p(v) \to p(k \mid \alpha, \Theta)$, and $q(v) \to p(j, k \mid \alpha, \Theta)$. The renumerator simplifies dramatically to the sample count. Then one takes N samples of k using Gibbs sampling from $p(k, j \mid \alpha, \Theta)$, and forms the estimate $\hat{p}(j \mid \alpha, \Theta)$

$$\frac{N}{\sum_n 1/p(j \mid k_n, \alpha, \Theta)}$$

This formula is a harmonic mean, the inverse of the mean of the inverses, suggested in [GS, BJ1].

Unfortunately, the convergence is not stable in general [CC], the variance of the approximation can be large (since k is finite discrete, and no probabilities are zero, it must be finite). In practice, one can see this because the importance weights $w_n = \frac{p(v_n)}{q(v_n)}/\sum_n \frac{p(v_n)}{q(v_n)}$ are mostly near zero and usually only one or two significantly nonzero. Thus the estimate is usually dominated by the least $p(j \mid k_n, \alpha, \Theta)$ seen so far. This makes it highly unstable.

3.3 Mean Field Approximation

The weakness in the last approach occurs because of the need to estimate the normalising constant, in the second form of importance sampling. Instead, use a proposal distribution $q(v)$ for which a normaliser is known. Then the estimate becomes

$$\frac{1}{N}\sum_n p(j, k_n \mid \alpha, \Theta)\frac{1}{q(k_n)} \tag{2}$$

Since the optimum $q(v)$ is proportional to $f(v)p(v)$ (when $f(v) \geq 0$), one could develop $q(v)$ by a Kullback-Leibler minimiser. So set $q(k) = \prod_{l < L} q_l(k_l)$ and find $q()$ to minimise

$$\mathcal{E}_{k \sim q(k)}\left[\log \frac{q(k)}{p(j, k_n \mid \alpha, \Theta)}\right]$$

This yields the system of rewrite rules [GB]

$$q_l(k_l) \propto \exp\left(\mathcal{E}_{\mathbf{k}_{-l}\sim\prod_{m\neq l} q_m(k_m)}\left[\log p(\mathbf{j}, \mathbf{k}_n \mid \boldsymbol{\alpha}, \boldsymbol{\Theta})\right]\right)$$

where $\mathbf{k}_{-l}$ is $\mathbf{k}$ with the entry k_l removed. The rewriting system will converge due to the general properties of Kullback-Leibler minimisation. This simpifies to (in the first proportion, note $C_{k'}$ is a function of $\mathbf{k}$ and thus includes k_l),

$$q_l(k_l) \propto \theta_{k_l,j_l} \exp\left(\sum_{k'} \mathcal{E}_{\mathbf{k}_{-l}\sim\prod_{m\neq l} q_m(k_m)}\left[\log \Gamma(C_{k'} + \alpha_{k'})\right]\right) ,$$

$$q_l(k) \propto \theta_{k,j_l} \exp\left(\mathcal{E}_{\mathbf{k}_{-l}\sim\prod_{m\neq l} q_m(k_m)}\left[\log(C'_k + \alpha_k)\right]\right) , \tag{3}$$

where $C'_k = C_k - 1_{k_l=k}$ (which is independent of k_l since its effect is removed). Now $\mathcal{E}_{u\sim p(u)}\left[g(u)\right]$ can be approximated by $g(\overline{u})$ or $g(\overline{u}) - \sigma_u^2\frac{1}{2\overline{u}^2}$, where $\overline{u}$ and σ_u^2 are the mean and variance by $p(u)$. Thus we have two options for the rewrite rules then, the simpler first order version is

$$q_l(k) \propto \theta_{k,j_l}\left(\mathcal{E}_{\mathbf{k}_{-l}\sim\prod_{m\neq l} q_m(k_m)}\left[C'_k\right] + \alpha_k\right)$$

$$\propto \theta_{k,j_l}\left(\sum_{m\neq l} q_m(k) + \alpha_k\right) \tag{4}$$

Note, after all this theory, we have derived an approach virtually identical to the iterated pseudo-counts importance sampler (IS-IP) of [WMSM]. That this method performed far better than its related importance sampling algorithms [WMSM, Figure 1] comes as no surprise, given its derivation here as a mean-field approximation to the optimal importance sampler. For the second order version we subtract the variance term, which is computed similarly. Since this is part of a larger approximation, either version could work to construct a proposal distribution.

3.4 Left-to-Right Samplers

Wallach [Wal] suggests a particular sampler that breaks the problem into a series of parts:

$$p(\mathbf{j} \mid \boldsymbol{\alpha}, \boldsymbol{\Theta}) = \prod_{l<L} p(j_l \mid j_1, \ldots, j_{l-1}, \boldsymbol{\alpha}, \boldsymbol{\Theta}) . \tag{5}$$

Each term is estimated seperately using vector samples $(k_1, \ldots, k_{l-1}) \sim p(k_1, \ldots, k_{l-1} \mid j_1, \ldots, j_{l-1}, \boldsymbol{\alpha}, \boldsymbol{\Theta})$:

$$p(j_l \mid j_1, \ldots, j_{l-1}, \boldsymbol{\alpha}, \boldsymbol{\Theta})$$

$$\approx \frac{1}{|\text{Sample}|} \sum_{(k_1,\ldots,k_{l-1})\in\text{Sample}} p(j_l \mid j_1, \ldots, j_{l-1}, k_1, \ldots, k_{l-1}, \boldsymbol{\alpha}, \boldsymbol{\Theta}) \tag{6}$$

where the probability in the mean in Equation (6) is calculated using

$$p(j_l \mid j_1, \ldots, j_{l-1}, k_1, \ldots, k_{l-1}, \boldsymbol{\alpha}, \boldsymbol{\Theta}) = \sum_k \theta_{k_l, j_l} p(k_l \mid k_1, \ldots, k_{l-1}, \boldsymbol{\alpha}, \boldsymbol{\Theta}) . \quad (7)$$

3.5   Left-to-Right Particle Sampler

Wallach's approach generates the vector samples $(k_1, \ldots, k_{l-1})$ for different l independently as follows:

1. For $l = 0, \ldots, L - 1$,
   (a) For $l' = 0, \ldots, l - 1$, resample $k_{l'}$ using
   $$k_{l'} \sim p(k_{l'} \mid j_1, \ldots, j_{l-1}, k_1, \ldots, k_{l'-1}, k_{l'+1}, \ldots, k_{l-1}, \boldsymbol{\alpha}, \boldsymbol{\Theta}) .$$
   (b) Sample k_l using $p(k_l \mid j_1, \ldots, j_l, k_1, \ldots, k_{l-1}, \boldsymbol{\alpha}, \boldsymbol{\Theta})$.
   (c) Record sample details using the current values $(k_0, \ldots, k_l)$ and Formula (7).

This is done R times as a so-called particle sampler. One particle's run generates L vectors from size 1 to L:

$$\{k_0\}, \{k_0, k_1\}, \{k_0, k_1, k_3\}, \{k_0, k_1, k_3, k_4\}, \ldots, \{k_0, \ldots, k_{L-1}\} .$$

All R particles takes $RL^2/2$ multinomial samples and generates RL vectors used to generate estimates for the L terms in Equation (5).

3.6   Left-to-Right Sequential Sampler

Alternatively, generate the samples sequentially, so instead of R independent operations,

1. For $l = 0, \ldots, L - 1$,
   (a) Repeat R times:
       i. For $l' = 0, \ldots, l$, resample $k_{l'}$ using
       $$k_{l'} \sim p(k_{l'} \mid j_1, \ldots, j_l, k_1, \ldots, k_{l'-1}, k_{l'+1}, \ldots, k_l, \boldsymbol{\alpha}, \boldsymbol{\Theta}) .$$
       ii. Record detail using the sample $(k_0, \ldots, k_l)$ and Formula (7).
   (b) Form the mean from the R samples to estimate Formula (6).
2. Form the estimate of Equation (5) from the L means.

This has the same complexity as the particle version but in contrast is easily seen to produce an unbiased estimate of Equation (5) as R approaches infinity for fixed L.

Lemma 1. *The left to right sampler gives unbiased estimates of $p(\boldsymbol{j} \mid \boldsymbol{\alpha}, \boldsymbol{\Theta})$ for sufficiently large sample size R.*

Proof. For sufficiently large R, the L estimates for $p(j_l \mid j_1, \ldots, j_{l-1}, \boldsymbol{\alpha}, \boldsymbol{\Theta})$ become independent, and thus the mean of their product is the product of their means. Since each of these are unbiased, their product is unbiased.   □

The particle sampler does not produce unbiased estimates regardless of sample size R because each particle still only runs a finite time.

4   Experiments

There are four different algorithms to compare:

Harmonic mean (HM): the early method know to be biased.
Mean-field importance sampler (MFI): the importance sampler of Equation (2) using the mean-field approximation built using rewrites rules of Equation (4).
Left-to-right particle sampler (LR): Wallach's method.
Left-to-right sequential sampler (LRS): Wallach's method modified to run sequentially.

Note the first two are linear in the document size, and the second two are quadratic.

Two different experiments are performed. The first does exact computation of the document likelihood formula for small K, L in single artificial cases. The second generates samples of artificial data with realistic K, L from a known, larger scale model taken from a real problem. This allows testing of the algorithms under conditions where the truth is known, and thus the results can be properly evaluated.

4.1   Comparison with Exact Calculation

For K^L in the trillions, the exact calculation is feasible. We therefore evaluate the four different sampling algorithms in the context of a specific topic model and a specific document. For this, we generate a model according to a Dirichlet posterior, and then generate a document according to the model, and then do the evaluation. C code for this is available from the DCA distribution[1].

The topic model has K topics for different values (usually 3,4,5), a vocabulary size of $J = 1000$ and a document length of $L = 12, 14, 16, 18$. The model, the Θ matrix is generated with each topics θ_k generated by a symmetric Dirichlet with parameters uniformly γ (varied below). For the document, we take a fixed length L and then generate word indexes $j_1, ..., j_{L-1}$ according to the model with Dirichlet prior on the components having $\alpha = 0.1$.

First, a calibration test is done. For fixed $L = 14$, $K = 4$ and a sample size of 200, we generate 100 different model-document pairs then run the different likelihood estimation algorithms and compare them with the exact likelihood. In the MFI algorithm, 10 full cycles of Equation (4) for the initial mean field approximator are run. Times for the computation (averaged over 1000 runs) are about 0.7 milliseconds for HM and the two MFIs, and 4.3 milliseconds for LR and LRS. The exact computation takes 5 minutes and 46 seconds, for a 2.16GHz Intel Core Duo machine.

The sample mean and standard-deviation are given in Table 1, along with the resultant Student-t value (for 99 df) for whether the estimate is unbiased (so the mean is zero). We see that LRS is the clear winner, consistently more precise,

---

[1] DCA is available from NICTA, and this small evaluator is in `doc/Approx`.

Table 1. Estimator Precision for L=14, K=4

Θ prior parameter $\gamma = 0.2$			
Method	Mean	Std.Dev.	St's t
HM	-0.3357	0.2345	-14.3
MFI (1st ord)	0.0018	0.0114	1.58
MFI (2nd ord)	0.0016	0.0204	0.815
LR	0.0032	0.0256	1.26
LRS	0.00072	0.0156	0.46

Θ prior parameter $\gamma = 0.5$			
Method	Mean	Std.Dev.	St's t
HM	-0.210	0.120	-17.5
MFI (1st ord)	0.00131	0.0347	0.377
MFI (2nd ord)	0.00624	0.0244	2.55
LR	0.0128	0.0429	2.98
LRS	0.00002	0.0233	-0.0079

Θ prior parameter $\gamma = 1.0$			
Method	Mean	Std.Dev.	St's t
HM	-0.109	0.0878	-12.4
MFI (1st ord)	0.0181	0.0668	2.70
MFI (2nd ord)	0.0261	0.0380	6.87
LR	0.000796	0.0502	1.58
LRS	0.00457	0.0317	1.44

Θ prior parameter $\gamma = 3.0$			
Method	Mean	Std.Dev.	St's t
HM	-0.0440	0.0819	-5.37
MFI (1st ord)	0.0694	0.0797	8.71
MFI (2nd ord)	0.0553	0.0646	8.55
LR	0.00926	0.0417	2.22
LRS	0.00510	0.0259	1.97

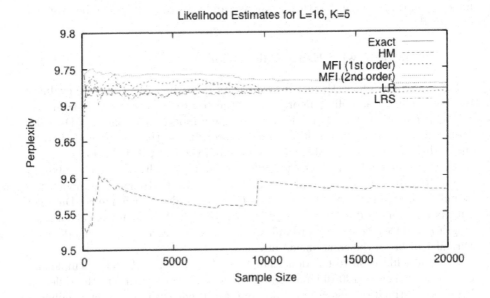

Fig. 1. Estimates for L=14, K=4 for increasing samples

where as all other methods are significantly different from the exact value at least once with a p-value of greater than 0.995[2]. LRS is mostly not significantly different at a p-value of 0.9. The two MFI approximations are comparable to Wallach's LR method. Timewise, LR and LRS are an order of magnitude slower for these comparable sample sizes.

---

[2] The cutoff t-value is 2.58 for 0.995 and 1.28 for 0.90.

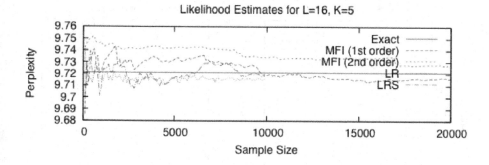

Fig. 2. Estimates for L=14, K=4 for increasing samples, closeup

Letting the different methods run out past 200 samples for a single model (sampled with $\gamma = 0.5$) plus document instance yields the plot given in Figures 1 and 2. This illustrates convergence to the exact value.

4.2   Comparison on Realistic Models

We built a topic model on a problem taken from a search engine news domain. In this domain, major stop words are removed, and less frequent words are removed leaving a total of $L = 50,182$ documents, $J = 28,251$ words and about 151 words on average per document. Then $K = 10$ and $K = 40$ topic models are built using standard LDA. Inspection shows these to be good looking models with clear separation between the topics and a clear semantics. From these known topic models, three data sets are generated according to the standard probabilistic LDA model: **s10** uses $K = 10$ and the existing number of words, $J = 28,251$; **s10s** uses $K = 10$ and a reduced $J = 5,000$ word count; and **s40** uses $K = 40$ and the existing number of words, $J = 28,251$. Thus three different data sets are created with known models and distinct topics and dictionary sizes.

Topic models were then built using vanilla LDA from varying subsets of the newly generated data sets, and their likelihood estimated on a hold out set of $10,000$ generated documents using the four different algorithms. Training set sizes used were $I = 1000, 2500, 5000, 10000, 20000, 30000, 40000$, and topics used were $K = 5, 8, 9, 10, 11, 12, 20, 40$ when the true $K = 10$ and $K = 20, 32, 35, 38, 40, 45, 50, 60$ when the true $K = 40$.

Variables are held fixed as much as possible, so when comparing MFI, LR and LRS at a given data set size, the same estimated model is used and the same hold out set is used to estimate likelihood. Note the HM method was not included in these comparisons because it is known to be poor. The LR and LRS methods were run with $R = 100$ particles/samples, and the MFI method was run with $R = 200$ samples. Note that LRS was also tested with $R = 2000$ samples, and the results where indistinguishable from that for $R = 100$ samples. For these estimates to be done on the 10000 test cases in the **s10** data set, MFI took 55 seconds and LR and LRS took 25 minutes and 30 seconds, approximately

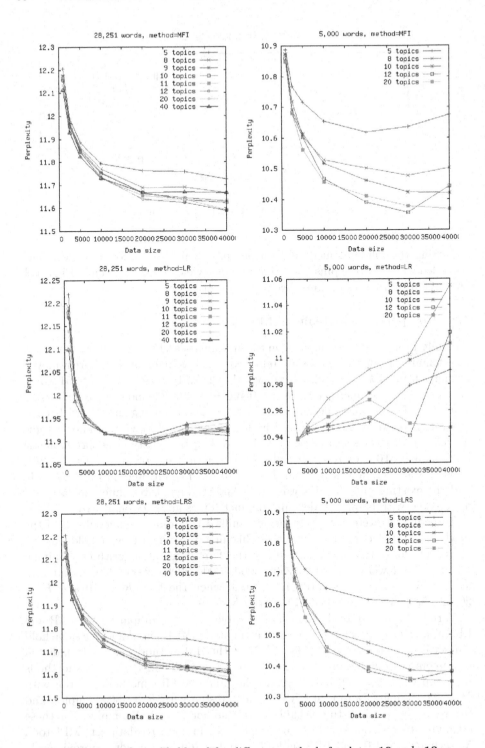

Fig. 3. Estimated test likelihood for different methods for data **s10** and **s10s**

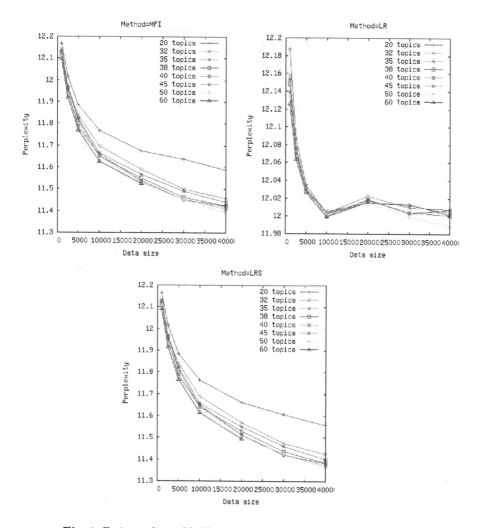

Fig. 4. Estimated test likelihood for different methods for data **s40**

20 times slower. The likelihood estimation for this is integrated into the DCA distribution[3].

Two views of the results are useful to look at. The first view compares the methods for different K as the data set sizes increases. Each method appears on a different plot. Two sets of plots are given, the first, Figure 3, shows the results for the **s10** data set with the large dictionary size, $J = 28,251$ and the **s10s** data set with the small dictionary size, $J = 5,000$. Like methods are presented side by

---

[3] DCA is available from NICTA, and this functionality is the "-X" option to command **mphier**. The methods HM, MFI (1st), LR and LRS with S samples correspond to using the flags "-XS,G", "-XS,I", "-XS,L", "-XS,M" respectively.

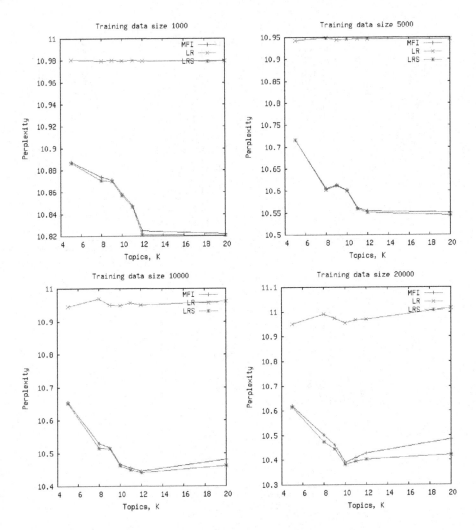

Fig. 5. Estimated test likelihood for different training set sizes on data **s10s**

side in the figure. The second set of plots, Figure 4, shows the results for the **s40** data set. Notice that the plots for the LR algorithm here indicates the perplexity starts to increase as data set sizes increase, in contrast to both MFI and LRS. This indicates a bias exists in the LR method.

The second view looks at how each method performs in selecting the "right" number of topics $K = 10$ for the **s10s** data set. The plots given in Figure 5 differ in the training sizes used, $I = 1000, 5000, 10000, 20000$. One can clearly see both MFI and LRS converging to the truth here, which has the "true" number of components at 10. The fast MFI method tracks LRS remarkably well. The LR method also does indicate the truth, but it is not as clear, and the distinction between different K is much finer, and thus harder to distinguish.

5   Discussion and Conclusion

A new method for estimating the likelihood of topic models has been developed that yielded significant improvements over those presented in [WMSM]. A second importance sampling method is revisited and shown to perform well in the kinds of model selection scenarios required in practice. Both methods, while similar to LR and IS-IP of Wallace *et al.*, come with an improved theory, and for one a proof that it can be used as a "gold standard". Our experiments also compare the approximations with an exact calculation, and demonstrate the use of the methods in a realistic model selection scenario using test set sizes typically required in practice (10, 000 versus 50 in [WMSM]), as well as a variety of training set sizes. This more rigorous testing showed that Wallach's left-to-right algorithm is slightly biased.

By converting Wallach's left-to-right algorithm for estimating likelihoods from a particle sampler to a sequential sampler (labelled LRS), the method becomes provably unbiased, and can thus be used as a gold standard for estimation with large enough number of samples. Moreover, the sequential sampler clearly picks the "right" number of topics using the tests sets, whereas Wallach's original particle samper does not do so as distinctly, and also provides a biased estimate of the likelihood. However, this LRS method is quadratic in the document size, and thus may not be realistic in practice where one wants to test thousands of documents.

A second method (labelled MFI) uses importance sampling with a proposal distribution based on a mean field approximation to the optimal importance sampler. This method is linear in the document size, and thus an order of magnitude faster than the other methods for comparable sample sizes, and while performing acceptably on the exact tests, seems to perform almost as well as LRS when applied to a collection of documents (where many document likelihood results are averaged over the collection). This suggests the second method can be used in place of LRS for efficiency.

Acknowledgements. NICTA is funded by the Australian Government as represented by the Department of Broadband, Communications and the Digital Economy and the Australian Research Council through the ICT Centre of Excellence program. Wray Buntine also acknowledges support from the EU project CLASS (IST project 027978), and thanks Helsinki Institute of Information Technology for providing some of the software and computer facilities on which the experimental work is based.

References

[AGvR]   Azzopardi, L., Girolami, M., van Risjbergen, K.: Investigating the relationship between language model perplexity and IR precision-recall measures. In: SIGIR 2003, pp. 369–370 (2003)

[BGJT]   Blei, D., Griffiths, T.L., Jordan, M.I., Tenenbaum, J.B.: Hierarchical topic models and the nested Chinese restaurant process. In: Thrun, S., Saul, L., Schölkopf, B. (eds.) Advances in Neural Information Processing Systems 16. MIT Press, Cambridge (2004)

[BJ1]   Buntine, W., Jakulin, A.: Applying discrete PCA in data analysis. In: UAI-2004, Banff, Canada (2004)

[BJ2]   Buntine, W.L., Jakulin, A.: Discrete components analysis. In: Saunders, C., Grobelnik, M., Gunn, S., Shawe-Taylor, J. (eds.) SLSFS 2005. LNCS, vol. 3940, pp. 1–33. Springer, Heidelberg (2006)

[BNJ]   Blei, D.M., Ng, A.Y., Jordan, M.I.: Latent Dirichlet allocation. Journal of Machine Learning Research 3, 993–1022 (2003)

[Can]   Canny, J.: GaP: a factor model for discrete data. In: SIGIR 2004, pp. 122–129 (2004)

[CC]   Carlin, B.P., Chib, S.: Bayesian model choice via MCMC. Journal of the Royal Statistical Society B 57, 473–484 (1995)

[GB]   Ghahramani, Z., Beal, M.J.: Propagation algorithms for variational Bayesian learning. In: NIPS, pp. 507–513 (2000)

[GS]   Griffiths, T.L., Steyvers, M.: Finding scientific topics. In: PNAS Colloquium (2004)

[GSBT]   Griffiths, T.L., Steyvers, M., Blei, D.M., Tenenbaum, J.B.: Integrating topics and syntax. In: Saul, L.K., Weiss, Y., Bottou, L. (eds.) Advances in Neural Information Processing Systems 17, pp. 537–544. MIT Press, Cambridge (2005)

[Hof]   Hofmann, T.: Probabilistic latent semantic indexing. In: Research and Development in Information Retrieval, pp. 50–57 (1999)

[LM]   Li, W., McCallum, A.: Pachinko allocation: DAG-structured mixture models of topic correlations. In: ICML 2006: Proc. of the 23rd Int. Conf. on Machine learning, pp. 577–584. ACM, New York (2006)

[LS]   Lee, D., Seung, H.: Learning the parts of objects by non-negative matrix factorization. Nature 401, 788–791 (1999)

[MLM]   Mimno, D., Li, W., McCallum, A.: Mixtures of hierarchical topics with Pachinko allocation. In: ICML 2007: Proceedings of the 24th international conference on Machine learning, pp. 633–640. ACM, New York (2007)

[NAXC]   Nallapati, R., Ahmed, A., Xing, E.P., Cohen, W.W.: Joint latent topic models for text and citations. In: Proc. of the 14th ACM SIGKDD Int. Conf. on Knowledge Discovery and Data Mining, Las Vegas, pp. 542–550. ACM, New York (2008)

[PSD]   Pritchard, J.K., Stephens, M., Donnelly, P.J.: Inference of population structure using multilocus genotype data. Genetics 155, 945–959 (2000)

[RZGSS]   Rosen-Zvi, M., Griffiths, T., Steyvers, M., Smyth, P.: The author-topic model for authors and documents. In: Proc. of the 20th Annual Conf. on Uncertainty in Artificial Intelligence (UAI 2004), Arlington, Virginia, pp. 487–494. AUAI Press (2004)

[Wal]   Wallach, H.: Structured Topic Models for Language. PhD thesis, University of Cambridge (2008)

[WMSM]   Wallach, H.M., Murray, I., Salakhutdinov, R., Mimno, D.: Evaluation methods for topic models. In: Bottou, L., Littman, M. (eds.) Proceedings of the 26th International Conference on Machine Learning, ICML 2009 (2009)

Conditional Density Estimation with Class Probability Estimators

Eibe Frank and Remco R. Bouckaert

Department of Computer Science, University of Waikato, New Zealand
{eibe,remco}@cs.waikato.ac.nz

Abstract. Many regression schemes deliver a point estimate only, but often it is useful or even essential to quantify the uncertainty inherent in a prediction. If a conditional density estimate is available, then prediction intervals can be derived from it. In this paper we compare three techniques for computing conditional density estimates using a class probability estimator, where this estimator is applied to the discretized target variable and used to derive instance weights for an underlying univariate density estimator; this yields a conditional density estimate. The three density estimators we compare are: a histogram estimator that has been used previously in this context, a normal density estimator, and a kernel estimator. In our experiments, the latter two deliver better performance, both in terms of cross-validated log-likelihood and in terms of quality of the resulting prediction intervals. The empirical coverage of the intervals is close to the desired confidence level in most cases. We also include results for point estimation, as well as a comparison to Gaussian process regression and nonparametric quantile estimation.

1   Introduction

In this paper we investigate methods for performing conditional density estimation using class probability estimators. Conditional density estimation makes it possible to quantify and visualize the uncertainty associated with the prediction of a continuous target variable. Given an observed vector of attribute values, a conditional density estimator provides an entire density function for the target variable, rather than a point estimate consisting of a single value. This function can then be visualized, or it can be summarized in prediction intervals that contain the true target value with a certain pre-specified probability.

As an example, consider the artificial data shown in Figure 1, consisting of two superimposed Mexican hats, each with Gaussian noise that exhibits constant variance. In this problem, there is a single attribute (shown on the x axis) that is used to predict the target variable (shown on the y axis). For each attribute value x, there is a conditional density function $f(y|x)$. Figure 2 shows conditional density functions for three values of x, namely 0, 3, and 8. These density functions are obviously unknown for real-world data. However, if they were available, we could use them to quantify predictive uncertainty, e.g. in the form of prediction intervals. The aim of conditional density estimation is to accurately estimate conditional density functions like these.

Z.-H. Zhou and T. Washio (Eds.): ACML 2009, LNAI 5828, pp. 65–81, 2009.
© Springer-Verlag Berlin Heidelberg 2009

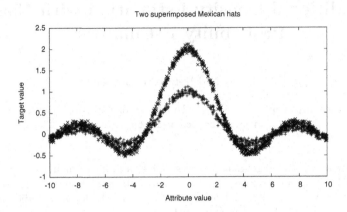

Fig. 1. Example data used to illustrate the output of CDE

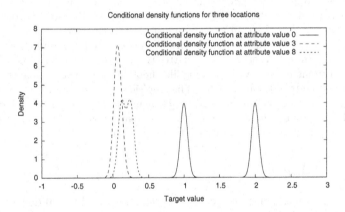

Fig. 2. Three conditional density functions for superimposed Mexican hats

Although information on predictive uncertainty is very useful in interactive applications of prediction methods, there has been comparatively little work on conditional density estimation in statistics [1] and machine learning [2] outside the area of Bayesian models for regression, e.g. Gaussian process regression [3]. A significant amount of related material can be found in the economics literature [4], but this work has a focus on time series problems. In machine learning, a number of publications discuss conditional density estimation for particular types of neural networks, see e.g. [5,6,7,8,9,10,11,12]. Excluding Bayesian regression and neural networks, work on conditional density estimation in machine learning appears to be rare. One exception is [13], which investigates methods for learning conditional density trees in the context of Bayesian networks. Another is [2], which describes a computationally efficient version of kernel conditional density estimation—a popular method in statistics [1].

The aim of this paper is to investigate a generic discretization-based technique for conditional density estimation that wraps around an existing machine learning algorithm—more specifically, a class probability estimator. A key advantage of generic techniques is that they can leverage existing implementations of machine learning algorithms in software suites like the Weka workbench [14]. It is well known that the most accurate algorithm for a particular prediction problem cannot in general be determined a priori and that experimentation with a collection of algorithms is necessary. Thus it is useful to have access to generic techniques that can be used to generate conditional density estimates in conjunction with existing class probability estimation techniques.

The idea of using class probability estimation for conditional density estimation is not new, and has been used in the context of modeling auction price uncertainty using boosting [15]. The contribution of this paper is that we show how to combine class probability estimation with univariate normal and kernel density estimators to improve on the histogram-based estimator used in [15]. Empirical evaluation is used to ascertain that progress has been made. We use two approaches for evaluation, both based on computing performance measures on independent test data. First, we compute log-likelihood values based on the predicted density estimates [9]. Second, we compute prediction intervals from the conditional density estimates based on pre-specified confidence levels, and measure their size and coverage [2]. The aim is to achieve small intervals with empirical coverage that is close to the desired confidence level.

The paper is structured is follows. In the next section, we discuss how class probability estimation can be combined with univariate density estimators to obtain conditional density estimates. Section 3 presents experimental results comparing the performance of histogram-based estimation with that obtained from normal and kernel estimators, based on two underlying class probability estimators. We also include results for point estimation, as well as a comparison to Gaussian process regression and nonparametric quantile estimation. Section 4 summarizes our findings and concludes the paper.

2   Conditional Density Estimation via Class Probabilities

We assume access to a class probability estimation scheme—e.g. an ensemble of class probability estimation trees—that can provide class probabilities $p(c|X)$ based on some labeled training data, where c is a class value and X an instance described by some attribute values. The basic idea is to discretize the continuous target values y into intervals that can be treated as class values c and use $p(c|X)$ to obtain a weight for each y, conditioned on X. A univariate density estimate— e.g. a normal density—constructed from these weighted target values in the training data, constitutes a conditional density estimate.

2.1   Three Density Estimators Based on Weighted Target Values

For the discussion that follows, we assume that the target variable has been discretized into non-overlapping intervals ("bins"), for example, by applying equal-frequency discretization to the target values in the training data.[1] Regardless of which discretization method is applied, once the target values have been discretized, a class probability estimator can be used to estimate $p(c|X)$. Note that the class values created in this fashion exhibit a natural order, and it may be advantageous to exploit this by choosing an appropriate underlying class probability estimator.

Let c_y be the bin (i.e. class) that contains the target value y and let $p(c_y|X)$ be the predicted probability of that class given X, which is obtained from the class probability estimator. Let n be the total number of target values in the training data and n_c be the number of target values in bin c.

We compute a conditional density estimate by weighting the target values in the training data and then using these weights in a standard univariate density estimator to obtain a conditional density estimate. A weight $w(y_i|X)$ for a particular target value y_i given an instance X is computed by "spreading" the predicted class probability for bin c_{y_i} across all $n_{c_{y_i}}$ target values in that bin:

$$w(y_i|X) = n \frac{p(c_{y_i}|X)}{n_{c_{y_i}}},$$

where the multiplier n ensures that the sum of weights is n.

The weight $w(y_i|X)$ can be viewed as an estimate of how likely it is that a future observed target value associated with X will be close to y_i, based on the class probability estimation model that has been inferred from the discretized training data. Given weights for all target values in the training data, we can then use a univariate density estimator, applied to these weighted values, to obtain a conditional density estimate $f(y|X)$.

The simplest estimator is the standard univariate normal estimator:

$$f_{normal}(y|X) = \mathcal{N}(y; \mu_X, \sigma_X^2),$$

where μ_X and σ_X^2 are the mean and variance respectively, computed based on the weighted target values; i.e. the mean is defined as $\mu_X = \frac{1}{n} \sum_{i=0}^{n} w(y_i|X) y_i$ and the variance as $\sigma_X^2 = \frac{1}{n} \sum_{i=0}^{n} w(y_i|X)(y_i - \mu_X)^2$.

Although the normal estimator is useful when the data is approximately normal, it is not flexible. A popular non-parametric alternative is a kernel density estimator. Using Gaussian kernels with kernel bandwidth σ_{kernel}, the weighted kernel estimator is:

$$f_{kernel}(y|X) = \frac{1}{n} \sum_{i=0}^{n} w(y_i|X) \mathcal{N}(y; y_i, \sigma_{kernel}^2)$$

The bandwidth parameter determines how closely the estimator fits the (weighted) data points. We use a data-dependent value based on the global

---

[1] Note that it is also possible to create a bin for each unique target value that occurs in the training data, assuming this is feasible given the computational resources that are available, and assuming an appropriate class probability estimator is applied.

(weighted) standard deviation and the number of data points, namely $\sigma_{kernel} = \frac{\sigma_X}{n^{1/4}}$. This is based on [16], which advocates a bandwidth chosen at $O(n^{-1/4})$.

Note that the kernel estimator is a "lazy" estimator and requires more computational effort than the normal estimator. However, the procedure for computing $f_{kernel}(y|X)$ can be sped up significantly using a binary search for the kernel closest to y and scanning the sorted list of kernels in both directions until the overall contribution from visiting additional kernels becomes negligible. Hence, the number of kernels that have to be evaluated to compute $f_{kernel}(y|X)$ is usually much smaller than n.

Another possible univariate density estimator, and perhaps the most obvious one in this context, is a histogram estimator based on the bins provided by the discretization. In this case, the density is assumed to be constant in each bin c and based on the bin's width r_c —the difference between the upper and lower boundary of the bin—and the class probability assigned to it:

$$f_{bins}(y|X) = \frac{p(c_y|X)}{r_{c_y}}$$

This estimator has been used for conditional density estimation in [15].

It is instructive to write the histogram estimator in a form analogous to the kernel estimator, based on weighted target values:

$$f_{bins}(y|X) = \frac{1}{n} \sum_{i=0}^{n} w(y_i|X) \frac{I(c_{y_i} = c_y)}{r_{c_y}},$$

where $I(a)$ takes on value 1 if the proposition a is true and 0 otherwise. This formulation shows that the histogram estimator can be viewed as a kernel estimator where the kernel function is identical for all target values in a bin, with kernel value $\frac{1}{r_{c_y}}$ inside the bin, and value 0 outside. Note that this means that potentially valuable information about the relative position of y with respect to the individual y_i in a bin is discarded.

Another drawback of the histogram estimator is that it can return density 0 if y falls into a bin that receives weight 0 or if it is located outside the range of all bins. To circumvent this problem we use the following adjusted version in our experiments:

$$f'_{bins}(y|X) = \frac{1}{n+2} \left(n f_{bins}(y|X) + \begin{cases} \frac{1}{max-min} : y \in [min, max] \\ \mathcal{N}(y; max, \sigma^2_{kernel}) : y > max \\ \mathcal{N}(y; min, \sigma^2_{kernel}) : y < min \end{cases} \right),$$

where min is the smallest bin boundary of all bins, and max the largest one. This means one data point is notionally spread out across the full $[min, max]$ range, and half a kernel function from the kernel estimator is attached to the left- and right-most bin respectively. Appropriate normalization ensures that the overall estimate integrates to one.

2.2   Examples

The Mexican hat data from Figure 1 can be used to illustrate the behavior of the three estimators discussed above. To this end, we discretized the 2,000 target values in Figure 1 into 20 bins using equal-frequency discretization and then applied a random-forest-based method, described in more detail in Section 3, to estimate class probabilities for the 20 discretization intervals.

The true conditional density function at attribute value 0 exhibits two well-separated peaks at target values 1 and 2 respectively (cf. Figure 2). Figure 3 shows the three different conditional density estimates for attribute value 0. It also shows an (unconditional) density estimate that was generated by applying a kernel density estimator to the unweighted target values. This latter estimate is labeled "prior kernel" in the figure and has a peak close to target value 0 because that is where most of the data in Figure 1 is located.

The figure shows that both the histogram estimator and the (conditional) kernel estimator reflect the two peaks in the true conditional density function quite accurately—as a unimodal estimator, the normal estimator is obviously not able to do so. However, a visual comparison of the estimates to the true conditional density shows that they do not model the height of the two peaks perfectly: they should be of the same height. This is due to the influence of discretization: the histogram estimate shows that the first peak is represented by two bars (corresponding to two discretization intervals) that together cover a wider range of target values than the single bar corresponding to the second peak. Thus the predicted class probability is spread across a wider range of target values and the height of the peak is reduced.

Let us now consider a situation where the two peaks in the true conditional density function are closer together. This is the case for attribute value 8 (cf. Figure 2). Figure 4 shows the conditional density estimates for this value. As before, the normal estimator is problematic, but less so as in the previous example. The kernel estimator models the two peaks quite well. The histogram estimator also has two main peaks, but additionally a third narrow peak that is a consequence of under-smoothing.

The data also exhibits attribute values where the normal estimator is appropriate, e.g. attribute value 3. The two Mexican hats intersect in the vicinity of this value, giving rise to a unimodal conditional density (cf. Figure 2). Figure 5 shows that the normal estimator yields the most accurate representation of the true density in this case. Both the histogram estimator and the kernel estimator overfit; however, the kernel estimator exhibits less under-smoothing.

2.3   Computing Prediction Intervals

An important application of conditional density estimation is the generation of prediction intervals. For a given confidence level α and an instance X, we would like to obtain an $\alpha\%$ prediction interval that contains the target value for X with probability α.

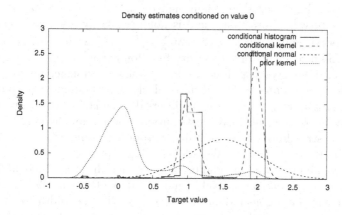

Fig. 3. Conditional density estimates for attribute value 0; prior estimate also included

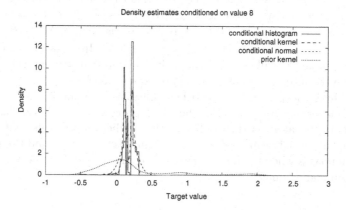

Fig. 4. Conditional density estimates for attribute value 8; prior estimate also included

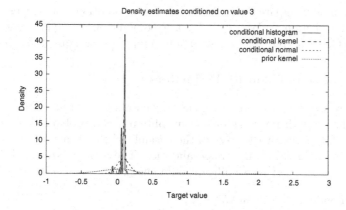

Fig. 5. Conditional density estimates for attribute value 3; prior estimate also included

Assuming we know the true conditional density $f(y|X)$, we can determine appropriate intervals—note that there is not necessarily a single interval—by choosing interval boundaries b_i such that $\sum_i \int_{b_i}^{b_{i+1}} f(y|X)dy = \alpha$, i.e. the area under the chosen segments of the density function equals α.

In practice, the true density function is not known and hence replaced by the estimators introduced above. This is particularly straightforward given the simple normal estimator $f_{normal}(y|X)$: the interval boundaries are easily obtained using the inverse of the cumulative distribution function for the normal density.

It is not so straightforward to obtain appropriate boundaries from a kernel density estimator $f_{kernel}(y|X)$. We adopt an approximate method, which assumes that the aim is to find interval boundaries such that the sum of all interval widths for the given confidence level is minimized. To this end, the range of the target variable is first segmented into 1,000 very small equal-width intervals. The density estimator is applied to the two end points of each interval and the two resulting density estimates are averaged. This average is then multiplied by the interval width to get an estimate of the area under the density function for that interval. The intervals are subsequently sorted in descending order according to their corresponding areas a_j, and the top k intervals chosen. More precisely, k is minimized subject to the constraint $\sum_{j=1}^{k} a_j \geq \alpha$. In most cases, many of the resulting intervals will be adjacent. Adjacent intervals are merged and the resulting merged intervals are output as prediction intervals.

Exactly the same process can also be applied in conjunction with the histogram estimator $f_{bins}(y|X)$. Using the same process in both cases facilitates a fair comparison based on the quality of the prediction intervals that are obtained. Like the kernel estimator, the histogram estimator can produce multiple prediction intervals when the density estimate is multi-modal (see, e.g., Figure 3 for a multi-modal estimate).

3   Experiments

In this section, we compare the performance of the conditional density estimators introduced in Section 2. We also evaluate point estimation performance and compare to Gaussian process regression and nonparametric quantile estimation.

3.1   Comparison of Density Estimators

We present results for two underlying class probability estimators: linear support vector machines, with Platt scaling [17] to obtain useful probability estimates,[2] and random forests, with 100 trees in the ensemble.[3] Performance estimates are obtained using 10 times 10-fold cross-validation. The corrected resampled paired t-test [18] is used to test for significant differences, at a 5% significance level.

---

[2] SMO with options -V 5 -M in Weka.

[3] To eliminate bias, each tree was grown from 66% of the training data, sampled w/o replacement, and the remainder used to estimate class probabilities.

Because a discretized regression target yields an ordinal variable, the method introduced in [19] was used to exploit this ordering information by creating multiple two-class problems. The base learners (i.e. SVMs and random forests) were applied to these two-class problems. This method was also used for conditional density estimation in [15], using an additional monotonizing step that we also applied for the results shown here. All results are based on the data from [20].

One method of measuring the performance of a conditional density estimator is to measure average log-likelihood (the average of $\log(f(y|X))$ across all test instances, where y is the actual target value in the test instance). We use base 2 logarithm. To make this measure more informative, we subtract from this the log-likelihood obtained for an unconditional kernel density estimator (i.e. the "prior kernel" in Figures 3, 4, and 5). A positive value means that the conditional estimator improves on the prior estimator.

Table 1 shows results obtained using random forests.[4] Scores are shown for all three density estimators from Section 2. For the first set of results, the target variable was discretized into 10 bins, using equal-frequency discretization; for the second set, it was discretized into 20 bins. For each set of results, statistical significance is measured with respect to the histogram-based method.

The results show that both the kernel estimator and the normal estimator improve on the simple histogram estimator used previously in [15]. There are only a handful of cases were the histogram estimator significantly outperforms the kernel estimator. In spite of the fact that the normal estimator is not flexible, it performs well compared to the histogram estimator. However, there are also a number of cases were it performs significantly worse and also worse than the kernel estimator, indicating that additional flexibility can be important.

Table 2 shows the results obtained using SVMs. These results are even more strongly in favor of the kernel estimator and the normal estimator, e.g. there is only a single case where the histogram estimator yields significantly higher improvement in log-likelihood vs. the prior estimator than the kernel estimator.

It is noteworthy that the effect of the number of discretization intervals used to obtain class probability estimates is relatively minor for the kernel estimator and the normal estimator in most cases. However, the likelihood scores indicate that increasing the number of intervals from 10 to 20 renders the histogram estimator more prone to overfitting.

In practice, conditional density estimation is often used to obtain prediction intervals. Hence, we also evaluate the quality of these prediction intervals. To this end, we measured empirical coverage of target values in the test data as well as average width of the predicted intervals. The aim is to obtain narrow intervals with empirical coverage close to the chosen confidence level. For the results shown here, we chose a 95% confidence level. Empirical coverage is expressed as the average percentage of target values in the test data that were in the range of the predicted intervals. Interval width is expressed relative to the full range of target values in the training data: a relative width of 100 means that the

---

[4] The meta and schlvote data are excluded from this and the next table, due to the high variance in the estimates. No significant differences in performance were obtained.

Table 1. Mean improvement in log-likelihood vs. prior estimator, using random forests

Dataset	10 bins			20 bins		
	Histogram	Kernel	Normal	Histogram	Kernel	Normal
auto93	0.21±0.79	0.90±0.28 ∘	0.75±0.64	-0.36±1.17	0.87±0.38 ∘	0.75±0.62 ∘
autoHorse	2.23±0.47	1.96±0.21 •	1.23±0.13 •	2.87±0.61	2.20±0.25 •	1.28±0.15 •
autoMpg	0.72±0.44	1.37±0.19 ∘	1.37±0.10 ∘	0.02±0.64	1.39±0.23 ∘	1.40±0.11 ∘
autoPrice	1.10±0.60	1.46±0.16 ∘	0.90±0.20	0.43±0.81	1.53±0.19 ∘	0.95±0.22
baskball	-1.33±0.85	0.02±0.39 ∘	0.43±0.66 ∘	-2.08±1.22	0.00±0.43 ∘	0.43±0.65 ∘
bodyfat	2.56±0.24	2.38±0.17 •	1.59±0.14 •	3.38±0.25	2.77±0.16 •	1.62±0.16 •
bolts	0.99±1.11	1.28±0.34	0.73±0.29	0.10±1.58	1.28±0.40 ∘	0.78±0.27
breastTumor	-0.78±0.60	-0.27±0.31 ∘	-0.13±0.14 ∘	-0.64±0.70	-0.26±0.25	-0.11±0.14 ∘
cholesterol	-0.99±0.43	-0.08±0.14 ∘	0.28±0.84 ∘	-1.68±0.72	-0.02±0.16 ∘	0.31±0.89 ∘
cleveland	0.48±0.22	0.64±0.19 ∘	-0.33±0.16 •	0.48±0.22	0.64±0.19 ∘	-0.33±0.16 •
cloud	0.39±0.95	1.11±0.31 ∘	0.74±0.32	-0.46±1.21	1.14±0.35 ∘	0.77±0.43 ∘
cpu	2.72±0.77	2.23±0.51 •	1.07±0.73 •	2.51±0.98	2.16±0.62	0.88±0.83 •
detroit	0.58±1.56	-0.86±5.55	0.16±2.03	0.35±1.82	-0.52±4.52	0.45±1.32
echoMonths	0.55±0.61	0.58±0.23	0.33±0.12	-0.03±0.83	0.59±0.25 ∘	0.33±0.13
elusage	-1.16±1.45	0.69±0.73 ∘	0.88±0.38 ∘	-1.89±1.51	0.48±1.00 ∘	0.87±0.49 ∘
fishcatch	1.89±0.71	2.25±0.23	1.55±0.15	1.66±0.76	2.37±0.24 ∘	1.54±0.17
fruitfly	-0.99±0.68	-0.13±0.15 ∘	-0.35±0.34 ∘	-1.96±0.98	-0.14±0.15 ∘	-0.33±0.34 ∘
gascons	0.67±1.46	1.43±0.50	1.24±0.33	0.23±1.90	1.59±0.57 ∘	1.40±0.36
housing	1.04±0.30	1.39±0.14 ∘	1.02±0.13	0.55±0.47	1.46±0.17 ∘	1.06±0.13 ∘
hungarian	-0.38±0.13	0.89±0.17 ∘	-1.12±0.26 •	-0.38±0.13	0.89±0.17 ∘	-1.12±0.26 •
longley	0.41±1.82	1.21±0.65	1.09±0.30	0.18±1.96	1.28±0.79	1.12±0.25
lowbwt	0.22±0.48	0.72±0.22 ∘	0.74±0.16 ∘	-0.33±0.67	0.73±0.22 ∘	0.76±0.16 ∘
mbagrade	-0.99±1.09	-0.01±0.43 ∘	0.00±0.43 ∘	-1.83±1.37	-0.04±0.47 ∘	0.03±0.44 ∘
pbc	0.09±0.30	0.38±0.12 ∘	0.23±0.10	-0.63±0.49	0.39±0.11 ∘	0.23±0.10 ∘
pharynx	-1.35±0.79	0.21±0.21 ∘	0.06±0.28 ∘	-2.34±0.91	0.23±0.21 ∘	0.06±0.25 ∘
pollution	-0.84±0.92	0.26±0.23 ∘	0.46±0.34 ∘	-1.56±1.05	0.24±0.22 ∘	0.44±0.34 ∘
pwLinear	0.24±0.58	1.00±0.20 ∘	0.94±0.13 ∘	-0.50±0.79	1.05±0.23 ∘	0.97±0.13 ∘
quake	-0.53±0.13	-0.22±0.09 ∘	-0.86±0.07 •	-0.54±0.13	-0.23±0.10 ∘	-0.86±0.07 •
sensory	-0.77±0.30	-0.13±0.21 ∘	-0.04±0.07 ∘	-0.76±0.31	-0.13±0.21 ∘	-0.04±0.07 ∘
servo	0.30±0.91	1.32±1.48 ∘	1.08±0.32 ∘	-0.90±1.15	1.46±0.75 ∘	1.09±0.32 ∘
sleep	-0.30±0.97	0.36±0.35 ∘	0.35±0.22 ∘	-0.89±1.07	0.35±0.36 ∘	0.33±0.25 ∘
strike	0.42±1.90	0.51±1.79	-1.30±2.93 •	-1.28±3.49	-0.15±3.39 ∘	-2.16±5.10
veteran	-0.35±1.30	0.39±0.83 ∘	-0.17±1.35	-1.41±1.22	0.30±0.45 ∘	-0.28±1.26 ∘
vineyard	-0.10±1.26	0.66±0.97 ∘	0.67±1.24	-0.28±1.43	0.49±1.18	0.73±1.15

∘/• statistically significant improvement/degradation

total interval width is equal to the full range of target values in the training data, yielding no useful information. For the results shown here, we used 10 discretization intervals. Results for 20 intervals are not shown due to lack of space, but yield comparable relative performance.

Table 3 shows results for random forests, and Table 4 those for linear SVMs. When interpreting these results, it is important to keep in mind that there is a trade-off between maximizing coverage and minimizing interval width: it is trivial to achieve high coverage by predicting very large intervals and vice versa. Hence, it is useful to pay attention in particular to those cases where a method dominates another one based on both criteria.

The results show that both the kernel estimator and the normal estimator often produce wider intervals than the histogram estimator. This is because they generally perform more smoothing, as could also be seen in the examples in Section 2.2. However, in contrast to the histogram estimator, they achieve empirical coverage close to, or above, the chosen 95% confidence level for almost all datasets, which is important for practical applications where the end user expects reliable prediction intervals. Moreover, there are several cases where the kernel estimator and the normal estimator dominate the histogram estimator

Table 2. Mean improvement in log-likelihood vs. prior estimator, using linear SVMs

Dataset	10 bins Histogram	10 bins Kernel	10 bins Normal	20 bins Histogram	20 bins Kernel	20 bins Normal
auto93	-0.59±1.32	0.88±0.70 o	1.12±0.65 o	-1.38±1.67	0.71±1.35 o	1.10±0.48 o
autoHorse	2.03±0.75	2.33±0.51	1.82±0.65	2.58±0.78	2.50±1.19	1.82±0.53 •
autoMpg	0.18±0.50	1.15±0.22 o	1.33±0.14 o	-0.39±0.64	1.14±0.25 o	1.37±0.14 o
autoPrice	0.71±0.69	1.36±0.24 o	1.24±0.21 o	0.31±0.79	1.35±0.34 o	1.34±0.26 o
baskball	-0.95±0.76	-0.03±0.44 o	0.43±0.59 o	-1.53±0.95	0.02±0.29 o	0.45±0.64 o
bodyfat	0.68±1.82	1.06±1.75 o	1.63±0.26	0.22±2.34	0.92±2.28 o	1.67±0.37
bolts	0.01±1.23	0.87±0.42 o	0.42±0.33	-0.85±1.33	0.82±0.51 o	0.42±0.32 o
breastTumor	0.08±0.29	-0.05±0.11	-0.04±0.06	0.37±0.35	-0.04±0.07 •	-0.04±0.06 •
cholesterol	-0.51±0.33	-0.02±0.10 o	0.27±0.85 o	-0.68±0.46	0.01±0.14 o	0.28±0.88 o
cleveland	0.42±0.26	0.67±0.27 o	-0.34±0.18 •	0.42±0.26	0.67±0.27 o	-0.34±0.18 •
cloud	-0.10±0.93	0.74±1.78	0.91±0.53 o	-1.13±1.07	0.82±0.44 o	0.87±0.51 o
cpu	0.97±0.94	2.03±0.55 o	1.81±0.70 o	0.24±1.25	2.05±0.66 o	1.98±0.77 o
detroit	0.14±2.96	0.69±2.18	0.88±0.83	-0.48±3.10	0.45±2.33	0.84±0.85
echoMonths	0.06±0.58	0.37±0.21	0.24±0.11	-0.35±0.80	0.36±0.23 o	0.24±0.12 o
elusage	-1.32±1.44	0.38±0.51 o	0.41±0.29 o	-1.75±1.48	0.35±0.56 o	0.46±0.32 o
fishcatch	1.26±0.83	2.15±0.39 o	2.04±0.30 o	0.60±0.94	2.16±0.39 o	2.12±0.30 o
fruitfly	-0.58±0.55	-0.01±0.05 o	-0.27±0.32	-1.07±0.78	-0.02±0.07 o	-0.28±0.32 o
gascons	0.23±1.94	0.85±1.99	1.08±0.76	-0.02±2.16	0.82±1.88	0.99±0.97
housing	0.98±0.30	1.21±0.19 o	0.95±0.45	0.68±0.38	1.22±0.21 o	1.01±0.42
hungarian	-0.36±0.16	1.06±0.22 o	-1.17±0.60 •	-0.36±0.16	1.06±0.22 o	-1.17±0.60 •
longley	-0.01±2.27	0.98±1.39 o	1.27±0.47	-0.19±2.07	1.14±1.25 o	1.29±0.47 o
lowbwt	-0.01±0.54	0.63±0.22 o	0.68±0.16 o	-0.28±0.60	0.62±0.24 o	0.68±0.15 o
mbagrade	-0.62±0.79	0.02±0.27 o	0.04±0.19 o	-1.06±0.93	0.02±0.21 o	0.03±0.21 o
pbc	-0.36±0.39	0.24±0.15 o	0.22±0.10 o	-1.00±0.49	0.25±0.15 o	0.24±0.10 o
pharynx	-0.40±0.49	0.33±0.19 o	0.12±0.24 o	-1.35±0.81	0.32±0.21 o	0.13±0.21 o
pollution	-1.16±1.43	0.07±1.01 o	0.52±0.35 o	-1.66±1.37	0.10±0.79 o	0.59±0.28 o
pwLinear	0.19±0.64	0.99±0.29 o	1.14±0.28 o	-0.51±0.88	0.96±0.37 o	1.14±0.32 o
quake	-0.19±0.04	-0.01±0.02 o	-0.86±0.05 •	-0.19±0.04	-0.02±0.03 o	-0.86±0.05 •
sensory	-0.47±0.19	-0.09±0.12 o	-0.17±0.03 o	-0.45±0.19	-0.09±0.12 o	-0.17±0.03 o
servo	0.31±1.00	1.44±0.30 o	0.90±0.27	-0.21±1.13	1.49±0.35 o	0.92±0.28 o
sleep	-0.42±1.02	0.42±0.40 o	0.43±0.27 o	-0.84±1.08	0.47±0.38 o	0.43±0.27 o
strike	-0.58±0.75	0.38±0.21 o	-0.97±0.68	-1.29±0.77	0.33±0.44 o	-0.97±0.81
veteran	-0.39±0.92	0.23±0.23 o	-0.57±1.09	-0.90±0.97	0.21±0.16 o	-0.57±1.17
vineyard	-0.36±1.31	0.21±1.03	0.63±0.99 o	-0.56±1.36	0.28±0.62 o	0.65±0.83 o

o/• statistically significant improvement/degradation

regarding both width and coverage (e.g. baskball, cholesterol, lowbwt, meta). The opposite is never the case. This is most likely due to the loss of location information in the histogram estimator. Considering the relative performance of the normal estimator and the kernel estimator, we can see that the former generally creates narrower intervals. As both exhibit acceptable coverage in most cases, the kernel estimator appears preferable.

3.2 Point Estimation Performance

The results so far show that the proposed methods can produce useful conditional density estimates and corresponding prediction intervals. However, it is important to consider the quality of the point estimates that they correspond to. To compute point estimates that are designed to minimize the squared error, we can compute the expected value of the target variable based on the conditional density estimate. In the case of the normal estimator and the kernel estimator, this is simply the weighted sum of the target values used to construct the estimator (i.e. the expected value is identical in both cases).

Table 3. Quality of prediction intervals, using random forests and 10 bins

Dataset	Coverage			Relative width		
	Histogram	Kernel	Normal	Histogram	Kernel	Normal
auto93	89.4± 9.2	97.4± 5.3 ∘	97.5± 5.1 ∘	55.3± 6.7	52.9± 6.0	55.3± 5.4
autoHorse	96.8± 4.3	97.8± 3.4	98.9± 2.2	35.4± 6.6	31.6± 4.3 ∘	35.6± 3.5
autoMpg	88.9± 5.2	95.3± 3.4 ∘	98.7± 1.8 ∘	33.8± 2.0	35.6± 2.0 ●	39.3± 1.6 ●
autoPrice	93.7± 6.0	98.6± 3.2 ∘	99.1± 2.5 ∘	38.8± 6.7	40.5± 6.7 ●	46.8± 4.3 ●
baskball	75.0±12.2	94.3± 7.6 ∘	95.4± 7.0 ∘	73.6± 3.5	62.3± 6.7 ∘	60.7± 5.9 ∘
bodyfat	98.8± 2.0	99.2± 1.8	99.2± 1.7	37.8± 2.4	37.6± 2.5	37.0± 2.3
bolts	88.0±15.7	100.0± 0.0 ∘	100.0± 0.0 ∘	66.0± 7.5	73.6± 9.8 ●	79.4± 8.7 ●
breastTumor	76.2± 7.7	94.0± 4.9 ∘	91.4± 5.2 ∘	59.6± 2.9	75.6± 3.6 ●	76.1± 3.3 ●
cholesterol	90.0± 5.3	94.7± 4.5 ∘	95.6± 4.0 ∘	68.7± 3.9	46.0± 6.2 ∘	47.9± 5.6 ∘
cleveland	90.1± 5.0	97.2± 3.1 ∘	97.3± 3.0 ∘	67.6± 3.2	84.6± 3.8 ●	98.2± 3.5 ●
cloud	90.0±10.1	96.1± 5.8	97.9± 4.1 ∘	43.6± 7.0	44.5± 7.0	50.0± 5.6 ●
cpu	96.2± 3.8	97.2± 3.9	98.5± 2.4	27.0± 7.9	17.8± 6.0 ∘	25.4± 6.2
detroit	85.5±32.8	87.5±32.1	87.5±32.1	87.2±10.7	88.4±33.0	80.7±28.6
echoMonths	87.8±10.4	96.5± 4.6 ∘	96.0± 5.2 ∘	71.5± 3.7	87.9± 2.7 ●	86.3± 3.0 ●
elusage	61.8±20.8	89.6±14.8 ∘	96.2± 9.3 ∘	38.0± 5.9	46.9± 8.3 ●	54.8± 7.0 ●
fishcatch	91.6± 7.7	98.2± 3.4 ∘	98.8± 3.0 ∘	26.1± 5.7	30.2± 5.9 ●	37.5± 4.3 ●
fruitfly	82.5± 9.8	93.1± 7.9 ∘	93.9± 7.6 ∘	70.3± 3.6	76.4± 5.7 ●	75.8± 4.3 ●
gascons	88.7±20.4	99.0± 7.0	100.0± 0.0	59.7± 6.3	63.6±11.2	67.0± 8.7 ●
housing	94.8± 3.0	98.7± 1.5 ∘	99.2± 1.3 ∘	41.5± 1.7	44.9± 2.1 ●	48.8± 2.0 ●
hungarian	55.8±10.3	100.0± 0.0 ∘	96.1± 3.5 ∘	78.7± 3.0	64.2± 4.7 ∘	139.2± 8.1 ●
longley	87.5±32.1	100.0± 0.0	100.0± 0.0	79.7± 8.4	97.7±21.1 ●	91.2±18.2 ●
lowbwt	89.9± 6.9	96.1± 4.8 ∘	96.7± 4.0 ∘	49.6± 1.8	46.8± 2.2 ∘	46.9± 2.1 ∘
mbagrade	69.1±20.2	90.8±10.1 ∘	90.1±10.5 ∘	67.7± 4.8	84.0± 5.2 ●	79.0± 5.7 ●
meta	87.6± 4.8	96.7± 2.2 ∘	96.4± 2.1 ∘	60.6± 7.1	6.7± 1.9 ∘	18.6± 2.7 ∘
pbc	91.9± 4.0	97.1± 2.7 ∘	96.3± 2.9 ∘	72.6± 1.5	79.1± 1.6 ●	80.5± 1.8 ●
pharynx	72.5±10.5	93.8± 5.4 ∘	94.4± 5.2 ∘	57.1± 3.5	69.9± 3.7 ●	72.7± 3.8 ●
pollution	79.5±15.1	94.8± 9.7 ∘	95.8± 8.3 ∘	80.5± 2.6	79.6± 4.3	72.4± 4.9 ∘
pwLinear	89.8± 6.8	98.1± 3.2 ∘	98.6± 2.8 ∘	54.8± 1.9	57.5± 2.0 ●	53.9± 2.2
quake	90.0± 2.1	93.8± 1.5 ∘	94.3± 1.5 ∘	46.3± 0.5	53.4± 0.6 ●	65.7± 0.6 ●
schlvote	87.0±18.4	91.9±15.8	95.4±13.2	54.1±14.8	42.8±16.4 ∘	51.8±16.8
sensory	86.5± 4.6	89.9± 4.1 ∘	95.8± 2.5 ∘	58.6± 4.5	56.7± 1.5	58.9± 1.4
servo	83.7± 9.0	96.2± 5.1 ∘	96.0± 4.5 ∘	30.3± 7.9	33.3± 8.8 ●	38.5± 9.2 ●
sleep	80.5±16.5	96.4± 8.9 ∘	94.3±10.0 ∘	74.5± 5.3	90.0± 5.3 ●	84.8± 6.4 ●
strike	89.2± 4.0	95.6± 2.8 ∘	97.4± 2.1 ∘	51.0± 5.9	13.9± 1.6 ∘	23.6± 2.5 ∘
veteran	85.1±11.1	94.9± 6.3 ∘	95.9± 5.6 ∘	51.9± 6.9	44.6± 5.5 ∘	54.6± 5.8
vineyard	81.6±16.9	93.7±10.3 ∘	94.0±10.1 ∘	56.8±10.9	59.7±10.8	54.9± 9.3

∘/● statistically significant improvement/degradation

Table 5 shows the root relative squared error for point estimates obtained this way, for the two base learners used previously (applied in conjunction with the wrapper for ordinal classification), based on 20 discretization intervals, and compared to two dedicated regression methods: 100 bagged unpruned regression trees and linear support vector machines for regression. These results show that the point estimates produced using regression by discretization are generally competitive with those produced by dedicated regression schemes.

3.3 Comparison to Gaussian Process Regression and Nonparametric Quantile Regression

In this section, we compare the performance of discretization-based density estimation to Gaussian process regression and techniques for quantile regression. To this end, we use SMO as the base learner, in conjunction with the ordinal wrapper and univariate kernel density estimation with 20 discretization bins. To yield comparable results, RBF kernels were used in SMO, and the γ parameter for the kernel and the C parameter for regularization were optimized using *in-*

Table 4. Quality of prediction intervals, using linear SVMs and 10 bins

Dataset	Coverage			Relative width		
	Histogram	Kernel	Normal	Histogram	Kernel	Normal
auto93	75.5± 14.0	92.8± 9.0 ○	96.4± 6.6 ○	35.8± 7.1	38.4± 6.5	42.1± 6.3 ●
autoHorse	91.5± 6.8	95.1± 5.1	96.2± 4.3 ○	21.6± 5.3	19.6± 4.1 ○	23.3± 4.3 ●
autoMpg	83.4± 6.3	94.8± 3.7 ○	97.9± 2.2 ○	37.6± 2.5	39.2± 2.6 ●	38.9± 2.5 ●
autoPrice	87.3± 8.4	95.7± 5.8 ○	98.4± 3.6 ○	32.2± 6.2	32.8± 5.3	36.2± 4.9 ●
baskball	83.4± 11.3	94.4± 7.0 ○	97.8± 4.3 ○	75.6± 3.6	63.4± 6.2 ○	61.7± 5.5 ○
bodyfat	91.4± 5.7	94.3± 4.7	96.6± 3.4 ○	27.9± 2.1	25.5± 1.3 ○	25.8± 1.3 ○
bolts	79.3± 20.1	98.0± 6.8 ○	99.0± 6.1 ○	55.4± 9.8	77.6±11.7 ●	88.7± 9.7 ●
breastTumor	90.0± 5.2	95.8± 4.4 ○	94.1± 4.1 ○	76.5± 3.0	85.6± 2.7 ●	80.5± 1.6 ●
cholesterol	94.6± 4.6	95.5± 4.5	95.9± 3.9	72.9± 2.9	46.3± 6.1 ○	48.5± 5.5 ○
cleveland	85.6± 6.9	96.5± 3.4 ○	94.2± 3.9 ○	56.0± 4.3	71.8± 5.1 ●	91.5± 4.7 ●
cloud	81.9± 10.8	94.8± 6.6 ○	96.1± 5.3 ○	31.5± 7.0	35.6± 7.0	37.8± 7.0 ●
cpu	87.5± 7.4	96.5± 4.3 ○	98.9± 2.1 ○	22.1± 7.0	16.6± 5.4 ○	20.1± 6.1
detroit	82.0± 35.2	90.0±28.4	93.0±24.6	78.3± 12.1	78.4±19.0	74.8±15.9
echoMonths	85.0± 8.9	96.5± 4.7 ○	94.9± 6.2 ○	70.2± 4.0	96.3± 2.2 ●	91.8± 3.6 ●
elusage	64.3± 22.8	91.0±11.8 ○	95.3± 9.3 ○	60.3± 6.3	83.6± 8.3 ●	80.6± 8.6 ●
fishcatch	86.8± 10.1	95.6± 5.8 ○	97.4± 3.9 ○	27.1± 5.0	27.0± 4.7	30.4± 5.2 ●
fruitfly	88.5± 9.5	93.4± 7.1	94.2± 7.1	75.4± 3.9	76.7± 5.2	75.3± 3.2
gascons	82.0± 23.7	91.7±18.9	92.8±17.9	53.8± 10.2	63.6±18.5 ●	62.9±18.0 ●
housing	89.5± 4.2	94.8± 2.9 ○	96.4± 2.4 ○	34.0± 1.8	35.9± 2.2 ●	39.3± 2.6 ●
hungarian	59.7± 12.2	99.7± 0.9 ○	94.5± 5.2 ○	76.4± 4.2	57.7± 6.0 ○	127.7±10.7 ●
longley	73.0± 38.5	94.5±22.4	98.5±11.1	66.3± 8.3	73.5±18.8	72.2±14.8
lowbwt	89.4± 7.2	95.9± 4.6 ○	96.9± 3.9 ○	52.1± 2.8	48.0± 2.7 ○	48.9± 2.7 ○
mbagrade	80.2± 14.5	97.1± 7.1 ○	92.4±10.4 ○	82.3± 4.9	97.7± 3.4 ●	90.9± 5.5 ●
meta	86.3± 7.2	98.2± 1.8 ○	99.4± 1.1 ○	65.5± 11.2	5.3± 1.1 ○	17.8± 2.9 ○
pbc	85.7± 5.7	96.1± 2.8 ○	96.7± 2.9 ○	70.4± 3.0	79.1± 3.0 ●	79.7± 3.2 ●
pharynx	83.3± 7.7	94.9± 4.8 ○	95.8± 4.5 ○	59.7± 3.4	70.3± 4.0 ●	75.6± 3.6 ●
pollution	76.3± 18.2	90.7±13.3 ○	91.8±12.0 ○	61.0± 6.9	64.6± 5.7	59.1± 5.2
pwLinear	84.5± 7.8	94.3± 5.0 ○	95.0± 5.3 ○	36.8± 3.0	40.0± 3.3 ●	39.8± 3.5 ●
quake	94.5± 1.4	98.0± 1.3 ○	94.6± 1.3	51.6± 1.0	58.1± 0.9 ●	67.5± 0.7 ●
schlvote	82.2± 20.6	89.0±18.2	93.9±14.3	56.2± 17.9	50.7±21.5	54.2±21.4
sensory	93.1± 3.5	95.2± 3.0 ○	96.3± 2.4 ○	72.1± 3.4	66.1± 1.9 ○	64.8± 1.5 ○
servo	81.0± 9.2	95.7± 5.4 ○	98.0± 3.6 ○	29.0± 7.2	32.8± 7.5 ●	40.6± 8.2 ●
sleep	79.4± 17.8	98.6± 4.9 ○	94.3±10.1 ○	68.8± 6.4	86.1± 7.8 ●	81.8± 6.5 ●
strike	85.5± 5.9	95.4± 3.2 ○	97.7± 2.3 ○	62.6± 14.4	20.6± 1.7 ○	30.1± 2.3 ○
veteran	88.6± 8.7	96.1± 5.6 ○	95.8± 6.0 ○	65.6± 4.9	51.1± 3.9 ○	60.7± 4.4 ○
vineyard	81.8± 19.2	91.8±13.1	92.5±11.4	56.1± 8.3	60.0±11.1	56.0±10.6

○/● statistically significant improvement/degradation

ternal 10-fold cross-validation, choosing from 10^i with $i \in \{-5, 4, ..., 4, 5\}$ for γ and 10^i with $i \in \{-2, -1, 0, 1, 2\}$ for C.[5]

Table 6 shows the relative performance of discretization-based density estimation and Gaussian process regression (GPR) with RBF kernels. Gaussian process regression was applied with inputs normalized to $[0, 1]$, as was SMO. The target was also normalized for GPR, to make choosing an appropriate noise level σ easier. Predictions were transformed back into the original range of the target variable. The noise level σ for GPR, and γ for the RBF kernel, were optimized using internal 10-fold cross-validation, with the same values for γ as in the case of SMO, and values 10^i with $i \in \{-2, -1.8, -1.6, ..., -0.4, -0.2, 0\}$ for σ.

The log-likelihood scores exhibit 11 significant differences, three in favor of GPR and eight against. Comparing the relative scores for GPR to those for the normal estimator in Table 2 shows that there are four cases where the gain for these techniques is well below zero while it is well above zero for the kernel-density-based

---

[5] To reduce runtime, the tolerance parameter in SMO was increased from 0.001 to 0.1, which did not significantly impact performance on the datasets investigated.

Table 5. Root relative squared error: regression by discretization (20 bins) vs. dedicated regression methods

Dataset	Random forests (20 bins)	Bagged regression trees	Linear SVMs (20 bins)	Linear SVMs for regression
auto93	58.6± 13.6	93.8± 21.4 o	53.1± 16.2	62.5± 18.8
autoHorse	33.3± 11.2	40.3± 11.7 o	30.6± 11.2	31.5± 8.7
autoMpg	37.0± 5.3	40.2± 7.2	38.1± 5.7	39.0± 5.6
autoPrice	37.8± 8.6	37.9± 11.3	39.5± 10.8	45.6± 8.9 o
baskball	84.5± 12.5	85.5± 21.3	84.3± 9.9	82.2± 17.9
bodyfat	22.4± 5.8	13.8± 11.0 •	29.1± 6.0	9.7± 12.7 •
bolts	33.2± 13.8	35.8± 29.3	49.0± 22.2	32.8± 23.1 •
breastTumor	100.1± 7.3	107.8± 11.6 o	97.9± 1.6	98.1± 7.7
cholesterol	99.0± 3.2	103.4± 10.0	99.6± 1.2	103.7± 9.2
cleveland	72.5± 6.8	74.1± 10.3	68.9± 9.1	71.5± 9.0
cloud	48.6± 14.0	45.7± 14.7	50.0± 12.9	37.2± 13.3 •
cpu	40.3± 17.5	27.2± 12.6	35.7± 19.1	29.2± 14.4
detroit	143.2±229.4	133.0±145.5	149.9±241.1	103.8±119.9
echoMonths	71.7± 11.5	76.3± 14.8 o	74.0± 9.5	74.1± 14.2
elusage	50.1± 17.1	57.5± 22.6 o	62.3± 14.4	60.5± 21.7
fishcatch	24.6± 7.5	20.8± 6.3	22.4± 5.9	25.8± 5.5
fruitfly	104.1± 5.7	116.9± 15.5 o	100.7± 2.6	102.4± 9.7
gascons	25.2± 10.2	24.2± 13.2	26.3± 10.6	24.1± 16.2
housing	41.6± 8.1	38.1± 8.6	45.2± 10.8	54.5± 11.2 o
hungarian	73.6± 10.3	78.0± 11.5 o	71.6± 11.5	88.6± 16.8 o
longley	49.9± 46.3	53.2± 71.6	56.1± 58.9	15.0± 14.0 •
lowbwt	61.5± 8.2	64.2± 10.3	64.0± 6.7	64.9± 10.9
mbagrade	97.6± 19.3	99.0± 23.4	96.2± 9.7	88.4± 23.8
meta	95.7± 22.2	157.1± 70.3 o	108.0± 34.3	81.6± 19.6
pbc	83.2± 5.9	83.0± 9.3	81.5± 6.0	82.4± 8.5
pharynx	78.5± 9.2	103.9± 6.7 o	76.2± 8.9	83.6± 13.3 o
pollution	78.0± 8.8	69.1± 17.9	69.3± 17.3	68.6± 25.8
pwLinear	46.2± 6.9	40.4± 7.8 •	42.0± 8.3	50.7± 10.1 o
quake	100.0± 1.7	100.9± 2.8	100.0± 0.2	107.7± 1.7 o
schlvote	77.0± 26.5	72.3± 29.4	91.5± 36.3	89.7± 39.4
sensory	86.7± 4.2	89.0± 6.9	95.2± 1.8	94.7± 5.9
servo	38.9± 14.4	34.3± 16.5	37.2± 14.5	62.7± 23.6 o
sleep	79.0± 18.3	79.6± 27.1	77.0± 22.8	74.7± 22.0
strike	79.8± 13.2	91.3± 15.6 o	85.2± 8.6	82.3± 12.4
veteran	90.7± 15.0	105.8± 26.4 o	92.3± 6.4	89.1± 13.7
vineyard	60.7± 23.5	66.1± 31.1	70.6± 23.0	72.6± 28.9

o/• statistically significant improvement/degradation

estimate: cleveland, hungarian, strike, and veteran. This indicates that it is not appropriate to model the predictive distribution with a normal density in these cases.

Considering the quality of the prediction intervals, there is one case where GPR performs significantly better than the discretization-based approach according to both coverage and width (bodyfat), and two where it is significantly worse (hungarian, quake). Ignoring significance, there are four additional cases where GPR performs better according to both statistics (auto93, autoMpg, cpu, lowbwt) and three more cases where it is worse (cleveland, echoMonths, servo). It is noticeable that the observed coverage is generally closer to the desired confidence level (95%) for the discretization-based approach.

Due to space constraints, we cannot show detailed results for the root relative squared error here. In most cases, the error is comparable. However, there are three very small datasets with 40 or less instances (bolts, gascons, longley) where GPR performs significantly better. It also performs significantly better on cpu and sensory, and significantly worse on servo.

Table 6. Comparison of discretization-based conditional density estimation (SVMs, kernel density estimator, 20 bins) with Gaussian process regression (GPR), using RBF kernels in both cases

Dataset	Mean improvement in log-likelihood		Coverage		Relative width	
	RBF SVMs (20 bins)	RBF GPR	RBF SVMs (20 bins)	RBF GPR	RBF SVMs (20 bins)	RBF GPR
auto93	0.68± 0.95	1.00± 0.77	92.70± 9.98	96.18± 6.93	38.43± 7.81	37.49± 6.32
autoHorse	2.63± 0.63	1.48± 1.80	94.46± 5.41	95.86± 4.48	16.20± 4.58	16.44± 2.69
autoMpg	1.29± 0.27	1.45± 0.23	93.56± 4.28	96.38± 3.61	32.71± 2.54	31.88± 2.72
autoPrice	1.27± 0.46	1.08± 0.44	92.83± 5.87	91.32± 6.98	28.46± 5.22	25.95± 4.85
baskball	-0.10± 0.56	0.50± 0.74	93.11± 8.55	97.61± 5.09	60.16± 6.45	64.79± 1.04 o
bodyfat	1.57± 1.77	1.74± 3.72	94.83± 4.49	98.41± 2.60 •	19.75± 2.78	8.92± 3.28 •
bolts	0.69± 0.55	-0.12± 2.37	95.50±10.88	84.00±22.90	79.29±13.09	41.66±14.30 •
breastTumor	-0.03± 0.09	-0.11± 0.13	95.52± 4.31	89.36± 7.13 o	83.83± 3.21	72.08±15.20 •
cholesterol	-0.01± 0.07	0.15± 0.93	95.09± 4.45	97.36± 4.08	46.19± 6.28	61.88± 4.89 o
cleveland	0.87± 0.40	-0.45± 0.24 o	96.60± 4.09	95.75± 5.61	63.73± 5.95	99.20±10.24 o
cloud	0.83± 0.40	0.61± 1.78	93.53± 7.00	92.63± 8.46	35.17± 7.15	26.48± 5.99 •
cpu	2.93± 0.67	2.76± 0.91	95.97± 4.94	99.32± 2.97	9.89± 4.18	7.88± 2.31
detroit	0.87± 1.84	0.26± 2.31	91.50±26.64	81.50±38.04	70.43±27.58	54.88±23.71
echoMonths	0.51± 0.30	0.22± 0.17 o	97.15± 4.85	93.69±10.10	80.39± 6.33	91.69±16.24
elusage	0.50± 0.52	0.74± 0.57	93.13±11.80	90.37±13.15	75.63±10.53	47.80± 6.23 •
fishcatch	2.67± 0.44	2.59± 0.33	95.57± 4.72	91.29± 6.82	16.84± 3.52	12.42± 2.23 •
fruitfly	-0.04± 0.17	-0.42± 0.31 o	92.75± 7.42	94.12± 8.08	72.57± 5.18	88.36±16.62 o
gascons	1.59± 0.82	3.60± 0.89 •	95.17±13.67	94.50±14.03	46.84±11.89	15.95±16.79 •
housing	1.42± 0.28	1.37± 0.30	93.76± 3.50	94.84± 2.76	28.48± 2.47	30.54± 3.91
hungarian	1.12± 0.24	-1.38± 0.27 o	99.69± 1.10	89.02± 8.07 o	53.80± 5.95	128.91±21.98 o
longley	0.95± 1.57	2.63± 1.90	95.00±19.46	90.50±27.24	71.34±20.75	17.22± 4.90 •
lowbwt	0.59± 0.25	0.70± 0.18	95.07± 5.18	95.86± 4.87	48.05± 2.79	40.97± 0.82 •
mbagrade	0.05± 0.25	0.10± 0.42	93.76± 8.69	95.36±11.11	95.62± 4.25	99.57± 8.31
meta	18.8±124.0	17.5±142.1	98.24± 1.92	99.45± 1.02 •	5.03± 1.33	54.84±51.38 o
pbc	0.27± 0.17	0.22± 0.17	95.36± 3.48	91.89± 4.41 o	76.51± 2.59	64.72± 1.85 •
pharynx	0.27± 0.25	0.16± 0.23	93.34± 5.68	94.62± 5.63	68.76± 3.78	70.25± 4.17
pollution	0.14± 0.48	0.65± 0.60 •	91.33±12.64	91.00±12.40	65.44± 7.37	55.93±13.61 •
pwLinear	0.93± 0.37	1.21± 0.30 •	92.45± 6.49	92.15± 6.64	38.50± 3.47	32.32± 5.72 •
quake	-0.03± 0.03	-0.87± 0.06 o	94.12± 1.34	94.12± 1.49 o	57.97± 0.74	62.42± 0.16 o
schlvote	0.15± 9.74	0.65± 5.81	89.00±17.88	96.92± 9.81	50.30±18.12	130.58±37.14 o
sensory	-0.04± 0.14	-0.10± 0.13	94.49± 3.36	95.90± 6.35	62.55± 3.00	63.96± 9.09
servo	1.65± 0.36	-0.07± 0.78 o	95.09± 5.38	94.97± 6.48	24.34± 6.18	45.41± 4.18 o
sleep	0.36± 0.49	0.42± 0.36	96.92± 7.79	88.21±13.08 o	85.20± 6.68	70.77±11.89 •
strike	0.51± 0.97	-1.09± 0.69 o	95.22± 3.03	98.24± 2.14 •	16.52± 2.55	35.67± 8.41 o
veteran	0.25± 0.28	-0.52± 0.90 o	95.04± 5.83	95.62± 5.97	49.17± 5.12	64.79± 5.69 o
vineyard	0.22± 0.91	0.89± 0.97	92.90±11.79	91.80±12.82	57.98±10.18	47.93± 8.98 •

o, • statistically significant improvement or degradation

We now consider the task of quantile estimation. Table 7 shows the pinball loss of our discretization-based method when used for this task (50% quantile), compared to two existing quantile regression techniques, including the nonparametric quantile estimation method NPQR, based on the results in [21]. For details of the pinball loss and the datasets used, see [21]. Note that the results from [21] are based on a single 10-fold cross-validation.

The approximate quantiles for our method were obtained from the weighted kernel density estimates in the same fashion as the interval boundaries used previously (see Section 2.3). The pinball loss was used as the performance estimate for the internal cross-validation employed to choose parameters for the SVMs.

The results show that, in spite of the fact that NPQR optimizes the pinball loss directly using quadratic optimization, discretization-based estimation generally yields highly competitive results. There are only two datasets where the pinball loss is clearly significantly worse: crabs and cpu. In both cases, given the small

Table 7. Comparison of pinball loss (50%-quantile) for proposed method (SVMs with RBF kernels, kernel density estimator, 20 bins), linear quantile regression (Linear QR), and nonparametric quantile regression (NPQR). Results for the latter two methods are reproduced from [21], with standard errors converted to standard deviations.

Dataset	RBF SVMs (20 bins)	Linear QR	NPQR
caution	23.71± 9.29	32.40± 8.73	22.56± 8.04
ftcollinssnow	42.87± 9.48	40.82± 8.85	39.08± 9.27
highway	23.42±10.42	45.39±21.12	25.33±10.86
heights	34.85± 2.27	34.50± 2.16	34.53± 2.16
sniffer	10.83± 2.55	12.78± 3.33	9.92± 2.82
snowgeese	14.23±10.57	13.85±10.38	18.50±14.88
ufc	21.91± 2.41	23.20± 2.85	21.22± 2.70
birthwt	37.75± 6.73	38.15± 5.88	37.19± 5.88
crabs	4.39± 1.08	2.24± 0.39	2.14± 0.36
GAGurine	16.06± 3.62	27.87± 4.38	14.57± 3.33
geyser	30.37± 5.43	32.50± 3.69	30.75± 4.20
gilgais	10.79± 1.94	16.12± 3.03	12.40± 1.98
topo	17.17± 6.73	26.51± 8.13	14.39± 4.95
BostonHousing	11.80± 2.14	17.50± 2.85	10.76± 1.83
CobarOre	40.88±15.84	41.93±15.60	39.29±20.07
engel	13.83± 3.21	13.72± 3.42	13.01± 2.55
mcycle	18.90± 4.52	37.88± 8.28	17.06± 4.26
BigMac2003	19.46±10.82	21.75± 8.55	17.89± 9.15
UN3	23.00± 5.50	26.32± 5.10	23.96± 5.52
cpus	6.13± 5.31	5.73± 3.12	1.06± 0.51

loss, a possible explanation for the relatively poor performance is the coarseness of the discretization: running the method using 40 discretization intervals yields a reduced pinball loss of 3.84 and 3.89 respectively.

4   Conclusions

This paper considered conditional density estimates based on combining class probability estimators with univariate density estimators. Our results indicate that kernel estimators are generally superior to normal estimators in this context, and that both are preferable to histogram-based estimation. We have presented results for both support vector machines and random forests as underlying class probability estimators. The proposed techniques yield useful prediction intervals as well as competitive point estimates. They also show promise when compared to Gaussian process regression and nonparametric quantile estimation.

The generic aspect of the approach discussed here is valuable because of the availability of a large number of class probability estimation schemes that are potentially applicable. An avenue for future work is to investigate the performance of other univariate density estimators in the context considered in this paper. For example, rather than using a kernel density estimator or a normal estimator, one could apply a mixture model.

References

1. Hyndman, R.J., Bashtannyk, D.M., Grunwald, G.K.: Estimating and visualizing conditional densities. Journal of Computational and Graphical Statistics 5(4), 315–336 (1996)

2. Holmes, M., Gray, A., Isbell, C.: Fast nonparametric conditional density estimation. In: Proc. 23rd Conf. on Uncertainty in AI. AUAI Press (2007)
3. Rasmussen, C.E., Williams, C.K.I.: Gaussian Processes for Machine Learning. MIT Press, Cambridge (2006)
4. Tay, A.S., Wallis, K.F.: Density forecasting: A survey. Journal of Forecasting 19, 235–254 (2000)
5. Neuneier, R., Ferdinand Hergert, W.F., Ormoneit, D.: Estimation of conditional densities: A comparison of neural network approaches. In: Proc. Int. Conf. on Artificial Neural Networks, pp. 689–692. Springer, Heidelberg (1994)
6. Bishop, C.M.: Neural networks for pattern recognition. Oxford UP, Oxford (1995)
7. Williams, P.M.: Using neural networks to model conditional multivariate densities. Neural Computation 8(4), 843–854 (1996)
8. Papadopoulos, G., Edwards, P., Murray, A.: Confidence estimation methods for neural networks: a practical comparison. IEEE Transactions on Neural Networks 12(6) (2001)
9. Carney, M., Pádraig Cunningham, J.D., Lee, C.: Predicting probability distributions for surf height using an ensemble of mixture density networks. In: Proc. 22nd Int. Conf. on Machine learning, pp. 113–120. ACM Press, New York (2005)
10. Stützle, E., Hrycej, T.: Numerical method for estimating multivariate conditional distributions. Computational Statistics 20(1), 151–176 (2005)
11. Carney, M., Cunningham, P.: Making good probability estimates for regression. In: Fürnkranz, J., Scheffer, T., Spiliopoulou, M. (eds.) ECML 2006. LNCS (LNAI), vol. 4212, pp. 582–589. Springer, Heidelberg (2006)
12. Carney, M., Cunningham, P.: Calibrating probability density forecasts with multi-objective search. In: Proc. 17th Europ. Conf. on AI, pp. 791–792. IOS Press, Amsterdam (2006)
13. Davies, S., Moore, A.: Interpolating conditional density trees. In: Proc. 18th Annual Conf. on Uncertainty in AI, pp. 119–127. Morgan Kaufmann, San Francisco (2002)
14. Witten, I.H., Frank, E.: Data Mining: Practical Machine Learning Tools and Techniques, 2nd edn. Morgan Kaufmann, San Francisco (2005)
15. Schapire, R.E., Stone, P., McAllester, D.A., Littman, M.L., Csirik, J.: Modeling auction price uncertainty using boosting-based conditional density estimation. In: Proc. 19th Int. Conf. on Machine Learning, pp. 546–553. Morgan Kaufmann, San Francisco (2002)
16. Hjort, N.L., Walker, S.G.: A note on kernel density estimators with optimal bandwidths. Statistics & Probability Letters 54, 153–159 (2001)
17. Platt, J.C.: Probabilistic outputs for support vector machines and comparison to regularized likelihood methods. In: Advances in Large Margin Classifiers, pp. 61–74. MIT Press, Cambridge (1999)
18. Nadeau, C., Bengio, Y.: Inference for the generalization error. Machine Learning 52, 239–281 (2003)
19. Frank, E., Hall, M.: A simple approach to ordinal classification. In: Flach, P.A., De Raedt, L. (eds.) ECML 2001. LNCS (LNAI), vol. 2167, pp. 145–156. Springer, Heidelberg (2001)
20. Frank, E., Trigg, L.E., Holmes, G., Witten, I.H.: Naive Bayes for regression (technical note). Machine Learning 41(1), 5–25 (2000)
21. Takeuchi, I., Le, Q.V., Sears, T.D., Smola, A.J.: Nonparametric quantile estimation. Journal of Machine Learning Research, 1231–1264 (2006)

Linear Time Model Selection
for Mixture of Heterogeneous Components

Ryohei Fujimaki, Satoshi Morinaga, Michinari Momma,
Kenji Aoki, and Takayuki Nakata

NEC Common Platform Software Research Laboratories
http://www.nec.co.jp/rd/en/datamining/index.html

Abstract. Our main contribution is to propose a novel model selection methodology, expectation minimization of description length (EMDL), based on the minimum description length (MDL) principle. EMDL makes a significant impact on the combinatorial scalability issue pertaining to the model selection for mixture models having types of components. A goal of such problems is to optimize types of components as well as the number of components. One key idea in EMDL is to iterate calculations of the posterior of latent variables and minimization of expected description length of both observed data and latent variables. This enables EMDL to compute the optimal model in linear time with respect to both the number of components and the number of available types of components despite the fact that the number of model candidates exponentially increases with the numbers. We prove that EMDL is compliant with the MDL principle and enjoys its statistical benefits.

Keywords: heterogeneous mixture model, model selection, minimum description length.

1 Introduction

Model selection is one of the most significant and challenging issues in machine learning. Its goal is not only to estimate the optimal parameter values for a given model but also to find the optimal model itself. One of the most common approaches is to select, from model candidates, a model which optimizes an information criterion such as the minimum description length (MDL) criterion [1], Akaike's information criterion (AIC) [2], minimum message length [3,4,5], etc.

The recent growing need for analyzing data sampled from a complicated generative process necessitates the use of a class of mixture models consisting heterogeneous types of components e.g. latent variable models [6], polynomial curves of various degrees, distributions with different dimensional dependencies, etc. Applications of such models cover a wide range of real problems such as image recognition, speech recognition, text/blog clustering, bioinfomatics, etc. Although it is necessary to specify appropriate types of components in order to make the best use of such models, it is usually impossible to know them in advance. Therefore, it is of crucial importance to simultaneously optimize both the

Z.-H. Zhou and T. Washio (Eds.): ACML 2009, LNAI 5828, pp. 82–97, 2009.

component types and parameters of each component. We refer to the class of such problems as model selection of heterogeneous mixture (MSHM) and possible types of components as "component seeds."

Naive approaches based on one of the above model selection criteria would rely on the exhaustive search in which all model candidates are examined. They are, however, impractical in MSHM because both the number of candidate models and the computational cost increase exponentially with the size of the search space (i.e., increases of both the maximum number of components and the number of component seeds).

One reasonable approach to reducing the computational cost is to apply a greedy search technique, such as the pattern search algorithm [7,8]. Although such an algorithm empirically achieves fast convergence, the computational cost in the worst case is of the same order as that of a full search (i.e., exponential time). Further, the cost for each step in the optimization is high when the number of the model candidates is large.

More sophisticated approaches based on the bayesian framework [9] and the lossy coding framework [10] have been proposed. Ghahramani and Beal [9] proposed a model selection method, on the basis of automatic relevance determination and the variational bayes method, and presented its effectiveness. However, since their optimization algorithm requires a temperature parameter and its scheduling heuristics, its complexity is not guaranteed to be small. Ma et al [10] proposed a model selection method on the basis of a "bottom-up" hierarchical clustering-like algorithm. Since its complexity is polynomial in both the number of component seeds and the number of data, it is difficult to apply their method to a large model set and a large data set. Further the Ma's algorithm can deal with only the data that consist of multiple Gaussian-like groups.

The main contribution of this paper is to propose a novel model selection methodology, expectation minimization of description length (EMDL), based on the MDL principle. We adopt the MDL criterion because, among the above criteria, it has numerous advantages including consistency, fast convergence rate for model estimation, achievement of the minimax regret, etc [11,12].

EMDL has a similar idea to the EM algorithm [13] and the Snob algorithm [3,4,5] in that it employs iterative steps of computing the posterior of latent variables, and minimizing the expected description length of the observed data and the latent variables (i.e., a "complete data set"). The key difference between EMDL and them is that, in each minimization step, EMDL optimizes the types of components as well as parameter vales for each component. As we explain later section, this makes linear time MSHM possible. In Section 3.4, we will show, as a theoretical justification, that the description length of observed data (i.e., a "incomplete data set") monotonically decreases in each iteration. This means that EMDL is compliant with the MDL principle and enjoys its statistical benefits.

The main significance of EMDL is that the time required for EMDL to compute an optimal model increases only linearly with the maximum number of components and model seeds, even though the number of candidate models grows exponentially. In addition, on the basis of the MDL principle, EMDL is able to be applied to arbitrary types of component seeds. To the best of our knowledge, this is the first theoretically-justifiable methodology for linear time MSHM which selects optimal types of components as well as the optimal number of components.

We evaluate EMDL via simulations based two types of heterogeneous mixture models. Computational cost for EMDL is practical and increases only linearly with the increases of the maximum number of components and the number of component seeds. Further, we show that EMDL outperforms naive "MDL + EM algorithm" approaches in terms of both model selection and parameter estimation performances.

2   Preliminaries

Let us denote a set of candidate models as $\mathcal{M} = \{M_i | i = 1, \ldots, |\mathcal{M}|\}$ where $|\circ|$ represents the number of elements in a set $\circ$. M_i represents a parametric family of probability distributions[1]: $M_i = \{P(X; \boldsymbol{\theta}^{M_i}) | \boldsymbol{\theta}^{M_i} \in \Theta_i\}$. Here X is a random variable in $\mathcal{X}$ and the function $P(\bullet; \star)$ is a probability density/mass function (pdf.) of $\bullet$ parameterized by $\star$. A parameter $\boldsymbol{\theta}^{M_i} = (\theta_1, \ldots, \theta_{K_{M_i}})$ is a element of a parameter set Θ_i where K_{M_i} is the dimensionality of $\boldsymbol{\theta}^{M_i}$ and is a function of M_i. We denote the maximum likelihood estimator of a parameter $\star$ as $\hat{\star}$.

2.1   MDL Principle

The fundamental idea behind the MDL principle is that, for given $\mathcal{M}$ and data $\boldsymbol{x}^N = \boldsymbol{x}_1, \ldots, \boldsymbol{x}_N$, we should select a model that compresses $\boldsymbol{x}^N$ most. The shortest description length of $\boldsymbol{x}^N$ with $\mathcal{M}$ is called the stochastic complexity.

Rissanen [1] proposed the two-part code to define the stochastic complexity. Let $M \in \mathcal{M}$ be a member of candidate models. The two-part code first encodes M itself and then encodes $\boldsymbol{x}^N$. The total description length $\ell(\boldsymbol{x}^N; M)$ is then defined as $\ell(\boldsymbol{x}^N; M) = \ell(\boldsymbol{x}^N | M) + \ell(M)$. Here $\ell(\boldsymbol{x}^N | M)$ and $\ell(M)$ represent the description length of $\boldsymbol{x}^N$ given the model M and one of M, respectively. Note that $\ell(\boldsymbol{x}^N; M)$ satisfies the Kraft's inequality [14] which is a necessary and sufficient condition of the existence of prefix encoding.

There exist several ways for defining $\ell(\boldsymbol{x}^N | M)$. A number of codes [11] can be constructed with the following asymptotic length[2] as:

$$\ell(\boldsymbol{x}^N | M) \sim \sum_{i=1}^{N} - \log P(\boldsymbol{x}_i; \hat{\boldsymbol{\theta}}^M) + \frac{K_M}{2} \log N, \tag{1}$$

---

[1] For notational simplicity, a superscription of $\star$ represents a model candidate of which $P(X; \star)$ is a element.

[2] The description length (1) is known to be rather redundant since the constant term which is ignored in (1) can be $\mathcal{O}(\log N)$. The description length of more refined "maximum-likelihood code" is presented by Rissanen [15], in which the redundant term is only $\mathcal{O}(1)$.

where, in the right side, the first term and the second term can be interpreted as the description length of x^N for the given parameter $\hat{\theta}^M$ (i.e., $\ell(x^N|\hat{\theta}^M)$) and the description length of the parameter $\hat{\theta}^M$ for the give model M (i.e., $\ell(\hat{\theta}^M|M)$). The description length $\ell(M)$ can be defined using an arbitrary manner which satisfies the Kraft's inequality $\sum_{M \in \mathcal{M}} 2^{-\ell(M)} \leq 1$. We will explain $\ell(M)$ for our problem in the latter section.

A MDL-based learning algorithm selects the optimal model $\bar{M}$ and its parameter value $\bar{\theta}^M$ which minimizes the description length as:

$$\bar{M}, \bar{\theta}^M = \arg \min_{M, \bar{\theta}^M} \left\{ \ell(x^N|\bar{\theta}^M) + \ell(\bar{\theta}^M|M) + \ell(M) \right\}. \tag{2}$$

MDL assumes that $\bar{\theta}^M$ is the maximum likelihood estimator. Therefore, in general, MDL-based algorithms for mixture models firstly estimate $\bar{\theta}^M$ using the EM algorithm for all M, and then obtain $\bar{M}$ which minimize the left size of the above equation [16].

Among a number of information criteria, the MDL criterion has some superiority in a number of respects; e.g. consistency, fast convergence rate for model estimation, achievement of the minimax regret, etc. For more detail about the MDL principle, see [11].

2.2   Mixture of Heterogeneous Components

Let us consider a parametric family of probability distributions $V_j = \{P(X; \phi^{V_j})| \phi^{V_j} \in \Phi_j\}$ where $\phi^{V_j} = (\phi_1^{V_j}, \ldots, \phi_{J_{V_j}}^{V_j})$ represents an element of a parameter set Φ_j corresponding to V_j, respectively. Note that each $P(X; \phi^{V_j})$ may have not only a different parameter dimensionality J_{V_j} but also a different functional form according to V_j. We then define "components seeds" as a set $\mathcal{S} = \{V_j | j = 1, \ldots, |\mathcal{S}|\}$.

We denote a set of models of mixture distributions with the above heterogeneous components as $\mathcal{H} = \{H_i | i = 1, \ldots, |\mathcal{H}|\}$. We here refer to $\mathcal{H}$ as model candidates[3]. For a given component seeds $\mathcal{S}$, a model candidate $H \in \mathcal{H}$ is defined as follows[4]:

$$\left\{ P(X; \theta) = \sum_{c=1}^{C} \pi_c P(X; \phi_c^{S_c}) \right\}, \tag{3}$$

where the number of mixed components C and types of components $S_c \in \mathcal{S}$ ($c = 1, \ldots, C$) are specified by H (i.e., $C = C(H)$ and $S_c = S_c(H)$), and π_c is the mixture ratio of the cth component. We here define π, ϕ and θ as $\pi = (\pi_1, \ldots, \pi_C)$, $\phi = (\phi_1^{S_1}, \ldots, \phi_C^{S_C})$ and $\theta = \{\pi, \phi\}$. Let us denote the dimensionality of $\phi_c^{S_c}$ as J_c. For example, if S_c is a one-dimensional gaussian, $J_c = 2$ (mean and variance).

---

[3] For example, if V_1 is a gaussian distribution and V_2 is a gamma distribution (and $|\mathcal{S}| = 2$), H_i is a mixture of gaussian and gamma distributions.

[4] Hereafter, if there is no danger of confuse, we omit the superscription H of θ^H for notational simplicity.

MSHM is defined as a class of problems in which we find an optimal model $\bar{H} \in \mathcal{H}$ and optimal parameter values $\bar{\theta}$ for $\bar{H}$. For MSHM, standard MDL-based model selection algorithms first estimate $\bar{\theta}$ for a candidate H using the EM algorithm [13], and then calculate the description length $\ell(\boldsymbol{x}^N; H)$. The optimal model $\bar{H}$ is one which achieves the minimum description length. Unfortunately, these algorithms suffer from exponentially increasing computational cost with increases of the maximum number of components $C_{\max}$ and $|\mathcal{S}|$.

3   Expectation Minimization of Description Length

3.1   Expected Description Length for Complete Data Set

Let us consider the latent variables $\boldsymbol{z}^N = \boldsymbol{z}_1, \ldots, \boldsymbol{z}_N$, where $\boldsymbol{z}_i = (z_{i1}, \ldots, z_{iC})$. $\boldsymbol{z}_i$ indicates the component assignment of $\boldsymbol{x}_i$ and $z_{ic} = 1$ if $\boldsymbol{x}_i$ is generated from the cth component and $z_{ic} = 0$ otherwise. A pair $(\boldsymbol{x}^N, \boldsymbol{z}^N)$ is a "complete data set" while $\boldsymbol{x}^N$ is an "incomplete data set". In MSHM, a joint distribution of $(\boldsymbol{x}_i, \boldsymbol{z}_i)$ for a given H is defined as $\prod_{c=1}^{C} (\pi_c P(\boldsymbol{x}_i; \phi_c^{S_c}))^{z_{ic}}$ where the marginal distribution with respect to $\boldsymbol{x}_i$ is a element of (3).

The description length of the complete data set $\ell(\boldsymbol{x}^N, \boldsymbol{z}^N; H)$ is described as $\ell(\boldsymbol{x}^N, \boldsymbol{z}^N; H) = \ell(\boldsymbol{x}^N|\boldsymbol{z}^N, H) + \ell(\boldsymbol{z}^N|H) + \ell(H)$. Here, $\ell(\boldsymbol{x}^N|\boldsymbol{z}^N, H)$ is the description length of $\boldsymbol{x}^N$, given $\boldsymbol{z}^N$ and H. $\ell(\boldsymbol{z}^N|H)$ is the description length of $\boldsymbol{z}^N$, given H.

We can compute $\ell(H)$ using the universal prior distribution for integers [17] as follows:

$$\ell(H) = \sum_{c=1}^{C} \log^* J_c + \log^*(C-1) + Q, \tag{4}$$

where Q is a constant term[5]. The dimensionality of $\hat{\boldsymbol{\pi}}$ is $C - 1$ because of the constraint $\sum_{c=1}^{C} \hat{\pi}_c = 1$.

On the basis of the code used for (1), we can compute the remaining terms as follows[6]:

$$\ell(\boldsymbol{x}^N|\boldsymbol{z}^N, H) = -\sum_{i=1}^{N} \sum_{c=1}^{C} z_{ic} \log P(\boldsymbol{x}_i; \hat{\phi}_c^{S_c}) + \sum_{c=1}^{C} \frac{J_{S_c}}{2} \log\Big(\sum_{i=1}^{N} z_{ic}\Big), \tag{5}$$

and

$$\ell(\boldsymbol{z}^N|H) = -\sum_{i=1}^{N} \sum_{c=1}^{C} z_{ic} \log \hat{\pi}_c + \frac{C-1}{2} \log N. \tag{6}$$

Note that, for $\ell(\boldsymbol{x}^N, \boldsymbol{z}^N; H)$, we can separately optimize a type of each component. This enables us to select the optimal model in linear time with respect to

---

[5] The function $\log^*$ is defined, for an integer K as $\log^* K = \log K + \log \log K + \cdots$, where only the positive terms are included in the sum.

[6] We can derive more refined description lengths on the basis of the maximum likelihood code [15]) in the same manner.

the number of component seeds since we do not need to consider the combinations of component seeds.

In practice, however, since we do not have values for the latent variables, we cannot compute $\ell(\boldsymbol{x}^N, \boldsymbol{z}^N; H)$. Instead, we compute the expected description length of a complete data set with respect to a given posterior distribution of latent variables $P(\boldsymbol{z}^N | \boldsymbol{x}^N, \boldsymbol{\theta}, H)$ which is described as follows:

$$P(\boldsymbol{z}^N | \boldsymbol{x}^N, \boldsymbol{\theta}, H) \propto \prod_{i=1}^{N} \prod_{c=1}^{C} \left(\pi_c P(\boldsymbol{x}_i; \hat{\boldsymbol{\phi}}_c^{S_c}) \right)^{z_{ic}}. \tag{7}$$

Hereafter, let $E_Z[\bullet]$ represent the expectation value of $\bullet$ with respect to $P(\boldsymbol{z}^N | \boldsymbol{x}^N, \boldsymbol{\theta}, H)$.

We can then derive the expectation of $\ell(\boldsymbol{x}^N, \boldsymbol{z}^N; H)$ as follows:

$$E_Z[\ell(\boldsymbol{x}^N, \boldsymbol{z}^N; H)] = -\sum_{i=1}^{N} \sum_{c=1}^{C} E_Z[z_{ic}] \log P(\boldsymbol{x}_i; \hat{\boldsymbol{\phi}}_c^{S_c}) + \sum_{c=1}^{C} \frac{J_{S_c}}{2} E_Z \left[\log \left(\sum_{i=1}^{N} z_{ic} \right) \right]$$
$$- \sum_{i=1}^{N} \sum_{c=1}^{C} E_Z[z_{ic}] \log \hat{\pi}_c + \frac{C-1}{2} \log N + \sum_{c=1}^{C} \log^* J_{S_c} + \log^*(C-1), \tag{8}$$

where we ignore the constant term Q. Note that the first and third terms is the same as the objective of the EM algorithm. A key trick of EMDL is the second term which represents "expected" parameter description length for complete data. This summed form does not have interaction between different components. As we will explain in the next section, this makes it possible to optimize the types of components, separately.

3.2 Optimization Algorithm

To search for an optimal model, we first fix the number of mixture components C. EMDL then optimizes types of components S_c and the parameter values $\boldsymbol{\phi}_c^{S_c}$ by iterating calculations of (7) and minimizations of (8). As we explain in Section 3.4, this iteration monotonically decreases the description length of an "incomplete data set" $\ell(\boldsymbol{x}^N; H^{(t)})$. Hereafter, let a value superscripted by (t) represent a tth iteration value. EMDL selects the optimal number of mixture components $\bar{C}$ with which $\ell(\boldsymbol{x}^N; H)$ is minimized.

Expectation Process. To minimize $E_Z^{(t)}[\ell(\boldsymbol{x}^N, \boldsymbol{z}^N; H)]$, we need to compute the expectation of the latent variable $E_Z^{(t)}[z_{ic}]$ and the expectation of the logarithm of the sum of latent variables $E_Z^{(t)}[\log(\sum_{i=1}^{N} z_{ic})]$ for all components ($c = 1, \ldots, C$).

We may easily compute $E_Z^{(t)}[z_{ic}]$ as follows:

$$E_Z^{(t)}[z_{ic}] = \frac{\pi_c^{(t-1)} P(\boldsymbol{x}^N; \boldsymbol{\phi}_c^{S_c^{(t-1)}})}{\sum_{c=1}^{C} \pi_c^{(t-1)} P(\boldsymbol{x}^N; \boldsymbol{\phi}_c^{S_c^{(t-1)}})}. \tag{9}$$

Since each $\boldsymbol{z}_i$ is independent of the others, time complexity for calculating all $E_Z^{(t)}[z_{ic}]$ $(i = 1, \ldots, N, c = 1, \ldots, C)$ will be $\mathcal{O}(N)$.

Exponential time with respect to N is required to compute $E_Z^{(t)}[\log(\sum_{i=1}^{N} z_{ic})]$ since we must consider 2^N sequences of $z_{1c}, \ldots, z_{Nc}$. To avoid this, we employ a fast approximation method[7], described in Section 3.3.

Minimization Process. Once $E_Z^{(t)}[z_{ic}]$ and $E_Z^{(t)}[\log(\sum_{i=1}^{N} z_{ic})]$ are calculated, we minimize $E_Z^{(t)}[\ell(\boldsymbol{x}^N, \boldsymbol{z}^N; H)]$. We can separately optimize a type for each component $S_c^{(t)}$ and its parameter $\hat{\boldsymbol{\phi}}_c^{S_c^{(t)}}$ by solving the following problem:

$$\arg \min_{S_c, \boldsymbol{\phi}_c^{S_c}} \left\{ -\sum_{i=1}^{N} E_Z^{(t)}[z_{ic}] \log P(\boldsymbol{x}_i; \boldsymbol{\phi}_c^{S_c}) + \frac{J_{S_c}}{2} E_Z^{(t)} \left[\log \left(\sum_{i=1}^{N} z_{ic} \right) \right] + \log^* J_{S_c} \right\}. \tag{10}$$

The mixture ratio is calculated as follows:

$$\hat{\pi}_c^{(t)} = \frac{\sum_{i=1}^{N} E_Z^{(t)}[z_{ic}]}{N}. \tag{11}$$

One of key ideas in EMDL is that we can separately search for an optimal component seed $S_c^{(t)}$ for each individual component. Therefore we do not need to take the combinations of S_c into account. This enables us to optimize S_c as well as C in linear time with respect to $|\mathcal{S}|$. Note that, for every iteration, EMDL solves a sub-model selection problem (10) and the dimensionality of parameters can change over the course of the iterations.

For convergence determination, we check the decrease in the description length $\ell(\boldsymbol{x}^N; H^{(t-1)}) - \ell(\boldsymbol{x}^N; H^{(t)})$. Note that it is difficult to use the difference between $\hat{\boldsymbol{\theta}}^{(t)}$ and $\hat{\boldsymbol{\theta}}^{(t-1)}$ for convergence determination because the dimensionality of $\hat{\boldsymbol{\theta}}^{(t)}$ changes over the course of the iterations.

Algorithm 1 presents a pseudo-code for EMDL. For each number of components C, we iterate the expectation process and the minimization process until convergence. In this iteration, the types of components and its parameter values are optimized. EMDL select the optimal number of components $\bar{C}$ with which $\ell(\boldsymbol{x}^N; H)$ is minimized. We denote the optimal model and its parameter as $\bar{H}$ and $\bar{\boldsymbol{\theta}}$. A disadvantage of EMDL over the Snob algorithm [3,4,5] is that EMDL

---

[7] $E_Z^{(t)}[\log(\sum_{i=1}^{N} z_{ic})]$ can diverge to infinity if the probability of $\sum_{i=1}^{N} z_{ic} = 0$ is not zero. In practice, we can assume the probability is zero for the following two reasons. First, with increasing of N, the probability converges to zero. Second, $\sum_{i=1}^{N} z_{ic} = 0$ means that no data is generated from the distribution of the cth components. On the basis of the MDL principle, the probability with which such a model would be selected is zero.

Algorithm 1. EMDL based Model Selection

1: **Input:** x^N, $\mathcal{S}$ and $C_{\max}$
2: Initialization : $\bar{H} \leftarrow$ NULL and $\bar{L} \leftarrow \infty$
3: **for** $C = 1, \dots, C_{\max}$ **do**
4:     $t \leftarrow 1$ and initialize $H^{(t)}$ and $\hat{\boldsymbol{\theta}}^{(t)}$.
5:     **repeat**
6:         $t \leftarrow t + 1$.
7:         Evaluate (9) and (12).
8:         **for** $c = 1, \dots, C$ **do**
9:             Calculate $S_c^{(t)}$, $\hat{\phi}_{S_c}^{(t)}$, $\hat{\pi}_c^{(t)}$ by (10) and (11).
10:         **end for**
11:         Update: $H^{(t)} \leftarrow \{C, S_1^{(t)}, \dots, S_C^{(t)}\}$ and $\hat{\boldsymbol{\theta}}^{(t)} \leftarrow \{\hat{\boldsymbol{\pi}}^{(t)}, \hat{\boldsymbol{\phi}}^{(t)}\}$.
12:         Evaluate $\ell(x^N; H^{(t)})$ and convergence.
13:     **until** $\ell(x^N; H^{(t)})$ converges
14:     **if** $\ell(x^N; H^{(t)}) < \bar{L}$ **then**
15:         $\bar{L} \leftarrow \ell(x^N; H^{(t)})$, $\bar{H} \leftarrow H^{(t)}$ and $\bar{\boldsymbol{\theta}} \leftarrow \hat{\boldsymbol{\theta}}^{(t)}$.
16:     **end if**
17: **end for**
18: **Output:** the optimal model $\bar{H}$ and parameter $\bar{\boldsymbol{\theta}}$

---

requires an additional input $C_{\max}$ to restrict the search space[8]. However, as the Snob algorithm requires linear computational time with respect to the number of components C in each iteration step, the required computational time of EMDL also increases linearly with $C_{\max}$. We may remove the additional input using a split and merge technique as with the Snob algorithm.

3.3 Fast Approximation Method

Let us denote $\mu_{ic} = E_Z[z_{ic}]$ and $\mu_c = \sum_{i=1}^{N} \mu_{ic}$. On the basis of the Taylor expansion around μ_c, we approximate $E_Z[\log(\sum_{i=1}^{N} z_{ic})]$ as follows:

$$E_Z[\log(\sum_{i=1}^{N} z_{ic})] \approx \sum_{m=0}^{M} b_m E_Z\left[\left(\sum_{i=1}^{N} z_{ic}\right)^m\right], \tag{12}$$

where b_m is appropriate coefficients and functions of μ_c, and M is a parameter to control the approximation accuracy.

The key of (12) is that we can calculate an arbitrary order moment in $\mathcal{O}(N)$ time on the basis of the following two properties: 1) $z_{ic}^n = z_{ic}$ ($n \geq 1$) and 2) $E_Z[z_{ic}z_{jc}] = \mu_{ic}\mu_{jc}$ ($i \neq j$). For example, we can derive the second moment r_{c2} as $r_{c2} = \mu_c^2 + \mu_c - \sum_{i=1}^{N} \mu_{ic}^2$.

---

[8] The Snob algorithm does not need the input $C_{\max}$ because it searches, in each iteration step, the optimal number of components using a split and merge technique.

3.4 Justification of EMDL

The Kraft's inequality [14] indicates[9] that the code for $\ell(\boldsymbol{x}^N; H)$ can be constructed on the basis of a joint probability distribution over $\boldsymbol{x}^N$, $\hat{\boldsymbol{\theta}}$ and H, which we denote as $P(\boldsymbol{x}^N, \hat{\boldsymbol{\theta}}, H)$. Then the following theorem holds, which are proved in a manner similar to that of the justification of the EM algorithm [13].

Theorem 1. *Each iteration defined in Section 3.2 monotonically decreases the description length of the incomplete data set $\ell(\boldsymbol{x}^N; H)$.*

Proof. The description length $\ell(\boldsymbol{x}^N; H) = -\log P(\boldsymbol{x}^N, \hat{\boldsymbol{\theta}}, H)$ can be rewrote as follow:

$$\ell(\boldsymbol{x}^N; H) = E_Z^{(t)}\left[-\log \frac{P(\boldsymbol{x}^N, \boldsymbol{z}^N, \hat{\boldsymbol{\theta}}, H)}{P(\boldsymbol{z}^N | \boldsymbol{x}^N, \hat{\boldsymbol{\theta}}, H)}\right]. \tag{13}$$

On the basis of the optimality with respect to $E_Z^{(t)}[\ell(\boldsymbol{x}^N, \boldsymbol{z}^N; H)]$ and the Jensen's inequality [14], we can prove Theorem 1, as $\ell(\boldsymbol{x}^N; H^{(t+1)}) \leq \ell(\boldsymbol{x}^N; H^{(t)})$. □

Under certain holomorphic conditions, all the limit points for the sequence $(H^{(t)}, \hat{\boldsymbol{\theta}}^{(t)})$ are stationary points in $\ell(\boldsymbol{x}^N; H)$. This means that EMDL is compliant with the MDL principle and enjoys its statistical benefits.

4 Experiments and Discussion

We investigated basic behaviors of EMDL on the basis of the following two types of intuitively-understandable heterogeneous mixture models. The first type, which we refer to as the Gaussian model, is a mixture of three dimensional Gaussian distributions with various dimensional dependencies. The second type, which we refer to as the Polynomial model, is a mixture of polynomial curves of various degrees.

In the experiments described below, all results represent the averages of ten runs. In each run, we randomly restarted EMDL and the EM algorithm ten times to avoid obtaining bad local minima. We denote the true model in each run as H^* and the number of components in H^* as C^*.

4.1 Evaluation of Computational Speed and Model Selection Performance

Comparison Methods. We evaluated here the computational speed and model selection performance of EMDL by comparing it with two other model selection methods, referred to here as "Full Search" and "Pattern Search". For fair

---

[9] Strictly speaking, it is not necessary for $P(\boldsymbol{x}^N, \hat{\boldsymbol{\theta}}, H)$ to be a probability distribution but it is necessary for it to be a subprobability distribution. We can, however, without loss of generality, assume it to be a probability distribution by normalizing it as $P(\boldsymbol{x}^N, \hat{\boldsymbol{\theta}}, H) / \sum_{\boldsymbol{x}^N, \hat{\boldsymbol{\theta}}, H} P(\boldsymbol{x}^N, \hat{\boldsymbol{\theta}}, H)$ since the normalization constant $\sum_{\boldsymbol{x}^N, \hat{\boldsymbol{\theta}}, H} P(\boldsymbol{x}^N, \hat{\boldsymbol{\theta}}, H)$ is not related to the optimization of $\hat{\boldsymbol{\theta}}$ and H.

comparisons, both of them conduct model selection on the basis of the MDL principle[10]. Full Search first estimates parameter values for all model candidates in $\mathcal{H}$, and then selects an optimal model. Pattern Search greedily searches for an optimal model in $\mathcal{H}$ using a pattern search algorithm [7,8]. In both of them, parameter values are estimated using the EM algorithm. Since the EM algorithm does not obtain the globally optimal parameter values, there is a possibility that Pattern Search will calculate MDL value for individual model candidates more than once, unlike Full Search, which calculates MDL values for individual model candidates only once. This means that both EMDL and Pattern Search have the potential to outperform Full Search.

Performance Evaluation Criteria. We used three criteria here to evaluate model selection performance: 1) description length $\ell(x^N; \bar{H})$; 2) R_H = ratio of the number of runs in which $\bar{H} = H^*$; and 3) R_C = ratio of the number of runs in which $\bar{C} = C^*$. Since $\ell(x^N; \bar{H})$ is the objective function of model selection, a lower $\ell(x^N; \bar{H})$ is preferred. R_H is used to evaluate whether the model selected in a given run is consistent with the true model. R_C is a criterion similar to R_H, but it is used to check the consistency of only the number of components.

True Model and Data Generation. In the Gaussian model evaluation, for each component of the true model, we set a mean vector value randomly taken from the range of $[-5, 5]$ and a covariance matrix element value from the range of $[0, 1]$. We also randomly set a dimensional dependency for each component. For this evaluation[11], $|\mathcal{S}| = 5$. We first compared EMDL, Full Search and Pattern Search for varying values of $C_{\max}$. This helps us understand how EMDL behaves with the increase in search space that accompanies an increasing $C_{\max}$.

In the Polynomial model evaluation, for each component of the true model, we randomly selected a polynomial curve function from the eight pre-defined polynomial functions shown in Fig. 1. Data with independent gaussian noise added were generated according to these polynomial curves. The variance of the gaussian noise for each component was randomly set from the range of $[1, 4]$. For this evaluation, we set $C_{\max}$ to five and compared the three methods for varying the maximum degree of estimated polynomial curves $D_{\max}$. This helps us understand how EMDL behaves with the increase in search space that accompanies an increasing the number of component seeds $|\mathcal{S}|$.

We chose $C^* = 5$ for the Gaussian model and $C^* = 4$ for the Polynomial model because at higher values the high computational cost of Full Search would make our comparisons impossible to achieve. For each model, we randomly set the

---

[10] Based on our motivation, we should evaluate that EMDL actually optimize the criterion in linear time without being trapped by bad local minima. For the purpose, comparisons among algorithms with that criterion are more appropriate than that with Ma's [10] or Ghahramani's [9].

[11] Five seeds: the single seed in which all dimensions are dependent on one another, the three possible seeds in which one or another of the three dimensions is independent of the other two, and the single seed in which all dimensions are independent of one another.

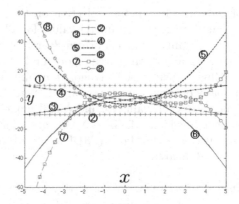

Fig. 1. Eight true components for the Polynomial model. ①:$y = 10$, ②:$y = -10$, ③:$y = 2x$, ④:$y = -2x$, ⑤:$y = 2x^2 - 3$, ⑥:$y = -2x^2 + 3$, ⑦:$y = 0.5x^3 - 1.5x^2 - 2x + 4$, ⑧:$y = -0.5x^3 + 1.5x^2 + 2x - 4$.

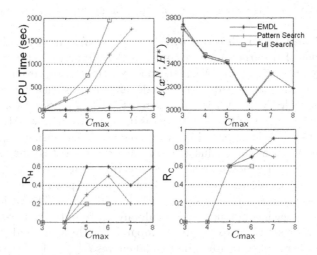

Fig. 2. Comparisons of computational speeds and model selection performance of EMDL, Pattern Search and Full Search in the Gaussian model experiments

true mixture ratio, and data were then randomly generated in accord with the true model for each run. In this experiment, we set $N = 500$.

Results and Discussion. Fig. 2 shows the results with respect to the Gaussian model. With increasing C_{max}, CPU time for both Pattern Search and Full Search increased quite rapidly (top left), while that for EMDL increased only linearly, and EMDL performed model selection much faster than the two others. Further, although $\ell(x^N; \bar{H})$ for all methods were roughly equivalent (top right), EMDL outperformed the others in terms of R_H and R_C (bottom left and right). These

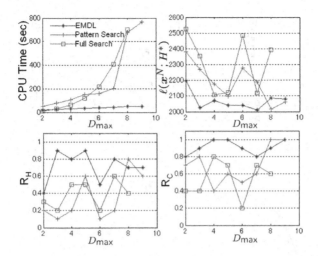

Fig. 3. Comparisons of computational speeds and model selection performance of EMDL, Pattern Search and Full Search in the Polynomial model experiments

results indicate that EMDL is resistant to bad local minima in the model space. Note that Full Search suffered from bad local minima and was inferior even to Pattern Search as we noted in Section 4.1 that it had the potential to be.

Fig. 3 shows results with respect to the Polynomial model. As with the Gaussian model, the CPU time for EMDL increased only linearly with D_{max}. Further, EMDL outperformed the other two methods in terms of $\ell(\boldsymbol{x}^N; \bar{H})$, R_H and R_C. In the case $D_{max} = 8$, Pattern Search outperformed EMDL because EMDL was trapped by a bad local minimum. The difference in the case $D_{max} = 10$ was not statistically significant. These perturbations will disappear with the larger number of runs.

Fig. 4 illustrates an example of model selection process for EMDL in a Polynomial model experiment. t, F_i and D_i represent, respectively, the number of iterations, the estimated ith polynomial curve, and the degree of F_i. The graph for $t = 1$ (top left) shows the results of random initialization. As the model selection proceeded, the estimated curves gradually proceeded toward a fit with the data. Results show the optimal degrees of the components ($D_1 \sim D_4$) frequently changed during the process and that EMDL successfully searched, simultaneously, for the optimal model and optimal parameter values. At $t = 20$ (bottom left), the model converged to the true model and the model selection process was completed.

4.2 Evaluation of Distribution Estimation Performance

We next evaluated parameter estimation performance of EMDL, that is, whether or not EMDL was more resistant to bad local optima in a parameter space than the EM algorithm. For this purpose, we compared log-likelihood values for

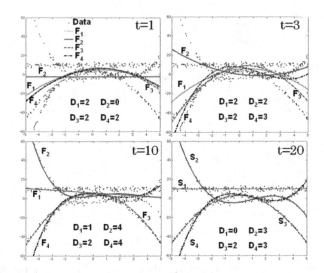

Fig. 4. Illustration of model selection process for EMDL in the Polynomial model experiment

training and test data with respect to data of both the Gaussian model and the Polynomial model on the basis of the two parameters: $\bar{\theta}$ and $\tilde{\theta}$. $\bar{\theta}$ is the parameter estimated using EMDL when $\bar{H} = H^*$. $\tilde{\theta}$ is the parameter for H^*, estimated using the EM algorithm. In this procedure, since both $\bar{\theta}$ and $\tilde{\theta}$ are parameter values for H^*, we can compare, in a fair manner, the distribution estimation performances of EMDL and the EM algorithm. In this experiment, data were generated in the same manner as in our previous experiments.

Table 1 shows the log-likelihood values for EMDL and the EM algorithm. In all cases (and for both training and test data), the log-likelihood values calculated using EMDL were higher than those calculated using the EM algorithm. This means that EMDL is able to estimate better parameter values than the EM algorithm and that EMDL performs well both as a parameter estimation method and as a model selection method.

This result can be interpreted as follows. Since EMDL simultaneously searches for the optimal model and its parameter values, it can be expected to be resistant to bad local optima, as is the case with the EM algorithm with split and merge operation [18]. While the EM algorithm stops when it reaches a local optimum in a parameter space, when EMDL reaches a local optimum of a parameter, it is able to avoid being caught by it by jumping to another model and to continue to search for parameter values.

4.3   Evaluation on UCI Datasets

We next evaluated whether or not EMDL outperformed Full Search and Pattern Search on Iris, Housing, Ecoli, Wine and Yeast datasets in the UCI reposi-

Table 1. Comparison of values of log-likelihood for training/test data calculated using EMDL and the EM algorithm

	GAUSSIAN MODEL		POLYNOMIAL MODEL	
	EMDL	EM	EMDL	EM
$C^* = 2$	-3238/-3259	-3482/-3505	-2372/-2405	-2372/-2405
$C^* = 3$	-3919/-3955	-3942/-3975	-2600/-2609	-2680/-2689
$C^* = 4$	-3837/-3837	-3999/-3988	-2819/-2855	-2845/-2883

tory [19]. Since the true models behind these data are unknown, we employed the description length $\ell(\boldsymbol{x}^N; \bar{H})$, cross-validated (CV) log-likelihood (10-fold), CPU time as evaluation criteria. For each run for each dataset, we set $C_{\max} = 5$ and optimized the Gaussian model on the basis of randomly chosen three features. Higher values for $C_{\max}$ and a Gaussian model with higher dimensionality were not available because of high computational costs for the comparison methods.

Table 2 shows the comparison of $\ell(\boldsymbol{x}^N; \bar{H})$, CV log-likelihood and CPU time for EMDL, Pattern Search (MDL), Full Search (MDL) and Pattern Search (AIC). EMDL could obtain the better (or at least competitive) solutions with respect to $\ell(\boldsymbol{x}^N; \bar{H})$ and CV log-likelihood with much smaller CPU time than the others. This result indicates that, in the case where we do not know an appropriate model, EMDL can find the appropriate model much faster than the others without reducing model selection accuracy (or even can obtain a better model.)

Table 2. Comparison of the averages of $\ell(\boldsymbol{x}^N; \bar{H})$/cross-validated likelihood/CPU time (sec) for EMDL, Pattern Search (MDL), Full Search (MDL) and Pattern Search (AIC). Since $\ell(\boldsymbol{x}^N; \bar{H})$ is the criterion for MDL, we omitted the value for AIC.

	EMDL	PATTERN (MDL)	FULL	PATTERN (AIC)
IRIS	416/-102/25	424/-108/345	420/-107/579	×/-112/473
HOUSING	1319/-763/24	1210/-925/310	1183/-928/532	×/-792/394
ECOLI	-605/165/39	-606/165/779	-605/166/829	×/165/510
WINE	590/-126/31	588/-125/548	588/-125/671	×/-138/453
YEAST	-4323/1147/81	-4331/1146/1469	-4334/1146/1533	×/1152/877

5 Concluding Remarks

This paper presented a novel model selection methodology for model selection for heterogeneous mixture (MSHM) problems, which we referred to as expectation minimization of description length (EMDL). EMDL employs iterative steps of computing the posterior of latent variables, and minimizing the expected description length of the complete data set. For MSHM, this enables EMDL to optimize types of components as well as the number of components in linear time with respect to both the maximum number of components and the number of component seeds. The experimental results based on UCI datasets and artificial datasets presented that EMDL outperforms common "MDL + the EM algorithm" approaches not only in terms of computational cost but also in terms of both model selection performance and parameter estimation performance.

Although we here applied EMDL to mixture of gaussian models and mixture of polynomial curves, it is applicable to the other mixtures of heterogeneous components such as latent variable models, factor analyzers, etc. We will formulate EMDL algorithms for these models and will compare them to the other algorithms (e.g., the Ghahramani's method for mixture of factor analyzers [9]).

References

1. Rissanen, J.: Modeling by shortest data description. Automatica 14, 465–471 (1978)
2. Akaike, H.: Information theory and an extension of the maximum likelihood principle. In: Petrov, B.N., Caski, F. (eds.) Proceedings of the 2nd International Symposium on Information Theory, pp. 267–281 (1973)
3. Wallace, C.S., Dowe, D.L.: Intrinsic classification by mml - the snob program. In: Proceedings of the 7th Australian Joint Conf. on Artificial Intelligence, pp. 37–44 (1994)
4. Wallace, C.S., Dowe, D.L.: Mml clustering of multi-state, poisson, von mises circular and gaussian distributions. Statistics and Computing 10, 73–83 (2000)
5. Edwards, R.T., Dowe, D.L.: Single factor analysis in mml mixture modelling. In: Wu, X., Kotagiri, R., Korb, K.B. (eds.) PAKDD 1998. LNCS, vol. 1394, pp. 96–109. Springer, Heidelberg (1998)
6. Bishop, C., Tipping, M.: A hierarchical latent variable model for data visualization. IEEE Transaction on Pattern Analysis and Machine Intelligence 20(3), 281–293 (1998)
7. Dennis, J.E., Torczon, V.J.: Derivative-free pattern search methods for multidisciplinary design problems. In: Proceedings of the 5th AIAA/ USAF/NASA/ISSMO Symposium on Multidisciplinary Analysis and Optimization (1994)
8. Momma, M., Bennett, K.P.: A pattern search method for model selection of support vector regression. In: Proceedings of the SIAM International Conference on Data Mining. SIAM, Philadelphia (2002)
9. Ghahramani, Z., Beal, M.J.: Variational inference for bayesian mixtures of factor analysers. In: Advances in Neural Information Processing Systems 12, pp. 449–455. MIT Press, Cambridge (2000)
10. Ma, Y., Derksen, H., Hong, W., Wright, J.: Segmentation of multivariate mixed data via lossy data coding and compression. IEEE Transactions on Pattern Analysis and Machine Intelligence 29(9), 1546–1562 (2007)
11. Grunwald, P.D., Myung, I.J., Pitt, M.A.: Advances in minimum description length. MIT Press, Cambridge (2005)
12. Yamanishi, K.: A learning criterion for stochastic rules. Machine Learning 9, 165–203 (1992)
13. Dempster, A.P., Laird, N.M., Rubin, D.B.: Maximum likelihood from imcomplete data via the em algorithm. Journal of the Royal Statistical Society B39(1), 1–38 (1977)
14. Cover, T.M., Thomas, J.A.: Elements of Information Theory. Wiley-Interscience, Hoboken (1991)
15. Rissanen, J.: Fisher information and stochastic complexity. IEEE Transaction on Information Theory 42(1), 40–47 (1996)

16. Tenmoto, H., Kudo, M., Shimbo, M.: MDL-based selection of the number of components in mixture models for pattern classification. In: Amin, A., Pudil, P., Dori, D. (eds.) SPR 1998 and SSPR 1998. LNCS, vol. 1451, pp. 831–836. Springer, Heidelberg (1998)
17. Rissanen, J.: Universal prior for integers and estimation by minimum description length. Annals of Statistics 11(2), 416–431 (1983)
18. Ueda, N., Nakano, R.: EM algorithm with split and merge operations for mixture models. IEICE transactions on information and systems E83-D(12), 2047–2055 (2000)
19. Asuncion, A., Newman, D.: UCI machine learning repository (2007)

Max-margin Multiple-Instance Learning
via Semidefinite Programming

Yuhong Guo

Department of Computer & Information Sciences
Temple University, Philadelphia, PA 19122, USA
yuhong@temple.edu

Abstract. In this paper, we present a novel semidefinite programming approach for multiple-instance learning. We first formulate the multiple-instance learning as a combinatorial maximum margin optimization problem with additional instance selection constraints within the framework of support vector machines. Although solving this primal problem requires non-convex programming, we nevertheless can then derive an equivalent dual formulation that can be relaxed into a novel convex semidefinite programming (SDP). The relaxed SDP has $\mathcal{O}(T)$ free parameters where T is the number of instances, and can be solved using a standard interior-point method. Empirical study shows promising performance of the proposed SDP in comparison with the support vector machine approaches with heuristic optimization procedures.

1 Introduction

Multiple-instance learning was introduced by Dieterich et al. [1] to solve a generalized supervised classification problem where the data set is composed of many *bags* such that each of them contains many instances and the class labels are associated with the bags, instead of individual instances. A bag is labeled as a *positive* bag if it contains at least one positive instance; otherwise it is labeled as a negative bag. Different from the standard supervised learning where all training instances are with known labels, the labels for individual instances in a positive bag is unknown in a multiple-instance learning problem, which makes the multiple-instance learning a much more challenging problem than the standard supervised classification.

Multiple-instance learning problems arise naturally from many application domains. One prominent example is the problem of drug activity prediction [1], where each molecule has a bag of different conformations, and the molecule qualified to make a drug has at least one conformation that could tightly bind to the target protein molecules. A second application is in content-based image retrieval or classification [2,3,4], where each image can be viewed as a bag of local subimages and one image is relevant with respect to one particular category if it has at least one relevant subimage. Another application is the problem of text categorization [5], where each document contains multiple passages over different

Z.-H. Zhou and T. Washio (Eds.): ACML 2009, LNAI 5828, pp. 98–108, 2009.

topics and a document is considered relevant regarding to one particular topic when it has one or more passages on this topic.

Motivated by these application challenges, multiple-instance learning has become one active research area in machine learning. A number of multiple-instance learning approaches have been developed in the literature, including special purpose algorithms using axis-parallel rectangular hypothesis [1], diverse density [3,6], kernel methods [7], support vector machines [5,8], ensemble methods [9], boosting methods [10], non-i.i.d. style methods [11] and etc.

In this paper, we propose to extend one popular classification method, maximum margin classification, to address the multiple-instance learning problem. Two maximum margin multiple-instance learning methods, *mi-SVM* and *MI-SVM*, based on support vector machines have been proposed in [5], mi-SVM for instance-level classification and MI-SVM for bag-level classification. The mi-SVM explicitly treats the instance labels in positive bags as unobserved hidden variables subject to constraints defined by their bag labels. In comparison, the MI-SVM aims to maximize the bag margin, which is defined as the margin of the most positive instance in case of positive bags, or the margin of the least negative instance in case of negative bags. However, due to the combinatorial nature of the formulated maximum margin problems, iterative heuristic procedures were used to conduct optimization in [5], which naturally suffer from the problem of local optima. Here we propose to formulate the multiple-instance learning as a combinatorial optimization problem of maximizing the classification margin with additional instance selection constraints. Like the mi-SVM, our approach can be categorized as an instance-level method. However, instead of figuring out the labels for all instances in positive bags, our approach selects only positive instances to use and ignore the negative ones in positive bags. The primal maximum margin formulation we developed is still non-convex, but we nevertheless can derive its equivalent dual formulation which can be relaxed into a convex semidefinite programming (SDP) problem by exploiting the Schur complement lemma. The relaxed SDP has $\mathcal{O}(T)$ free parameters where T is the number of instances, and can be solved using a standard interior-point method. Our empirical study shows promising performance of the proposed SDP in comparison with the support vector machine approaches with heuristic optimization procedures.

The remainder of this paper is organized as follows. After establishing the preliminaries and notations in Section 2, we present our maximum margin formulation for multiple-instance learning in Section 3. In Section 4, we derive an equivalent dual formulation and show it can be finally relaxed to yield a convex semidefinite programming problem which allows a global solution to be computed. Experimental results are reported in Section 5. We finally conclude the paper in Section 6.

2 Preliminaries

Since our approach is based on support vector machines (SVMs), we will first establish the background knowledge of SVMs as well as establish the notation we will use. Assume we are given labeled training instances $(\mathbf{x}_i, y_i), \cdots, (\mathbf{x}_T, y_T)$, where each instance is assigned to one of two classes $y_i \in \{-1, +1\}$. The goal of a SVM is to find the linear discriminant function $f_{\mathbf{w},b}(\mathbf{x}) = \mathbf{w}^\top \phi(\mathbf{x}) + b$ that achieves maximum margin (*separation*) between the two classes in $\phi(\mathbf{x})$ space. Note here $\phi(\mathbf{x})$ denotes the general feature vector produced from the original feature vector $\mathbf{x}$, and it is introduced to cope with nonlinear classification. The standard primal soft margin SVM is formulated as follow

$$\min_{\mathbf{w}, b, \boldsymbol{\xi},} \quad \frac{1}{2}\|\mathbf{w}\|^2 + C\boldsymbol{\xi}^\top \mathbf{e} \tag{1}$$
$$\text{s.t.} \quad y_i(\mathbf{w}^\top \phi(\mathbf{x}_i) + b) \geq 1 - \xi_i, \forall_{i=1}^T$$
$$\boldsymbol{\xi} \geq 0$$

where slack variables $\boldsymbol{\xi}$ are introduced to cope with noisy instances and the non-separability of the training data; C is a parameter that controls the tradeoff between the separation margin and the misclassification error; b is a parameter to control the bias; and $\mathbf{e}$ denotes the vector of all 1 entries. For the simplicity reason, we will use the same $\mathbf{e}$ notation to denote any vectors with all 1 entries later in the paper. The length of the $\mathbf{e}$ vector for each of its appearance can be determined from the context. By introducing Lagrangian multipliers and following the standard procedure, an equivalent dual formulation for SVM in (1) can be obtained

$$\max_{\boldsymbol{\alpha}} \quad \boldsymbol{\alpha}^\top \mathbf{e} - \frac{1}{2}\boldsymbol{\alpha}^\top (K \circ \mathbf{yy}^T)\boldsymbol{\alpha} \tag{2}$$
$$\text{s.t.} \quad 0 \leq \boldsymbol{\alpha} \leq C, \ \boldsymbol{\alpha}^\top \mathbf{y} = 0$$

where $\boldsymbol{\alpha}$ denotes the vector of dual variables; K denotes the $T \times T$ kernel matrix formed from the inner products of feature vectors $\Phi = [\phi(\mathbf{x}_1), \cdots, \phi(\mathbf{x}_T)]$ such that $K = \Phi^\top \Phi$; $\mathbf{y}$ denotes the label vector, such that $\mathbf{y} = [y_1, \cdots, y_T]^\top$; $A \circ B$ denotes componentwise matrix multiplication.

In the following section, we will extend the standard SVM to address the problem of multiple-instance learning.

3   Max-margin Multiple-Instance Learning

Different from the standard supervised learning scenario where a label is assigned to each training instance, in multiple-instance learning, a label is assigned to a bag of instances. A bag is labeled as a positive bag if it contains at least one positive instance; otherwise it is labeled as a negative bag, which means all instances in negative bags are negative instances. Thus the difficulty for extending any standard supervised learning methods to address the multiple-instance learning problem lies in that the labels for instances in positive bags are unknown. Moreover, different from the standard semi-supervised learning scenario, we need to guarantee that at least one instance from each positive bag gets a positive label.

The *mi-SVM* proposed in [5] views the instance labels in positive bags as hidden variables, and maximizes a soft margin criterion jointly over discriminant model parameters and possible label assignments while taking an extra checking step to enforce that at least one instance gets a positive label for each positive bag. However, there are usually a lot of ambiguities with regard to the label assignments over the hidden variables. On the other hand, we can obtain many confirmed negative instances from negative bags. Based on these observations, we propose to select only a set of positive instances from positive bags to use together with the negative instances from negative bags for multiple-instance learning. Our intuition is to incorporate only the most useful information into the model learning process while avoiding the unnecessary ambiguities. Specifically, we propose to formulate the multiple-instance learning as a combinatorial optimization problem that maximizes a soft SVM margin criterion jointly over both the model parameters and the instance selection variables. The instance selection variables are used to choose the most informative positive instances from the positive bags such that when the selected positive instances are incorporated into the proposed maximum margin model, the soft margin criterion can be maximumly optimized. Below we will present this joint optimization model in detail. Moreover, we will show later that this optimization model with instance selection variables can lead to a simple SDP formulation.

Assume we are given a multiple-instance training set with N bags of instances $\{\mathbf{B}_1, \cdots, \mathbf{B}_N\}$, where the first N_p bags are positive bags, following by N_n negative bags such that $N_p + N_n = N$. Assume each bag $\mathbf{B}_i$ contains t_i instances such that $\sum_i^{N_p} t_i = T_p$, $\sum_{i=N_p+1}^{N} t_i = T_n$ and $T_p + T_n = T$. Our maximum margin multiple-instance learning can be formulated as follows

$$\min_{\boldsymbol{\eta}} \min_{\mathbf{w}, b, \boldsymbol{\xi}} \frac{1}{2}\|\mathbf{w}\|^2 + C\boldsymbol{\xi}^\top[\boldsymbol{\eta}; \mathbf{e}] \tag{3}$$

$$\text{s.t.} \quad y_i(\mathbf{w}^\top \boldsymbol{\phi}(\mathbf{x}_i) + b) \geq 1 - \xi_i, \ \forall_{i=1}^{T}$$

$$\boldsymbol{\xi} \geq 0$$

$$\boldsymbol{\eta} \in \{0,1\}^{T_p \times 1}$$

$$\mathbf{A}\boldsymbol{\eta} \geq \mathbf{e}$$

where $\boldsymbol{\eta}$ denotes the vector of instance selection binary variables; $y_i = 1$ for $i = 1, \cdots, T_p$ and $y_i = -1$ for $i = 1 + T_p, \cdots, T$; $\mathbf{A}$ is a $N_p \times T_p$ binary matrix such that

$$\mathbf{A} = \begin{bmatrix} ones(1, t_1), & zeros(1, t_2), & \cdots, & zeros(1, t_{N_p}) \\ zeros(1, t_1), & ones(1, t_2), & \cdots, & zeros(1, t_{N_p}) \\ \vdots & \vdots & \vdots & \vdots \\ zeros(1, t_1), & zeros(1, t_2), & \cdots, & ones(1, t_{N_p}) \end{bmatrix}$$

and all the other notations are same as introduced before. Note that the constraint $\mathbf{A}\boldsymbol{\eta} \geq \mathbf{e}$ is used to guarantee that at least one positive instance from each positive bag will be selected. Given fixed $\boldsymbol{\eta}$, the optimization problem (3) will become a standard SVM optimization problem over a training set formed by the negative instances from negative bags and the positive instances selected using η from positive bags.

The minimization problem (3) we formulated is a NP-hard combinatorial optimization problem. In order to obtain an efficient convex optimization, we first need to derive its dual formulation.

Proposition 1. *For fixed η, the inner minimization problem in (3), that is*

$$\min_{\mathbf{w},b,\boldsymbol{\xi}} \frac{1}{2}\|\mathbf{w}\|^2 + C\boldsymbol{\xi}^\top[\eta;\mathbf{e}] \tag{4}$$
$$s.t. \quad y_i(\mathbf{w}^\top\boldsymbol{\phi}(\mathbf{x}_i)+b) \geq 1-\xi_i, \ \forall_{i=1}^T$$
$$\boldsymbol{\xi} \geq 0$$

is equivalent to the following dual maximization problem

$$\max_{\boldsymbol{\alpha}} \quad \boldsymbol{\alpha}^\top\mathbf{e} - \frac{1}{2}\boldsymbol{\alpha}^\top(K \circ \mathbf{y}\mathbf{y}^T)\boldsymbol{\alpha} \tag{5}$$
$$s.t. \quad 0 \leq \boldsymbol{\alpha} \leq C[\eta;\mathbf{e}]$$
$$\boldsymbol{\alpha}^\top\mathbf{y} = 0$$

Proof. The proof is simple. Note the minimization problem in (4) is a slightly modified version of the standard SVM optimization in (1). The only difference lies in that the $\boldsymbol{\xi}$ in the objective function of (4) is weighted by a vector $[\eta;\mathbf{e}]$. Thus following the standard procedure for deriving a dual formulation of SVMs, an equivalent dual formulation (5) can be obtained.   ∎

Exploiting Proposition 1, the minimization problem (3) can be rewritten into the following equivalent min-max optimization problem by simply replacing the inner minimization of (3) with its equivalent dual formulation (5)

$$\min_{\eta}\max_{\boldsymbol{\alpha}} \quad \boldsymbol{\alpha}^\top\mathbf{e} - \frac{1}{2}\boldsymbol{\alpha}^\top(K \circ \mathbf{y}\mathbf{y}^T)\boldsymbol{\alpha} \tag{6}$$
$$s.t. \quad 0 \leq \boldsymbol{\alpha} \leq C[\eta;\mathbf{e}]$$
$$\boldsymbol{\alpha}^\top\mathbf{y} = 0$$
$$\eta \in \{0,1\}^{T_p\times1}$$
$$\mathbf{A}\eta \geq \mathbf{e}$$

Although the dual optimization problem in (6) does not provide a convex solution immediately, it provides a foundation for further reformulation.

4   Semidefinite Programming

In this section, we will reformulate the min-max optimization problem (6) obtained in the previous section to finally get a convex semidefinite programming problem which can provide a global solution without local optima.

Theorem 1. *The combinatorial min-max optimization problem in (6) is equivalent to the following minimization problem*

$$\min_{\eta,\mu,\lambda,\epsilon,\delta} \quad \delta \tag{7}$$

$$s.t. \quad \begin{pmatrix} K \circ \mathbf{y}\mathbf{y}^\top & (\mathbf{e}+\mu-\lambda+\epsilon\mathbf{y}) \\ (\mathbf{e}+\mu-\lambda+\epsilon\mathbf{y})^\top & 2\delta - 2C\lambda^\top[\eta;\mathbf{e}] \end{pmatrix} \succeq 0$$

$$\mu \geq 0$$

$$\lambda \geq 0$$

$$\eta \in \{0,1\}^{T_p \times 1}$$

$$\mathbf{A}\eta \geq \mathbf{e}$$

Proof. The min-max optimization problem (6) can be equivalently rewritten as

$$\min_{\eta} \quad \delta \tag{8}$$

$$s.t. \quad \delta \geq \max_{\alpha} \alpha^\top \mathbf{e} - \frac{1}{2}\alpha^\top(K \circ \mathbf{y}\mathbf{y}^\top)\alpha$$

$$0 \leq \alpha \leq C[\eta;\mathbf{e}]$$

$$\alpha^\top\mathbf{y} = 0$$

$$\eta \in \{0,1\}^{T_p \times 1}$$

$$\mathbf{A}\eta \geq \mathbf{e}$$

Below we will express the constraint $\delta \geq \max_\alpha \alpha^\top\mathbf{e} - \frac{1}{2}\alpha^\top(K \circ \mathbf{y}\mathbf{y}^\top)\alpha$ as a linear matrix inequality for given η.

First define the Lagrangian of the maximization problem (5) by

$$\mathcal{L}(\alpha,\mu,\lambda,\epsilon,\delta) = \alpha^\top\mathbf{e} - \frac{1}{2}\alpha^\top(K \circ \mathbf{y}\mathbf{y}^\top)\alpha + \mu^\top\alpha + \lambda^\top(C[\eta;\mathbf{e}] - \alpha) + \epsilon\alpha^\top\mathbf{y}$$

where $\mu \geq 0$, $\lambda \geq 0$ and $\epsilon \in \mathbb{R}$. By duality [12], we have

$$\max_{\alpha} \min_{\mu,\lambda,\epsilon,\delta} \mathcal{L}(\alpha,\mu,\lambda,\epsilon,\delta) = \min_{\mu,\lambda,\epsilon,\delta} \max_{\alpha} \mathcal{L}(\alpha,\mu,\lambda,\epsilon,\delta).$$

Then the inner maximization over α, $\max_\alpha \mathcal{L}(\alpha,\mu,\lambda,\epsilon,\delta)$, can be easily solved by determining a critical point, since $\mathcal{L}(\alpha,\mu,\lambda,\epsilon,\delta)$ is concave in α. By setting $\partial\mathcal{L}/\partial\alpha = 0$, we obtain $\alpha = (K \circ \mathbf{y}\mathbf{y}^\top)^{-1}(\mathbf{e}+\mu-\lambda+\epsilon\mathbf{y})$. Substituting this into $\mathcal{L}(\alpha,\mu,\lambda,\epsilon,\delta)$, we can form the following dual problem of (5)

$$\min_{\mu,\lambda,\epsilon,\delta} \quad C\lambda^\top[\eta;\mathbf{e}] + \frac{1}{2}(\mathbf{e}+\mu-\lambda+\epsilon\mathbf{y})^\top(K \circ \mathbf{y}\mathbf{y}^\top)^{-1}(\mathbf{e}+\mu-\lambda+\epsilon\mathbf{y}) \tag{9}$$

$$s.t. \quad \mu \geq 0$$

$$\lambda \geq 0$$

This implies that for any δ, the constraint $\delta \geq \max_\alpha \alpha^\top\mathbf{e} - \frac{1}{2}\alpha^\top(K \circ \mathbf{y}\mathbf{y}^\top)\alpha$ holds if and only if there exist $\mu \geq 0, \lambda \geq 0$ and ϵ such that

$$\delta \geq C\lambda^\top[\eta;\mathbf{e}] + \frac{1}{2}(\mathbf{e}+\mu-\lambda+\epsilon\mathbf{y})^\top(K \circ \mathbf{y}\mathbf{y}^\top)^{-1}(\mathbf{e}+\mu-\lambda+\epsilon\mathbf{y})$$

or equivalently using the Schur complement lemma [12] such that

$$\begin{pmatrix} K \circ \mathbf{y}\mathbf{y}^\top & (\mathbf{e}+\boldsymbol{\mu}-\boldsymbol{\lambda}+\epsilon\mathbf{y}) \\ (\mathbf{e}+\boldsymbol{\mu}-\boldsymbol{\lambda}+\epsilon\mathbf{y})^\top & 2\delta - 2C\boldsymbol{\lambda}^\top[\boldsymbol{\eta};\mathbf{e}] \end{pmatrix} \succeq 0 \tag{10}$$

Substituting this into (8) yields (7). ∎

However, the minimization problem (7) is still not convex for two reasons. First, there is a bilinear term $\boldsymbol{\lambda}^\top[\boldsymbol{\eta};\mathbf{e}]$ in the matrix inequality constraint (10), which makes the constraint non-convex. Second, the existence of binary constraints over variables $\boldsymbol{\eta}$ makes the overall optimization problem a combinatorial optimization. We thus need to solve these two issues to obtain an efficient convex optimization problem.

For the problem of bilinear term, we notice that $\boldsymbol{\lambda}^\top[\boldsymbol{\eta};\mathbf{e}] = \boldsymbol{\lambda}_{1:T_p}^\top\boldsymbol{\eta}+\boldsymbol{\lambda}_{T_p+1:T}^\top\mathbf{e}$, and $2\boldsymbol{\lambda}_{1:T_p}^\top\boldsymbol{\eta} = (\boldsymbol{\lambda}_{1:T_p}+\boldsymbol{\eta})^\top(\boldsymbol{\lambda}_{1:T_p}+\boldsymbol{\eta}) - \boldsymbol{\lambda}_{1:T_p}^\top\boldsymbol{\lambda}_{1:T_p} - \boldsymbol{\eta}^\top\boldsymbol{\eta}$. Note that since $\boldsymbol{\eta}$ is a vector of binary variables, thus $\boldsymbol{\eta}^\top\boldsymbol{\eta} = \boldsymbol{\eta}^\top\mathbf{e}$. Now we introduce two new variables g and h such that $g = (\boldsymbol{\lambda}_{1:T_p}+\boldsymbol{\eta})^\top(\boldsymbol{\lambda}_{1:T_p}+\boldsymbol{\eta})$ and $h = \boldsymbol{\lambda}_{1:T_p}^\top\boldsymbol{\lambda}_{1:T_p}$. For simplicity reason, we also let $\mathbf{u}$ denote the constant vector $[\mathbf{0};\mathbf{e}]$, where $\mathbf{0}$ is a vector of all 0 entries, such that $\boldsymbol{\lambda}^\top\mathbf{u} = \boldsymbol{\lambda}^\top[\mathbf{0};\mathbf{e}] = \boldsymbol{\lambda}_{T_p+1:T}^\top\mathbf{e}$. Therefore

$$2\boldsymbol{\lambda}^\top[\boldsymbol{\eta};\mathbf{e}] = g - h - \boldsymbol{\eta}^\top\mathbf{e} + 2\boldsymbol{\lambda}^\top\mathbf{u}.$$

Substituting this back to the matrix inequality constraint in (7), we obtain an equivalent optimization problem

$$\min_{\boldsymbol{\eta},\boldsymbol{\mu},\boldsymbol{\lambda},\epsilon,\delta,g,h} \quad \delta \tag{11}$$

$$\text{s.t.} \quad \begin{pmatrix} K \circ \mathbf{y}\mathbf{y}^\top & (\mathbf{e}+\boldsymbol{\mu}-\boldsymbol{\lambda}+\epsilon\mathbf{y}) \\ (\mathbf{e}+\boldsymbol{\mu}-\boldsymbol{\lambda}+\epsilon\mathbf{y})^\top & 2\delta - C(g-h-\boldsymbol{\eta}^\top\mathbf{e}+2\boldsymbol{\lambda}^\top\mathbf{u}) \end{pmatrix} \succeq 0$$

$$g = (\boldsymbol{\lambda}_{1:T_p}+\boldsymbol{\eta})^\top(\boldsymbol{\lambda}_{1:T_p}+\boldsymbol{\eta})$$

$$h = \boldsymbol{\lambda}_{1:T_p}^\top\boldsymbol{\lambda}_{1:T_p}$$

$$\boldsymbol{\mu} \geq 0$$

$$\boldsymbol{\lambda} \geq 0$$

$$\boldsymbol{\eta} \in \{0,1\}^{T_p\times1}$$

$$A\boldsymbol{\eta} \geq \mathbf{e}$$

Now we have successfully got rid of the bilinear term from the matrix inequality constraint. However, two new quadratic equality constraints have been introduced. In order to obtain a convex optimization, we need to relax the two quadratic equality constraints into inequality constraints

$$g - (\boldsymbol{\lambda}_{1:T_p}+\boldsymbol{\eta})^\top(\boldsymbol{\lambda}_{1:T_p}+\boldsymbol{\eta}) \geq 0$$

$$h - \boldsymbol{\lambda}_{1:T_p}^\top\boldsymbol{\lambda}_{1:T_p} \geq 0$$

Using quadratic inequality constraints to replace corresponding equality constraints is a typical relaxation technique used in the literature, e.g. [13], towards obtaining convex semidefinite approximations. Here the inequality constraints above can then be rewritten equivalently into convex linear matrix inequality constraints according to the Schur complement lemma [12]

$$\begin{pmatrix} I & (\boldsymbol{\lambda}_{1:T_p} + \boldsymbol{\eta}) \\ (\boldsymbol{\lambda}_{1:T_p} + \boldsymbol{\eta})^\top & g \end{pmatrix} \succeq 0 \tag{12}$$

$$\begin{pmatrix} I & \boldsymbol{\lambda}_{1:T_p} \\ \boldsymbol{\lambda}_{1:T_p}^\top & h \end{pmatrix} \succeq 0 \tag{13}$$

Finally replacing the two equality quadratic constraints in (11) with the relaxed constraints (12) and (13) and relaxing the integer constraints over $\boldsymbol{\eta}$ into continuous constraints $0 \leq \boldsymbol{\eta} \leq 1$, we obtain a relaxed optimization problem

$$\min_{\boldsymbol{\eta}, \boldsymbol{\mu}, \boldsymbol{\lambda}, \epsilon, \delta, g, h} \quad \delta \tag{14}$$

$$\text{s.t.} \quad \begin{pmatrix} K \circ \mathbf{y}\mathbf{y}^\top & (\mathbf{e} + \boldsymbol{\mu} - \boldsymbol{\lambda} + \epsilon\mathbf{y}) \\ (\mathbf{e} + \boldsymbol{\mu} - \boldsymbol{\lambda} + \epsilon\mathbf{y})^\top & 2\delta - C(g - h - \boldsymbol{\eta}^\top\mathbf{e} - 2\boldsymbol{\lambda}^\top\mathbf{u}) \end{pmatrix} \succeq 0$$

$$\begin{pmatrix} I & (\boldsymbol{\lambda}_{1:T_p} + \boldsymbol{\eta}) \\ (\boldsymbol{\lambda}_{1:T_p} + \boldsymbol{\eta})^\top & g \end{pmatrix} \succeq 0$$

$$\begin{pmatrix} I & \boldsymbol{\lambda}_{1:T_p} \\ \boldsymbol{\lambda}_{1:T_p}^\top & h \end{pmatrix} \succeq 0$$

$$\boldsymbol{\mu} \geq 0, \quad \boldsymbol{\lambda} \geq 0$$

$$0 \leq \boldsymbol{\eta} \leq 1, \quad \mathbf{A}\boldsymbol{\eta} \geq \mathbf{e}$$

The problem in (14) is a convex optimization problem, more specifically, a semidefinite programming problem. It has $\mathcal{O}(T)$ free parameter in the SDP cone and $\mathcal{O}(T)$ linear inequality constraints, that involves a worst-case computational complexity of $\mathcal{O}(T^{4.5})$. It is much more efficient than the SDP problems formulated for semi-supervised support vector machines such as in [14] which has $\mathcal{O}(T^2)$ free parameter in the SDP cone. Our SDP problem can be efficiently solved by using an interior-point method [13] implemented in some optimization packages, such as SeDuMi [15]. In our experiments, we used the Yalmip interface [16] together with the optimization engine of SeDuMi to solve this semidefinite programming problem.

After the training process, we obtain continuous optimal $\boldsymbol{\eta}^*$ values. We then use a heuristic rounding procedure to recover the discrete binary values $\hat{\boldsymbol{\eta}}$ by enforcing the constraints $\mathbf{A}\hat{\boldsymbol{\eta}} \geq \mathbf{e}$ while minimizing $\boldsymbol{\lambda}_{1:T_p}^{*\top}\hat{\boldsymbol{\eta}}$. (See the objective function of (9).) After recovering the discrete $\hat{\boldsymbol{\eta}}$ values, the target maximum margin discriminant function can be learned by solving the optimization problem (4) or its dual (5).

5   Experimental Results

We have conducted experiments on various data sets to evaluate the proposed semidefinite programming approach, comparing with the two maximum margin approaches, mi-SVM and MI-SVM, proposed in [5]. In our experiments, in order to reduce the ambiguity of the problem, we used equality constraints $\mathbf{A}\eta = \mathbf{e}$ instead of the inequality constraints for the proposed SDP. This implies that we only select one most promising positive instance from each positive bag. The C parameters used for each approach are selected based on results obtained using one random training/test split of the data.

5.1 Musk Data Sets

We first conducted experiments using the benchmark Musk data sets for multiple-instance learning, Musk1 and Musk2. The Musk data sets are produced for the task of drug activity prediction, and have been described in detail in [1]. The two data sets, Musk1 and Musk2, consist of instances describing different conformations of various molecules. A bag is defined as a set of all conformations for one molecule. A positive bag has at least one instance, that is one conformation of the molecule, that can bind well to a target protein.

We conducted experiments by randomly selecting 4/5 of the bags in Musk1 (1/8 in Musk2) as training data and keeping the remaining as test data. The experiments were repeated 10 times and the average test accuracies are reported in Table 1. One can see that the proposed SDP approach returns the best result among the three methods on data set Musk2. However, on Musk1, mi-SVM gives the best accuracy value and the bag-level method, MI-SVM, gives the weakest result. This might indict that on Musk1 data set it is helpful to incorporate more instances to build the classification model.

Table 1. Classification accuracy results on the Musk data sets(%)

Data Set	#Bags	#Data	SDP	MI-SVM	mi-SVM
Musk1	92	476	69.5	69.0	**71.6**
Musk2	102	6598	**61.3**	58.9	59.7

5.2 Corel Image Data Sets

We have also conducted experiments on the corel image data sets used in [5]. Here an image is viewed as a bag, which consists of a set of instances, that is segments, characterized by color, texture and shape descriptors. We used the three data sets constructed in [5]: Elephant, Fox and Tiger. The problem is to determine whether a given animal is present in an image. For each data set, we conducted experiments by randomly sampling 3/5 of the bags as training data and keeping the remaining as test data. The test accuracy results reported in Table 2 are averages over 10 repeated runs. In this case, the proposed SDP approach outperforms the other two methods on both Fox and Tiger data sets. However, MI-SVM presents a better test accuracy than SDP on the Elephant data set.

Table 2. Classification accuracy results on the Corel data sets(%)

Data Set	#Bags	#Data	SDP	MI-SVM	mi-SVM
Elephant	200	1391	74.8	**76.7**	70.8
Fox	200	1320	**56.8**	52.3	55.0
Tiger	200	1220	**73.6**	71.9	69.4

5.3 Text Categorization

Finally, we conducted experiments for text categorization using the text data sets generated from the publicly available TREC9 data set in [5]. In a multiple-instance learning setting, each document of the data set corresponds to a bag, where the instances in the bag are overlapping passages splitted from the document, consisting of 50 words in length. For each data set, we randomly sampled 1/3 of the data as training set and kept the remaining as test set. We repeated this process 10 times and the average results are reported in Table 3. Evidently here the SDP approach presents a more consistent advantage over the other two SVM methods. On the 7 data sets, the SDP has only been slightly overperformed by mi-SVM on TREC3. These results suggest that better performance can be gained by pursuing convex global optimization.

Table 3. Classification accuracy results on the TREC9 text sets(%)

Data Set	#Bags	#Data	SDP	MI-SVM	mi-SVM
TREC1	400	3224	**92.7**	92.5	85.8
TREC2	400	3344	**75.1**	74.4	63.4
TREC3	400	3246	74.3	73.2	**74.6**
TREC4	400	3391	**77.7**	76.9	72.8
TREC7	400	3367	**72.5**	70.9	63.8
TREC9	400	3300	**59.9**	55.0	59.0
TREC10	400	3453	**74.4**	73.8	67.8

6 Conclusion and Future Work

We have presented a novel maximum margin semidefinite programming approach for multiple-instance learning. Comparing to two other maximum margin approaches based on heuristic procedures that suffer from local optima, our convex approach can be solved using a global optimization method. Unlike the semi-supervised SDP method proposed in the literature which has $\mathcal{O}(T^2)$ parameters, our SDP has only $\mathcal{O}(T)$ parameters and can be solved more efficiently. The empirical results reported in the experimental section suggest that the proposed SDP can yield more promising results than the two locally optimized nonconvex maximum margin methods.

Although the SDP approach proposed in this paper selects only positive instances from positive bags to use, it is still an instance-level method. Extending it to get a bag-level approach is an interesting future work we are considering.

References

1. Dietterich, T., Lathrop, R., Lozano-Perez, T.: Solving the multiple-instance problem with axis-parallel rectangles. Artificial Intelligence Journal 89 (1997)
2. Carson, C., Thomas, M., Belongie, S., Hellerstein, J., Malik, J.: Blobworld: A system for region-based image indexing and retrieval. In: Huijsmans, D.P., Smeulders, A.W.M. (eds.) VISUAL 1999. LNCS, vol. 1614, pp. 509–517. Springer, Heidelberg (1999)
3. Maron, O., Ratan, A.: Multiple-instance learning for natural scene classification. In: Proceedings of the International Conference on Machine Learning (1998)
4. Zhang, Q., Goldman, S., Yu, W., Fritts, J.: Content-based image retrieval using multiple-instance learning. In: Proceedings of the International Conference on Machine Learning (2002)
5. Andrews, S., Tsochantaridis, I., Hofmann, T.: Support vector machines for multiple-instance learning. In: Advances in Neural Information Processing Systems (2002)
6. Zhang, Q., Goldman, S.: EM-dd: An improved multiple-instance learning technique. In: Advances in Neural Information Processing Systems (2001)
7. Gartner, T., Flach, P., Kowalczyk, A., Smola, A.: Multi-instance kernels. In: Proceedings of the International Conference on Machine Learning (2002)
8. Mangasarian, O., Wild, E.: Multiple instance classification via successive linear programming. Journal of Optimization Theory and Applications 137 (2008)
9. Zhou, Z., Zhang, M.: Ensembles of multi-instance learners. In: Lavrač, N., Gamberger, D., Todorovski, L., Blockeel, H. (eds.) ECML 2003. LNCS (LNAI), vol. 2837, pp. 492–502. Springer, Heidelberg (2003)
10. Andrews, S., Hofmann, T.: Multiple instance learning via disjunctive programming boosting. In: Advances in Neural Information Processing Systems (2003)
11. Zhou, Z., Sun, Y., Li, Y.: Multiple-instance learning by training instances as non-i.i.d. samples. In: Proceedings of the International Conference on Machine Learning (2009)
12. Lanckriet, G., Cristianini, N., Bartlett, P., Ghaoui, L., Jordan, M.: Learning the kernel matrix with semidefinite programming. Journal of Machine Learning Research 5 (2004)
13. Boyd, S., Vandenberghe, L.: Convex Optimization. Cambridge U. Press, Cambridge (2004)
14. Xu, L., Neufeld, J., Larson, B., Schuurmans, D.: Maximum margin clustering. In: Advances in Neural Information Processing Systems (2004)
15. Sturm, J.: Using SeDuMi 1.02, a MATLAB toolbox for optimization over symmetric cones. Optimization Methods and Software 11 (1999)
16. Lofberg, J.: YALMIP: a toolbox for modeling and optimization in MATLAB. In: Proceedings of the CACSD Conference (2004)

A Reformulation of Support Vector Machines
for General Confidence Functions

Yuhong Guo[1] and Dale Schuurmans[2]

[1] Department of Computer & Information Sciences
Temple University
www.cis.temple.edu/~yuhong
[2] Department of Computing Science
University of Alberta
www.cs.ualberta.ca/~dale

Abstract. We present a generalized view of support vector machines that does not rely on a Euclidean geometric interpretation nor even positive semidefinite kernels. We base our development instead on the *confidence matrix*—the matrix normally determined by the direct (Hadamard) product of the kernel matrix with the label outer-product matrix. It turns out that alternative forms of confidence matrices are possible, and indeed useful. By focusing on the confidence matrix instead of the underlying kernel, we can derive an intuitive principle for optimizing example weights to yield robust classifiers. Our principle initially recovers the standard quadratic SVM training criterion, which is only convex for kernel-derived confidence measures. However, given our generalized view, we are then able to derive a principled relaxation of the SVM criterion that yields a convex upper bound. This relaxation is *always* convex and can be solved with a linear program. Our new training procedure obtains similar generalization performance to standard SVMs on kernel-derived confidence functions, but achieves even better results with indefinite confidence functions.

1 Introduction

Support vector machines were originally derived from purely geometric principles [13,1]: given a labeled training set, one attempts to solve for a consistent linear discriminant that maximizes the minimum Euclidean distance between any data point and the decision hyperplane. Specifically, given $(\mathbf{x}_1, y_1), ..., (\mathbf{x}_t, y_t)$, $y \in \{-1, +1\}$, the goal is to determine a $(\mathbf{w}, b)$ such that $\min_i y_i(\mathbf{w}^\top \mathbf{x}_i + b)/\|\mathbf{w}\|$ is maximized. Vapnik [13] famously proposed this principle and formulated a convex quadratic program for efficiently solving it. With the addition of slack variables the dual form of this quadratic program can be written

$$\min_{\boldsymbol{\alpha}} \frac{1}{2} \sum_{ij} \alpha_i \alpha_j y_i y_j \mathbf{x}_i^\top \mathbf{x}_j - \sum_i \alpha_i \quad \text{subject to} \quad 0 \leq \boldsymbol{\alpha} \leq \beta, \ \boldsymbol{\alpha}^\top \mathbf{y} = 0 \quad (1)$$

where the dual variables $\boldsymbol{\alpha}$ behave as weights on the training examples.

One of the key insights behind the support vector machine approach is that the training vectors appear only as inner products in both training and classification, and

Z.-H. Zhou and T. Washio (Eds.): ACML 2009, LNAI 5828, pp. 109–119, 2009.
© Springer-Verlag Berlin Heidelberg 2009

therefore can be abstracted away by a general kernel function. In this case, the kernel function, $k(\mathbf{x}_i, \mathbf{x}_j)$, simply reports inner products $\langle \phi(\mathbf{x}_i), \phi(\mathbf{x}_j) \rangle$ in some arbitrary feature (Hilbert) space. Combining the kernel abstraction with the ν-SVM formulation of [12,4] one can re-express (1) in the more general form

$$\min_{\boldsymbol{\alpha}} \; \boldsymbol{\alpha}^\top (K \circ \mathbf{y}\mathbf{y}^\top)\boldsymbol{\alpha} \quad \text{subject to} \quad 0 \le \boldsymbol{\alpha} \le \beta, \; \boldsymbol{\alpha}^\top \mathbf{y} = 0, \; \boldsymbol{\alpha}^\top \mathbf{e} = 1 \qquad (2)$$

where K is the *kernel matrix*, $K_{ij} = \langle \phi(\mathbf{x}_i), \phi(\mathbf{x}_j) \rangle$, the matrix $\mathbf{y}\mathbf{y}^\top$ is the *label matrix*, the vector $\mathbf{e}$ consists of all 1's, and $\circ$ denotes componentwise matrix multiplication (Hadamard product).

Although (2) appears to be a very general formulation of the weight training problem for $\boldsymbol{\alpha}$, it is in fact quite restrictive: for (2) to be convex, the combined matrix $K \circ \mathbf{y}\mathbf{y}^\top$ must be positive semidefinite, implying that K itself must be conditionally positive semidefinite.[1] Thus, it is commonly assumed that support vector machines should be applied on conditionally positive semidefinite kernels K.

Although the restriction to conditional positive semidefiniteness might not appear onerous, it is actually problematic in many natural situations. First, as [11] notes, verifying that a putative kernel function $k(\cdot, \cdot)$ is conditionally positive semidefinite can be a significant challenge. Second, as many authors note [2,3,7,9,10,11,14] using indefinite kernels and only approximately optimizing (2) can often yield similar or even better results than using conventional positive semidefinite kernels. (A frequently used example is the hyperbolic tangent "kernel" $\tanh(a\langle \mathbf{x}_i, \mathbf{x}_j \rangle + b)$.) Third, adding conditional positive semidefiniteness as a constraint causes difficulty when attempting to *learn* a kernel (similarity measure) directly from data.

In fact, it is this third difficulty that is the main motivation for this research. We are interested in *learning* similarity measures from data that we can then use to train accurate classifiers. One can easily devise natural ways of doing this (we elaborate on one approach below), but unfortunately in these cases ensuring positive semidefiniteness ranges from hard to impossible. To date, most successful attempts at learning conditionally positive semidefinite kernels have been reduced to taking convex combinations of known conditionally positive semidefinite kernels [8,5]. But we would like to consider a wider range of techniques for learning similarities, and therefore we seek to generalize (2) to exploit general similarity matrices. Our goal is to develop an efficient weight optimization procedure for $\boldsymbol{\alpha}$ that does not require a positive semidefinite matrix $K \circ \mathbf{y}\mathbf{y}^\top$, while still preserving the desirable generalization and sparseness properties achieved by standard SVM training.

Below in Section 2 we show how the standard kernel classifier can be generalized to consider more abstract *confidence functions* $c(y_i y_j | \mathbf{x}_i \mathbf{x}_j)$ that play the same role as the usual kernel-label combination $y_i y_j k(\mathbf{x}_i, \mathbf{x}_j)$. We then briefly outline some natural approaches for learning confidence functions from data in Section 3. The approach we propose there is very simple, but effective. Nevertheless, it suffers from the drawback of not being able to ensure a positive semidefinite matrix for optimization. Section 4

---

[1] A symmetric matrix K is conditionally positive semidefinite if $\mathbf{z}^\top K \mathbf{z} \ge 0$ for all $\mathbf{z}$ such that $\mathbf{z}^\top \mathbf{e} = 0$. K need only be conditionally positive semidefinite to ensure $K \circ \mathbf{y}\mathbf{y}^\top$ is positive semidefinite because of the assumption $\boldsymbol{\alpha}^\top \mathbf{y} = 0$. That is, if $(\boldsymbol{\alpha} \circ \mathbf{y})^\top \mathbf{e} = \boldsymbol{\alpha}^\top \mathbf{y} = 0$, then we immediately obtain $\boldsymbol{\alpha}^\top (K \circ \mathbf{y}\mathbf{y}^\top)\boldsymbol{\alpha} = (\boldsymbol{\alpha} \circ \mathbf{y})^\top K(\boldsymbol{\alpha} \circ \mathbf{y}) \ge 0$.

then outlines our main development. Given the general confidence function viewpoint, we derive an α-weight optimization procedure from intuitive, strictly *non*-geometric first principles. The first procedure we derive simply recovers the classical quadratic objective, but from a new perspective. With this re-derivation in hand, we are then able to formulate a novel relaxation of the standard SVM objective that is both principled while also being guaranteed to be convex. Finally, in Section 6 we present experimental results with this new training principle, showing similar performance to standard SVM training with standard kernel functions, but obtaining stronger performance using indefinite confidence functions learned from data.

2   Confidence Function Classification

Our goal is to develop a learning and classification scheme that can be expressed more abstractly than the usual formulation in terms of $y_i y_j k(x_i x_j)$. We do this via the notion of a *confidence function*, $c(y_i y_j | x_i x_j)$, which expresses a numerical confidence that the label pair $y_i y_j$ is in fact correct for the input pair x_i and x_j. A large confidence value expresses certainty that the label pair is correct, while a small value correspondingly expresses a lack of confidence that the label pair is correct (or certainty that the label pair is wrong). We make no other assumptions about the confidence function, although it is usually presumed to be symmetric: $c(y_i y_j | x_i x_j) = c(y_j y_i | x_j x_i)$.

Although the notion of a pairwise confidence function might seem peculiar, it is in fact exactly what the $y_i y_j k(x_i, x_j)$ values provide to the SVM. In particular, if we make the analogy $c(y_i y_j | x_i x_j) = y_i y_j k(x_i, x_j)$ and assume $y \in \{-1, +1\}$, one can see that $y_i y_j k(x_i, x_j)$ behaves as a simple form of confidence function: the value is relatively large if either $y_i = y_j$ and x_i and x_j are similar under the kernel k, or if $y_i \neq y_j$ and x_i and x_j are *dissimilar* under the kernel. We therefore refer to the matrix $C = K \circ yy^\top$ as the *confidence matrix*.

Proposition 1. *If the entries of the confidence matrix $K \circ yy^\top$ are strictly positive, then the training data is linearly separable in the feature space defined by K.*

This proposition clearly shows that high confidence values translate into an accurate classifier on the training data. In fact, it is confidences, not similarities, that lie at the heart of support vector machines: The SVM methodology can be recast strictly in terms of confidence functions, abstracting away the notion of a kernel entirely, without giving up anything (except the Euclidean geometric interpretation). To illustrate, consider the standard SVM classifier: Assuming a vector of training example weights, α, has already been obtained from the quadratic minimization (2), the standard classification rule can be rewritten strictly in terms of confidence values

$$\hat{y} = \text{sign}\left(\left(\sum_j \alpha_j y_j k(x, x_j)\right) + b\right) = \arg\max_y \left(\left(\sum_j \alpha_j y y_j k(x, x_j)\right) + by\right)$$

$$= \arg\max_y \left(\left(\sum_j \alpha_j c(y y_j | x x_j)\right) + by\right) \quad (3)$$

Thus a test example $\mathbf{x}$ is classified by choosing the label y that exhibits the largest weighted confidence when paired against the training data.

Quite obviously, the SVM training algorithm itself can also be expressed strictly in terms of a confidence matrix over training data.

$$\min_{\boldsymbol{\alpha}} \; \boldsymbol{\alpha}^\top C \boldsymbol{\alpha} \quad \text{subject to} \quad 0 \leq \boldsymbol{\alpha} \leq \beta, \; \boldsymbol{\alpha}^\top \mathbf{y} = 0, \; \boldsymbol{\alpha}^\top \mathbf{e} = 1 \qquad (4)$$

This is just a rewriting of (2) with the substitution $C = K \circ \mathbf{y}\mathbf{y}^\top$, which does not change the fact that the problem is convex if and only if C is positive semidefinite. However, the formulation (4) is still instructive. Apparently the SVM criterion is attempting to re-weight the training data to *minimize* expected confidence. Why? Below we argue that this is in fact an incorrect view of (4), and suggest that, alternatively, (4) can be inter-preted as attempting to *maximize* the robustness of the classifier against changes in the training labels. With this different view, we can then formulate an alternative training criterion—a relaxation of (4)—that still attempts to maximize robustness, but is convex for *any* confidence matrix C. This allows us to advance our goal of *learning* confidence functions from data, while still being able to use SVM training of the example weights without having to ensure positive semidefiniteness.

Before turning to the interpretation and relaxation of (4), we first briefly discuss approaches that can be explored to learning confidence functions.

3   Learning Confidence Functions

There are many natural ways to consider learning a confidence function $c(y_i y_j | \mathbf{x}_i \mathbf{x}_j)$ from training data. A straightforward approach is to explore any known similarity learn-ing techniques to learn an arbitrary kernel matrix and then obtain a confidence function by combining the y terms. Alternatively, one can also learn the confidence function di-rectly. One of such simple techniques has been explored in [6]. Given training labels, one can just straightforwardly learn to predict label pairs $y_i y_j$ given their corresponding input vectors $\mathbf{x}_i$ and $\mathbf{x}_j$. Concretely, given examples $(\mathbf{x}_1, y_1), ..., (\mathbf{x}_t, y_t)$, it is easy to form the set of training pairs $\{(\mathbf{x}_i \mathbf{x}_j, y_i y_j)\}$ from the original data, which doubles the number of input features and squares the number of training labels and classes. (Sub-sampling can always be used to reduce the size of this training set.) Given such pairwise training data, standard probabilistic models can be learned to predict the probability of a label pair given the input vectors.

Many standard probabilistic methods, especially discriminative methods, can be used for learning pairwise predictors. For example, the logistic regression classifiers can be trained to maximize the conditional likelihood $P(y_i y_j | \mathbf{x}_i \mathbf{x}_j)$ of the observed label pairs given the conjoined vector of inputs $\mathbf{x}_i \mathbf{x}_j$. Once learned, a pairwise model would clas-sify test inputs $\mathbf{x}$ by maximizing the product $\hat{y} = \arg\max_y \prod_j P(y y_j | \mathbf{x}\mathbf{x}_j)$.[2] Clearly, this is equivalent to using a confidence function $c(y_i y_j | \mathbf{x}_i \mathbf{x}_j) = \log P(y_i y_j | \mathbf{x}_i \mathbf{x}_j)$ and classifying with respect to uniform example weights $\boldsymbol{\alpha}$. Surprisingly, [6] obtained good classification results with this simple approach. In this paper, we are interested in the

---

[2] [6] also considered other techniques for classification, including correlating the predictions on the test data in a transductive manner, but we do not pursue these extensions here.

connection to support vector machines and attempt to improve the basic method by optimizing the α-weights.

Note that, as a confidence measure, using a log probability model, $\log P(y_i y_j | \mathbf{x}_i \mathbf{x}_j)$, is a very natural choice. It can be trained easily from the available data, and performs quite well in practice. However, using a log probability model for a confidence function raises a significant challenge: $\log P(y_i y_j | \mathbf{x}_i \mathbf{x}_j)$ is always non-positive and therefore any confidence matrix C it produces, since it is strictly non-positive, cannot be positive semidefinite. This raises the same difficulty one faces with non-positive semidefinite kernels, which motivates us to reformulate the quadratic optimization criterion (4), so that convexity can be achieved more generally while preserving the effective generalization properties.

4   Optimizing Training Example Weights: An Alternative View

Given a confidence classifier (3) it is natural to consider adjusting the training example weights α to improve accuracy. At first glance, the quadratic minimization criterion (4) used by SVMs appears to be adjusting the example weights to *minimize* the confidence of the training labels y_i. However, we can argue that this interpretation is misleading. In fact, standard kernel-based confidence functions have a special property that masks a key issue: how confidences change when a training label is flipped. For the classifier (3), it is not the absolute confidence that counts, but rather the *relative* confidences between the correct label and the incorrect label. That is, we would like the confidence of a correct label to be larger than the confidence of a wrong label. For kernel-based confidence functions it turns out that the relationship between the relative confidences is greatly restricted.

Observation 1. *Let $\overline{y}$ denote a label flip, $-y$. If $c(y_i y_j | \mathbf{x}_i \mathbf{x}_j) = y_i y_j k(\mathbf{x}_i \mathbf{x}_j)$ then*

$$\sum_j \alpha_j c(y y_j | \mathbf{x} \mathbf{x}_j) + by = -\sum_j \alpha_j c(\overline{y} y_j | \mathbf{x} \mathbf{x}_j) - b\overline{y} \tag{5}$$

However, the relationship (5) is obviously not true in general. For example, it is violated by any probabilistic confidence function defined by $c(y_i y_j | \mathbf{x}_i \mathbf{x}_j) = \log P(y_i y_j | \mathbf{x}_i \mathbf{x}_j)$.

Thus for kernel-based confidence functions, the confidence in the opposite label is always just the negation of the confidence in the current label.

We now show how the concept of minimizing sensitivity to label flips on the training data recovers the classical SVM training criterion. Consider a training example $(\mathbf{x}_i, y_i)$ and an arbitrary current set of weights α. The current confidence in the training label y_i is

$$\left(\sum_j \alpha_j c(y_i y_j | \mathbf{x}_i \mathbf{x}_j) \right) + by_i$$

Now consider the change in the confidence in y_i that would result if a single training label y_k was actually mistaken. That is, if the current value of y_k is incorrect and should have been given the opposite sign, then the mistake we are making in y_i's confidence is

$$\Delta_k c(y_i) = \alpha_k c(y_k y_i | \mathbf{x}_k \mathbf{x}_i) - \alpha_k c(\overline{y_k} y_i | \mathbf{x}_k \mathbf{x}_i) \qquad (6)$$

$$= 2\alpha_k c(y_k y_i | \mathbf{x}_k \mathbf{x}_i) \qquad (7)$$

Note that (7) holds only under the kernel-based restriction (5), but in this special case the confidence penalty is just twice the original confidence. If (5) does not hold, then (6) can be used. The sum of the local confidence changes measures the overall sensitivity of the classification of training label y_i to possible mislabelings of other data points

$$\Delta c(y_i) = \sum_k \Delta_k c(y_i) \qquad (8)$$

The smaller this value, the less likely y_i is to be misclassified due to a mislabeling of some other data point. That is, the sensitivity to label flips should be minimized if the classifier is to be made more robust. Nevertheless, there might be a trade-off between the sensitivities of different training examples. Therefore, as a final step, we consider minimizing the overall *weighted* sensitivity of the training labels. This yields the minimization objective

$$\sum_i \alpha_i \Delta c(y_i) = \sum_i \alpha_i \sum_k \alpha_k \left(c(y_k y_i | \mathbf{x}_k \mathbf{x}_i) - c(\overline{y_k} y_i | \mathbf{x}_k \mathbf{x}_i) \right) \qquad (9)$$

$$= 2 \sum_{ik} \alpha_i \alpha_k c(y_k y_i | \mathbf{x}_k \mathbf{x}_i) \qquad (10)$$

Again, (10) only holds under the kernel-based restriction (5), but if this is violated, the more general form (9) can be used.

Therefore, if we are using a kernel-based confidence function, we recover exactly the same training criterion as the standard SVM (4). What is interesting about this derivation is that it does not require any reasoning about Euclidean geometry or even feature (Hilbert) spaces. The argument is only about adjusting the example weights to reduce the sensitivity of the classifier (3) to any potential mistakes in the training labels. That is, from the perspective of α-weight optimization, it is only the reflection property (5) and the desire to minimize sensitivity to mislabeled training examples that yields the same minimization objective as standard SVMs. The remaining constraints in (4) are also easily justified in this context: It is natural to assume that the example weights form a convex combination, and therefore $0 \leq \alpha$, $\alpha^\top e = 1$. It is also natural to preserve class balance in the re-weighting, hence $\alpha^\top y = 0$. Finally, as a regularization principle, it makes sense to limit the magnitude of the largest weights so that too few examples do not dominate the classifier, hence $\alpha \leq \beta$.

Of course, re-deriving an old criterion from alternative principles is not a significant contribution. However, what is important about this perspective is that it immediately suggests principled alternatives to the SVM criterion that can still reduce sensitivity to potential training label changes. Our goal is to reformulate the objective to avoid a quadratic form, since this prevents effective optimization on indefinite confidence functions, which are the confidence functions we are most interested in (Section 3). It turns out that just a minor adjustment to (10) yields just such a procedure.

Given the goal of minimizing the sensitivity to training label changes, previously we sought to minimize the *weighted* sensitivity, using the same weights being optimized, which leads to the quadratic form (10) and the optimization problem (4). However, rather than minimize weighted sensitivity, one could instead be more conservative and attempt to minimize the *maximum* sensitivity of any label in the training set. That is, we would like to adjust the example weights so that the worst sensitivity of any training label y_i to potential mislabelings of other examples is minimized. This suggestion immediately yields our proposal for a new training criterion

$$\min_{\boldsymbol{\alpha}} \max_i \sum_k \alpha_k \Big(c(y_k y_i | \mathbf{x}_k \mathbf{x}_i) - c(\overline{y_k} y_i | \mathbf{x}_k \mathbf{x}_i) \Big) \tag{11}$$

$$\text{subject to} \quad 0 \le \boldsymbol{\alpha} \le \beta, \; \boldsymbol{\alpha}^\top \mathbf{y} = 0, \; \boldsymbol{\alpha}^\top \mathbf{e} = 1$$

Once $\boldsymbol{\alpha}$ has been optimized, the offset b can be chosen to minimize training error.

Proposition 2. *The objective (11) is convex for any confidence function c, and moreover is an upper bound on (4).*

The proof of this proposition is obvious. The minimization objective is a maximum of linear functions of $\boldsymbol{\alpha}$, and hence is convex. Given the constraints $0 \le \boldsymbol{\alpha}, \boldsymbol{\alpha}^\top \mathbf{e} = 1$ we immediately have $\max_i f(i) \ge \sum_i \alpha_i f(i)$.
   As a practical matter, (11) can be solved by a simple linear program

$$\min_{\boldsymbol{\alpha}, \delta} \delta \text{ subject to } \delta \ge \sum_k \alpha_k \Big(c(y_k y_i | \mathbf{x}_k \mathbf{x}_i) - c(\overline{y_k} y_i | \mathbf{x}_k \mathbf{x}_i) \Big) \quad \forall i$$

$$0 \le \boldsymbol{\alpha} \le \beta, \; \boldsymbol{\alpha}^\top \mathbf{y} = 0, \; \boldsymbol{\alpha}^\top \mathbf{e} = 1 \tag{12}$$

This formulation produces a convex relaxation of the ν-SVM criterion for *any* confidence function $c(y_i y_j | \mathbf{x}_i \mathbf{x}_j)$, and provides an alternative option for using indefinite kernels.

5   Related Work

Most work related to our proposed approach concerns learning SVMs with indefinite kernels [10,2,3,11]; although in this paper we address the slightly more general problem of learning from confidence functions—a superset of confidences derived from indefinite kernels.
   As noted, if the kernel matrix is not conditionally positive semidefinite, the standard SVM training problem (2) becomes non-convex, hence hard to optimize. To overcome this difficulty, some authors have suggested using the indefinite kernel directly but instead solve an approximate form of the SVM training problem [11,9,7]. Most, however, propose to transform the indefinite kernel into a positive semidefinite kernel and then apply standard SVM training. Such transformation methods include "denoise" (neglect the negative eigenvalues), "flip" (flip the sign of the negative eigenvalues) and "shift" (shift all eigenvalues by a positive constant to make all positive); see [14] for details. A limitation of using such simple transformation methods is that valuable information about the data can be lost in the transformation process. Therefore, more recently, a number of papers have begun to pursue a middle ground strategy, where the original

indefinite kernel is used in training, while the SVM objective is perturbed as little as possible to maintain a convex objective [10,2,3]. These papers proceed by fixing the original indefinite kernel K_0 and then modifying the training objective as follows.

$$\max_{\alpha} \min_{K} \quad \alpha^\top e - \frac{1}{2}\alpha^\top (K \circ yy^\top)\alpha + \rho\|K - K_0\|^2 \tag{13}$$
$$\text{subject to} \quad \alpha^\top y = 0, \ 0 \leq \alpha \leq \beta, \ K \succeq 0$$

We provide some comparison of our approach to these different techniques below.

6 Experimental Results

We implemented the new weight optimization scheme based on linear programming (12) and compared it both to standard SVM quadratic minimization (4) and to a simple approach of using uniform weights. In addition, for *indefinite* confidence measures, we compared our proposed method to the existing kernel transformation methods mentioned in Section 5 above. Note that training based on (4) can only be efficiently performed when the kernel is positive semidefinite. Furthermore, note that although using uniform weights appears to be naive, for high quality confidence measures such as those learned by training reasonable probability models, uniform weighting can still achieve highly competitive generalization performance—a fact that will be revealed below.

We compared these different weight optimization schemes on a variety of confidence functions, including those defined by standard positive semidefinite kernels (linear dot product and RBF), as well as the sigmoid kernel tanh, and two probabilistic confidence models trained using naive Bayes and logistic regression respectively. In our result tables below, we will denote the results produced by the proposed weight optimization scheme (12) as $\alpha 1$, for the uniform weighting scheme as $\alpha 0$, and for the standard SVM quadratic minimization scheme (4) as $\alpha 2$. We compared the test accuracy of these various algorithms on a set of two-class UCI data sets. All of our experimental results are averages over 5 times repeats with training size equal to 100 or 4/5ths of the data size.

In the first study we compared how the different weight optimization schemes performed using the positive semidefinite confidence functions determined by the linear and RBF kernels respectively. Table 1 shows that the new weight optimization scheme (12) achieves comparable generalization performance to standard quadratic training (4). The uniform weighting strategy is generally inferior in the linear kernel case (being dominated on all data sets except placing second on *flare* and *german*). Although uniform weighting performs relatively better in the RBF kernel case, the results are still not comparable to the linear and quadratic weighting schemes. The reason is that the confidence functions are only weakly informative here, and simply averaging them still yields a sensitive classifier. It is encouraging to note that our convex relaxation retains most of the benefit of the original quadratic objective in this case.

We also compared our proposed approach to existing methods for learning with indefinite kernels. Table 2 shows the comparative results we obtained using indefinite kernels produced by tanh functions. Interestingly, the straightforward *denoise* and *flip* transformations yield very good results using tanh kernels, while *shift* produces very

Table 1. Comparison of the test accuracy of different α-weight optimizers on UCI data sets using the positive semidefinite confidence functions defined by linear (L) and RBF ($k(\mathbf{x}_i, \mathbf{x}_j) = \exp(-\|\mathbf{x}_i - \mathbf{x}_j\|^2)$) kernels. Here, $\alpha0$ denotes the results for uniform weighting; $\alpha1$ for our proposed linear optimization (12); and $\alpha2$ for standard quadratic SVM optimization (4).

	L-α0	L-α1	L-α2	RBF-α0	RBF-α1	RBF-α2
australian	0.613	0.815	0.843	0.729	0.784	0.713
breast	0.928	0.969	0.970	0.947	0.971	0.962
cleve	0.617	0.794	0.802	0.817	0.819	0.792
corral	0.571	0.900	0.929	0.907	0.936	1.000
crx	0.546	0.850	0.837	0.649	0.770	0.677
diabetes	0.651	0.754	0.705	0.741	0.705	0.730
flare	0.829	0.829	0.760	0.800	0.815	0.829
german	0.700	0.705	0.666	0.498	0.628	0.700
glass2	0.756	0.781	0.819	0.892	0.867	0.864
heart	0.780	0.824	0.835	0.764	0.794	0.774
hepatitis	0.813	0.863	0.863	0.725	0.825	0.825
mofn-3-7	0.779	0.781	0.805	0.704	0.776	0.857
pima	0.651	0.763	0.736	0.772	0.701	0.694
vote	0.615	0.931	0.935	0.881	0.913	0.875
average	0.703	0.826	0.822	0.773	0.807	0.807

Table 2. Comparison of the test accuracy of different methods using indefinite kernels produced by $\tanh$ ($k(\mathbf{x}_i, \mathbf{x}_j) = tanh(0.001 \cdot \mathbf{x}_i\mathbf{x}_j - 1)$). Here, the column headings refer to the various techniques reviewed in Section 5, except that $\alpha1$ refers to our proposed method based on (12).

	denoise	flip	shift	reg-svm	α1
australian	0.852	0.852	0.554	0.782	0.844
breast	0.967	0.967	0.650	0.965	0.969
cleve	0.778	0.767	0.541	0.798	0.763
corral	0.893	0.893	0.429	0.871	0.879
crx	0.842	0.841	0.429	0.828	0.879
diabetes	0.758	0.759	0.651	0.759	0.699
flare	0.829	0.698	0.829	0.828	0.829
german	0.722	0.723	0.700	0.726	0.695
glass2	0.768	0.765	0.540	0.781	0.781
heart	0.827	0.827	0.553	0.837	0.815
hepatitis	0.875	0.875	0.813	0.813	0.825
mofn-3-7	1.000	1.000	0.779	0.816	0.813
pima	0.756	0.763	0.651	0.766	0.735
vote	0.935	0.935	0.615	0.928	0.921
average	0.843	0.833	0.624	0.821	0.818

poor performance. The sophisticated *reg-svm* technique, based on (13) [10,2,3], requires much more involved training, yet produces very similar results to the simple linear optimization scheme we propose.

Table 3. Comparison of the test accuracy of different α-weight optimizers on UCI data sets, using the indefinite confidence functions produced by pairwise naive Bayes (NB), and logistic regression (LR). Here, $\alpha 0$ denotes the results for uniform weighting; and $\alpha 1$ for our proposed linear optimization (12).

	LR-$\alpha 0$	LR-$\alpha 1$	NB-$\alpha 0$	NB-$\alpha 1$
australian	0.850	0.850	0.846	0.847
breast	0.959	0.933	0.972	0.969
cleve	0.802	0.804	0.840	0.846
corral	0.871	0.893	0.900	0.900
crx	0.845	0.845	0.842	0.836
diabetes	0.744	0.741	0.761	0.755
flare	0.820	0.820	0.831	0.821
german	0.719	0.718	0.714	0.713
glass2	0.806	0.813	0.873	0.895
heart	0.813	0.812	0.822	0.827
hepatitis	0.900	0.900	0.925	0.938
mofn-3-7	0.911	0.903	0.887	0.924
pima	0.743	0.743	0.758	0.759
vote	0.918	0.918	0.919	0.912
average	0.836	0.835	0.849	0.853

Another interesting test of the method is on using *learned* indefinite confidence functions, such as those determined by an estimated probability model $\log P(y_i y_j | \mathbf{x}_i \mathbf{x}_j)$. In these cases, the quadratic objective is non-convex and cannot be solved by standard quadratic optimizers. However, as mentioned, our relaxation remains convex. [11] suggests more sophisticated approach for training in these cases, but their methods are substantially more technical than the simple technique proposed here. Table 3 shows the results of our linear weight optimization procedure and the uniform weighting on the indefinite confidence functions. Noticeably the probabilistic (trained) confidence functions yield better results than the ones produced in the previous tables by standard SVMs with both linear and RBF kernels. Moreover, even uniform weighting already achieves good results for these confidence functions, since the confidence functions obtained using LR and NB are very informative.

The main benefit of the new approach is the ability to reliably optimize example weights for a wider range of confidence functions. We believe this is a useful advantage over SVM training because most natural confidence functions, in particular learned confidence functions, are not usually positive semidefinite but have a wider potential for generalization improvement over using fixed kernels, as our results suggest.

7   Conclusion

We have introduced a simple generalization of support vector machines based on the notion of a *confidence function* $c(y_i y_j | \mathbf{x}_i \mathbf{x}_j)$. This view allows us to think of SVM training as attempting to minimize the sensitivity of the classifier to perturbations of the training labels. From this perspective, we can not only re-derive the standard SVM objective without appealing to Euclidean geometry, we can also devise a new training objective that is convex for arbitrary, not just positive semidefinite, confidence functions. Of course, other optimization objectives are possible, and perhaps superior ones could still be developed. An important research direction is to develop a generalization theory for our relaxed training procedure that is analogous to the theory that has already been developed for SVMs.

Acknowledgments. The authors gratefully acknowledge support from Temple University, AICML, NSERC, MITACS and the Canada Research Chairs program.

References

1. Boser, B., Guyon, I., Vapnik, V.: A training algorithm for optimal margin classifiers. In: Proceedings 5th Annual Conference on Computational Learning Theory, COLT (1992)
2. Chen, J., Ye, J.: Training SVM with indefinite kernels. In: Proceedings of the International Conference on Machine Learning, ICML (2008)
3. Chen, Y., Gupta, M., Recht, B.: Learning kernels from indefinite similarities. In: Proceedings of the International Conference on Machine Learning, ICML (2009)
4. Crisp, D.J., Burges, C.J.C.: A geometric interpretation of v-SVM classifiers. In: Advances in Neural Information Processing Systems, NIPS 2000 (2000)
5. Cristianini, N., Shawe-Taylor, J., Elisseeff, A., Kandola, J.: On kernel target-alignment. In: Advances in Neural Information Processing Systems, NIPS 2001 (2001)
6. Guo, Y., Greiner, R., Schuurmans, D.: Learning coordination classifiers. In: Proceedings of the 19th International Joint Conference on Artificial Intelligence, IJCAI 2005 (2005)
7. Haasdonk, B.: Feature space interpretation of SVMs with indefinite kernels. IEEE Transactions on Pattern Analysis and Machine Intelligence 27(4) (2005)
8. Lanckriet, G., Cristianini, N., Bartlett, P., Ghaoui, L., Jordan, M.: Learning the kernel matrix with semidefinite programming. Journal of Machine Learning Research 5 (2004)
9. Lin, H.-T., Lin, C.-J.: A study on sigmoid kernels for SVM and the training of non-PSD kernels by SMO-type methods (2003)
10. Luss, R., d'Aspremont, A.: Support vector machine classification with indefinite kernels. In: Advances in Neural Information Processing Systems, NIPS 2007 (2007)
11. Ong, C.S., Mary, X., Canu, S., Smola, A.J.: Learning with non-positive kernels. In: Proceedings of the International Conference on Machine Learning, ICML (2004)
12. Schoelkopf, B., Smola, A., Williamson, R., Bartlett, P.: New support vector algorithms. Neural Computation 12(5), 1207–1245 (2000)
13. Vapnik, V.: Statistical Learning Theory. Wiley, Chichester (1998)
14. Wu, G., Chang, Y., Zhang, Z.: An analysis of transformation on non-positive semidefinite similarity matrix for kernel machines. In: Proceedings of the International Conference on Machine Learning, ICML (2005)

Robust Discriminant Analysis Based on Nonparametric Maximum Entropy

Ran He, Bao-Gang Hu, and Xiao-Tong Yuan

National Laboratory of Pattern Recognition,
Institute of Automation,
Chinese Academy of Sciences
95 Zhongguancun Donglu, Beijing 100190, China
{rhe,bghu,xtyuan}@nlpr.ia.ac.cn

Abstract. In this paper, we propose a Robust Discriminant Analysis based on maximum entropy (MaxEnt) criterion (MaxEnt-RDA), which is derived from a nonparametric estimate of Renyi's quadratic entropy. MaxEnt-RDA uses entropy as both objective and constraints; thus the structural information of classes is preserved while information loss is minimized. It is a natural extension of LDA from Gaussian assumption to any distribution assumption. Like LDA, the optimal solution of MaxEnt-RDA can also be solved by an eigen-decomposition method, where feature extraction is achieved by designing two Parzen probability matrices that characterize the within-class variation and the between-class variation respectively. Furthermore, MaxEnt-RDA makes use of high order statistics (entropy) to estimate the probability matrix so that it is robust to outliers. Experiments on toy problem , UCI datasets and face datasets demonstrate the effectiveness of the proposed method with comparison to other state-of-the-art methods.

1 Introduction

Feature extraction plays an important role in machine learning and computer vision. It is a common preprocessing step to learn a low-dimensional subspace from the raw input variables which might be strongly relevant and redundant [1]. From different viewpoints, there are two major categories of feature extraction: unsupervised and supervised.

In unsupervised feature extraction, the data class labels are unknown. The low-dimensional representation is learned by minimizing reconstruction error or preserving structural information of data. A best-known unsupervised method is principal component analysis (PCA) [2]. To deal with the Gaussian assumption problem in PCA, some important extensions of PCA are developed by locally optimizing the weighted scatter matrix, including locality preserving projections (LPP) [3], Laplacian PCA [4] and etc.

In supervised feature extraction [5][6] [7], the information of class labels is used to learn a low-dimensional subspace by maximizing the class differences. The linear discriminant analysis (LDA) [8] is the most representative one. It is a widely used in classification tasks due to its computational simplicity. Despite its wide use, LDA assumes that data distribution of each class is Gaussian. Thus, this will be certainly hard to make LDA adapt to data under non-Gaussian distribution. Among algorithms for solving this problem, nonparametric estimation is a most popularly used technique

Z.-H. Zhou and T. Washio (Eds.): ACML 2009, LNAI 5828, pp. 120–134, 2009.

and nonparametric LDAs have therefore been developed. The main difference between nonparametric LDAs and LDA is in that nonparametric LDAs introduce nonparametric within-class scatter and between-class scatter. Nonparametric LDA (NDA) [9], marginal Fisher analysis (MFA) [10] and linear Laplacian discrimination (LLD) [11] are most representatives of nonparametric methods. Although those nonparametric methods are useful for solving non-Gaussian data, they are often designed based on heuristic strategy and the learned subspace depends on the structure of training set. When there is noise, the learned subspace may be biased.

Another well-known supervised feature extraction category is information theoretic learning (ITL) [12], where the mutual information (or entropy) is selected as the objective. Feature extraction is achieved by directly maximizing the quadratic mutual information between the class label and the features [13]. Gaussian [14] and Parzen window [15] probability density functions are used to estimate mutual information. In [16], feature extraction is implemented by solving a minimum entropy problem (or maximizing the information potential) and an iterative algorithm based on half-quadratic is proposed. The experimental results demonstrate that the methods based on ITL have potential ability to discover principal curve of the data and are robust to noise [17][16]. However, the ITL based methods are often solved by iterative algorithms and hence have relatively high computation complexity.

In this paper, the maximum entropy (MaxEnt) criterion, which provides a natural means to process information in the form of constraints [18], is introduced in linear feature extraction. A MaxEnt Robust Discriminant Analysis (MaxEnt-RDA) is proposed where entropy is defined as the Renyi's quadratic entropy. The MaxEnt distribution is estimated by a nonparametric Parzen window density estimator. As a result, the calculation of differential entropy becomes a summation over pair wise interactions of the data. The proposed MaxEnt-RDA has several interesting perspectives: (1) it utilizes entropy maximum as objective and hence has a clearly theoretical foundation. The high order statistic information of data are preserved during feature extraction. And it is robust to noise. (2) it can be solved by an eigen-decomposition method which avoids iterative calculation of entropy. The optimal solution consists of the principal eigenvectors of two probability matrices corresponding to MaxEnt distribution. (3) Its assumption of distribution is free so that it can effectively capture the underline distribution of multimodal data statistics and deal with non-Gaussian distribution data.

The remainder of this paper is outlined as follows. The concepts of LDA and MFA are briefly reviewed in Section 2. The theoretical properties of MaxEnt objective function and MaxEnt-RDA are described in Section 3. We evaluate our method on UCI machine learning datasets and face datasets in Section 4. Finally, we conclude the paper in Section 5.

2   LDA and MFA

Suppose that we have a matrix $X = [x_1, \ldots, x_n]$ of n d-dimensional samples and a projection matrix $U = [u_1, \ldots, u_m]$ whose columns constitute the bases of the m-dimensional subspace. There are c classes in the data set X and each class C_j has n_j samples. According to definitions in [19], sample mean for all of the classes and for each class can be defined as,

$$\mu = \frac{1}{n}\sum_{i=1}^{n} x_i \quad \mu_j = \frac{1}{n_j}\sum_{x_j \in C_j} x_j \tag{1}$$

and the covariance matrices as:

$$\widehat{\Sigma} = \frac{1}{n}\sum_{i=1}^{n}(x_i - \mu)(x_i - \mu)^T \tag{2}$$

$$\widehat{\Sigma}_j = \frac{1}{n_j}\sum_{x_j \in C_j}(x_j - \mu_j)(x_j - \mu_j)^T \tag{3}$$

We define the scatter matrices S_T, S_W and S_B by,

$$S_T = n\widehat{\Sigma} = X(I - W_t)X^T \tag{4}$$

$$S_W = \sum_{j=1}^{c} n_j\widehat{\Sigma}_j = \sum_{j=1}^{c} X_j(I - W_j)X_j^T = X(I - W_w)X^T \tag{5}$$

$$S_B = S_T - S_W \tag{6}$$

where W_t is a $n \times n$ matrix with all the elements equal to $1/n$, W_j is a $n_j \times n_j$ matrix with all the elements equal to $1/n_j$, and W_w reads the diagonal block form of $W_w = diag(W_1, \ldots, W_c)$. S_W and S_B are often called within-class scatter matrix and between-class scatter matrix respectively. The LDA tries to solve the following maximum problem:

$$J_{LDA}(u) = \frac{u^T S_T u}{u^T S_W u} \sim \frac{u^T S_B u}{u^T S_W u} \tag{7}$$

where u is a d-dimension vector. It is easy to show that the vector u that maximizes J_{LDA} must satisfy

$$S_T u = \lambda S_W u \tag{8}$$

If S_W is nonsingular, the solution of LDA can be obtained by a conventional eigenvalue problem:

$$(X(I - W_w)X^T)^{-1}X(I - W_t)X^T u = \lambda u \tag{9}$$

Since LDA assumes that data distribution of each class is Gaussian, LDA could not adapt to data under non-Gaussian distribution. MFA is an important extension of LDA from viewpoint of Graph Embedding. Two graphs in MFA are designed to characterize the within-class compactness matrix and the between-class separability, respectively. The within-class compactness matrix of MFA is defined as

$$S_w^{MFA} = \sum_{i=1}^{n} \sum_{x_i \in N_{k_1}(x_j) \text{ or } x_j \in N_{k_1}(x_i)} \|u^T x_i - u^T x_j\|^2 \tag{10}$$

$$= 2u^T X(D_w^{MFA} - W_w^{MFA})X^T u$$

$$(D_w^{MFA})_{ii} = \sum_{j \neq i} (W_w^{MFA})_{ij}$$

$$(W_w^{MFA})_{ij} = 1 \quad if\ x_j \in N_{k_1}(x_i)\ or\ x_i \in N_{k_1}(x_j); 0\ else$$

where $N_{k_1}(x_i)$ means the k_1 nearest neighbors in the same class of the sample x_i and D_w^{MFA} is a diagonal matrix. The between-class compactness matrix of MFA is defined as

$$S_b^{MFA} = \sum_{i=1}^{n} \sum_{x_i \in P_{k_1}(x_j) \ or \ x_j \in P_{k_1}(x_i)} ||u^T x_i - u^T x_j||^2 \qquad (11)$$

$$= 2u^T X (D_b^{MFA} - W_b^{MFA}) X^T u$$

$$(D_b^{MFA})_{ii} = \sum_{j \neq i} (W_b^{MFA})_{ij}$$

$$(W_w^{MFA})_{ij} = 1 \quad if \ x_j \in P_{k_2}(x_i) \ or \ x_i \in P_{k_2}(x_j); \quad 0 \ else$$

Where $P_{k_2}(x_i)$ is a set of data pairs that are the k_2 nearest pairs among $\{(x_i, x_j), x_i \in C_i, x_j \notin C_i\}$ and D_b^{MFA} is a diagonal matrix. MFA provides a new approach to deal with Gaussian assumption in LDA and thus can improve accuracy of classification rates in non Guassian distribution applications such as face recognition.

However, a main drawback of MFA is that MFA depends on the structure of the data points near the boundary of different classes. When there is noise, the performance of MFA will decrease (see experiment 4.3).

3   MaxEnt Robust Discriminant Analysis

Renyi's quadratic entropy of a dataset X with Probability Density Function (PDF) $f_X(x)$ is defined by

$$H(X) = -\log \int f_X^2(x) dx \qquad (12)$$

If Parzen window method is used to estimate the P.D.F., $f_X(x)$ can be obtained as

$$\widehat{f}_{X;\sigma}(x) = \frac{1}{n} \sum_{i=1}^{n} G(x - x_i, \sigma) \qquad (13)$$

where $G(x - x_i, \sigma)$ is the Gaussian kernel with bandwidth σ

$$G(x - x_i, \sigma) = \frac{1}{\sqrt{2\pi}\sigma} \exp(-\frac{(x - x_i)^2}{2\sigma^2}) \qquad (14)$$

Substitute $f_X(x)$ in (12) with (13), the estimate of entropy by Parzen method can be obtained as [13]:

$$H(X) = -\log(\frac{1}{n^2} \sum_{i=1}^{n} \sum_{j=1}^{n} G(x_j - x_i, \sigma)) \qquad (15)$$

In supervised linear feature extraction, one considers the following constraint MaxEnt problem:

$$\max_{U} H(U^T X) \quad s.t. \ H(U^T X|C) = c_1 \ \& \ U^T U = I \qquad (16)$$

where conditional entropy $H(X|C)$ is defined as [20]:

$$H(X|C) = \sum_{j=1}^{c} p(C_j)H(X|C = C_j) \tag{17}$$

When the formula of $f_X(x)$ is given, the MaxEnt distribution in (16) is only relative to the subspace U, i.e., the MaxEnt objective is a function of subspace U. The different subspace will give a different system state measured by entropy. We expect that in the subspace U entropy of all data is maximized so that the loss of information is minimized, meanwhile entropy of each class (measured by conditional entropy in (17)) is nearly invariant so that the structure of each class is not broken. After the dimension reduction, the information loss is minimized meanwhile the structural information of individual class is preserved. Since the distribution is estimated by Parzen window method, it can model the data's distribution more accurately. Note that the orthogonal constraint is necessary and important. It is in accord with the idea that the system of coordinates carries no information [18].

Theorem 1. *The optimal solution for (16) is given by the following eigenvector problem*

$$XL_t(u)X^T u = \lambda X L_w(u) X^T u \tag{18}$$

where

$$L_t(u) = D^t(u) - W^t(u) \tag{19}$$

$$L_w(u) = D^w(u) - W^w(u) \tag{20}$$

$$W_{ij}^t(u) = \frac{2G(u^T x_i - u^T x_j, \sigma)}{\sigma^2 \sum_{i=1}^{n} \sum_{j=1}^{n} G(u^T x_i - u^T x_j, \sigma)}$$

$$W_{ij}^w(u) = I_{(c_i = c_j)} p(c_i) \frac{2G(u^T x_i - u^T x_j, \sigma)}{\sigma^2 \sum_{i=1}^{n_i} \sum_{j=1}^{n_i} G(u^T x_i - u^T x_j, \sigma)}$$

$$D_{ii}^t(u) = \sum_{j=1}^{n} W_{ij}^t(u), \quad D_{ii}^w(u) = \sum_{j=1}^{n} W_{ij}^w(u)$$

(D^t and D^w are diagonal matrices)

Proof. We follow the standard theory of numerical optimization and applying the Lagrangian factor on (16) (Here we only consider the first constraint). The PDF $f_X(x)$ is defined in (13), we have:

$$J_H(u) \triangleq -\log \frac{1}{n^2} \sum_{i=1}^{n} \sum_{j=1}^{n} G(u^T x_i - u^T x_j, \sigma)$$

$$+ \lambda(\sum_{j=1}^{c} p(C_j) \log \frac{1}{n_j^2} \sum_{k=1}^{n_j} \sum_{l=1}^{n_j} G(u^T x_k - u^T x_l, \sigma) - c_1) \tag{21}$$

where the Lagrangian multiplier λ for enforcing the first constraints. The KKT condition for optimal solution specifies that the gradient of $J_H(u)$ must be zero:

$$\frac{\partial J_H(u)}{\partial u} = \sum_{i=1}^{n}\sum_{j=1}^{n}(W_{ij}^t(u) - \lambda W_{ij}^w(u))((u^T x_i - u^T x_j)x_i^T)^T$$

$$= XL_t(u)X^T u - \lambda X L_w(u)X^T u = 0$$

The above equation gives the fixed point relation

$$X L_t(u)X^T u = \lambda X L_w(u)X^T u$$

Intuitively, the optimal u is the eigenvectors of (18).

In Theorem 1, we can find that the format of solution of MaxEnt is similar to that of Graph Embedding. They all can be solved by eigen-decomposition method. $W_{ij}^w(U)$ can be seen as a within-class scatter matrix, which represents interactions between pairs of samples inside each class. $W_{ij}^t(U)$ consists of interactions between all pairs of samples, regardless of class information. Given a U, $W_{ii}^w(U)$ (or $W_{ii}^t(U)$) is an approximate of probability contribution on x_i under the jth Parzen estimate, and $D_{ii}^w(U)$ (or $D_{ii}^t(U)$) is an approximate of probability value on x_i under the Parzen estimate. Hence, we denote and $W_{ij}^w(U)$ and $W_{ij}^t(U)$ as within-class and between-class Parzen probability matrices respectively.

Recent theoretical results [21][16] illustrate that there is a close relationship between entropy objective and robust estimators [22]. Algorithms based on information theoretic objectives often can significantly improve the robustness [21][16]. For MaxEnt-RDA, if there are outliers that are significantly faraway from the rest of the data points, those outliers will obtain small values in within-class and between-class Parzen probability matrices and hence they will less affect the objective.

Look at (18) in Theorem 1, we recognize that it is still hard to find a closed form for this problem, because this eigenvalue problem is nonlinear. Hence the objective of $J_H(u)$ is still difficult to be directly optimized. In next section, we introduce an approximate strategy for optimization of the problem.

3.1   Algorithm of MaxEnt-RDA

From now on, we derive an algorithm to solve the MaxEnt problem. From theorem 1, we can learn that solution of MaxEnt-RDA can be given by an eigen-decomposition method. However, the eigen-decomposition problem in (18) is nonlinear and the solution is dependent on U in a non-trivial way. The estimate of PDF is performed on the reduced dimension instead of original input feature space. To address this problem, we approximate each Gaussian kernel term by its first-order Taylor expansion. Since expanding $exp(-z)$ at z_0 leads to $exp(-z) \approx exp(-z_0) - exp(-z_0)(z - z_0)$, let $z = \frac{||U^T x_i - U^T x_j||^2}{\sigma^2}$ and $z_0 = \frac{||x_i - x_j||^2}{\sigma^2}$, we have

$$G(U^T x_i - U^T x_j, \sigma) \approx -G(x_i - x_j, \sigma)||U^T x_i - U^T x_j||^2 + const \quad (22)$$

Substitute (22) into (16), we finally reduce the MaxEnt objective to a constraint graph embedding problem:

$$\max tr(U^T X L_t(I) X^T U) \quad s.t. \quad tr(U^T X L_w(I) X^T U) = c_1 \ \& \ U^T U = I \quad (23)$$

where we remove the $log(x)$ function because $log(x)$ is a convex function. Since the $X L_t(I) X^T$ and $X L_w(I) X^T$ are symmetric and semi-positive definite, by using the Lagrangian technique, it is easy to prove that the solution is given by

$$X L_t(I) X^T U = \Lambda X L_w(I) X^T U \quad (24)$$

where Λ is a diagonal matrix whose diagonal elements are m largest eigenvalues. Compared with (18) and (24), we can learn that, in the Taylor approximation solution, we assume that the MaxEnt distribution on the subspace U and original input feature are quite similar so that $L_t(U) \approx L_t(I)$ and $L_w(U) \approx L_w(I)$. The merit of this solution based on Taylor expansion is that it is a unique and global solution.

The bandwidth σ is an important parameter in MaxEnt-RDA, which is used in Parzen estimate of $f_X(x)$. The bandwidth controls all properties of the estimator [21]. Considering the theoretical analysis of nonparametric entropy estimators [15], we set the bandwidth as a factor of average distance between projected samples:

$$\sigma^2 = \frac{1}{sn^2} \sum_{i=1}^{n} \sum_{j=1}^{n} (x_i - x_j)^2 \quad (25)$$

where s is a scale factor.

The detailed description of MaxEnt-RDA is summarized as Algorithm 1.

Input: data matrix X, desired dimensionality m, bandwidth parameter σ
Output: orthonormal matrix U

1. Project the data set into the PCA subspace by retaining $n - c$ dimensions. Let W_{PCA} denote the transformation matrix of PCA.
2. Construct the approximated within-class and between-class Parzen probability matrices on the PCA subspace according to (20) and (19).
3. Finding the optimal projection direction U^* by solving (24). U^* is given by the eigenvectors corresponding to m largest eigenvalues.
4. Output the final linear projection direction as $U = W_{PCA} * U^*$

Algorithm 1. MaxEnt-RDA

3.2   Relation to Mutual Information and LDA

By Lagrangian factor method, we can rewrite (16) as

$$\max_U J_I(U) = H(U^T X) - \lambda_W (H(U^T X | C) - c_1) \quad s.t. \quad U^T U = I \quad (26)$$

When λ_W is set to 1, $J_I(U)$ becomes the mutual information which has been used as the objective in [15][14].

Theorem 2. *If the data is Gaussian distributed and $\lambda_W = 1$, the solution of LDA gives of a lower bound of MaxEnt in (26):*

$$J_I(u) \geq \tfrac{1}{2} \log(J_{LDA}(u)) \tag{27}$$

Proof. Since $\frac{d}{2} \log 2\pi + \frac{d}{2} = \sum\limits_{j=1}^{c} p(C_j)(\frac{d}{2} \log 2\pi + \frac{d}{2})$, we have

$$
\begin{aligned}
J_I(u) &= H(u^T X) - H(u^T X | C) \\
&= \frac{1}{2} \log u^T \Sigma u - \frac{1}{2} \sum_{j=1}^{c} p(C_j) \log(u^T \Sigma_j u) \\
&\geq \frac{1}{2} \log u^T \Sigma u - \frac{1}{2} \log(\sum_{j=1}^{c} \frac{n_j}{n} u^T \Sigma_j u) \\
&= \frac{1}{2} \log(\frac{1}{n} u^T S_T u) - \frac{1}{2} \log(\sum_{j=1}^{c} \frac{n_j}{n}(\frac{1}{n_j} u^T S_j u)) \\
&= \frac{1}{2} \log(\frac{\frac{1}{n} u^T S_T u}{\frac{1}{n} u^T S_W u})
\end{aligned}
$$

According to the definition of S_B in (6), we have

$$J_I(u) \geq \frac{1}{2} \log(\frac{u^T S_B u}{u^T S_W u}) = \frac{1}{2} \log(J_{LDA})$$

From theorem 2, we can learn that if the data is Gaussian distributed and $\lambda_W = 1$, LDA provides a lower bound of the MaxEnt problem. This theorem also gives a theoretical explanation that LDA subspace is often selected as an initial guess of optimal subspace in gradient ascend method based ITL algorithms.

Theorem 3. *The maximum entropy in (26) is bounded and nonincreasing when $0 \leq \lambda_W \leq 1$, i.e.,*

$$0 \leq J_I(U_F^T) \leq J_I(U) \tag{28}$$

where $U \in R^{d \times d}$ is an orthonormal matrix, $U_F^T : R^d \to R^m$ and $m < d$.

Proof. Since $H(X) \geq H(X|C) \geq \lambda_W H(X|C)$ [20], we have

$$H(U_F^T X) - \lambda_W H(U_F^T X | C) \geq 0 \tag{29}$$

Let $U_B \in R^{d \times (d-m)}$ be a matrix which is the complement subspace of U_F, define the matrix U as

$$U^T X = [U_F^T X \quad U_B^T X], \quad U = [U_F \quad U_B] \tag{30}$$

Since U is orthonormal matrix and H(X) is differential entropy, it follows that $H(X) = H(U^T X)$ [20], we have

$$
\begin{aligned}
J_I(U) &= H(X) - \lambda_W H(X|C) \\
&= H(U^T X) - \lambda_W H(U^T X|C) \\
&= H(U_F^T X\, U_{\bar{F}}^T X) - \lambda_W H(U_F^T X\, U_{\bar{F}}^T X|C) \\
&= H(U_F X) + H(U_{\bar{F}} X|U_F X) - \lambda_W H(U_F X|C) - \lambda_W H(U_{\bar{F}} X|U_F X, C) \\
&= H(U_F X) - \lambda_W H(U_F X|C) + (H(U_{\bar{F}} X|U_F X) - \lambda_W H(U_{\bar{F}} X|U_F X, C)) \\
&\geq H(U_F X) - \lambda_W H(U_F X|C)
\end{aligned}
$$

Theorem 3 states that the maximum entropy of any orthonormal subspace of the original feature space is bounded and nonincreasing. Furthermore, because the entropy $H(X)$ is a concave function of $f_X(x)$, there is at least a maxima solution of (26).

It is clear that MaxEnt-RDA is a nature extension of LDA from Gaussian distribution assumption to any distribution assumption. Without the prior information on data distributions, the between-class variance of MaxEnt-RDA can better characterize the separability of different classes than the between-class variance as in LDA. Compared with $L_t(U)$ and S_T, we can learn that although the distribution assumptions of MaxEnt-RDA and LDA are different, their solutions have similar matrix format.

According to (26), when λ_W is set to 1, the objective of MaxEnt-RDA becomes the mutual information. Different from previous ITL algorithms, we propose an approximate algorithm to solve the MaxEnt problem. This eigen-decomposition method not only can save computation cost but also has a unique and global solution.

Recently, linear regression based discriminant methods [23][24][25] are proposed in subspace learning for computational convenience. They use multinomial class indicator matrix [26][23] as the regression targets. Theoretical results show that the linear regression based method is equal to LDA in some conditions [24][25]. Since the mean square criterion in linear regression is sensitive to outliers and non-Gaussian noise [21], the linear regression based discriminant methods are often sensitive to outliers [16]. Compared regression based methods, MaxEnt-RDA is robust to outliers due to its MaxEnt objective.

4   Experiments

In this section, we applied the proposed MaxEnt-RDA algorithm to several real-world pattern recognition problems and compared its performance with PCA, LDA and MFA.

4.1   A Toy Problem

In the toy problem, a two-class problem is discussed. The data for each class is non-Gaussian distribution. As shown in Fig. 1, class one has 100 2D points from a bimodal Gaussian distribution (blue triangles), with centers at (-3.5, 3.5) and (3.5, 1.0); class two has also 100 2D points (red circles) which are drawn from a bimodal Gaussian distribution with centers at (3.5, -3.5) and (0.5, 1.0). The scale factor s in (25) of MaxEnt-RDA was set to 8.

The solid line in Fig. 1 represents the derived classification hyperline for MaxEnt-RDA; and the dashed line is the optimal classification hyperline for LDA. It is obvious

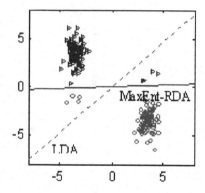

Fig. 1. A toy problem: comparative optimal classification hyperlines for a two-class bimodal Gaussian distribution

Table 1. UCI datasets used in the experiments

data set	Dimension	Classes	Instances
Derm	34	6	366
Glass	9	6	214
Isolet	617	26	1559
Mfeat	216	10	2000

that optimal hyperline of MaxEnt-RDA can perfectly separate the two-class data. But LDA fails to find an optimal direction in the non-Gaussian case. Given a suitable scale factor s in the bandwidth σ, MaxEnt-RDA will learn a correct hyperline. It provides an efficient means for discovering non-Gaussian structures in the data.

4.2   Classification on UCI Dataset

We applied MaxEnt-RDA to four UCI data sets in UCI machine learning repositories [27]. Tab. 1 shows a brief summary of the data sets used in this experiment, which have been used in many feature extraction studies [13] [14][15]. For each data set, we performed 10-fold cross validation (CV) 10 times and computed the average correct classification rate. Each dimension of raw data was normalized using the means and the standard variances. The Nearest-Neighbor [19] algorithm based on Euclidean distance, which is commonly applied as the classifier in feature extraction, is utilized as the classifier. The facial images in the second row of Fig. 3 illustrate examples of the noisy image.

Fig. 2 shows the average correct classification rates of each data set with various numbers of extracted features. For "Glass" dataset, the number of extracted features m is varied from 2 to $d - 1$. For data sets with a high input dimension such as the "Derm", "Ionosphere" and "Isolet" data sets, the number of extracted features in the Figure 2 was truncated at 30 for a clear view. The final reduced dimension after LDA is $c - 1$.

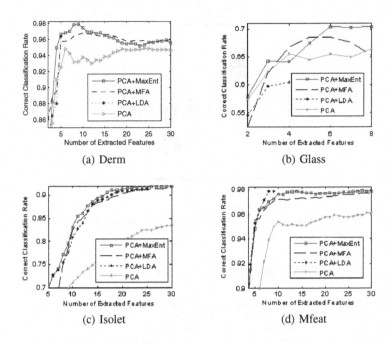

Fig. 2. Classification accuracy for UCI data sets

The results in Fig. 2 show that MaxEnt-RDA achieves the highest correct classification rate on four datasets. In Fig.2 (a) and (b), the highest correct classification rate of MaxEnt-RDA is clearly higher than that of other methods. In Fig. 2(c), MaxEnt-RDA, LDA and MFA are all perform significantly better than PCA. Although the results of MaxEnt-RDA, LDA and MFA are very close, MaxEnt-RDA still slightly outperforms LDA and MFA. In Fig. 2(d), MaxEnt-RDA outperforms MFA; and the highest correct classification rate of MaxEnt-RDA is slightly larger that of LDA. The results on UCI datasets further demonstrate that MaxEnt-RDA provides an efficient method for discriminant feature extraction.

4.3   Robustness to Noise

In real face recognition system, it is necessary to label large amount of training facial images to learn a robust classifier. But during collection of facial images, two types of noise may be occur. The one is mislabeling noise and the other is image noise. The mislabeling noise occurs when we mislabel one person's facial image to another; and the image noise occurs when the person's face is close to border of the camera so that we can only crop part of the face.

In this experiment, we use CMU PIE face database [28] to validate different methods under above two noise. The facial images are collected from a subset of CMU PIE face database that contains more than 40,000 facial images of 68 subjects. These still images are acquired across different poses, illuminations and facial expressions. We choose the

Fig. 3. Cropped facial images and their corresponding noisy images on CMU PIE database

five near frontal poses (C05, C07, C09, C27, C29) and use all the images under different illuminations and expressions, thus we get 170 images for each individual. All facial images are manually aligned and cropped. The size of each cropped image is 32×32. For each individual (20, 30 ,40) images are randomly selected for training and the rest are used for testing. And 5%, 10%,20% facial images per person is randomly selected as noisy samples. To eliminate statistical deviations, all experiments are averaged over 50 random splits and the mean as well as the standard deviation are reported. The Nearest-Center [19] algorithm based on Euclidean distance is used as the classifier.

Table 2. Comparison of different algorithms on mislabeling noise (average correct classification rate $\pm$ standard deviation). Best of all results are highlighted in bold.

	20 train			
	0%	5%	10%	20%
PCA + LDA	87.9 ± 0.71	87.1± 0.75	86.2 ± 0.82	84.4 ± 0.87
PCA + MFA	90.9 ± 0.58	88.8± 0.75	86.9 ± 0.90	83.4 ± 0.94
PCA + MaxEnt-RDA	**91.9** ± 0.56	**90.6**± 0.64	**89.3** ± 0.79	**86.6** ± 0.78
	30 train			
PCA + LDA	90.7 ± 0.37	90.2 ± 0.47	89.6 ± 0.50	88.4 ± 0.50
PCA + MFA	92.2 ± 0.53	91.0 ± 0.63	89.4 ± 0.57	86.8 ± 0.73
PCA + MaxEnt-RDA	**94.0** ± 0.31	**93.2** ± 0.40	**92.4** ± 0.40	**90.7** ± 0.43
	40 train			
PCA + LDA	92.1 ± 0.44	91.8 ± 0.44	91.5 ± 0.43	90.5 ± 0.55
PCA + MFA	93.0 ± 0.46	92.0 ± 0.53	90.7 ± 0.61	88.4 ± 0.71
PCA + MaxEnt-RDA	**94.9** ± 0.33	**94.5** ± 0.36	**94.0** ± 0.37	**92.8** ± 0.45

Tab. 2 shows the results of different methods under mislabeling noise. MaxEnt-RDA achieves the highest average correct classification rate on all testing datasets. Furthermore, its standard deviation is also smaller than two other methods. This means that MaxEnt-RDA can learn a stable result under mislabeling noise. Since the mislabeling noise changes the relationship of neighborhood near the boundary of different classes, MFA's performance decreases rapidly when more mislabeling samples are added into the training set. When 20% mislabeling samples are added, average correct classification rate of MFA is even lower than that of LDA. Since MaxEnt-RDA tries to estimate a MaxEnt distribution measured by Parzen window, it can efficiently deal with the mislabeling noise so that it performs better than other methods.

Tab. 3 shows the results of different methods under noisy images. Similar to mislabeling noise, when noise occurs, all methods' correct classification rates decrease. But it seems that all methods are more robust to image noise than mislabeling noise. MaxEnt-RDA still achieves the highest average correct classification rate, and its standard deviation is smaller than two other methods. Experimental results illustrate that MaxEnt-RDA is robust against training noise for both features (image) and labels.

Table 3. Comparison of different algorithms on image noise(average correct classification rate ± standard deviation). Best of all results are highlighted in bold.

	20 train			
	0%	5%	10%	20%
PCA + LDA	87.9 ± 0.71	87.5 ± 0.69	87.4 ± 0.65	87.0 ± 0.73
PCA + MFA	90.9 ± 0.58	89.5 ± 0.67	89.2 ± 0.68	88.7 ± 0.69
PCA + MaxEnt-RDA	**91.9 ± 0.56**	**91.2 ± 0.55**	**91.0 ± 0.57**	**90.5 ± 0.58**
	30 train			
PCA + LDA	90.7 ± 0.37	90.4 ± 0.39	90.3 ± 0.39	90.0 ± 0.40
PCA + MFA	92.2 ± 0.53	91.5 ± 0.52	91.2 ± 0.54	90.9 ± 0.51
PCA + MaxEnt-RDA	**94.0 ± 0.31**	**93.5 ± 0.33**	**93.3 ± 0.33**	**93.0 ± 0.31**
	40 train			
PCA + LDA	92.1 ± 0.44	91.9 ± 0.47	91.8 ± 0.44	91.6 ± 0.47
PCA + MFA	93.0 ± 0.46	92.4 ± 0.45	92.2 ± 0.50	91.9 ± 0.46
PCA + MaxEnt-RDA	**94.9 ± 0.33**	**94.6 ± 0.39**	**94.5 ± 0.37**	**94.2 ± 0.38**

5   Conclusion

A MaxEnt-RDA method is proposed for discriminant feature extraction based on the MaxEnt criterion. It utilizes Renyi's quadratic entropy as objective and hence has a clearly theoretical foundation. An eigen-decomposition algorithm is proposed to approximately solve the MaxEnt problem based on Taylor expansion. The proposed method bases on the Parzen window estimate of MaxEnt distribution and hence can effectively overcome the limitations of the traditional LDA algorithm in data distribution assumption. And it is robust against image noise and mislabeling noise. Experiments on UCI datasets and face databases have demonstrated the superiority of MaxEnt-RDA compared with traditional discriminant methods.

Acknowledgement

This work was supported in part by NSF of China (No. 60275025) and MOST of China (No. 2007DFC10740).

References

1. Guyon, I., Elissee, A.: An introduction to variable and feature selection. Journal of Machine Learning Research 3, 1157–1182 (2003)
2. Jolliffe, I.T.: Principal component analysis. Springer, New York (1986)
3. He, X., Yan, S., Hu, Y., Niyogi, P., Zhang, H.: Face recognition using laplacianfaces. IEEE Transactions on Pattern Analysis and Machine Intelligence 27(3), 328–340 (2005)
4. Zhao, D., Lin, Z., Tang, X.: Laplacian pca and its applications. In: International Conference on Computer Vision (2007)
5. Zhu, M., Martinez, A.M.: Subclass discriminant analysis. IEEE Transactions on Pattern Analysis and Machine Intelligence 28(8), 1274–1286 (2006)
6. Hamsici, O.C., Martinez, A.M.: Bayes optimality in linear discriminant analysis. IEEE Transactions on Pattern Analysis and Machine Intelligence 30(4), 647–657 (2008)
7. Tao, D.C., Li, X.L., Wu, X.D., Maybank, S.J.: Geometric mean for subspace selection. IEEE Transactions on Pattern Analysis and Machine Intelligence 31(2), 260–274 (2009)
8. Belhumeur, P., Hespanha, J., Kriegman, D.: Eigenfaces vs. fisherfaces: recognition using class specific linear projection. IEEE Transactions on Pattern Analysis and Machine Intelligence 19(1), 711–720 (1997)
9. Fukunaga, K.: Statistical pattern recognition. Academic Press, London (1990)
10. Yan, S., Xu, D., Zhang, B., Zhang, H., Yang, Q., Lin, S.: Graph embedding and extensions: a general framework for dimensionality reduction. IEEE Transactions on Pattern Analysis and Machine Intelligence 29(1), 40–51 (2007)
11. Zhao, D., Lin, Z., Xiao, R., Tang, X.: Linear laplacian discrimination for feature extraction. In: IEEE conference on Computer Vision and Pattern Recognition (2007)
12. Principe, J., Xu, D., Iii, J.W.F.: Information-theoretic learning, http://www.cnel.ufl.edu/bib/./pdf_papers/chapter7.pdf
13. Torkkola, K.: Feature extraction by nonparametric mutual information maximization. Journal of Machine Learning Research 3, 1415–1438 (2003)
14. Nenadic, Z.: Information discriminant analysis: feature extraction with an information-theoretic objective. IEEE Transactions on Pattern Analysis and Machine Intelligence 29(8), 1394–1407 (2007)
15. Hild II, K.E., Erdogmus, D., Torkkola, K., Principe, C.: Feature extraction using information-theoretic learning. IEEE Transactions on Pattern Analysis and Machine Intelligence 28(9), 1385–1392 (2006)
16. Yuan, X., Hu, B.: Robust feature extraction via information theoretic learning. In: International Conference on Machine Learning (ICML 2009), Montreal, Canada (2009)
17. Rao, S., Liu, W., Principe, J., de Medeiros Martins, A.: Information theoretic mean shift algorithm. In: Machine Learning for Signal Processing (2006)
18. Caticha, A., Giffin, A.: Updating probabilities. In: The 26th International Workshop on Bayesian Inference and Maximum Entropy Methods, Paris, France (2006)
19. Duda, R.O., Hart, P.E., Stork, D.G.: Pattern classification. Wiley-Interscience, Hoboken (2000)
20. Cover, T., Thomas, J.: Elements of Information Theory, 2nd edn. John Wiley, New Jersey (2005)
21. Liu, W., Pokharel, P.P., Principe, J.C.: Correntropy: Properties and applications in non-gaussian signal processing. IEEE Transactions on Signal Processing 55(11), 5286–5298 (2007)
22. Huber, P.: Robust statistics. Wiley, Chichester (1981)
23. Baek, J., Son, Y.S.: Local linear logistic discriminant analysis with partial least square components. In: Li, X., Zaïane, O.R., Li, Z.-h. (eds.) ADMA 2006. LNCS (LNAI), vol. 4093, pp. 574–581. Springer, Heidelberg (2006)

24. Ye, J.: Least squares linear discriminant analysis. In: International Conference on Machine Learning, ICML (2007)
25. Cai, D., He, X., Han, J.: Spectral regression for efficient regularized subspace learning. In: International Conference on Computer Vision, pp. 1–7 (2007)
26. Hastie, T., Tibshirani, R., Friedman, J.: The elements of statistical learning: Data mining, inference, and prediction. Springer, Heidelberg (2001)
27. Newman, D., Hettich, S., Blake, C., Merz, C.: Uci repository of machine learning databases (1998), http://www.ics.uci.edu/mlearn/MLRepository.html
28. Sim, T., Baker, S., Bsat, M.: The cmu pose, illumination, and expression database. IEEE Transactions on Pattern Analysis and Machine Intelligence 25, 1615–1618 (2005)

Context-Aware Online Commercial Intention Detection

Derek Hao Hu[1,*], Dou Shen[2], Jian-Tao Sun[3], Qiang Yang[1], and Zheng Chen[3]

[1] Hong Kong University of Science and Technology
{derekhh,qyang}@cse.ust.hk
[2] Microsoft Research
doushen@microsoft.com
[3] Microsoft Research Asia
{jtsun,zhengc}@microsoft.com

Abstract. With more and more commercial activities moving onto the Internet, people tend to purchase what they need through Internet or conduct some online research before the actual transactions happen. For many Web users, their online commercial activities start from submitting a search query to search engines. Just like the common Web search queries, the queries with commercial intention are usually very short. Recognizing the queries with commercial intention against the common queries will help search engines provide proper search results and advertisements, help Web users obtain the right information they desire and help the advertisers benefit from the potential transactions. However, the intentions behind a query vary a lot for users with different background and interest. The intentions can even be different for the same user, when the query is issued in different contexts. In this paper, we present a new algorithm framework based on skip-chain conditional random field (SCCRF) for automatically classifying Web queries according to *context-based online commercial intention*. We analyze our algorithm performance both theoretically and empirically. Extensive experiments on several real search engine log datasets show that our algorithm can improve more than 10% on F1 score than previous algorithms on commercial intention detection.

1 Introduction

The rapid development of World Wide Web has impacted almost every aspect of our daily life and more and more activities happen on the Internet. Among these activities, one important kind is commercial activities, which form an ecosystem and attract a lot of players.In this ecosystem, the behaviors of Web users play a critical role. The behaviors include shopping online, or conducting online research for actual deals. As we are aware of, most Web users start their online behaviors by submitting a Web query to a search engine. Therefore, accurately understanding the intentions behind the issued queries is of great importance to

---

* This work was done when Derek Hao Hu was an intern at Microsoft Research Asia.

Z.-H. Zhou and T. Washio (Eds.): ACML 2009, LNAI 5828, pp. 135–149, 2009.

the mentioned ecosystem. In this paper, we focus on detecting the commercial intentions of Web queries, which is not thoroughly studied yet as the general query intentions studied in [2,11,9,6].

Detecting Online Commercial Intention (OCI) from Web queries is not trivial, considering the following three difficulties. The first difficulty is that many queries are very short. [5] studied an Excite search-service transaction log and showed that approximately 93% of the Web queries contained less than 4 terms. It is extremely hard to derive user intention solely based on the queries. The second difficulty is that a Web query often has multiple meanings and hence is ambiguous. For example, the word "jaguar" has dozens of meanings, which can either mean an animal or a kind of luxury cars, or others[1]. The third difficulty is that the intention of a Web query can vary given different contexts. For example, even if "jaguar" takes the meaning of being a kind of luxury car, it either encodes a commercial intention (when the user wants to buy a car), or non-commercial intention (where the user just wants to find some luxury car pictures).

[4] first defined the notion of OCI and provided a non-context-aware approach to detect OCI in Web queries. The authors formalize the problem as a binary classification problem to decide whether a search query is intended for commercial purposes such as intending to buy a product or finding product information as in the research stage. The proposed solution, based on query enrichment through search engines and traditional text classification techniques [4], solves the first difficulty as we discussed. However, their method cannot tackle the second and third difficulties. In this paper, we propose a new algorithm to analyze the commercial intention of Web queries, taking the contextual information of the submitted queries into consideration. With these information provided, our method can provide more accurate predictions for commercial intention of Web users.

Figure 1 shows the workflow of our algorithm. We consider a newly asked query and then consider two kinds of features, one is the generalized OCI intention degree which extracts features from top result pages when this query is issued to the search engine. The other is the historical similarity feature, which takes past queries into consideration. It first detects all the similar queries in the user's personal query log up to a specific length. And then it computes the semantic similarity kernel function by using query expansion techniques as external information sources. Next these two feature functions are used in a skip-chain conditional random field model (SCCRF). A skip-chain conditional random field model adds possible links between non-adjacent nodes, in contrast to the more commonly used linear chain conditional random field, where only adjacent nodes are connected. The reason for us to use SCCRF is to grab the connection between the query labels (commercial or non-commercial) of non-adjacent nodes since simply connecting edges between adjacent nodes would not accumulate enough information. Finally, the queries are classified as being commercial or non-commercial. Details are presented in Section 3.

---

[1] http://en.wikipedia.org/wiki/Jaguar_%28disambiguation%29

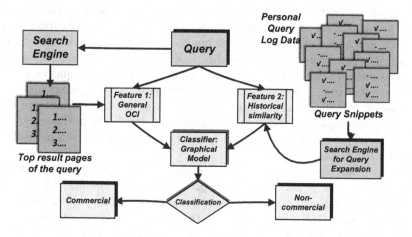

Fig. 1. Framework of our context-based OCI detection algorithm

Detecting OCI from contextual queries may appear to be similar to the context-aware query classification problem recently proposed in [3]. However, how to integrate contextual information in OCI detection is completely different from context-aware query classification. In context-aware query classification, one can learn much knowledge from direct relationship between adjacent queries as well as the QC taxonomy information. In OCI detection, we can only have limited knowledge from adjacent query relationships since we only have two categories. Furthermore, we do not have a taxonomy from whose structure we can mine useful information. Therefore, we proposed a different approach from [3] to define context-aware features and accumulate useful information from the surrounding queries.

The rest of the paper is organized as follows. We show some related works on general OCI detection and query classification methods. We then present our solution in Section 3 for detecting context-based OCI. In Section 4, we compare our approaches against the baseline method in [4] using real query log data from AOL and Live Search. Finally, in Section 5, we give the conclusion of this paper and describe some possible future research directions.

2 Related Work

A machine learning-based approach for predicting OCI based on Web pages or queries was proposed in [4]. When detecting OCI from Web pages, the traditional approach of document classification is used, where we are given a training set D consisting of training Web pages $\langle d_j, C_i \rangle$, $i = 1$ or 2 and $1 \leq j \leq |D|$, where C_1 means the Web page has commercial intention and C_2 means the Web page has non-commercial intention. The Web pages are represented by the Vector Space Model. The keywords are extracted from both the content texts and tag attributes of all the labeled Web pages in the training data.

The problem of Web query classification is closely related to the OCI detection problem, although current research works on query classification do not tackle this problem. Currently, works on query classification can be split into two groups, one is classifying queries according to query types, such as informational or navigational or transactional[2,7]; and the other is classifying queries according to the query topic, such as "computers/hardware" or "computers/software" [1]. However, as mentioned in [4], OCI follows an independent dimension compared to query topic classification or query type classification. Therefore, the methodologies for these two types of query classification can not be used directly for OCI detection.

One paper that is particularly related to this work is context-aware query classification [3]. In that paper, the authors present a query classification approach also based on conditional random fields and aiming at mining useful information from user sessions. Our work differs from that paper in several aspects.

In [3], the authors defined several kinds of "context-aware" features based on direct association between adjacent labels, and also taxonomy-based association between adjacent labels. However, when we aim to detect online commercial intention, such features cannot be simply borrowed to work in this scenario.

Firstly, in [3], they use a linear-chain conditional random field model and simply calculate the number of occurrences of a pair of adjacent query labels as the direct association between adjacent labels. It's easy to see that in the problem of OCI detection, we can only form four pairs of adjacent query labels and therefore the knowledge we can gain from using such information is very limited. To solve this problem, we propose to apply the skip-chain conditional random field to better capture the information hidden between nonadjacent, but related queries and we will also demonstrate its usefulness in the experimental results section.

Secondly, another part of contextual features is the taxonomy-based association, where the structure of the taxonomy is exploited to mine useful information between "sibling" categories. In particular, some transitions between categories may not occur in the training data and therefore simply using the number of observed transitions between adjacent query labels may not reflect the distribution in real-world applications. Therefore, in [3], the association between two sibling categories are considered to be stronger than non-sibling categories. Nevertheless, in the problem of OCI detection, such a taxonomy does not exist since we only have two categories instead of the multiple categories available in general QC. We also cannot aim to use taxonomy based association to provide contextual features.

Therefore, both the two parts of contextual features proposed in [3] cannot be applied to the OCI detection problem, which makes our problem and approach highly different from [3]. Furthermore, in our experiments section, we do not use any information about the user clicked URLs since that information is missing from our query log data. Such information is available and is defined as a major part of "context" in [3], which further shows the differences between our work and [3].

3   Context-Based OCI Detection

3.1   Overview

In this section, we describe our algorithm for context-based online commercial intention detection from personal query logs.

We first make the basic assumption on context-based OCI detection: a search engine has access to the query log of a specific user, or at least the query log from this same IP address. Here we assume that a query log (or clickthrough log in some literature) consists of a set of queries associated with the user-clicked search result pages or snippets. Stated formally, the query log is a set Q where each element is at least a triple $\langle U, T, Q, [C] \rangle$, where U indicates the user ID, or any other information (e.g. IP address) that can differentiate one user from another, T indicates the time where this query is issued and Q indicates the string of the Web query. Other elements may also be added to this query log so that more information will be encoded, such as in the case of a clickthrough log where we have C, which indicates what pages or URLs the Web user actually clicked on or the snippets of the clicked pages. In the following, we refer to the log data consistently as query logs, and do not consider the existence of C since including these contents would be straightforward.

The assumption on user identification can often be satisfied in real world, where search engine companies record the query logs of different registered users or their IP addresses for differentiating them from each other. When the personal query log data are available, for each incoming query, we can use the information from the personal query log to find similar queries that has strong correlation with the new query by the same user or user group.

3.2   Modeling Query Logs via CRF

Since we want to take a sequence of queries instead of a single query as our input when we train the classifier, we find that the problem fits well with conditional random field as our graphical model for context-based OCI detection. Conditional Random Field, which was first proposed by Lafferty et al.[8], is widely used in relational learning which directly models the conditional distribution $p(\mathbf{y}|\mathbf{x})$. In this paper, we use a variant of the widely used linear-chain CRF model, the skip-chain CRF proposed in [12], to model the context-based OCI issue for the following reasons. Firstly, skip-chain CRF has deep roots in natural language processing area (NLP). In NLP, the problem of Named Entity Recognition (NER) has similarities with the context-based OCI detection problem, which needs to model the correlation between non-consecutive identical words in the text. Secondly, being a probabilistic graphical model, skip-chain CRF has its advantage in modeling uncertainty in a natural and convenient way. Thirdly, the key issue in skip-chain CRF is how to add skip edges. Semantic similarities, which represents how similar two queries are in terms of their intended meaning such as category of targeting pages or products, are the major issues we consider when creating skip edges in the CRF model. Based on the above reasons, we believe that skip-chain CRF would be a model appropriate for handling context-based OCI detection.

The main advantage of a skip-chain CRF (SCCRF) model over the commonly used linear-chain CRF models is that the skip-chain CRF model has an additional type of potential, which is represented using long-distance edges. Formally, we build a SCCRF model as follows. Assume that the original personal query log length is N. In order to acquire training data consisting of personal query logs with each length as L, we can build a training set with cardinality $(N - L + 1)$, where each data instance consists of L consecutive queries in the original personal query log, i.e. each personal query log starts with the query item $1, 2, \ldots, N - L + 1$ and has a length of L. In our experiments of Section 4, we will empirically evaluate the classifier performance with different values of L.

Graphical models like CRF or skip-chain CRF directly represents the conditional distribution $p(\mathbf{y}|\mathbf{x})$, where in our context-based OCI detection problem, $\mathbf{y}$ indicates the target label of each query as being commercial or non-commercial and $\mathbf{x}$ states an observed personal query log of length L. We let x_t be the observed t^{th} query in the personal query log, and let y_t be a random variable to indicate the OCI value inferred from the tth query, in the final setting if the y_t inferred is larger than 0.5, we assume that the given query has commercial intention. This parameter setting of threshold follows that of [4].

The conditional random field model is represented by a factor graph, which is a bipartite graph $G = (V, F, E)$ in which a variable $v_s \in V$ is connected to a factor node $\Psi_A \in F$ if v_s is an argument to Ψ_A. For skip-chain CRF, it is essentially a linear-chain CRF with additional long-distance edges between queries x_i and x_j such that $f(x_i, x_j) > \theta$ (Refer to Figure 2 for an illustration). θ is a parameter that can be tuned to adjust the confidence of such correlations between different queries. In our experiments, we will evaluate the effect of changing the parameter θ on the classifier accuracy.

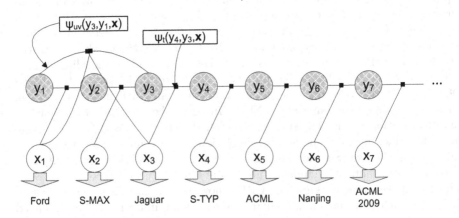

Fig. 2. Illustration of the SCCRF model

For an observation sequence $\mathbf{x}$, let $\mathcal{I} = \{u, v\}$ be the set of all pairs of queries for which there are skip edges (all edges except edges connecting adjacent queries) connected with each other. The probability of a label sequence $\mathbf{y}$ given an observation activity sequence $\mathbf{x}$ is:

$$p(\mathbf{y}|\mathbf{x}) = \frac{1}{Z(x)} \prod_{t=1}^{n} \Psi_t(y_t, y_{t-1}, \mathbf{x}) \prod_{(u,v) \in I} \Psi_{uv}(y_u, y_v, \mathbf{x}). \tag{1}$$

In the above Equation 1, Ψ_t are the potential functions for linear-chain edges and Ψ_{uv} are the factors over the skip edges (Also refer to Figure 2 for illustration). $Z(x)$ is the normalization factor. We define the potential functions Ψ_t and Ψ_{uv} in Equation 2 and Equation 3 as:

$$\Psi_t(y_t, y_{t-1}, \mathbf{x}) = \exp\left(\sum_k \lambda_{1k} f_{1k}(y_t, y_{t-1}, \mathbf{x}, t)\right) \tag{2}$$

$$\Psi_{uv}(y_u, y_v, \mathbf{x}) = \exp\left(\sum_k \lambda_{2k} f_{2k}(y_u, y_v, \mathbf{x}, u, v)\right) \tag{3}$$

λ_{1k} are the parameters of the linear-chain template and λ_{2k} are the parameters of the skip-chain template. Each of the potential functions factorize according to a set of features f_{1k} or f_{2k}. We will describe our choice of feature functions later.

Learning the weights λ_{1k} and λ_{2k} for the skip-chain CRF model can be achieved by maximizing the log-likelihood of the training data, which requires the computation of calculating partial derivative and optimization techniques. We omit the detailed algorithm of inference and parameter estimation of the skip-chain CRF model. Interested readers can consult [8,12] for technical details.

3.3   Modeling Semantic Similarities between Queries

We now specify how to calculate the feature functions in our CRF settings. Two kinds of feature functions must be computed. One is the "query-intention" pair, the other is the "query-query" pair. For the first function, it is easy to follow the traditional IR techniques, where we can first choose some keywords from the query snippets or the top landing pages, count the occurrence of these words and then learn the corresponding weight. In this paper, we used the OCI value of top 10 result pages calculated from the baseline method[2] [4] as the features representing the "query-intention" pair.

The remaining issue is how to define a good "query-query" similarity function that can measure the semantic similarity of different search queries. An accurate function that can reflect the inherent semantic similarity or correlation between

---

[2] http://adlab.msn.com/OCI/OCI.aspx

different queries will substantially improve the overall accuracy. Due to the inherent feature of the length of queries, the traditional method of word-based similarity metric cannot be used. Therefore, we use an approach, based on the idea of query expansion, to measure the similarity of short text snippets the considered query pairs.

Our first expansion, called the "first-order" query expansion, is to compute the cosine similarity of query snippets. Given two queries Q_1 and Q_2, we first issue these two queries into the search engine and then retrieve the result pages of the corresponding two queries and the m query snippets. Then we combine these m query snippets as one document and consider the cosine similarity between them. In other words, we first compute the TFIDF vector of the two documents, denote them as A and B, and then compute: $\theta = \arccos \frac{A \cdot B}{\|A\|\|B\|}$.

However, although cosine similarity is a traditional similarity metric, it may not satisfy our need to measure the semantic similarity between different queries since the query snippets are rather short and the document pairs will not contain common terms. Also even if we consider the cases of the two snippets containing the same terms, it may not mean that these same terms mean the same thing in different contexts of different query snippet documents. So measuring similarity based on word terms is not a good choice for our problem. In our experiment section 4, we will show that our "first-order" query expansion does not perform very well, compared to the "second-order" query expansion we used.

Our second expansion goes a step further, compared to the "first-order" query expansion, and we call the idea of "second-order" query expansion based on the pages we get when we issue the query snippets again into the search engine. In this way, the information we get from this particular query snippet is increased. This step is similar to the solution given in [10]. We consider the maximum similarity between these query snippets and take the value as the value of feature function between "query-query" pairs in the SCCRF model.

Stated formally, we have $f(y_u, y_v, x) = \max_{1 \leq i \leq m, 1 \leq j \leq m} g(S_{ui}, S_{vj}))$, $g(S_{ui}, S_{vj})$ is defined as the value of similarities between query snippet S_{ui} and S_{vj}, where S_{ui} is the ith query snippet when we issue the uth query into the search engine, S_{vj} is defined similarly.

We use a kernel function to compute the semantic similarities of given query pairs based on the query expansion framework. Let S_x represent a query snippet. First, we get more expanded information of S_x by using the idea of query expansion. We input this query snippet S_x into the search engine and the top n returned Web pages are retrieved, say, $p_1, p_2, \ldots, p_n$. Then we compute the TFIDF term vector v_i for each Web page p_i. For each v_i, it is truncated to only include its m highest weighted terms, where $m = 50$, is used as a balance between evaluation efficiency and expressive power.

Then we let $C(x)$ be the centroid of L_2 normalized vectors: $C(x) = \frac{1}{n} \sum_{i=1}^{n} \frac{v_i}{\|v_i\|_2}$. Finally, we compute $QE(x)$, $QE(x)$ is the L_2 normalization of $C(x)$: $QE(x) = \frac{C(x)}{\|C(x)\|_2}$. The kernel function of query snippets $K(x, y) = QE(x) \cdot QE(y)$. It can be observed that $K(x, y)$ is a valid kernel function.

Therefore, after defining the corresponding feature functions between "query-intention" pairs and "query-query" pairs, we have enough information to build the SCCRF model from the training data. When testing, a new query arrives, and we take the past $L-1$ query into consideration, which forms together a personal query log with L subsequent queries, and then label the query sequence $\mathbf{y}$ and take the last element from the vector y_L as the label: commercial / non-commercial of this query. We further describe our algorithm workflow in Algorithm 1.

Algorithm 1. Algorithm description for context-based online commercial intention

Input: N is the length of a query log, where each query item is represented by $\{x_i, y_i\}$, where x_i is the ith query and y_i is the corresponding i^{th} label for x_i. Q, which is a newly asked query.

Output: P, which is the probability for Q as being commercial intended.

Assumption: Assume all the queries in the personal query log we considered here are issued by the same user or user group.

Parameters: θ and L, which suggests the confidence parameter for us to add the skip edges and the length of the personal query log training data, correspondingly.

1: **for** $i = 1$ to $N - L + 1$ **do**
2:      Initialize the ith training data as empty.
3:      **for** $j = 0$ to $L - 1$ **do**
4:           Add the $(i+j)^{th}$ query x_{i+j} to the ith training data.
5:      **end for**
6: **end for**
7: **for** $i = 1$ to N **do**
8:      Issue the query x_i to the search engine to get the top P landing pages. P can be tuned to reflect more information from landing pages. To simplify, we set $P = 10$ in our experiments.
9:      Compute the corresponding OCI value of these landing pages from the baseline method [4].
10:      Use these values as features for f_1.
11: **end for**
12: Train the corresponding SCCRF model from the training set created.
13: **for** $i = N - L + 2$ to N **do**
14:      Add the query x_i to the test personal query log.
15: **end for**
16: Add the query Q to the test personal query log. Now it contains L terms.
17: **for** $i = 1$ to L **do**
18:      **for** $j = 1$ to $i - 1$ **do**
19:           Compute the semantic similarity of T_i and T_j, i.e. $K(T_i, T_j) = QE(T_i) \cdot QE(T_j)$ as defined.
20:           **if** $K(T_i, T_j) > \theta$ **then**
21:                Add a skip edge between y_i and y_j, corresponding to the feature function $f_2(y_i, y_j, \mathbf{x})$.
22:           **end if**
23:      **end for**
24: **end for**

---

4   Empirical Evaluation

In order to validate the effectiveness of our algorithm, we compare our algorithm to the baseline method proposed in [4]. Several parameters occur in our algorithm, θ, which appears in the skip-chain Conditional Random Field model to mark the confidence of the nonadjacent edges we created. The larger the θ is, we are more confident of the semantic similarity of the edges. Furthermore, another parameter is the parameter L, which is the length of each personal query log in the training data set, also it means when two queries have a distance more than L, we will not consider their semantic similarities because otherwise if the two queries have distances larger than L, they may not be in the same searching session and have no temporal correlation between each other, even though they may have high semantic similarities. In our experiment, we will empirically evaluate how our classifier performance will be affected when we tune these two parameters.

4.1   Description of Datasets

We use two datasets in our experiment, the first is a publicly released AOL query log dataset[3]. The AOL query log data consists of around 20M Web queries collected from around 650,000 Web users, where the data is sorted by anonymous User ID and sequentially arranged. Another query log data we acquired is from Live Search and the original query log data is collected in March 2008. The Live Search query log data consists of around 450M distinct web queries from around 2.5M different query search sessions.

The dataset format of the two querylog datasets are rather similar. Each item of the query log datasets includes {AnonID, Query, QueryTime, ItemRank, ClickURL}. AnonID is an anonymous user ID number. Query is the query issued by the user, case shifted with most punctuation removed. QueryTime is the time at which the query was submitted for search. ItemRank was the rank of the item on which they clicked, if the user clicked on a search result. ClickURL is the domain portion of the URL in the clicked result, if the user had clicked on a search result.

In this paper we do not use the clicked URL information (ClickURL), since this information is often relatively sparse in the query log. Another reason for us not to consider clickthrough information in computing the semantic similarities of the SCCRF model is the infeasibility of getting the clicked URL in most cases. For example, in the AOL query log dataset we only have access to the clicked domains, not the clicked webpages. Therefore we settled on expanding the query information by a "second-order" expansion instead.

Because both the AOL dataset and the Live Search dataset are rather huge and do not contain any label of commercial intention, we had randomly chosen 100 users who had submitted at least 100 queries from both AOL and Live Search datasets and had manually labeled 100 consecutive queries for each of the 100

---

[3] http://www.gregsadetsky.com/aol-data/

users we had selected. Three labelers labeled the 20K queries we had chosen. Each labeler is told to take the surrounding queries as well as clicked URLs into account to determine the commercial intention degree of the labeled query. Each query is labeled as being "commercial", "non-commercial" or "unable to determine". A query is labeled as "commercial" if and only if at least two out of three labelers mark it as "commercial" and it's similar for "non-commercial" queries. We also delete some invalid queries in the query log. The distribution of commercial intention in the queries of the Web users we've chosen above is shown is the following Table 1. In all we had acquired 9,553 queries in AOL dataset, where 1,247 queries are labeled as commercial and the rest labeled as non-commercial; also we labeled 9674 queries in Live Search dataset, where 936 queries are labeled as commercial and 8,738 queries as non-commercial.

Table 1. OCI distribution of the selected datasets

Labeler	AOL Commercial	AOL Non-commercial	Live Commercial	Live Non-Commercial
1	1238	8627	919	8819
2	1430	8435	1025	8713
3	1117	8748	973	8765
Sum	1247	8306	936	8738

In the query log data, we first performed preprocessing to remove all invalid queries. We then divide each user's query into ten pairs of training and test data, by first choosing a random number between a certain size interval as the size of the training data while keeping the rest of the data for the user as the test data. We repeat this process ten times so as to obtain average results. For all the experiments in this section, we use precision, recall and F1-measure as the evaluation metric.

4.2 Performance of Baseline Classifier

For the baseline method, we use the classifier which is now currently available on the Web[4], following the work in [4], The parameter chosen in the Website is the best-tuned so we just compare the performance of our algorithm with this result. The classification result is in the following Table 2.

Table 2. Baseline Classifier Performance

Dataset	Precision	Recall	F1-Measure
AOL	0.817	0.796	0.806
Live Search	0.802	0.836	0.809

---

[4] http://adlab.msn.com/OCI/OCI.aspx

4.3 Varying the Confidence Parameter θ

We then analyze how different parameters of θ and l will affect our algorithm performance. When we vary the confidence paramter θ, another important objective is to verify the usefulness of skip edges created between non-adjacent queries and that skip-chain CRF beats the linear chain CRF in the OCI detection scenario.

We first set $l = 1000$ and tune different parameters on θ, we get the following result in Table 3. We run the experiments 10 times with different values of p, accuracy and variance are recorded in the following table. Column with header "AOL (Variance)" is the accuracy and variance of our algorithm performance on our selected AOL query log dataset and "Live Search (Variance)" is the accuracy and variance of our algorithm performance on Live Search dataset. Here we first set L, which is the length of the CRF model, as 50.

Table 3. Algorithm performance with varying parameter θ

θ	AOL (Variance)	Live Search (Variance)
$\theta = 0.01$	0.863 (0.002)	0.872 (0.003)
$\theta = 0.02$	0.887 (0.005)	0.878 (0.003)
$\theta = 0.04$	0.892 (0.003)	0.881 (0.004)
$\theta = 0.08$	0.901 (0.005)	0.893 (0.002)
$\theta = 0.1$	**0.913 (0.002)**	0.901 (0.004)
$\theta = 0.2$	0.912 (0.005)	**0.908 (0.003)**
$\theta = 0.4$	0.902 (0.004)	0.883 (0.006)
$\theta = 0.8$	0.871 (0.003)	0.852 (0.008)
Baseline	0.806	0.809

From Table 3, it is noteworthy to see that our proposed algorithm for context-based online commercial intention *always* performs better than the baseline approach, suggesting that taking contextual information, especially surrounding queries in a query session would possibly help. We can see that for different user queries, different values of θ may lead to best generalization ability. Also it's reasonable that the generalization ability will be best when θ is neither too big nor too small.

When θ is too small, different queries that may not be so relevant will be linked towards each other and hence noise is added to the edges between "query-query" pairs. When θ is large enough, classification accuracy will drop rapidly. This is due to the fact that large values of θ will be a too strict criteria between different queries, in that merely no skip link will be created. Therefore, large values of θ would create too few skip edges and then we cannot model the interleaving processes of user search behaviors through a CRF model which is very similar to a linear chain. This verifies our claim in Section 2 that using a linear-chain conditional random field and simply model the relationship between adjacent queries cannot effectively model the commercial intention of queries as a skip-chain CRF will do.

4.4   Varying Training Data Length L

In the next experiment, we test whether different parameters of L will lead to large variance in classification accuracy. From the result in the earlier experiment, here we empirically set θ as 0.1. We get the following result in Table 4.

Table 4. Algorithm performance with varying parameter L

L	AOL (Variance)	Live Search (Variance)
$L = 5$	0.872 (0.010)	0.871 (0.013)
$L = 10$	0.893 (0.011)	0.878 (0.010)
$L = 15$	0.882 (0.009)	0.891 (0.005)
$L = 20$	0.901 (0.005)	0.891 (0.003)
$L = 25$	0.910 (0.004)	0.897 (0.007)
$L = 30$	**0.913 (0.002)**	0.901 (0.004)
$L = 40$	0.909 (0.003)	**0.903 (0.005)**
$L = 50$	0.905 (0.003)	0.902 (0.003)
Baseline	0.806	0.809

Again, our experimental results show that the algorithm performance is rather steady although it typically suggests that it's better to set the history length L at around 30, which would achieve the best performance so far. And even if we only set L as 5, i.e. consider the history submitted queries up to 5, it would still perform better than the baseline approach.

4.5   Analysis of Training Time

In this subsection we will show the training time of our proposed approach with varying lengths of L. Since we have calculated the accuracy as well as variance from a "ten-fold wise" appoach, we would show the total time for us to train the model. Thus, the average time to train a CRF model of length L would be the time shown divided by 10. The result is shown in the following Table 5. Time is measured in seconds.

The result is rather promising. Even when we consider history length up to 50, the computational time is rather quick, only around 15 seconds to train a CRF model for ten times. Such a model is undoubtedly a good fit for real-world usage.

4.6   Comparison between Query Expansion Methods

Finally, we verify our claim that our way of "second-order" query expansion will perform better than "first-order" query expansion. We choose the settings $\theta = 0.1$ and $L = 30$ in "second-order" query expansion and test them against the "first-order" query expansion. The result is shown in Table 6.

Table 5. Training time with varying lengths of L on a Pentium Core 2 Dual 2.13GHz CPU

L	AOL Time	Live Search Time
$L = 5$	1.7s	1.7s
$L = 10$	3.0s	4.1s
$L = 15$	4.9s	5.2s
$L = 20$	6.2s	6.8s
$L = 25$	9.0s	10.2s
$L = 30$	11.1s	11.7s
$L = 40$	14.0s	14.1s
$L = 50$	15.3s	16.3s

Table 6. Comparison of first-order query expansion vs. second-order query expansion

Dataset	Baseline	First-Order	Second-Order
AOL	0.806	0.825 (0.007)	0.913(0.002)
Live Search	0.809	0.826 (0.006)	0.901(0.004)

The above table shows our use of "second-order" query expansion can substantially improve the quality over "first-order" query expansion alone. However, due to the fact that computing the semantic similarities through the "second-order" query expansion approach might be slow. It would be a tradeoff to use first-order query expansion instead, i.e. decrease the computational time while sacrifice some prediction accuracies.

5   Conclusion

In this paper, we have presented a new algorithm based on contextual information to solve the problem of context-based online commercial intention detection. Our work follows from the intuitive motivation that in the same session of a personal query log, semantic similarities and the correlation between surrounding queries can help improve the overall classification accuracy. Similar assumptions are also made in [3]. However, the context-aware features defined in that paper cannot be simply borrowed to tackle the OCI detection problem. Therefore, we exploited a skip chain CRF model to model the problem as a collective classification problem. We use an algorithm based on query expansion to consider the semantic similarity between queries, using query snippets as a major source of information. Our experiment on the AOL query log dataset as well as the Live Search Clickthrough Log Dataset shows that our algorithm can effectively improve the accuracy of context-based OCI detection.

Acknowledgement

Qiang Yang and Derek Hao Hu thank the support of Microsoft Research Project MRA07/08.EG01.

References

1. Beitzel, S.M., Jensen, E.C., Frieder, O., Grossman, D.A., Lewis, D.D., Chowdhury, A., Kolcz, A.: Automatic web query classification using labeled and unlabeled training data. In: SIGIR 2005, pp. 581–582 (2005)
2. Broder, A.Z.: A taxonomy of web search. SIGIR Forum 36(2), 3–10 (2002)
3. Cao, H., Hu, D.H., Shen, D., Jiang, D., Sun, J.-T., Chen, E., Yang, Q.: Context-aware query classification. In: SIGIR 2009 (2009)
4. Dai, H.K., Zhao, L., Nie, Z., Wen, J.-R., Wang, L., Li, Y.: Detecting online commercial intention (oci). In: WWW 2006, pp. 829–837 (2006)
5. Jansen, B.J.: The effect of query complexity on web searching results. Information Research 6(1) (2000)
6. Jansen, B.J., Booth, D.L., Spink, A.: Determining the user intent of web search engine queries. In: WWW 2007, pp. 1149–1150 (2007)
7. Kang, I.-H., Kim, G.-C.: Query type classification for web document retrieval. In: SIGIR 2003, pp. 64–71 (2003)
8. Lafferty, J.D., McCallum, A., Pereira, F.C.N.: Conditional random fields: Probabilistic models for segmenting and labeling sequence data. In: ICML 2001, pp. 282–289 (2001)
9. Li, X., Wang, Y.-Y., Acero, A.: Learning query intent from regularized click graphs. In: SIGIR 2008, pp. 339–346 (2008)
10. Sahami, M., Heilman, T.D.: A web-based kernel function for measuring the similarity of short text snippets. In: WWW 2006, pp. 377–386 (2006)
11. Shen, D., Sun, J.-T., Yang, Q., Chen, Z.: Building bridges for web query classification. In: SIGIR 2006, pp. 131–138 (2006)
12. Sutton, C.A., Rohanimanesh, K., McCallum, A.: Dynamic conditional random fields: factorized probabilistic models for labeling and segmenting sequence data. In: ICML 2004 (2004)

Feature Selection via Maximizing Neighborhood Soft Margin

Qinghua Hu, Xunjian Che, and Jinfu Liu

Harbin Institute of Technlology, Harbin, China
huqinghua@hit.edu.cn

Abstract. Feature selection is considered to be a key preprocessing step in machine learning and pattern recognition. Feature evaluation is one of the key issues for constructing a feature selection algorithm. In this work, we propose a new concept of neighborhood margin and neighborhood soft margin to measure the minimal distance between different classes. We use the criterion of neighborhood soft margin to evaluate the quality of candidate features and construct a forward greedy algorithm for feature selection. We conduct this technique on eight classification learning tasks. Compared with the raw data and other three feature selection algorithms, the proposed technique is effective in most of the cases.

1 Introduction

Feature selection plays an important role in a number of machine learning and pattern recognition tasks [1,2,3]. A great number of candidate features are provided to a learning algorithm for producing a complete characterization of a classification task. However, it is often the case that majority of the candidate features are irrelevant or redundant to the learning task, which will reduce the performance of the employed learning algorithm. The learning accuracy and training speed may be significantly deteriorated by these superfluous features [4]. So it is of fundamental importance to select the relevant and necessary features in the preprocessing step.

A lot of algorithms for feature selection have been developed and discussed in the last decade[5,6,7]. One of the key issues in constructing a feature selection algorithm is evaluating the quality of candidate features [8,9]. An optimal criterion should naturally relate the Bayes error rate of classification in the feature space. However, computing Bayes error rates requires detailed knowledge of the class probability distribution, whereas in practice class probabilities are unknown. One has to estimate these probabilities by making use of a finite size of samples, which is very difficult especially when dealing with highly dimensional feature spaces. Quite commonly, we focus on the design of performance measures to determine the relevance between features and decision. Distance, correlation, mutual information, consistency and dependency are usually considered as feasible alternatives. Mutual information is widely applied to characterize the relevance between categorical attributes and classification decisions [10]. Wang introduced

Z.-H. Zhou and T. Washio (Eds.): ACML 2009, LNAI 5828, pp. 150–161, 2009.

an axiomatic framework for feature selection based on mutual information [1]. A dependency-based feature selection algorithm was proposed, where dependency is defined as the ratio of so-called positive region in the rough set theory over the whole set of samples [11]. The samples with the same attribute values and different decisions are called classification boundary. However, the rate of positive region is not an effective estimate of classification accuracy. According to the Bayes rule, the samples with the same feature values will be classified as belonging to the majority class. Therefore, only the samples in the minority classes are misclassified in this case. Based on this observation, Dash and Liu introduced the measure of consistency and employed it to evaluate the quality of features where consistency is treated as the ratio of the samples which can be recognized with the Bayes rule [12]. Along this direction, a set of criteria were developed. From another viewpoint, classification margin was introduced to evaluate features in recent years.

It was derived that a classifier producing great class margin will get good generalization ability. The margin reflects the confidence of classifiers with respect to its decision. A set of algorithms based on margin have been proposed for evaluating and selecting features. In 2002, Crammer et al. gave two ways to define the margin of a sample in terms of a classifier rule: sample margin and hypothesis margin [13]. In 2004, Gilad-Bachrachy et al. introduced hypothesis margin to evaluate features and developed two algorithms, called Simba and G-flip, for selecting features [20]. In 2006, Sun and Li [14] showed that the famous feature evaluating algorithm Relief and its variant ReliefF [15] could also be considered as a margin based feature estimator. And then an iterative version of Relief (I-Relief) was introduced. In 2008, Li and Lu introduced a distance learning scheme based on loss-margin of nearest neighbor classification for ranking features [16].

Given a learning task, the theory about soft margin shows there should be a tradeoff between margin and training error rate. The training error rate would rise if we enlarge the classification margin. Dependency in neighborhood rough sets approximately measures the ratio of training accuracy, while the margin reflects the confidence of classification. One naturally expects that he can get an optimal subset of features which derives high classification accuracies and a great classification margin.

All of the techniques of Simba, G-flip, Reflief and I-Relief evaluate features with the average hypothesis margin, regardless of classification accuracies. In fact, we can also get a feature space where the average hypothesis margin is large, but the classification accuracy is very low. So we should consider both the classification accuracy and margin in evaluating features. In this work, we introduce the idea hidden in soft-margin support vector machines into neighborhood rough sets and propose a neighborhood soft-margin based feature evaluating and selecting technique. This criterion integrates the classification loss (characterized with neighborhood boundary) and margin (characterized with the size of neighborhood) to reflect the classification quality in the corresponding feature subspace.

2   Preliminary Knowledge

2.1   Soft Margin Support Vector Machine

We are given a set of sample points of the form $\mathbb{S} = \{(x_i, y_i)|x_i \in \mathbb{R}^N, y_i \in \{-1, +1\}\}_{i=1}^n$, where y_i is either +1 or -1, indicating the class to which the point x_i belongs. Each sample is a N-dimensional real vector. A maximum-margin hyperplane is required for dividing the points into one of the classes. Any hyperplane can be written as the set of points x satisfying

$$w \cdot x - b = 0 \tag{1}$$

where $w \cdot x$ is the dot product of w and x . The vector is a normal vector. The parameter $b/||w||$ determines the offset of the hyperplane from the origin along the normal vector w.

We want to choose the w and b to maximize the margin, or distance between the parallel hyperplanes that are as far apart as possible while still separating the data. These hyperplanes can be described by the equations

$$w \cdot x - b = 1 \quad and \quad w \cdot x - b = -1.$$

By using geometry, we know the distance between these two hyperplanes is $2/||w||$, so we can maximize margin by minimizing $||w||$. This can be transformed to a quadratic programming optimization problem. More clearly:

$$Minimize \; \frac{1}{2}||w||^2, \; subject \; to \; y_i(wx_i - b) \geq 1, \; i = 1, 2, \cdots, n. \tag{2}$$

This problem can now be solved by standard quadratic programming techniques and programs [17].

In 1995, Cortes and Vapnik suggested a modified maximum margin idea that allows for mislabeled examples. If there exists no hyperplane that can split two class of examples, a soft margin method will choose a hyperplane that splits the examples as cleanly as possible, while still maximizing the distance to the nearest cleanly split examples. The method introduces slack variables ξ_i, which measures the degree of misclassification of the datum x_i.

$$y_i(w \cdot x_i - b) \geq 1 - \xi_i, \quad i = 1, 2, \cdots, n \tag{3}$$

The objective function is then increased by a function which penalizes non-zero ξ_i, and the optimization becomes a trade off between a large margin, and a small error penalty. If the penalty function is linear, the optimization problem can be rewritten as

$$\frac{1}{2}||w||^2 + C\sum_i \xi_i, \; subject \; to \; y_i(w \cdot x_i - b) \geq 1 - \xi_i \; i = 1, 2, \cdots, n \tag{4}$$

where C is a constant for cost of the constrain violation. This constraint along with the objective of minimizing $||w||$ can be solved using Lagrange multipliers.

2.2   Neighborhood Rough Sets

Definition 1. *Given a set* $\mathbb{S}$ *of samples described with features F,* Δ *is a distance function on* $\mathbb{S}$. δ *is a positive constant, then the neighborhood of sample x is defined as* $\delta(x) = \{x_i | \Delta(x, x_i) \leq \delta\}$.

Usually, Euclidean distance is used if the samples are described with numerical features. However, some other functions can also be found for complex tasks [9].

The relation $\mathcal{N}$ of neighborhood divides the samples into a collection of subsets of samples, where $\mathcal{N}(x, y) = 1$ if $y \in \delta(x)$; otherwise, $\mathcal{N}(x, y) = 0$. We call $(\mathbb{S}, \mathcal{N})$ a neighborhood approximation space.

Definition 2. *Given* $(\mathbb{S}, \mathcal{N})$ *and an arbitrary subset* $X \subseteq \mathbb{S}$ *of samples, the lower and upper approximations of X in are defined as*

$$\underline{\mathcal{N}}X = \{x \in U | \delta(x) \subseteq X\}, \overline{\mathcal{N}}X = \{x \in U | \delta(x) \cap X \neq \emptyset\}$$

Definition 3. *Given* $(\mathbb{S}, \mathcal{N})$, $\mathbb{S}$ *is partitioned into subsets* $d_1, d_2, \ldots, d_m$ *with the decision attribute Y. Then the lower and upper approximations of classification in* $(\mathbb{S}, \mathcal{N})$ *are defined as*

$$\underline{\mathcal{N}}C = \bigcup_{i=1}^{m} \underline{\mathcal{N}}d_i; \overline{\mathcal{N}}C = \bigcup_{i=1}^{m} \overline{\mathcal{N}}d_i,$$

correspondingly, the approximation boundary of classification is defined as

$$BN_{\mathcal{N}}C = \overline{\mathcal{N}}C - \underline{\mathcal{N}}C$$

It is easy to show that $\overline{\mathcal{N}}C = \mathbb{S}$. So $BN_{\mathcal{N}}C = \mathbb{S} - \underline{\mathcal{N}}C$.

Definition 4. *Given* $(\mathbb{S}, \mathcal{N})$ *and the decision attribute Y, the neighborhood dependency of Y on the set of features F is computed with*

$$\gamma_{\mathcal{N}}(Y) = \frac{|\underline{\mathcal{N}}C|}{|\mathbb{S}|},$$

where $|A|$ *is the cardinality of A.*

We say decision Y depends on features F with degree γ. We say the classification is δ neighborhood consistent or δ neighborhood separable if $\gamma_{\mathcal{N}}(Y) = 1$.

3   Evaluating Feature Quality with Soft-Margin Like Technique

As to a linearly separable task, a large margin classifier can be illustrated as Fig. 4. The real margin is $2/||w||$ in this case.

Definition 5. *If the distance between the hyperplane and a sample is less than* δ *, we say that the two classes of samples is* δ *linearly inseparable[18].*

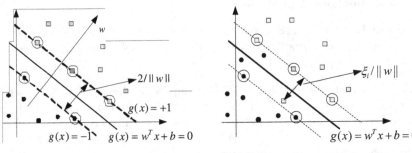

(a) Linear support vector machine

(b) Soft margin support vector machine

Fig. 1. Soft margin support vector machine

Obviously, as to Fig. 1(a), the task is δ linearly inseparable if $\delta > 1/||w||$. In fact, most of the tasks are linearly inseparable even if we set $\delta = 0$, as shown in Fig. 1 (b). The soft margin support vector machines were designed to deal with inseparable tasks by allowing for some mislabeled examples.

In filter based feature selection, we do not know the classifying hyperplane. So we can not compute the margins of samples in this case. Now, we consider samples near the hyperplane. It is easy to see that these samples are close to samples with different labels.

Theorem 1. *Given a set of samples $\mathbb{S}$ for classification learning, assume $\mathbb{S}$ is divided into two classes$\{y_+, y_-\}$ based on the decision attribute, where y_+ and y_- are denoted the samples with label +1 and -1, respectively. If the samples are δ neighborhood separable, we have that $\Delta(y_+, y_-) > \delta$, where $\forall x_i \in y_+$, $\forall x_j \in y_-$, $\Delta(y_+, y_-) = \min \Delta(x_i, x_j)$. In this case, we say the neighborhood margin of the classification task is no less than δ.*

It is easy to get from Fig. 2 that the task is 2ε neighborhood separable if it is ε linearly separable with respect to linear SVM because the least distance l between two samples with different classes is no less than 2ε.

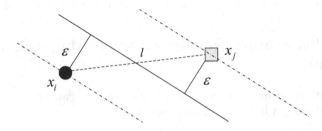

Fig. 2. Relation between *delta* neighborhood separable and ε linearly separable

If a task is δ neighborhood separable, then the distance between two samples with different classes is greater than δ. Namely given $\forall x_i \in y_+$ and $\forall x_j \in y_-$, we have $\Delta(x_i, x_j) > \delta$. It is easy to show that the task is 2δ neighborhood inseparable if it is δ linearly inseparable with respect to linear SVM because the least distance of two samples of different labels away from the hyperplane is less than δ.

One expects the classification task is separable at a large size of neighborhood. In this case, the margin between samples in different classes is large enough for discriminating them. So we can construct a new criterion for evaluating the quality of feature spaces based on neighborhood. However, just as pointed out above that we are usually confronted with tasks which are of little margins partly due to the complexity of tasks or noisy samples.

Figure 3 shows a classification task with a noisy sample, where samples in class 1 are denoted by $\bullet$ and samples in class 2 are marked with $\square$. We see that x_1 is the closest one to the second class if we do not consider x_2. In this case the task is δ_1 neighborhood separable. However, if there are some samples like x_2, the margins between different classes of samples are significantly reduced. The task is δ_2 neighborhood separable if x_2 is considered. As δ_2 is far less than δ_1, then the quality of the feature subspace gets much worse than the case that x_2 is not considered. The analysis shows that margin based feature evaluation is sensitive to noisy samples.

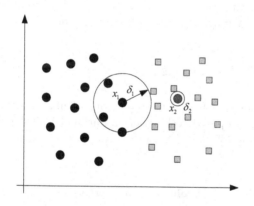

Fig. 3. Classification task with a noisy sample

In order to reduce the influence of noisy samples, we introduce the idea of soft-margin support vector machines by trading off between the size of neighborhood and the size of boundary set. Following the optimization objective function in soft-margin support vector machine, we give a new measure for evaluating features:

$$\min_{\delta} \frac{1}{2\delta^2} + \lambda\|B\| \tag{5}$$

where δ is the size of neighborhood, $||B||$ is the size of the boundary set with respect to δ, and λ is a nonnegative real number to reflect the weight of margin and size of the boundary set. We call this criterion neighborhood soft-margin.

Given a set of samples and parameter δ, we can compute the boundary set B. Then we get the quality of features. As to a certain subset of features, we can try a series of values for parameter δ, compute the neighborhood soft margins and get the maximal value of neighborhood soft-margin as the final output. Let $NSM = \frac{1}{2\delta^2} + \lambda||B||$. As we know that the size of the boundary set B increases with parameter δ. Assume $\delta = 0.01$, $\lambda = 1$, and $||B|| = 0$, then $NSM = 5000$. However, in some cases, we change $\delta = 0.1$, we get $||B|| = 20$, then $NSM = 250$. Thus, we should take 0.1 as the soft margin of the classification task although there are 20 samples in the margin. In the experiments, we try δ from 0.01 to $0.25 \times \sqrt{N}$ with step 0.01, where N is the number of features.

We construct a greedy algorithm to select features based on the proposed measure. We first compute the neighborhood soft margin of each feature and select the feature f_1 with the largest margin. Then we find a feature f_2 to get the largest margin in the subspace of f_1 and f_2, and so on until all the candidate features are selected. We get a rank of features by this procedure. Then we can check the effectiveness of the first m features, where m is specified by the users.

By introducing the idea in the soft-margin support vector machines, the neighborhood soft-margin technique is naturally not sensitive to data noise. Thus the proposed technique is robust. Moreover, this technique uses the neighborhood margin to evaluate candidate features. This maybe lead to an effective solution to feature selection. We will test this technique in the next section.

4   Experimental Analysis

Eight data were collected from UCI repository of machine learning databases [19] for testing the proposed technique, as outlined in Table 1. The numbers of samples vary from 178 to 2310, and the numbers of features are between 13 and 34.

Table 1. Data description

Data	Sample	Feature	Class
German	1000	20	2
Heart	270	13	2
Hepatitis	155	19	2
Horse	368	22	2
Iono	351	34	2
Segmentation	2310	18	7
Wdbc	569	30	2
Wine	178	13	3

In order to test the effectiveness of the proposed technique, we consider the distribution of distance between samples with different classes. Take the wine data as an example. We compute the distances between samples with different classes in feature space $A = \{1, 3, 7, 8, 10, 13\}$, where the feature values are transformed to the unit interval $[0, 1]$. Figure 4 (a) gives the histogram of distance distribution. We can see that most of the distances between different classes is greater than 0.2. However, we also see there are several samples whose distance is less than 0.2 and the minimal distance is 0.1644. Figure 4 (b) gives the histogram of distance distribution computed in subspace $B = \{1, 2, 5, 7, 10, 13\}$. The minimal margin is 0.2169. Comparing the minimal margin, we get that subspace B is much better than A. At the same time, we also compute the classification accuracies of 3-nearest neighbor classifier in these subspaces and the derived accuracies are both 0.97 ± 0.04. This fact shows the measure of crisp margin can not reflect the real ability of features. Soft margin should be introduced.

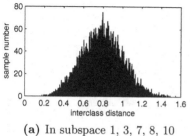

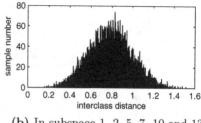

(a) In subspace 1, 3, 7, 8, 10   (b) In subspace 1, 2, 5, 7, 10 and 13

Fig. 4. Distance distribution between samples with different classes

Before evaluating features with neighbohood soft margin, we should specify the paramter λ. We try a set of values for this parameter and find $\lambda = 4$ is a good choice for most of the tasks. So we set $\lambda = 4$ in the following experiments. We evaluate the candidate features with neighbohood soft margin and get a rank for each classification task. Then we compute the classification accuracy of the first m features with KNN and RBF-SVM classifiers based on 10-fold crossvalidation, where $m = 1, 2, 3 \cdots$. Then we get a series of classification accuracies for each task, as shown in Figure 5.

In the same time, we also conduct feature selection on these data sets with correlation (CFS) [10], consistency (C) [12] and neighborhood rough sets (NRS) based algorithm [9].

The optimal numbers of features selected with these algorithms are given in Table 2. As a whole, we see that NSM-KNN selects the least features. For some sets, such as hepatitis, horse, iono and segmentation, just several features are selected. NSM-SVM selects almost the same number of features as consistency based algorithm. Correlation and neighborhood rough sets based algorithms get the most features.

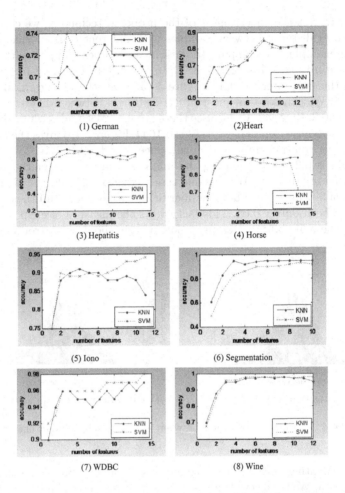

Fig. 5. Variation of classification accuracies with number of features

Table 2. Number of features selected with different algorithms

Data	Raw	NSM-KNN	NSM-SVM	CFS	C	NRS
German	20	7	**3**	5	11	11
Heart	13	8	8	**7**	10	9
Hepatitis	19	**4**	7	7	**4**	6
Horse	22	**4**	5	8	**4**	8
Iono	34	**4**	9	14	7	9
Segmentation	18	**3**	9	7	10	13
Wdbc	30	12	9	11	**7**	12
Wine	13	7	**5**	11	5	6
Total	**169**	**49**	**55**	**70**	**58**	**74**

Table 3. KNN accuracy of features selected with different algorithms ($k = 3$)

Data	Raw	NSM	CFS	C	NRS
German	0.69	0.73 ↑	0.70	0.72	0.71
Heart	0.82	0.83 ↑	0.83	0.84	0.83
Hepatitis	0.87	0.93 ↑	0.88	0.90	0.90
Horse	0.90	0.91 ↑	0.92	0.88	0.90
Iono	0.84	0.91 ↑	0.86	0.88	0.88
Segmentation	0.95	0.95−	0.95	0.95	0.95
Wdbc	0.97	0.97−	0.97	0.97	0.95
Wine	0.95	0.98 ↑	0.97	0.96	0.97

Table 4. RBF-SVM accuracy of features selected with different algorithms

Data	Raw	NSM	CFS	C	NRS
German	0.70	0.74 ↑	0.71	0.71	0.71
Heart	0.81	0.86 ↑	0.81	0.80	0.81
Hepatitis	0.84	0.91 ↑	0.90	0.90	0.91
Horse	0.72	0.91 ↑	0.92	0.83	0.89
Iono	0.94	0.93 ↓	0.95	0.93	0.94
Segmentation	0.92	0.93 ↑	0.91	0.91	0.93
Wdbc	0.98	0.97 ↓	0.97	0.96	0.97
Wine	0.98	0.98−	0.99	0.97	0.99

Then we compare the classification accuracies of these feature subsets in Tables 3 and 4, where "raw" denotes the classification accuracy of the raw data sets, NSM is the classification accuracy of features selected with neighborhood soft margin, and CFS, C and NRS are accuracies derived with features selected with correlation, consistency and neighborhood rough sets based algorithms, respectively.

In Tables 3 and 4, markers ↑, ↓ and − means the classification accuracy increase, decrease and does not change, respectively. Compared with KNN accuracies derived from the raw data and NSM, we see that six of the eight tasks are improved and two tasks keep invariant although more than two thirds of features are removed, as shown in Table 2. As to RBF-SVM, five tasks are improved and two tasks are worsened a little; one task keeps invariant. These results show that the proposed technique is effective for feature selection in most of the cases.

Comparing the proposed algorithm with other techniques of feature selection, we also see that NSM is competent. Especially for KNN classifiers, NSM is almost the best one on all the classification tasks except data set horse. As to RBF-SVM, NSM also get good results in majority of samples.

5 Conclusions

Feature selection is considered to be an important preprocessing step in machine learning and pattern recognition. Feature evaluation is a key issue when constructing an algorithm for feature selection. In this work, we propose a new concept of neighborhood margin and neighborhood soft margin, which reflects the minimal distance between different classes.

We use the criterion of neighborhood soft margin to evaluate the quality of candidate features and construct a forward greedy algorithm for feature selection. We conduct this technique on eight classification learning tasks. Compared with the raw data and other three feature selection algorithms, the proposed technique is effective in most of the cases.

Neighborhood soft margin can be looked as a local approximation strategy for feature evaluation. The connection between neighborhood soft margin, sample margin and hypothesis margin is not discussed in this paper. We will show a systematic analysis on these concepts and give risk estimation on neighborhood soft margin in the future.

Acknowledgement

This work is supported by National Natural Science Foundation of China under Grant 60703013 and Development Program for Outstanding Young Teachers in Harbin Institute of Technology under Grant HITQNJS.2007.017.

References

1. Wang, H., Bell, D., Murtagh, F.: Axiomatic Approach to Feature Subset Selection Based on Relevance. IEEE Transactions on Pattern Analysis and Machine Intelligence 21, 271–277 (1999)
2. Yu, L., Liu, H.: Efficient Feature Selection via Analysis of Relevance and Redundancy. Journal of Machine Learning Research 5, 1205–1224 (2004)
3. Guyon, I., Elisseeff, A.: An introduction to variable and feature selection. Journal of Machine Learning Research 3, 1157–1182 (2003)
4. Liu, H., Yu, L.: Toward integrating feature selection algorithms for classification and clustering. IEEE Transactions on Knowledge and Data Engineering 17, 491–502 (2005)
5. Kohavi, R., John, G.H.: Wrappers for feature subset selection. Artificial Intelligence 97, 273–324 (1997)
6. Abe, S., Thawonmas, R., Kobayashi, Y.: Feature selection by analyzing class regions approximated by ellipsoids. IEEE Transactions on Systems, Man, and Cybernetics-Part C: Applications and Reviews 28, 282–287 (1998)
7. Neumann, J., Schnorr, C., Steidl, G.: Combined SVM-based feature selection and classification. Machine learning 61, 129–150 (2005)
8. Ho, T.K., Basu, M.: Complexity measures of supervised classification problems. IEEE Trans. Pattern Analysis and Machine Intelligence 24, 289–300 (2002)
9. Hu, Q.H., Yu, D., Liu, J.F., Wu, C.: Neighborhood rough set based heterogeneous feature subset selection. Information sciences 178, 3577–3594 (2008)

10. Hall, M.A.: Correlation-based feature selection for discrete and numeric class machine learning. In: Proc. 17th Int'l. Conf. Machine Learning, pp. 359–366 (2000)
11. Hu, Q.H., Yu, D.R., Xie, Z.X.: Neighborhood classifier. Expert Systems with Applications 34, 866–876 (2008)
12. Dash, M., Liu, H.: Consistency-based search in feature selection. Artificial Intelligence 151, 155–176 (2003)
13. Crammer, K., Gilad-Bachrach, R., Navot, A., Tishby, N.: Margin analysis of the LVQ algorithm. In: Proc. 17th Conference on Neural Information Processing Systems (2002)
14. Sun, Y., Li, J.: Iterative RELIEF for Feature Weighting. In: Proc. 23rd Int'l. Conf. Machine Learning, pp. 913–920 (2006)
15. Kononenko, I.: Estimating Attributes: Analysis and Extensions of RELIEF. In: Bergadano, F., De Raedt, L. (eds.) ECML 1994. LNCS, vol. 784, pp. 171–182. Springer, Heidelberg (1994)
16. Li, Y., Lu, B.-L.: Feature selection based on loss-margin of nearest neighbor classification. Pattern Recognition (2008), doi:10.1016/j.patcog. 2008.10.011
17. Cortes, C., Vapnik, V.: Support-Vector Networks. Machine Learning 20, 273–297 (1995)
18. Vapnik, V.: Statistical learning theory. Wiley, NY (1998)
19. Blake, C.L., Merz, C.J.: UCI repository of machine learning databases (1998), http://www.ics.uci.edu/~mlearn/MLRepository.html
20. Gilad-Bachrach, R., Navot, A., Tishby, N.: Margin based feature selection- theory and algorithm. In: Proceddings of 21st international conference on machine learning, Banff, Canada (2004)

Accurate Probabilistic Error Bound for Eigenvalues of Kernel Matrix

Lei Jia and Shizhong Liao

School of Computer Science and Technology
Institute of Knowledge Science and Engineering
Tianjin University, Tianjin 300072, P. R. China
ljia@tju.edu.cn, szliao@tju.edu.cn

Abstract. The eigenvalues of the kernel matrix play an important role in a number of kernel methods. It is well known that these eigenvalues converge as the number of samples tends to infinity. We derive a probabilistic finite sample size bound on the approximation error of an individual eigenvalue, which has the important property that the bound scales with the dominate eigenvalue under consideration, reflecting the accurate behavior of the approximation error as predicted by asymptotic results and observed in numerical simulations. Under practical conditions, the bound presented here forms a significant improvement over existing non-scaling bound. Applications of this theoretical finding in kernel matrix selection and kernel target alignment are also presented.

1 Introduction

Kernel methods [1] are a class of algorithms which have proven to be very powerful and versatile for machine learning problems. Central to kernel methods is the kernel matrix, which is built by evaluating a Mercer kernel on all pairs of samples of the training set. It is well known that such matrices have a particularly nice spectral decomposition, having full set of eigenvectors which are orthogonal and only positive eigenvalues.

A number of learning algorithms rely on estimating spectral data on training samples and using this data as input to further analyze. For example, in principal component analysis (PCA) the subspace spanned by the first k singular vectors is used to give a k dimensional model of the data with minimal residual, hence forming a low dimensional representation of the data for analysis or clustering [2]. In ridge regression the coordinates are shrunk with respect to the orthonormal basis formed by the principal components of samples. Coordinate with respect to the principal component with a smaller variance is shrunk more. To extend these approaches to their non-linear versions, the kernel trick has been introduced. Then the learning procedure can be applied in kernel-defined feature spaces in what has become known as kernel principal component analysis (KPCA) [3] and kernel ridge regression (KRR) [4]. Most importantly the spectral data are estimated on kernel matrix instead.

Z.-H. Zhou and T. Washio (Eds.): ACML 2009, LNAI 5828, pp. 162–175, 2009.

The use of these techniques raises two questions. First, it is well known that when an algorithm uses inner product of two matrixes $\mathbf{M}^\top\mathbf{M}$, it tends to be unstable [5]. Unfortunately, a kernel matrix is equivalent to the inner product of two matrixes. Hence the first question is how does the kernel matrix computing impact the stability of kernel methods. Ng et al. [6] have undertaken a study of the sensitivity of the principal eigenvector to perturbations of the connection matrix. Martin [7] has studied the numerical unstable problem in kernel PCA computing. He has also provided an example where the kernel matrix computing results in numerical instability, in the case of linearly separable data in SVM. Following and extending their results, Jia and Liao [8] proposed a notion of perturbation stability to quantify the instability arising from computing eigenvalues on kernel matrix. These works solve the convergence of spectral properties of kernel matrices completely.

The second question, which motivated this research reported here, is how reliably the spectral data can be estimated from kernel matrix. The convergence of spectra of kernel matrices has attracted much attention, cf. Williams and Seeger [9], Bengio et al. [10], Shawe-Taylor et al. [11], and Zwald and Blanchard [12]. In general, the case of kernel matrices is a special case of the results obtained in Koltchinskii [13] and Koltchinskii and Giné [14], where even the distributions of the eigenvalues and spectral projections are investigated.

These results fail to correctly reflect the fact that the approximation error scales with the magnitude of the dominate eigenvalue. For small eigenvalues, the absolute bound is overly pessimistic. A more accurate bound has to scale with the magnitude of the dominate eigenvalue. This observation is particularly important for the kernel functions employed in machine learning which typically have slowly decaying eigenvalues. We will derive a refinement of the convergence result which shows that the variance in the estimate depends on the magnitude of the dominate eigenvalue. The resulting estimates of the approximation errors are consequently much tighter than previous results.

The rest of this paper is organized as follows. In section 2 we give background knowledge and main techniques that required to derive inequalities. We present our main theoretical results in section 3. Section 4 studies two applications of these findings. We end in section 5 with conclusions and future works.

Notation. We use lowercase boldface letters $\mathbf{x}$, $\mathbf{z}$ etc. for column vectors, and uppercase boldface letters $\mathbf{M}$, $\mathbf{S}$ etc. for matrices. We use $\lambda_i(\mathbf{M})$ for the eigenvalues of matrix $\mathbf{M}$. If $\mathbf{M}$ is evident from the context we drop it from the parenthesis. Without losing generalization, we assume $\lambda_1 \geq \cdots \geq \lambda_m$ throughout, i.e. λ_1 is the dominate eigenvalue.

2   Preliminaries

We first introduce some background knowledge and then the main techniques we will use to derive inequalities. Central to kernel methods is the kernel matrix, which is built by evaluating a Mercer kernel on all pairs of samples of the training set. Mercer kernels are defined as follows.

Definition 1 (Mercer Kernel). *Let μ be a probability measure on $\mathcal{X}$, and $\mathcal{H}_\mu(\mathcal{X}) \subset l^2$ the associated Hilbert space. Given a sequence $(\pi_i)_{i \in \mathbb{N}} \in l^1$ with $\pi_i \geq 0$, and an orthogonal family of unit norm functions $(\psi_i)_{i \in \mathbb{N}}$ with $\psi_i \in \mathcal{H}_\mu(\mathcal{X})$, the associated Mercer kernel is*

$$k(\mathbf{x}, \mathbf{z}) = \sum_{i=1}^{\infty} \pi_i \psi_i(\mathbf{x}) \psi_i(\mathbf{z}).$$

The numbers π_i will be called the eigenvalues of the kernel and ψ_i its eigenfunctions.

In practical situations, one will often use a Mercer kernel function where the above sum can be computed in closed form, and the expansion itself is in fact unknown. An example for such a kernel function is the radial basis kernel function (rbf-kernel) $k(\mathbf{x}, \mathbf{z}) = \exp\left(-\gamma \|\mathbf{x} - \mathbf{z}\|^2\right)$. We have called the numbers π_i eigenvalues and the functions eigenfunctions ψ_i. Actually, these are the eigenvalues and eigenfunctions of the integral operator associated with k defined by

$$T_k f(\mathbf{x}) = \int_{\mathcal{X}} k(\mathbf{x}, \mathbf{z}) f(\mathbf{z}) \, d\mu(\mathbf{z}).$$

Proposition 1. *The π_i and ψ_i occurring in the definition of a uniformly bounded Mercer kernel function k are the eigenvalues and eigenfunctions of T_k.*

Proof. We compute $T_k \psi_i(\mathbf{x})$

$$T_k \psi_i(\mathbf{x}) = \int_{\mathcal{X}} k(\mathbf{x}, \mathbf{z}) \psi_i(\mathbf{z}) \, d\mu(\mathbf{z}) = \int_{\mathcal{X}} \sum_{j=1}^{\infty} \pi_j \psi_j(\mathbf{x}) \psi_j(\mathbf{z}) \psi_i(\mathbf{z}) \, d\mu(\mathbf{z})$$

$$= \sum_{j=1}^{\infty} \pi_j \psi_j(\mathbf{x}) \langle \psi_j, \psi_i \rangle = \pi_i \psi_i(\mathbf{x}).$$

This concludes the proof.   □

With the quadrature method, we can discretize the operator T_k to the operator

$$\hat{T}_k f(\mathbf{x}) = \sum_{i=1,\dots,m} w_i k(\mathbf{x}, \mathbf{x}_i) f(\mathbf{x}_i).$$

It can be shown that the eigenvalues of $\hat{T}_k$ are the same as the ones of the matrix $\mathbf{KW}$, where $\mathbf{K}$ is the kernel matrix obtained by evaluating the kernel function on all pairs of object samples $\mathbf{x}_i, \mathbf{x}_j$, i.e. $\mathbf{K}$ is the $n \times n$ square matrix with entries $[\mathbf{K}]_{ij} = k(\mathbf{x}_i, \mathbf{x}_j)$, $\mathbf{W}$ is the diagonal matrix containing the weights w_i of the quadrature rule on the diagonal. The simplest quadrature rule is the Riemann sum in case of a uniform distribution and the weights w_i are given by $1/m$, hence $\mathbf{KW} = \frac{1}{m}\mathbf{K}$. As the number of samples m tends to infinity, certain properties of the kernel matrix show a convergent behavior, i.e., $\lambda_i\left(\frac{1}{m}\mathbf{K}\right)$ converge to the eigenvalues of the integral operator T_k with kernel function k with respect to the probability measure μ of the object samples $\mathbf{x}_i$:

$$\mathbb{E}_{\mathcal{X}} \left[\frac{1}{m} \lambda_i (\mathbf{K}) \right] = \pi_i,$$

where the $\mathbb{E}_{\mathcal{X}}$ is the expectation operator under selection of the sample. This relationship allows to study the properties of kernels with infinite expansions.

We will make use of the following theorems due to Ostrowski [15] bounding the perturbation on eigenvalues of a matrix.

Theorem 1 (Ostrowski). *Let* $\mathbf{M}$ *be a Hermitian* $m \times m$ *matrix, and* $\mathbf{S}$ *an* $m \times m$ *matrix. Then, for each* $1 \leq i \leq m$, *there exists an nonnegative real* θ_i *with* $\lambda_m (\mathbf{SS}^{\mathrm{H}}) \leq \theta_i \leq \lambda_1 (\mathbf{SS}^{\mathrm{H}})$, *such that*

$$\lambda_i (\mathbf{SMS}^{\mathrm{H}}) = \theta_i \lambda_i (\mathbf{M}). \tag{1}$$

We will also use the following results due to McDiarmid characterizing the concentration of an estimate to its expectation.

Theorem 2 (McDiarmid). *Let* $x_1, \dots, x_m$ *be independent random variables taking values in a set* $\mathcal{A}$, *and assume that* $f : \mathcal{A}^m \to \mathbb{R}$, *and* $f_k : \mathcal{A}^{m-1} \to \mathbb{R}$ *satisfy for* $1 \leq k \leq m$

$$\sup_{x_1, \dots, x_m} |f(x_1, \dots, x_m) - f_k(x_1, \dots, x_{k-1}, x_{k+1}, \dots, x_m)| \leq c_k,$$

then $\forall \epsilon > 0$,

$$P\{|f(x_1, \dots, x_m) - \mathbb{E}f(x_1, \dots, x_m)| > \epsilon\} \leq 2 \exp \left(\frac{-2\epsilon^2}{\sum_{i=1}^m c_k^2} \right).$$

Using McDiarmid's inequality, Shawe-Taylor et al. have shown that with increasing sample size, the solution concentrates around its expectation, such that the eigenvalues become concentrated as well.

Theorem 3 (Shawe-Taylor et al.). *Let* $\mathcal{S} = \{\mathbf{x}_1, \dots, \mathbf{x}_m\}$, $\mathbf{x}_i \in \mathbb{R}^n$, *be a sample set of* m *points drawn according to a distribution* μ *on* $\mathbb{R}^n$. *Given Mercer kernel function* $k(\mathbf{x}, \mathbf{z})$. *Let* $\mathbf{K} \in \mathbb{R}^{m \times m}$ *be a kernel matrix on* $\mathcal{S}$, $[\mathbf{K}]_{ij} = k(\mathbf{x}_i, \mathbf{x}_i)$. *Then* $\forall \epsilon > 0$,

$$P\left\{ \left| \frac{1}{m} \lambda_i (\mathbf{K}) - \mathbb{E}_{\mathcal{S}} \left[\frac{1}{m} \lambda_i (\mathbf{K}) \right] \right| > \epsilon \right\} \leq 2 \exp \left(\frac{-2\epsilon^2 m}{R^4} \right), 0 \leq i < m.$$

where $R^2 = \max_{\mathbf{x} \in \mathbb{R}^n} k(\mathbf{x}, \mathbf{x})$.

This result fails to correctly reflect the fact that the approximation error scales with the magnitude of the dominate eigenvalue. For small eigenvalues, the absolute bound is overly pessimistic.

For kernels with an infinite number of non-zero eigenvalues, k can be decomposed into a degenerate kernel $k^{[r]}$ and an error function e^r given a truncation point r [16]:

$$k^{[r]}(\mathbf{x}, \mathbf{z}) = \sum_{i=1}^{r} \pi_i \psi_i(\mathbf{x}) \psi_i(\mathbf{z}),$$

$$e^r(\mathbf{x}, \mathbf{z}) = k(\mathbf{x}, \mathbf{z}) - k^{[r]}(\mathbf{x}, \mathbf{z}).$$

The kernel matrices induced by $k^{[r]}$ and e^r will be denoted by $\mathbf{K}^{[r]}$ and $\mathbf{E}^r$, respectively, such that $\mathbf{E}_m^r = \mathbf{K} - \mathbf{K}^{[r]}$. Based on these notions, Braun [16] has proposed a deterministic bound which also holds for non-random choices of points $\mathbf{x}_1, \ldots, \mathbf{x}_m$.

Theorem 4 (Braun). *For* $1 \leq r \leq m$, $1 \leq i \leq m$,

$$|\lambda_i - \pi_i| \leq \pi_m \left\| \mathbf{C}_m^r \right\| + \pi_r + \left\| \mathbf{E}_m^r \right\|,$$

with $\mathbf{C}_m^r = {\mathbf{\Psi}_m^r}^\top \mathbf{\Psi}_m^r - \mathbf{I}_r$, $\mathbf{\Psi}_m^r$ *is the* $n \times r$ *matrix with entries* $[\mathbf{\Psi}_m^r]_{il} = \frac{1}{\sqrt{n}} \psi_l(\mathbf{x}_i)$.

Note that this bound requires the knowledge of the true eigenvalues and has a slower than stochastic rate. This limits the practical applicability of these bounds. In this paper, we will try to derive a probabilistic error bound which reflects the true behavior of the approximate eigenvalues better.

3   Main Results

The previous section outlined the relatively well-known perspective that we now apply to obtain the concentration results for the eigenvalues of kernel matrix.

Theorem 5. *Let* $\mathcal{S} = \{\mathbf{x}_1, \ldots, \mathbf{x}_m\}$, $\mathbf{x}_i \in \mathbb{R}^n$, *be a sample set of* m *points. Given Mercer kernel function* $k(\mathbf{x}, \mathbf{z})$. *Let* $\mathbf{K} \in \mathbb{R}^{m \times m}$ *be a kernel matrix on* $\mathcal{S}$, $[\mathbf{K}]_{ij} = k(\mathbf{x}_i, \mathbf{x}_j)$, $\hat{\mathbf{K}} \in \mathbb{R}^{m-1 \times m-1}$ *be a kernel matrix on* $\mathcal{S} \setminus \{\mathbf{x}_k\}$, $[\hat{\mathbf{K}}]_{ij} = k(\mathbf{x}_i, \mathbf{x}_j)$. *Then for* $1 \leq i < m$,

$$\lambda_i(\hat{\mathbf{K}}) = \theta_i \lambda_i(\mathbf{K}), 0 \leq \theta_i \leq 1.$$

Proof. Consider the matrix $\hat{\mathbf{K}}' \in \mathbb{R}^{m \times m}$ created by adding a zero-row (column) to k-th row (column) of $\hat{\mathbf{K}}$. It is easy to get that

$$\lambda_i(\hat{\mathbf{K}}') = \begin{cases} \lambda_i(\hat{\mathbf{K}}), i = 1, \ldots, m-1, \\ 0, i = m. \end{cases} \tag{2}$$

Define diagonal matrix $\mathbf{S} \in \mathbb{R}^{m \times m}$

$$\mathbf{S}_{ii} = \begin{cases} 1 , i = 1, \ldots, k-1, k+1 \ldots, m, \\ 0 , i = k. \end{cases}$$

then $\hat{\mathbf{K}}' = \mathbf{S}\mathbf{K}\mathbf{S}^{\top}$.

Note that $\lambda_1 \left(\mathbf{S}\mathbf{S}^{\top}\right) = 1$, $\lambda_m \left(\mathbf{S}\mathbf{S}^{\top}\right) = 0$. Using theorem 1 we have

$$\lambda_i(\hat{\mathbf{K}}') = \theta_i \lambda_i \left(\mathbf{K}\right), 0 \leq \theta_i \leq 1. \tag{3}$$

Using formulas (2) and (3) we have for $1 \leq i < m$,

$$\lambda_i(\hat{\mathbf{K}}) = \theta_i \lambda_i \left(\mathbf{K}\right), 0 \leq \theta_i \leq 1, \tag{4}$$

which concludes the proof.   □

Theorem 6. *Let* $\mathcal{S} = \{\mathbf{x}_1, \ldots, \mathbf{x}_m\}$, $\mathbf{x}_i \in \mathbb{R}^n$, *be a sample set of m points drawn according to a distribution μ on $\mathbb{R}^n$. Given Mercer kernel function $k(\mathbf{x}, \mathbf{z})$. Let* $\mathbf{K} \in \mathbb{R}^{m \times m}$ *be a kernel matrix on* $\mathcal{S}$, $[\mathbf{K}]_{ij} = k(\mathbf{x}_i, \mathbf{x}_i)$. *Then* $\forall \epsilon > 0$,

$$P\left\{\left|\frac{1}{m}\lambda_i \left(\mathbf{K}\right) - \mathbb{E}_{\mathcal{S}}\left[\frac{1}{m}\lambda_i \left(\mathbf{K}\right)\right]\right| > \epsilon\right\} \leq 2\exp\left(\frac{-2\epsilon^2 m}{\theta^2 \lambda_1 \left(\mathbf{K}\right)^2}\right), 0 \leq i < m,$$

where $0 \leq \theta \leq 1$.

Proof. Let $\hat{\mathbf{K}} \in \mathbb{R}^{m-1 \times m-1}$ be a kernel matrix on $\mathcal{S} \setminus \{\mathbf{x}_k\}$. Using theorem 5 we have

$$\lambda_i \left(\mathbf{K}\right) - \lambda_i(\hat{\mathbf{K}}) = \lambda_i \left(\mathbf{K}\right) (1 - \theta_i) \leq \lambda_1 \left(\mathbf{K}\right) (1 - \theta_i) \leq \lambda_1 \left(\mathbf{K}\right) \left(1 - \theta^{[k]}\right),$$

where

$$\theta^{[k]} = \min_{i=1,\ldots,m-1} \frac{\lambda_i(\hat{\mathbf{K}})}{\lambda_i \left(\mathbf{K}\right)}.$$

Then

$$\sup_{\mathcal{S}}\left|\frac{1}{m}\lambda_i \left(\mathbf{K}\right) - \frac{1}{m}\lambda_i(\hat{\mathbf{K}})\right| \leq \theta\frac{1}{m}\lambda_1 \left(\mathbf{K}\right),$$

where

$$\theta = \max_{k=1,\ldots,m} \left(1 - \theta^{[k]}\right).$$

Hence, we have $\theta \lambda_1(\mathbf{K})/m$ for all i and we obtain from an application of McDiarmid' theorem 2 that

$$P\left\{\left|\frac{1}{m}\lambda_i \left(\mathbf{K}\right) - \mathbb{E}_{\mathcal{S}}\left[\frac{1}{m}\lambda_i \left(\mathbf{K}\right)\right]\right| > \epsilon\right\} \leq 2\exp\left(\frac{-2\epsilon^2 m}{\theta^2 \lambda_1 \left(\mathbf{K}\right)^2}\right), \tag{5}$$

which concludes the proof.   □

This result is expressed in terms of 'concentration' inequality. Concentration means that the probability of a random empirical estimate deviating from its mean can be found as an exponentially decaying function of that deviation.

Remark 1. Note that besides dominate eigenvalue λ_1, the error bound presented here depends on the parameter θ. This seems to be an undesirable feature, as this requires extra computational efforts. However, we have adopted a more theoretical approach in this work with the goal to understand the underlying principles which permit the derivation of scaling bounds for individual eigenvalues. One could then use these results to construct statistical tests to estimate.

Remark 2. We can also loosen the bound $\theta\lambda_1(\mathbf{K})$ to be λ_1 by setting θ to 1. For general kernel matrix, λ_1 is bounded by $\lambda_1 \leq \|\mathbf{K}\|_1 = \max_j \sum_{i=1}^{m} |[\mathbf{K}]_{ij}|$. For non negative kernel matrix, say the RBF kernel matrix, λ_1 is bounded by $\min_j r_j(\mathbf{K}) \leq \lambda_1 \leq \max_j r_j(\mathbf{K})$, where $r_j(\mathbf{K}) = \sum_{i=1}^{m}[\mathbf{K}]_{ji}$.

Corollary 1. *Let* $\mathcal{S} = \{\mathbf{x}_1, \ldots, \mathbf{x}_m\}$, $\mathbf{x}_i \in \mathbb{R}^n$, *be a sample set of m points drawn according to a distribution* μ *on* $\mathbb{R}^n$. *Given Mercer kernel function* $k(\mathbf{x}, \mathbf{z})$. *Let* $\mathbf{K} \in \mathbb{R}^{m \times m}$ *be a kernel matrix on* $\mathcal{S}$, $[\mathbf{K}]_{ij} = k(\mathbf{x}_i, \mathbf{x}_j)$. *Then* $\forall \epsilon > 0$,

$$\mathrm{P}\left\{\left|\frac{1}{m}\lambda_i(\mathbf{K}) - \mathbb{E}_{\mathcal{S}}\left[\frac{1}{m}\lambda_i(\mathbf{K})\right]\right| > \epsilon\right\} \leq 2\exp\left(\frac{-2\epsilon^2 m}{\rho^2}\right), 0 \leq i < m,$$

where $\rho = \min\{\theta\lambda_1(\mathbf{K}), R^2\}$.

Proof. This corollary can be easily concluded from theorem 6 and 3.   □

Remark 3. As R^2 is defined on whole $\mathbb{R}^n$, it is usually intractable to compute R^2 for all kernels under all input space, unless we can guarantee that $k(\mathbf{x}, \mathbf{x})$ is bounded. This limits the practical applicability of this bounds. Thus $\theta\lambda_1(\mathbf{K})$ is of more practical use.

Consider the following example similar to [17].

Example 1. We construct a Mercer kernel using an orthogonal set of functions and a sequence of eigenvalues. To keep the example simple, consider Legendre polynomials $\psi_n(x)$, which are orthogonal polynomials on $[-1, 1]$. We take the first 20 polynomials, and set $\lambda_i = \exp(-i)$. Then,

$$k(\mathbf{x}, \mathbf{z}) = \sum_{i=0}^{19} \upsilon_i e^{-i} \psi_i(\mathbf{x})\psi_i(\mathbf{z})$$

defines a Mercer kernel, where $\upsilon_i = 1/(2i + 1)$ are normalization factors such that ψ_i have unit norm with respect to the probability measure induced by $\mu([a, b]) = |b - 1|/2$. Figure 1 shows the approximation errors (box plots) and the true eigenvalues (dot line) for the largest 20 eigenvalues. The dashed line plots the bound $\theta\lambda_1(\mathbf{K})$ on the approximation error as observed on the data, and the solid line plots the possible bound R^2. It is easy to see from Figure 1 that the proposed bound is tighter than existing one. This is consistent with theoretical analysis.

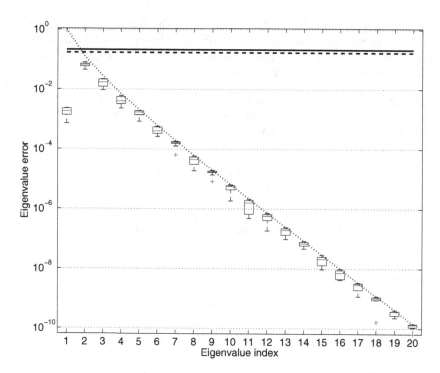

Fig. 1. The approximation errors (box plots) and the true eigenvalues (dot line) for the largest 20 eigenvalues. The dashed line plots the bound $\theta\lambda_1(\mathbf{K})$ on the approximation error as observed on the data, and the solid line plots the possible bound R^2.

Corollary 2. *Let the assumptions in corollary 1 hold. If the condition number h of kernel matrix $\mathbf{K}$ satisfies $h \leq 1/\theta$, then $\theta\lambda_1(\mathbf{K}) \leq R^2$.*

Proof. Since condition$(\mathbf{K}) = h$, then

$$\frac{m\lambda_1(\mathbf{K})}{h} \leq \sum_{i=1}^{m} \lambda_i(\mathbf{K}) = \mathrm{Tr}(\mathbf{K}) = \sum_{i=1}^{m} k(\mathbf{x}_i, \mathbf{x}_i) \leq mR^2.$$

Thus

$$\lambda_1(\mathbf{K}) \leq hR^2.$$

As $0 \leq \theta \leq 1$ and $h \leq 1/\theta$, we have

$$\theta\lambda_1(\mathbf{K}) \leq R^2.$$

This concludes the proof. □

Remark 4. Corollary 2 provides a condition that $\theta\lambda_1$ in place of R^2 gives a tighter bound. Specifically, if the condition number of a kernel matrix $\mathbf{K}$ is small enough (approaches to 1 in the limit), we can approximately have $\theta\lambda_1 \leq R^2$. In this situation, the obtained bound is better than the existing one.

Remark 5. However, if additional information on the kernel, or the parameter θ is fixed to a particular value, the bounds R^2 could be improved leading to more accurate bounds. We next discuss a case where R^2 can be computed precisely and is guaranteed to be smaller than $\theta\lambda_1$.

Corollary 3. *Let the assumptions in corollary 1 hold. If k is a translate invariant kernel, i.e. $k(\mathbf{x}, \mathbf{z}) = r(\|\mathbf{x} - \mathbf{z}\|)$, where $r(\cdot)$ is an arbitrary real function, suppose $\theta = 1$, then $R^2 \le \theta\lambda_1(\mathbf{K})$.*

Proof. Since

$$R^2 = \max_{X \in \mathbb{R}^n} k(X, X) = r(0),$$

then

$$\mathrm{Tr}(\mathbf{K}) = \sum_{i=1}^{m} k(\mathbf{x}_i, \mathbf{x}_i) = \sum_{i=1}^{m} \lambda_i(\mathbf{K}) = mr(0).$$

Further more, we have $\lambda_1 \ge \cdots \ge \lambda_n$, then

$$\lambda_1(\mathbf{K}) \ge \frac{1}{m} \sum_{i=1}^{m} \lambda_i(\mathbf{K}) \ge r(0) \ge R^2,$$

which concludes the proof.   □

4   Applications

In this section, we present two applications of the above results in two active research themes in kernel methods: kernel matrix selection and kernel target alignment.

4.1   Kernel Matrix Selection

The problem of assessing the quality of a kernel matrix is central to the theory of kernel methods, and deeply related to the problem of model selection, a classical theme in machine learning. In a practical kernel-based learning procedure, the kernel matrix may be partly or fully "uncertain" because of a crude knowledge of it. All that is known about the kernel matrix is that it belongs to a given class $\mathcal{K}$. Our goal is to choose particular kernel matrix that meets certain criterion form $\mathcal{K}$. A kernel matrix can be decomposed into two parts: its set of eigenvectors and its spectrum (set of eigenvalues). We will fix the eigenvectors and tune the spectrum from the data. For a kernel matrix $\mathbf{K}_0 = \mathbf{V}\mathbf{U}\mathbf{V}^\top$ and $c > 0$ we consider the spectral class of $\mathbf{K}_0$, given by

$$\mathcal{K} = \left\{\mathbf{K}_j | \mathrm{Tr}(\mathbf{K}_j) = c, \mathbf{V}^\top \mathbf{K}_j \mathbf{V} = \mathbf{\Lambda}, 1 \le j \le t\right\}$$
$$= \left\{f_j(\mathbf{K}_0) | \mathrm{Tr}(f(\mathbf{K}_0)) = c, f_j \in C\left(\mathbb{R}_0^+\right), 1 \le j \le t\right\},$$

where $\mathbf{V}$ is an orthogonal matrix, $\mathbf{\Lambda}$ is a diagonal matrix.

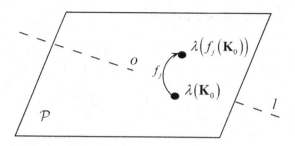

Fig. 2. The eigenvalues of $f_j(\mathbf{K}_0)$ are in a hyperplane $\mathcal{P}$ in $\mathbb{R}_+^m$. l is the normal vector of $\mathcal{P}$ that passes through the origin, O is the foot.

Remark 6. Notice that this class can be considered as the convex hull of the matrices $c\mathbf{v}_i\mathbf{v}_i^\top$ where $\mathbf{v}_i$ are the eigenvectors (columns of $\mathbf{V}$). For all $f_j(\mathbf{K}_0)$ in $\mathcal{K}$, $f_j(\mathbf{K}_0)$ has the same eigenspace.

As $f_j(\mathbf{K}_0) = \mathbf{V}f_j(\mathbf{\Lambda})\mathbf{V}^\top$, we have $\lambda_i(f_j(\mathbf{K}_0)) = f_j(\lambda_i)$. The eigenvalues of $f_j(\mathbf{K}_0)$ are actually constrained in a hyperplane $\mathcal{P}$ in $\mathbb{R}_+^m$, $\mathcal{P}: \sum_{i=1}^m p_i = c$, $p_i \in \mathbb{R}_+$ (see Figure 2). The f_j maps the point $(\lambda_1, \ldots, \lambda_m)$ to the point $(f_j(\lambda_1), \ldots, f_j(\lambda_m))$.

Theorem 7. *Fix the parameter θ and class $\mathcal{K}$, if there exists a f^* that maps $(\lambda_1, \ldots, \lambda_m)$ to the foot O (see Figure 2), or phrased differently, the condition number of $f^*(\mathbf{K}_0)$ approaches to 1, then the eigenvalues of $f^*(\mathbf{K}_0)$ have the tightest probabilistic error bound.*

Proof. Since $\lambda(f^*(\mathbf{K}_0)) \to O$, then

$$\lambda_1(f^*(\mathbf{K}_0)) = \lambda_2(f^*(\mathbf{K}_0)) = \cdots = \lambda_m(f^*(\mathbf{K}_0)) = \frac{1}{m}c.$$

Thus

$$\lambda_1(f^*(\mathbf{K}_0)) \le \lambda_1(f_j(\mathbf{K}_0)).$$

Using (5), we can conclude the proof.   □

4.2   Kernel Target Alignment

The kernel target alignment (KTA) [18] defined as follows is a quantitative measure of the agreement between a kernel and a learning task. By adapting a kernel to the labels and vice versa, a series of methods for clustering and transduction, kernel combination and selection can be derived.

Definition 2 (Alignment [18]). *The alignment of a kernel k_1 with a kernel k_2 is the quantity*

$$A(k_1, k_2) = \frac{\langle k_1, k_2 \rangle_p}{\sqrt{\langle k_1, k_1 \rangle_p \langle k_2, k_2 \rangle_p}}, \tag{6}$$

where $\langle f, g \rangle_p = \int_{\mathcal{X}^2} f(\mathbf{x}, \mathbf{z}) g(\mathbf{x}, \mathbf{z}) dP(\mathbf{x}) P(\mathbf{z})$.

One can consider the special case where $k_2 = y(\mathbf{x}) \otimes y(\mathbf{x})$. Then the alignment between a kernel and a target $y(\mathbf{x})$ can be expressed as

$$A(y) = \frac{\langle y \otimes y, k \rangle_p}{\|y \otimes y\|_p \|k\|_p} = \frac{\langle y \otimes y, k \rangle_p}{\|k\|_p},$$

where $\|y \otimes y\|_p^2 = \int_{\mathcal{X}^2} y(\mathbf{x}) y(\mathbf{z}) y(\mathbf{x}) y(\mathbf{z}) dP(\mathbf{x}) P(\mathbf{z}) = \left(\int_{\mathcal{X}} y(\mathbf{x})^2 dP(\mathbf{x}) \right)^2 = 1$.

In all real problems the value of the alignment $A(y)$ is impossible to estimate. One can, however, obtain an empirical alignment from sample set.

Definition 3 (Empirical Alignment [18]). *Given sample set* $S = \{(\mathbf{x}_i, y_i)\}_{i=1}^m$, $\mathbf{x}_i \in \mathbb{R}^n$, $y_i \in \{-1, +1\}$, *Mercer kernel function* $k(\mathbf{x}, \mathbf{z})$. *Let* $\mathbf{K} \in \mathbb{R}^{m \times m}$ *be a kernel matrix on* S, $[\mathbf{K}]_{ij} = k(\mathbf{x}_i, \mathbf{x}_i)$. *Then the kernel target alignment of* $\mathbf{K}$ *is*

$$A(\mathbf{K}) = \frac{\langle \mathbf{K}, \mathbf{Y}\mathbf{Y}^\top \rangle_{\mathsf{F}}}{\sqrt{\langle \mathbf{Y}\mathbf{Y}^\top, \mathbf{Y}\mathbf{Y}^\top \rangle_{\mathsf{F}} \langle \mathbf{K}, \mathbf{K} \rangle_{\mathsf{F}}}} = \frac{\mathbf{Y}^\top \mathbf{K} \mathbf{Y}}{m \|\mathbf{K}\|_{\mathsf{F}}}, \tag{7}$$

where $\mathbf{Y} = (y_1, \ldots, y_n)^\top$, $\langle \cdot, \cdot \rangle_{\mathsf{F}}$ *is Frobinues inner product,* $\| \cdot \|_{\mathsf{F}}$ *is the norm induced by* $\langle \cdot, \cdot \rangle_{\mathsf{F}}$.

A crucial property of the alignment for practical application is given the following theorem showing that it can be reliably estimated from its empirical estimate. This theorem makes use of theorems 2 and 5 to derive a probabilistic finite sample size bound on the approximation error of KTA.

Theorem 8. *Given a sample set* $S = \{(\mathbf{x}_i, y_i)\}_{i=1}^m$, $\mathbf{x}_i \in \mathbb{R}^n$, $y_i \in \{-1, +1\}$. *For a Mercer kernel function* $k(\mathbf{x}, \mathbf{z})$ *with feature vectors of norm 1, let* $\mathbf{K} \in \mathbb{R}^{m \times m}$ *be a kernel matrix on* S, $[\mathbf{K}]_{ij} = k(\mathbf{x}_i, \mathbf{x}_i)$. *Then* $\forall \epsilon > 0$

$$P\{|A(\mathbf{K}) - A(y)| > \epsilon\} \leq 2 \exp\left(\frac{-2\epsilon^2 (m-1)^2}{mC^2(\theta)} \right),$$

where $C(\theta) = |A(\mathbf{K})| \theta^{-1} \left(m - (m-1)\theta + \frac{2m-1}{\|\mathbf{K}\|_{\mathsf{F}}} \right)$, $\theta \in [0, 1]$.

Proof. Let $\hat{\mathbf{K}}$ be the kernel matrix on $S \setminus \{(\mathbf{x}_k, y_k)\}$, $[\hat{\mathbf{K}}]_{ij} = k(\mathbf{x}_i, \mathbf{x}_i)$. Using theorem 5 we have

$$\|\hat{\mathbf{K}}\|_{\mathsf{F}} = \sqrt{\sum_{i=1}^m \lambda_i^2(\hat{\mathbf{K}})} = \sqrt{\sum_{i=1}^m \theta_i^2 \lambda_i^2(\mathbf{K})}$$

$$\geq \sqrt{\sum_{i=1}^m \theta^{[k]^2} \lambda_i^2(\mathbf{K})} = \theta^{[k]} \|\mathbf{K}\|_{\mathsf{F}}, \tag{8}$$

where

$$\theta^{[k]} = \min_{i=1,\ldots,m-1} \frac{\lambda_i(\hat{\mathbf{K}})}{\lambda_i(\mathbf{K})}.$$

Let $\hat{\mathbf{Y}} = (y_1, \ldots, y_{k-1}, y_{k+1} \ldots, y_m)^\top$, then

$$\left| A(\mathbf{K}) - A(\hat{\mathbf{K}}) \right|$$

$$= \left| \frac{\langle \mathbf{K}, \mathbf{Y}\mathbf{Y}^\top \rangle_{\mathsf{F}}}{m \|\mathbf{K}\|_{\mathsf{F}}} - \frac{\langle \hat{\mathbf{K}}, \hat{\mathbf{Y}}\hat{\mathbf{Y}}^\top \rangle_{\mathsf{F}}}{(m-1)\|\hat{\mathbf{K}}\|_{\mathsf{F}}} \right|$$

$$\leq \left| \frac{\langle \mathbf{K}, \mathbf{Y}\mathbf{Y}^\top \rangle_{\mathsf{F}}}{m \|\mathbf{K}\|_{\mathsf{F}}} - \frac{\langle \mathbf{K}, \mathbf{Y}\mathbf{Y}^\top \rangle_{\mathsf{F}}}{(m-1)\|\hat{\mathbf{K}}\|_{\mathsf{F}}} \right| + \left| \frac{\langle \mathbf{K}, \mathbf{Y}\mathbf{Y}^\top \rangle_{\mathsf{F}}}{(m-1)\|\hat{\mathbf{K}}\|_{\mathsf{F}}} - \frac{\langle \hat{\mathbf{K}}, \hat{\mathbf{Y}}\hat{\mathbf{Y}}^\top \rangle_{\mathsf{F}}}{(m-1)\|\hat{\mathbf{K}}\|_{\mathsf{F}}} \right|$$

$$\leq \left| \langle \mathbf{K}, \mathbf{Y}\mathbf{Y}^\top \rangle_{\mathsf{F}} \right| \left(\frac{1}{(m-1)\|\hat{\mathbf{K}}\|_{\mathsf{F}}} - \frac{1}{m\|\mathbf{K}\|_{\mathsf{F}}} \right) + \frac{2m-1}{(m-1)\|\hat{\mathbf{K}}\|_{\mathsf{F}}}$$

$$\leq |A(\mathbf{K})| \left(\frac{m}{(m-1)\theta^{[k]}} - 1 \right) + \frac{2m-1}{(m-1)\theta^{[k]}\|\mathbf{K}\|_{\mathsf{F}}}$$

$$= \frac{|A(\mathbf{K})|}{(m-1)\theta^{[k]}} \left(m - (m-1)\theta^{[k]} + \frac{2m-1}{|\langle \mathbf{K}, \mathbf{Y}\mathbf{Y}^\top \rangle_{\mathsf{F}}|} \right)$$

$$= \frac{1}{m-1} C\left(\theta^{[k]} \right),$$

where the first inequality is derived from the absolute value inequality, the second inequality is derived from the assumption that feature vectors are of norm 1, the third inequality is derived from inequality (8). Thus

$$\sup_{k=1,\ldots,m} \left| A(\mathbf{K}) - A(\hat{\mathbf{K}}) \right| \leq \frac{1}{m-1} C(\theta).$$

where

$$\theta = \max_{k=1,\ldots,m} \left(1 - \theta^{[k]} \right).$$

Hence, we have $c_i = C(\theta)/(m-1)$ for all i and we obtain from an application of McDiarmid' theorem 2 that

$$P\left\{ |A(\mathbf{K}) - A(y)| > \epsilon \right\} \leq 2 \exp\left(\frac{-2\epsilon^2(m-1)^2}{mC^2(\theta)} \right),$$

which concludes the proof.   □

Remark 7. If one wants to obtain unbiased convergence, it is necessary to slightly modify the definition of $A(y)$ by removing the diagonal, since for finite samples it biases the expectation by receiving too large a weight. With this modification $A(y)$ in the statement of the theorem becomes the 'true alignment'. Using the definition of true alignment it is possible to prove that asymptotically as m tends to infinity the empirical alignment as defined above converges to the true alignment. However the change is not significant.

5   Conclusion

We have proposed a probabilistic error bound for eigenvalues of kernel matrix. This bound scales with the dominate eigenvalue under consideration, leading to conditionally more accurate bound for single eigenvalues than previously known bound, in particular for slowly decaying eigenvalues. Two applications of this theoretical finding have also been presented.

Note that the resulting bound needs the knowledge of the parameter θ, which requires extra computational efforts compared with existing non-scaling bounds. Thus, efficient calculating methods for θ need to be investigated in further works. Attentions of applying this bound to other themes in kernel methods are also deserved.

Acknowledgment

This work was supported in part by Natural Science Foundation of China under Grant No. 60678049 and Natural Science Foundation of Tianjin under Grant No. 07JCYBJC14600. The authors would like to thank the four anonymous reviewers for their detailed comments that helped to improve the quality of the paper.

References

1. Shawe-Taylor, J., Cristianini, N.: Kernel Methods for Pattern Analysis. Cambridge University Press, Cambridge (2005)
2. Shawe-Taylor, J., Cristianini, N., Kandola, J.: On the concentration of spectral properties. In: Dietterich, T.G., Becker, S., Ghahramani, Z. (eds.) Advances in Neural Information Processing Systems 14, pp. 511–517. MIT Press, Cambridge (2002)
3. Schölkopf, B., Smola, A., Muller, K.: Kernel principal component analysis. In: Schölkopf, B., Burges, C., Smola, A. (eds.) Advances in Kernel Methods: Support Vector Machines, pp. 327–352. MIT Press, Cambridge (1998)
4. Saunders, C., Gammerman, A., Vovk, V.: Ridge regression learning algorithm in dual variables. In: Proceedings of the 15th International Conference on Machine Learning, pp. 24–27 (1998)
5. Trefethen, L., Bau, D.: Numerical Linear Algebra. SIAM, Philadelphia (1997)
6. Ng, A.Y., Zheng, A.X., Jordan, M.I.: Link analysis, eigenvectors and stability. In: Proceedings of the 7th International Joint Conference on Artificial Intelligence, pp. 903–910 (2001)
7. Martin, S.: The numerical stability of kernel methods. Technical report, Sandia National Laboratory, New Mexico (2005)
8. Jia, L., Liao, S.: Perturbation stability of computing spectral data on kernel matrix. In: The 26th International Conference on Machine Learning, Workshop on Numerical Mathematics on Machine Learning (2009)
9. Williams, C., Seeger, M.: The effect of the input density distribution on kernel-based classifiers. In: Langley, P. (ed.) Proceedings of the 17th International Conference on Machine Learning, pp. 1159–1166. Morgan Kaufmann, San Francisco (2000)

10. Bengio, Y., Vincent, P., Paiement, J., Delalleau, O., Ouimet, M., Le Roux, N.: Spectral clustering and kernel PCA are learning eigenfunctions. Technical Report TR 1239, University of Montreal (2003)
11. Shawe-Taylor, J., Williams, C.K., Cristianini, N., Kandola, J.: On the eigenspectrum of the gram matrix and the generalization error of kernel-PCA. IEEE Transactions on Information Theory 51(7), 2510–2522 (2005)
12. Zwald, L., Blanchard, G.: On the convergence of eigenspaces in kernel principal component analysis. In: Weiss, Y., Schölkopf, B., Platt, J. (eds.) Advances in Neural Information Processing Systems 18, pp. 1649–1656. MIT Press, Cambridge (2006)
13. Koltchinskii, V.: Asymptotics of spectral projections of some random matrices approximating integral operators. Progress in Probability 43, 191–227 (1998)
14. Koltchinskii, V., Giné, E.: Random matrix approximation of spectra of integral operators. Bernoulli 6(1), 113–167 (2000)
15. Horn, R.A., Johnson, C.R.: Matrix Analysis. Cambridge University Press, Cambridge (1985)
16. Braun, M.L.: Accurate error bounds for the eigenvalues of the kernel matrix. Journal of Machine Learning Research 7, 2303–2328 (2006)
17. Braun, M.L.: Spectral Properties of the Kernel Matrix and their Relation to Kernel Methods in Machine Learning. PhD thesis, Bonn University (2005)
18. Cristianini, N., Kandola, J., Elisseeff, A., Shawe-Taylor, J.: On kernel-target alignment. Journal of Machine Learning Research 1, 1–31 (2002)

Community Detection on Weighted Networks: A Variational Bayesian Method

Qixia Jiang, Yan Zhang, and Maosong Sun

State Key Laboratory on Intelligent Technology and Systems
Tsinghua National Laboratory for Information Science and Technology
Department of Computer Science and Technology, Tsinghua University, Beijing 100084, China
qixia.jiang@gmail.com, zhang-y-05@mails.tsinghua.edu.cn,
sms@mail.tsinghua.edu.cn

Abstract. Massive real-world data are network-structured, such as social network, relationship between proteins and power grid. Discovering the latent communities is a useful way for better understanding the property of a network. In this paper, we present a fast, effective and robust method for community detection. We extend the constrained Stochastic Block Model (conSBM) on weighted networks and use a Bayesian method for both parameter estimation and community number identification. We show how our method utilizes the weight information within the weighted networks, reduces the computation complexity to handle large-scale weighted networks, measure the estimation confidence and automatically identify the community number. We develop a variational Bayesian method for inference and parameter estimation. We demonstrate our method on a synthetic data and three real-world networks. The results illustrate that our method is more effective, robust and much faster.

1 Introduction

In recent years, the application of data mining to real-life problems has led to a realization that more and more real data are network-structured. Such data involve massive relational information among objects. Given such data, people often hope to divide all the vertices within the networks into several latent communities. So we can suppress the complexity of the full description and better catch and understand the properties of these networks. It has been shown that models that use clustering of vertices are inferior to those models that find communities based on connectivity [1].

The methodology of finding communities in networks (community detection) comes from the physics community, and the algorithms for finding these communities are deterministic. These algorithms focus on optimizing an energy-based cost function. Such a cost function is always defined with fixed parameters over possible community assignments of nodes [2, 3]. A notable work proposed by Newman and Girvan [3] introduces *modularity* as a posterior measure of network structure. Modularity measures inter- and not inter- connectivity, which is with reference to the Null model[1].

---

[1] Null model is a statistical method and computer simulation for analyzing patterns in nature. All the parameters in Null model are zero except the intercept. It is also known as the *intercept-only* model.

Z.-H. Zhou and T. Washio (Eds.): ACML 2009, LNAI 5828, pp. 176–190, 2009.
© Springer-Verlag Berlin Heidelberg 2009

Modularity is influential in the community detection literatures and gains success in many applications [4, 5].

A potential problem of modularity is the assumption that all the relations in the network have been observed (complete-observation assumption). However, relational data in real world are always full of omissions, errors, and uncertainty. To overcome the shortcomings above, massive attentions are recently turned to the study of probabilistic generative models. So, Stochastic Block Model (SBM) [6,7,8,9,10,11,12,13,14,15,16] is proposed. The flexibility of this kind of models makes them be successfully applied in many domains, such as bioinformatics [9], document analysis [10, 11], image processing [12] and social networks [13].

However, much previous work on SBM [13, 14, 15, 16] is infeasible in community detection task, especially when the network is large-scale or weighted. This is mainly because:

1. *Information Loss.* The weight of relations reflects the importance of the interaction between objects. This important information is lost in those models that only consider binary relation [6, 13]. They just utilize the presence or absence information of relations and ignore the importance of them, which has been proven to be infeasible in many real-world applications [15, 17]. Under this consideration, we hope to utilize the weight information within the weighted networks to improve the performance of community detection;

2. *Computation Complexity.* Some algorithms [13, 14] model the interaction patterns between communities. In SBM, interaction patterns between communities are modeled as the probability of connecting two nodes within the different (or the same) communities, i.e. $p(e_{ij}|$ node i in the community g, node j in the community h), where e_{ij} is the edge between the node i and the node j. In this paper, we argue that if we focus on detecting the latent communities, it is unnecessary to model the interaction patterns. We only need to model the densities of intra- and inter- connectivity for nodes in the same or not the same communities, i.e. $p(e_{ij}|$ the nodes i and j are in the same community) and $p(e_{ij}|$ the nodes i and j are not in the same community). This can reduce the computation complexity and the performance is also comparable.

3. *Robustness and Accuracy.* Much previous work [13, 14, 15, 17, 16] uses maximum likelihood (ML) estimation to find the optimum model parameters. However, ML is a point estimation method, such that we can not measure the estimation quality or confidence. Furthermore, in much previous work [13, 14], the number of communities are estimated by Bayesian Information Criterion (BIC) and Integrated Classification Likelihood (ICL). Both BIC and ICL are suitable for single-peaked likelihood functions well-approximated by Laplace integration and studied in the large-N limit [6]. However, for a SBM, the assumption of a single-peaked function is invalidated by the underlying symmetries of the latent variables. So we hope to identify the number of communities automatically and obtain the estimation quality.

Motivated by above three problems, we extend the constrained Stochastic Block Model (conSBM) [6] on weighted networks. To utilize the weight information within the weighted networks, we model the densities of intra- and inter- connectivity for nodes by different distributions according to the different edge types. To obtain the better estimation of model parameters and the community number, our method allows to include the prior belief on both the model parameters and the community number. The inductive prior belief leads to two benefits: (1) We can use a uniform Bayesian estimation framework to simultaneously learn the model parameters and identify the community number; (2) Allowing a prior distribution over model parameters, we can incorporate expectation as another parameter estimate as well as variance information as a measure of estimation quality or confidence. Furthermore, to handle the large-scale weighted networks, unlike some previous work models the complex interaction patterns, we only model the densities of intra- and inter- connectivity for nodes in the same or not the same communities. This can speed up the algorithm and the performance is also comparable. And we develop a variational Bayesian method [18] for inference and parameter estimation.

In short, the contributions of our work can be summarized as follows:

- utilizing the weight information within the weighted networks to improve the performance of community detection;
- identifying the number of communities and estimating the model parameters in one uniform Bayesian framework and measuring the estimation quality;
- reducing the computation complexity to handel large-scale weighted networks.

We evaluate our method on the synthetic data and three real datasets, including a gene dataset, a scientist collaboration network and a word co-occurrence network. We also compare our approach with some previous methods. The results show that our method is much faster, more effective and robust.

The rest of the paper is organized as follows. Section 2 introduces preliminaries. In Section 3, our method is presented. The variational Bayesian approach for our method is given in Section 4. Section 5 presents the empirical results. Section 6 concludes the paper.

2   Preliminaries

Community detection is one important task of many relational data analysis problems. These relational data can be represented as a weighted graph $G = \{V, E\}$, directed or undirected, where $V = \{v_1, v_2, \ldots, v_N\}$ is the node set and $E = \{e_1, e_2, \ldots, e_M\}$ is the edge set. The size of the sets V, E is N and M respectively. The elements in the node set V are the vertices or objects of networks. The elements in the edge set E represent the relationships between the corresponding objects, and each edge e_m can be represented as a measurement taken on a pair of nodes (v_i, v_j), i.e. $e_m \equiv e_{ij}$, $m \in [1, M]$ and $i, j \in [1, N]$. In particular, if the graph is directed, the edge e_{ij} means that the object v_i interacts to or response to the object v_j, while if the graph is undirected, the edge e_{ij} means the co-occurrence of the objects v_i and v_j. The edges can take any type of values, i.e. a real number, a class label (multinomial), a positive integer, and so on. The value of the edge e_{ij} represents the importance of the relationship between corresponding two nodes v_i and v_j. For example, if e_{ij} takes a positive real number ($e_{ij} > 0$) means: (1) *presence information*. There exists an relation between such two nodes; (2) *importance information*. The weight reflects the importance of this relationship. Optionally, such a weighted graph can also be represented as an $N \times N$ dimensional adjacency matrix $\boldsymbol{E}$.

Given such a weighted graph, we hope to find a fast, robust and effective probabilistic model to detect the latent communities.

3   Model Formulation

Given a network, we often wish to model this network. So we can suppress the complexity of the full description and better catch and understand the properties of such a network. An effective approach is to divide the nodes into different latent communities, which is also called clustering or community detection. In this section, we will describe our approach, which is an extension of constrained Stochastic Block Model(conSBM) on weighted networks. We will also show the relationships with some previous work.

3.1   Constrained Stochastic Block Model

We assume there exist K latent communities within the N-node network. We hope to divide these N nodes into such K communities. We define $\sigma_i \in \{1, \ldots, K\}$ to be the community indicator of the i-th node. To detect the latent communities, we use the constrained Stochastic Block Model (conSBM), which consists of a multinomial distribution over the community assignments with the weights $\pi_\mu \equiv p(\sigma_i = \mu | \pi)$. After each node instantiates its community assignment, we will connect these N nodes to form the whole network. One important principle is that the nodes within the same community tend to be connected and have larger weight while the nodes within the different communities tend not to be connected and have smaller weight. The relationship between the nodes v_i and v_j is associated with a random variable e_{ij}, whose value reflects the importance of this relationship. Given the community assignments $\boldsymbol{\sigma}$, the edges are supposed to be independent. The value of an edge e_{ij} is assigned according to whether the nodes v_i and v_j are within the same community. Specifically, if two nodes are within the same community, the corresponding edge e_{ij} follows a distribution $f(\boldsymbol{\theta}_1)$, and if they are not within the same community, the edge e_{ij} follows a distribution $f(\boldsymbol{\theta}_0)$[2]. Note that the form of the distribution $f(\cdot)$ varies with the type of edges within the networks. Details are discussed in Section 3.2. The complete generative progress is as follows:

- sample the number of communities $K \sim p(K)$
- sample a K dimensional membership vector $\boldsymbol{\pi} \sim p_D(\boldsymbol{\alpha})$
- sample $\boldsymbol{\theta}_1 \sim p(\boldsymbol{\beta_1})$
- sample $\boldsymbol{\theta}_0 \sim p(\boldsymbol{\beta_0})$
- for each node $v_i \in V$
    - sample a specific community assignment $\sigma_i \sim p_M(\boldsymbol{\pi})$
- for each pair of nodes $(v_i, v_j) \in V \times V$
    - sample edge $e_{ij} \sim f(\boldsymbol{\theta}_{\delta_{\sigma_i \sigma_j}})$

---

[2] For simpleness, we assume $f(\cdot)$ is a distribution with a finite dimensional parameters.

where $p(K)$ is the prior belief on the number of communities, $p_D(\cdot)$ is a Dirichlet, $p_M(\cdot)$ is a Multinomial, $p(\beta_1)$ and $p(\beta_0)$ are the prior distributions of intra- and inter-community connection parameters θ_1 and θ_0 respectively, and $\delta_{\sigma_i,\sigma_j}$ is the *Kronecker delta*, which equals to one whenever $\sigma_i = \sigma_j$ and zero otherwise. This generative progress defines a (log) joint probability distribution:

$$\log p(E, \sigma, \pi, \theta_1, \theta_0 | \alpha, \beta_1, \beta_0) = \log p(\pi | \alpha) + \log p(\theta_1 | \beta_1) + \log p(\theta_0 | \beta_0)$$
$$+ \sum_{i>j} \delta_{\sigma_i \sigma_j} \log f(e_{ij} | \theta_1) + \sum_{i>j}(1 - \delta_{\sigma_i \sigma_j}) \log f(e_{ij} | \theta_0) + \sum_{\mu} \log \pi_{\mu} \sum_{i} \delta_{\sigma_i \mu} \quad (1)$$

The generative progress shows that our method is defined on the undirected weighted networks. However, our method is also suitable when the networks are directed. Equation 1 still holds for directed weighted networks, simply replacing the sum over $i > j$ by a sum over $i \neq j$.

3.2 Edge Types

To complete our method, we will discuss some different edge types in this section. These types are common in many real-world applications. We should emphasize that the type of edges determines the form of distribution $f(\cdot)$. Furthermore, for simplicity of operation, we assume the prior distributions of intra- and inter- community connection parameters are the conjugate distributions of $f(\cdot)$.

Bernoulli Distribution. In many applications, people only utilize the presence and absence information of an edge, i.e. Internet, protein-protein interaction network, power grid, etc. In such networks, the edges are assumed to follow a Bernoulli distribution:

$$e_{ij} | \sigma_i, \sigma_j \sim f(e_{ij} | \theta) = \theta_{\delta_{\sigma_i \sigma_j}}^{e_{ij}} (1 - \theta_{\delta_{\sigma_i \sigma_j}})^{1-e_{ij}}, e_{ij} \in \{0, 1\} \quad (2)$$

The conjugate distribution of Bernoulli is Beta distribution.

Multinomial Distribution. The relationship between nodes sometimes is a class label or assignment. For example, in social network, the edge between two people informs one existing relation: friends, colleagues, roommates and so on. So we assume such edges follow a multinomial distribution:

$$e_{ij} | \sigma_i, \sigma_j \sim f(e_{ij} | \theta_{\delta_{\sigma_i \sigma_j}}) = \prod_{k=1}^{K} \theta_{\delta_{\sigma_i \sigma_j}, k}^{\delta_{e_{ij} k}}, e_{ij} \in \{1, \ldots, K\} \quad (3)$$

The conjugate distribution of Multinomial is Dirichlet distribution.

Poisson Distribution. In collaboration networks, the value of an edge between two scientists always reflects their collaboration times. So we model the edges in these networks as Poisson variables:

$$e_{ij} | \sigma_i, \sigma_j \sim f(e_{ij} | \lambda_{\delta_{\sigma_i \sigma_j}}) = \frac{\lambda_{\delta_{\sigma_i \sigma_j}}^{e_{ij}} e^{-\lambda_{\delta_{\sigma_i \sigma_j}}}}{e_{ij}!}, e_{ij} \in \{0, 1, 2, 3, \ldots\} \quad (4)$$

The parameter $\theta_{\delta_{\sigma_i\sigma_j}} \equiv \lambda_{\delta_{\sigma_i\sigma_j}}$ can be understood as the mean of the collaboration times. The conjugate distribution of Poisson is Gamma distribution, $\Gamma(\alpha_\Gamma, \beta_\Gamma)$.

Gaussian Distribution. In traffic network or airport network, the edge e_{ij} is valued according to the number of passengers traveling from the city or the airport v_i to v_j. So we consider the edge following a Gaussian distribution:

$$e_{ij}|\sigma_i, \sigma_j \sim f(e_{ij}|\mu_{\delta_{\sigma_i\sigma_j}}, \Sigma^2_{\delta_{\sigma_i\sigma_j}}) = \frac{1}{\Sigma_{\delta_{\sigma_i\sigma_j}}\sqrt{2\pi}} e^{-\frac{(x-\mu_{\delta_{\sigma_i\sigma_j}})^2}{2\Sigma^2_{\delta_{\sigma_i\sigma_j}}}} , e_{ij} \in \mathbb{R} \quad (5)$$

the parameter $\theta_{\delta_{\sigma_i\sigma_j}}$ is $\{\mu_{\delta_{\sigma_i\sigma_j}}, \Sigma^2_{\delta_{\sigma_i\sigma_j}}\}$. For the Gaussian distribution with unknown mean and variance, the conjugate prior for the joint distribution of μ and Σ^2 is the normal inverse-gamma (Γ) distribution (i.e. normal-inverse-χ^2 N-Inv-$\chi^2(\mu_0, \Sigma_0^2/k_0; v_0, \Sigma_0^2)$).

3.3 Community Number Identification

The community number estimation is to choose the number of blocks, K. This is also considered as a special case of model selection problems. Many model selection strategies exist for hierarchical models, such as Bayesian Information Criterion (BIC) and Integrated Classification Likelihood(ICL). Both BIC and ICL are suitable for single-peaked likelihood functions well-approximated by Laplace integration and studied in the large-N limit. However, for a SBM the assumption of a single-peaked function is invalidated by the underlying symmetries of the latent variables. In our method, we consider the community number as one model parameter. We allow a prior distribution over the community number, so that we can use the Bayesian method to identify the community number. Such a problem can also be stated as follows: given a weighted graph $G = \{V, E\}$, to detect the most probable number of communities, $K^* = \arg max_K p(K|E)$. If we are lack of enough knowledge about the number of communities, we prefer a sufficient weak prior $p(K)$, such that maximizing $p(K|E) \propto p(E|K)p(K)$ is equivalent to maximizing $p(E|K)$.

3.4 Comparison with Previous Work

In this section, we compare our approach with previous work and point out the key difference between our approach and previous work.

Some work, such as M. Mariadassou and S. Robin's [14] and ICMc [16], proposes a mixture model to uncover the latent structure in weighted networks. The key difference between our approach and theirs can be concluded as follows:

1. *Interaction Patterns Modeling.* Interaction patterns are the relationships between communities. In M. Mariadassou and S. Robin's method, the probability of an edge between two nodes is defined by K^2 probability distributions F. Each distribution f_{gh} denotes the probability of connecting a node in the community g and a node in the community h. However, in the community detection task, people always ignore

the interaction patterns and focus on dividing all the nodes into K latent communities. So we just need to use two distributions to model the intra- and inter- community connection densities respectively, which is much faster in learning progress. In short, the first difference is that interaction patterns are community-specific in M. Mariadassou and S. Robin's approach and ICMc, while they are globally-shared in our method.

2. *Parameter Estimation.* Unlike a direct estimation of model parameters in M. Mariadassou and S. Robin's approach, we allow a prior distribution over the parameter set $\{\theta_1, \theta_0\}$. So our method can not only encode the maximum (a posteriori) value of data-generated parameters but also incorporate expectation as another parameter estimate as well as variance information as a measure of estimation quality or confidence.

3. *Community Number Estimation.* In M. Mariadassou and S. Robin's approach, they use the Integrated Classification Likelihood (ICL) criterion to choose the number of communities. The ICL criterion is suggested for single-peaked likelihood functions well-approximated by Laplace integration and studied in the large-N limit. However, for a SBM the assumption of a single-peaked function is invalidated by the underlying symmetries of the latent variables, i.e. nodes are distinguishable and modules indistinguishable. Thus we choose the community number by the Bayesian estimation.

Our method can be seen as an extension of Hoffman et al.'s work [6] on the weighted networks. They only model binary relations, which ignores the value of an edge and only utilizes the presence and absence information of edges. However, the value of an edge has been proven to be very important and useful in many real-world applications, such as social networks, biological networks, etc. So their work can be considered as a special case of our method when the networks are binary.

Our method can also be explained as a general case of the Potts model by defining $\mathcal{H} = -\log p(E, \sigma | \pi, \theta_1, \theta_0)$. Ignoring the constant, we can get the following form:

$$\mathcal{H} = -\sum_{i>j} \{\log f(e_{ij}|\theta_1) - \log f(e_{ij}|\theta_0)\} \cdot \delta_{\sigma_i \sigma_j} + \sum_{\mu} (-\log \pi_\mu) \cdot \sum_i \delta_{\sigma_i \mu} \quad (6)$$

Equation 6 is a Potts model Hamiltonian with constants $\log f(\cdot|\theta_1)$ and $\log f(\cdot|\theta_0)$, and $-\log \pi_\mu$ is the chemical potential.

4   Variational Bayesian Approach

The Bayesian estimation is to calculate the posterior of likelihood according to Bayes's rule:

$$p(\theta|E) = \frac{p(E|\theta) \cdot p(\theta)}{p(E)} \quad (7)$$

where E is the observations and θ is the model hyper-parameter. So the key problem is to calculate the likelihood. The likelihood is the marginal probability of observations E, which requires an integral over all the hidden variables.

$$p(E|\alpha, \beta_1, \beta_0)$$
$$= \int_{\Pi} \int_{\Theta} \int_{\Theta} \sum_{\sigma} p(E, \sigma, \pi, \theta_1, \theta_0 | \alpha, \beta_1, \beta_0) d\pi d\theta_1 d\theta_0 \tag{8}$$

which is intractable to compute in a closed form [13, 10]. In this section, we develop an approximation algorithm based on a variational method [18] for inferring the posterior distribution of the membership of each node. The basic idea behind variational methods is to posit a variational distribution $q(\pi, \sigma, \theta_1, \theta_0)$ on the hidden variables and then fit those parameters so that the variational distribution is close to the true posterior in *Kullback-Leibler* divergence. This corresponds to maximizing a lower bound, $\mathcal{L}(\gamma, \Phi, \eta_1, \eta_0; \alpha, \beta_1, \beta_0)$, which can be obtained on the log probability of the observations with *Jensen*'s inequality.

$$\log p(E|\alpha, \beta_1, \beta_0) \geq \mathcal{H}(q) + E_q[\log p(E, \sigma, \pi, \theta_1, \theta_0)]$$
$$= \mathcal{L}(\gamma, \Phi, \eta_1, \eta_0; \alpha, \beta_1, \beta_0) \tag{9}$$

where $\mathcal{H}(q)$ is the entropy of variational distribution q. Note that the distance between log-likelihood and this lower bound $\mathcal{L}$ is the *Kullback-Leibler* divergence from variational distribution q to the true posterior distribution. So minimizing the *KL* divergence from the q to the true posterior distribution is therefor equivalent to maximizing the low bound on log-likelihood.

4.1 Variational Bayes

To get a tractable bound on $\log p(E|\alpha, \beta_1, \beta_0)$, we prefer a fully factorized variational distribution so that the optimization is easy to operate. We specify the variational distribution q as the mean-field fully-factorized family:

$$q(\pi, \sigma, \theta_1, \theta_0 | \gamma, \Phi, \eta_1, \eta_0) = q(\theta_1 | \eta_1) \cdot q(\theta_0 | \eta_0) \cdot q_D(\pi | \gamma) \cdot \prod_{i}^{N} q_M(\sigma_i | \phi_i) \tag{10}$$

where q_D is a Dirichlet and q_M is a Multinomial.

Tightening the lower bound $\mathcal{L}(\gamma, \Phi, \eta_1, \eta_0; \alpha, \beta_1, \beta_0)$ leads to the following updates for free parameters Φ and γ. For each $i \in [1 \ldots N]$ and $\mu \in [1 \ldots K]$,

$$\phi_{i\mu}^* \propto \exp\left\{ \sum_{j \neq i} \left(E_q[\log f(e_{ij}|\theta_1)] - E_q[\log f(e_{ij}|\theta_0)] \right) \phi_{j\mu} + E_q[\log \pi_\mu] \right\} \tag{11}$$

$$\gamma_\mu^* = \sum_{i=1}^{N} \phi_{i\mu} + \alpha_\mu \tag{12}$$

The updates for free variational parameters $\{\eta_1, \eta_0\}$ vary with the forms of the distributions $f(\cdot)$s. For example, when $f(\cdot)$ is a Bernoulli, the updates for η_1 are $\eta_1^* = \beta_1 + \sum_{\mu=1}^{K} \sum_{i>j} \phi_{i\mu} \phi_{j\mu} e_{ij}$, and when $f(\cdot)$ is a multinomial, the updates for η_1 are $\eta_{1\mu}^* = \beta_{1\mu} + \sum_{\mu=1}^{K} \sum_{i>j} \phi_{i\mu} \phi_{j\mu} \delta_{e_{ij}\mu}$. The variational Bayesian approach is summarized in Figure 1.

Procedure. `Variational Bayesian Approach`

---

initialize $\boldsymbol{\Phi}^0$ randomly;
initialize $\gamma_{i,k}^0 = \frac{2N}{K}$ for all i, k;
initialize $\eta_1{}^0, \eta_0{}^0$;
repeat
    for $i = 1$ *to* N **do**
        update $\boldsymbol{\phi}_i{}^t$;
        normalize $\boldsymbol{\phi}_i{}^t$ to sum to 1;
    end
    update $\eta_1{}^t, \eta_0{}^t$ and $\boldsymbol{\gamma}^t$;
    $t \leftarrow t + 1$;
until *convergence* ;

Fig. 1. Variational Bayesian Approach

5   Experiments

In this section, we present the experimental results on the synthetic data and three real datasets including a gene regulations in A.Thaliana, a scientist collaboration network and a word co-occurrence network. These experimental results demonstrate the benefits of our method. We compare our approach with other related work on weighted networks.

5.1   Synthetic Data

We construct a binary network with different number of communities. The structure of such a network is shown in Figure 2(a). In each community, there exist four nodes and they are fully-connected. Figure 2(b) presents the comparison of our method with the

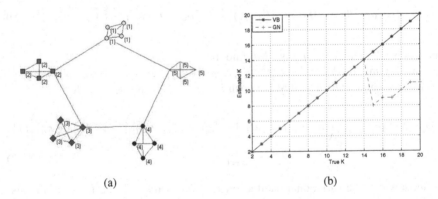

(a)

(b)

Fig. 2. Synthetic Data. a) The structure of synthetic network. The communities are designated by color and shape. Nodes within the same community are fully connected. b) The estimated number of communities (K). The diagonal means consistently identifying all the correct number of modules.

Girvan-Newman modularity [1] in the resolution limit test [19]. We can see that our approach can consistently identifies the correct number of modules, while the Girvan-Newman modularity fails when the number of true communities is large.

5.2   Gene Regulations in A. Thaliana

In this section, we will evaluate our approach on the gene regulations in A. Thaliana. The dataset is the partial correlations between the expression levels of 800 genes in various conditions [20]. We specify the partial correlations as an 800×800 adjacency matrix as in Figure 3(a), where each entry is a partial correlation coefficient. A partial correlation coefficient quantifies the correlation between two variables (e.g. gene expression levels) when conditioning on one or several other variables. A white point in

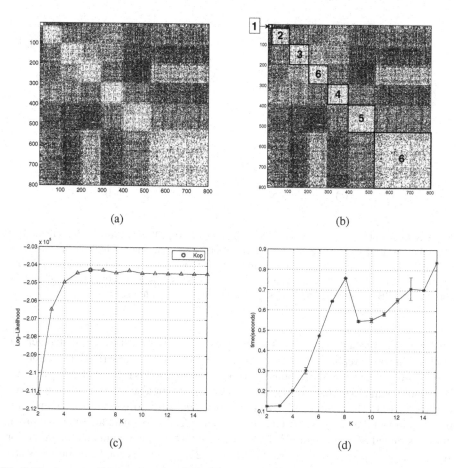

(a)   (b)

(c)   (d)

Fig. 3. The gene regulations in A. Thaliana. a) A adjacency matrix represents partial correlations between the expression levels of 800 genes in various conditions; b)The result of our approach; c) The log-likelihood for each K. The peak indicates the number of latent communities; d) The running time of our method with the different values of K. Note that our method only spends $7.221(\pm0.525)$ seconds on the total computational time for $K = 2, \ldots, 15$.

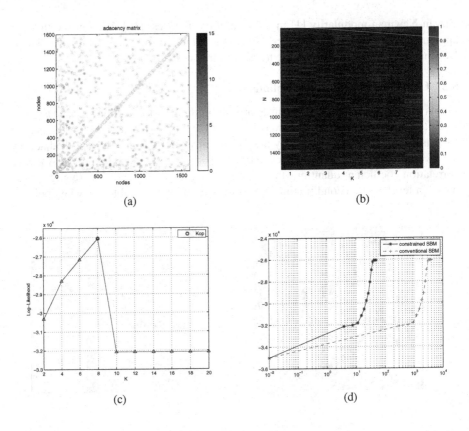

Fig. 4. Science collaboration network. a) The adjacency matrix. The gray level reflects the weight of a relationship; b) The structure discovered by our approach. The red indicates membership of each node to the corresponding community; c) The log-likelihood for each K. The peak indicates the number of latent communities $K^* = 8$; d) The running time of our method and the conventional SBM. Note that the X axis (seconds) is in the logarithm scale.

Figure 3(a) means a positive relation between two genes (a positive correlated genes), while a blue point represents a negative one. We set the $f(\cdot)$ as described in Section 3.1 to be a two dimensional Multinomial distribution. The result is presented in Figure 3(b). We should emphasize that M. Mariadassou and S. Robin also evaluate their approach on this dataset. They improperly split the community 6 into two separate communities, and in fact they identify seven communities. We also estimate the running time of our constrained SBM in Figure 3(d). We can see that the total computational time for $K = 2, \ldots, 15$ classes on a standard PC is $7.221(\pm 0.525)$ seconds, while M. Mariadassou and S. Robin report that their approach spends 1 hour [21].

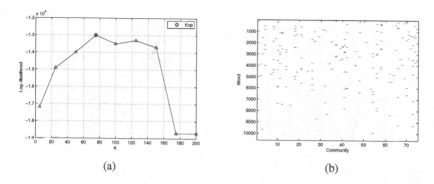

(a)   (b)

Fig. 5. a) Log-likelihood for each K. We measure the number of blocks on the X axis and the log-likelihood on dataset EAT on the Y axis. The peak indicates the number of latent blocks, $K^* = 75$; b) The structure discovered by our approach. The gray scale reflects the community membership of each node.

Table 1. An illustration of 10 communities from a 75-community solution for the *FreeAssoc* dataset. Each community is illustrated with the top 10 words that have the highest probability conditioned on that community.

c10:royalty	c70:speak	c75:emotion	c50:degree	c52:animal
king	speak	alone	great	bird
monarchy	talk	happy	superior	insect
kingdom	say	guilt	broad	grasshopper
queen	discussion	sad	narrow	parrot
princess	topic	joy	height	eagle
prince	communicate	lonely	compact	robin
royal	conversation	isolated	giant	dove
monarch	communication	regret	long	seagull
royalty	gossip	shame	vast	pelican
emperor	discuss	despair	magnitude	quail
c60:road	**c54:bread**	**c33:astronomy**	**c29:book**	**c66:vegetable**
street	nut	earth	book	cauliflower
alley	roll	asteroid	story	cabbage
road	almond	meteor	text	lettuce
way	butter	comet	novel	coleslaw
lane	bread	Saturn	publisher	celery
map	doughnut	universe	page	tomato
pavement	dough	galaxy	library	carrots
track	barley	stars	shelf	zucchini
avenue	bun	planets	handing	vagetable
highway	loaf	Mars	reader	turnip

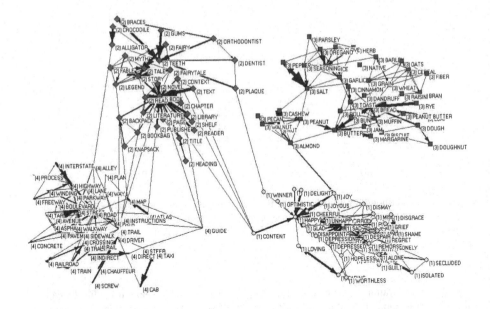

Fig. 6. A graphical representation of examples of 4 communities. The inferred communities are designated by color and shape. Each node corresponds to a word. And each arc represents a relationship between two words. The width of an arc reflects its weight.

5.3   Scientist Collaboration Network

We use another weighted network to evaluate our approach, which is a co-authorship network of scientists working on network theory and experiment [22]. Such a weighted network contains a total of 1,589 scientists and 2,742 relationships. The weight of an edge is a real number. We make the weight to be a positive integer and ensure that each edge takes a value form 0 to 15. So we set the generative distribution $f(\cdot)$ is a Poisson. Figure 4(a) is the adjacency matrix. Figure 4(b) represents the structure estimated by our approach. We also compare the running time of our method with conventional Poisson SBM. The result as in Figure 4(d) shows that our method is much faster.

5.4   Word Co-occurrence Network

To evaluate whether our method can handle large-scale networks, we test constrained SBM on *FreeAssoc* dataset [23]. *FreeAssoc* is a directed network with 10,617 vertices and 72,168 arcs. The weight on an arc represents the times that word v_i is associated with word v_j.

More than 6,000 participants produced nearly three-quarters of a million responses to 5,019 stimulus words. Participants were asked to write the first word that came to mind that was meaningfully related or strongly associated to the presented word on the blank shown next to each item. For example, if given BOOK_____, they might write READ on the blank next to it. This procedure is called a discrete association task because each participant is asked to produce only a single associate to each word. We detect the most probable number of communities as described in Section 3.3. The result is shown in Figure 5(a), which indicates the most probable number of communities is 75. Figure 5(b) represents the structure discovered by our method. The gray level indicates the community assignment to corresponding community. Table 1 illustrates examples of 10 communities (out of 75) inferred by constrained SBM. Each community is illustrated by top 10 word most likely to be generated conditioned on that community. Figure 6 is the graphical representation of examples of four communities, designated by color and shape. Each node corresponds a word whose probability conditioned on that community is above 80%. We can see that the words within the same community are semantic relative.

6 Conclusion

We have proposed a new method to detect the latent communities within the weighted networks. Our method is an extension of the Constrained Stochastic Block Model (conSBM) on weighted networks. In our method, we model the densities of intra- and inter- connectivity for nodes in the same or different modules. This leads to the reduction of computation complexity, so our method can handle the large-scale networks. Furthermore, we use the Bayesian estimation method to estimate model parameters and identify the number of communities. This allows us to include the priori belief on model parameters and the community number. Bayesian estimation can not only encode the maximum value of the data-generated model parameters but also measure the estimation quality or confidence.

In simulations, our model performs better than the notable community detection algorithm, Girvan-Newman algorithm, when identifying the number of communities. In experiments, we compare our method with some state-of-art methods on two real-world weighted networks, a gene correlation network and a collaboration network. The results show that our method is more effective, robust and much faster. We also evaluate our method on a word co-occurrence network with 10,617 words and 72,168 relations. We show that our model can discover meaningful communities, in which the words are semantic relative. All the results have demonstrated that our approach is more effective, robust and faster.

Acknowledgement

This work is supported by the National Science Foundation of China under Grant No. 60621062, 60873174 and the National 863 High-Tech Project under Grant No. 2007AA01Z148.

References

1. Girvan, M., Newman, M.: Community structure in social and biological networks. Proc. Natl. Acad. Sci. USA 99(12), 7821–7826 (2002)
2. Reichardt, J., Bornholdt, S.: Statistical mechanics of community detection. Phys. Rev. E 74 (2006)

3. Newman, M., Girvan, M.: Finding and evaluating community structure in networks. Phys. Rev. E 69 (2004)
4. Newman, M.: Detecting community structure in networks. The European Physical Journal B 38, 321–330 (2004)
5. Danon, L., Duch, J., Diaz-Guilera, A., Arenas, A.: Comparing community structure identification. Journal of Statistical Mechanics: Theory and Experiment 29(09) (2005)
6. Hoffman, J.M., Wiggin, C.H.: A bayesian approach to network modularity. Phys. Rev. Lett. 100 (2008)
7. Holland, P.W., Leinhardt, S.: Local structure in social networks. Sociological Methodology, 1–45 (1975)
8. Nowichi, K., Snijders, T.: Estimation and prediction for stochastic block-structures. Journal of the American Statistical Association 96, 1077–1087 (2001)
9. Airoldi, E., Blei, D., Fienberg, S., Xing, E.: Mixed membership stochastic block models for relational data with application to protein-protein interactions. In: ENAR (2006)
10. Blei, D., Ng, A., Jordan, M.: Latent dirichlet allocation. Journal of Machine Learning Research 3, 993–1022 (2003)
11. Chang, J., Blei, D.: Relational topic models for document networks. Artificial Intelligence and Statistics (2009)
12. Li, F.F., Perona, P.: A bayesian hierarchical model for learning natural scene categories. IEEE Computer Vision and Pattern Recognition (2005)
13. Airoldi, E., Blei, D., Fienberg, S., Xing, E.: Mixed membership stochastic blockmodels. Journal of Machine Learning Research 9, 1981–2014 (2008)
14. Mariadassou, M., Robin, S.: Uncovering latent structure in valued graphs: a variational approach. SSB-RR 10 (2007)
15. Shan, H., Banerjee, A.: Bayesian co-clustering. Techique Report (2008)
16. Sinkkonen, J., Aukia, J., Kaski, S.: Component models for large networks (2008)
17. Zhang, H., Qiu, B., Lee Giles, C., Foley, H.C., Yen, J.: An ldea-based community structure discovery approach for large-scale social networks. In: IEEE International Conference on Intelligence and Security Informatics (2007)
18. Jordan, M., Ghahramani, Z., Jaakkola, T., Saul, L.: An introduction to variational methods for graphical models. Machine Learning 37 (1999)
19. Kumpula, J., Saramäki, J., Kaski, K., Kertesz, J.: Limited resolution and multiresolution methods in complex network community detection. Eur. Phys. J. B 56, 41 (2007)
20. Opgen-Rhein, R., Strimmer, K.: Inferring gene dependency network from genomic longitudinal data: a functinonal data approach. REVSTAT 4, 53–65 (2006)
21. Robin, S., Mariadassou, M.: Uncovering latent structure in valued graphs: a variational approach, http://carlit.toulouse.inra.fr/MSTGA/Reunion_nov2008/Stephane.pdf
22. Newman, M.: Finding community structure in networks using the eigenvectors of matrices. Phys. Rev. E 74(3) (2006)
23. Nelson, D., McEvoy, C., Schreiber, T.: The university of south florida word association, rhyme, and word fragmentnorms, University of South Florida, Tampa (1999) (unpublished manuscript)

Averaged Naive Bayes Trees: A New Extension of AODE

Mori Kurokawa[1], Hiroyuki Yokoyama[1], and Akito Sakurai[2]

[1] KDDI R&D Laboratories, Inc., 2-1-15 Ohara, Fujimino-shi, Saitama, 356-8502 Japan
{mo-kurokawa,yokoyama}@kddilabs.jp
[2] Keio University, 3-14-1 Hiyoshi, Kouhoku-ku, Yokohama-shi, Kanagawa, 223-8522 Japan
sakurai@ae.keio.ac.jp

Abstract. Naive Bayes (NB) is a simple Bayesian classifier that assumes the conditional independence and augmented NB (ANB) models are extensions of NB by relaxing the independence assumption. The averaged one-dependence estimators (AODE) is a classifier that averages ODEs, which are ANB models. However, the expressiveness of AODE is still limited by the restricted structure of ODE. In this paper, we propose a model averaging method for NB Trees (NBTs) with flexible structures and present experimental results in terms of classification accuracy. Results of comparative experiments show that our proposed method outperforms AODE on classification accuracy.

Keywords: naive Bayes, augmented naive Bayes, averaged one-dependence estimators, naive Bayes trees, model averaging.

1 Introduction

Automatic classification is a key technique for retrieving useful information from large amount of data on users and Web contents accumulated on many websites. Automatic classification is usually implemented in two steps: 1) feature extraction and 2) model construction by analysing dependencies between features and classes. Feature extraction is a domain-specific problem. In contrast, model construction is a generic problem formulated as a problem of determining a function, called a classifier, that maps the instantiation of features $(A_1, \cdots, A_n) = (a_1, \cdots, a_n)$ to one of the classes $c \in \{c_1, \cdots, c_m\}$.

In this study, we aimed to improve classification accuracy by accurately modeling the dependencies between features and classes. Among the many previously proposed classifiers, averaged one-dependence estimators (AODE) is a classifier that averages one-dependence estimators (ODEs), which are extensions of NB models, and it is a state-of-the-art classifier in terms of classification accuracy. However, as ODEs still have limitations on its structure, there is scope for improvement in classification accuracy of the AODE. Therefore, we introduce naive Bayes trees (NBTs), which have more flexibility than ODEs, and propose a model averaging method for NBTs instead of ODEs.

The rest of this paper is organized as follows: In Section 2, related classification techniques are described. In Section 3, our proposed method is described. In Section 4, experimental settings are described. In Section 5, experimental results are described. Finally, in Section 6, the paper is concluded.

Z.-H. Zhou and T. Washio (Eds.): ACML 2009, LNAI 5828, pp. 191–205, 2009.

2 Related Classification Techniques

Bayesian classifiers or Bayesian network classifiers[1] are probabilistic models called Bayesian networks (BN) that are applied to the classification problem. Owing to the recent enhancements in computational power that have enabled fast execution of complex probabilistic computation, many Bayesian classifiers have been proposed and applied to various datasets.

The BN model is a probabilistic model expressed as a directed acyclic graph (DAG) where a node represents a random variable and an arc[1] represents the dependency between random variables. In a DAG that has an arc $X \leftarrow Y$, the node Y is called a parent node of X and X is called a child node of Y. Here, the graph expresses the conditional independence between X and the other child nodes of Y given their parent node Y.

Bayesian classifiers express the dependency between the class variable and feature variables $\{C, \boldsymbol{A}\}$. Let a feature variable A_k have the class variable C and some other feature variables $Pa(A_k) \subset \boldsymbol{A} \setminus A_k$ as its parents; then the joint probability of the class variable and feature variables is decomposed as follows:

$$P(C, \boldsymbol{A}) = P(C) \prod_k P(A_k | C, Pa(A_k)) \tag{1}$$

Although the problem of learning an optimal BN model structure was proved to be NP-hard by Chickering et al.[2], many approaches have been proposed for learning BN models.

In classification when a feature value vector $\boldsymbol{a}$ is given, the class $\hat{c}$ to be assigned is calculated to maximize the posterior probability $P(c|\boldsymbol{a})$ as follows:

$$\hat{c} = \underset{c \in C}{\operatorname{argmax}} P(c|\boldsymbol{a}) = \underset{c \in C}{\operatorname{argmax}} \frac{P(c, \boldsymbol{a})}{\sum_{c \in C} P(c, \boldsymbol{a})} = \underset{c \in C}{\operatorname{argmax}} P(c, \boldsymbol{a}) \tag{2}$$

2.1 Naive Bayes

Naive Bayes (NB) (shown in Fig. 1, upper left panel) is a simple Bayesian classifier that assumes the conditional independence between features, given that the class value is known. While classifying a newly observed instance with an unknown class, the class $\hat{c}$ to be assigned is calculated to maximize the joint probability of features and a class $P(c, \boldsymbol{a})$ as follows:

$$\hat{c} = \underset{c \in C}{\operatorname{argmax}} P(c, \boldsymbol{a}) \tag{3}$$

$$P(c, \boldsymbol{a}) = P(c) \prod_{k=1}^n P(a_k | c) \tag{4}$$

NB does not require too many computational resources and has been applied to many applications such as spam filters[3].

---

[1] In this paper, we refer to the link between variables in a BN graph as an arc, the link between variables in a decision tree (DT) graph as a branch, and the course from the root node to a node in a DT graph as a path.

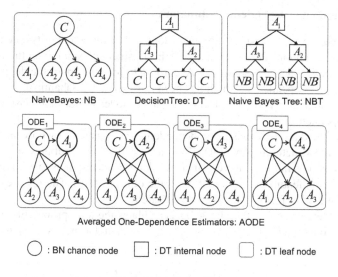

Averaged One-Dependence Estimators: AODE

Fig. 1. Related classifiers

2.2 Decision Tree

A DT (shown in Fig. 1, upper middle panel) is a tree-structured classifier, where an internal node represents a feature, a branch represents a value of the parent node, and a leaf node represents a class. In training, each internal node divides training data into the subset by feature values, and the feature of the internal node is decided on the basis of the maximum information gain criteria of the C4.5 algorithm[4]. Once the training data is divided into the subset on every path, a newly observed instance with an unknown class is classified to the class with the maximum frequency in the subset of data at the leaf node.

2.3 Naive Bayes Tree

An NB tree (NBT)[5] (shown in Fig. 1, upper right panel) is a hybrid classification technique that combines DT and NB to compensate for the weak probability estimation of the DT. In training of NBT, the NB classifiers, which are located at the leaf nodes, are trained using the subsets of the data assigned to their corresponding leaf nodes. The data are recursively divided into the subsets at the internal nodes on the paths from the root to the leaves. The division into the subsets at each internal node is based on the values of the feature assigned to the node. The feature of each internal node to be selected is the one that would give the maximum and above-threshold improvement in generalization accuracy, which is estimated by cross validation (CV). In classification, a newly observed instance with an unknown class is classified to the class with the maximum a posteriori probability calculated by the NB classifier at the leaf node. In Kohavi's experiments[5], the NBT shows higher classification accuracy than NB and DT for many datasets.

2.4 Averaged One-Dependence Estimators

Augmented NB (ANB) classifiers are extensions of NB by relaxing its conditional independence assumption. There are many ANB classifiers depending on the manner of relaxing the independence assumption. According to the concept of the degree of alleviation, called $\mathcal{N}$-dependence, which has been introduced by Sahami[6], the Bayesian classifier in which each node has at most $\mathcal{N}$ parent nodes besides the class variable is called $\mathcal{N}$-dependence.

Tree-augmented NB (TAN) is a one-dependence ANB classifier whose arcs between features are tree-structured. An AODE[7] is a classifier that combines specific TANs called ODEs (shown in Fig. 1, lower panel). Each ODE corresponds to each feature and the corresponding feature is the parent of all the other features. The feature that is the parent of all the other features is called super parent.

In model averaging of ODEs, the AODE calculates the joint probability of features and classes using each ODE $P_j(c, a)(j = 1, 2, \cdots, n)$ and averages them as follows:

$$\bar{P}(c, a) = \frac{1}{n} \sum_{j=1}^{n} P_j(c, a) \tag{5}$$

In classification, the class to which a newly observed instance is to be assigned is calculated to maximize the averaged joint probability.

2.5 Weightily Averaged One-Dependence Estimators

A weightily AODE[8] calculates normalized mutual information $I(C; A_j)$ between the class variable C and the super parent A_j and uses it to assign a weight to ODE_j in model averaging, as follows:

$$\bar{P}(c, a) = \sum_{j=1}^{n} w_j P_j(c, a) \tag{6}$$

$$w_j = \frac{I(C; A_j)}{\sum_j I(C; A_j)} \tag{7}$$

$$I(C; A_j) = \sum_{i=1}^{N} P(c^i, a_j^i) \log \frac{P(c^i, a_j^i)}{P(c^i)P(a_j^i)} \tag{8}$$

Mutual information between the class variable and a feature is often used in feature selection[9] and it expresses the degree of dependency between the class variable and a feature.

3 Proposed Method

Model averaging is a promising method for constructing a classifier with higher accuracy by averaging results of many accurate classifiers. To ensure improvement in accuracy, the diversity as well as accuracy of the member classifiers are important, as

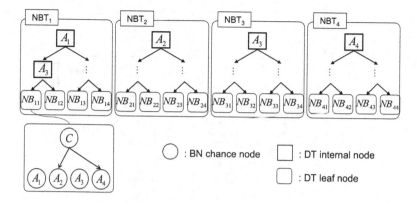

Fig. 2. NBTs to be averaged

described by Tang et al. in their paper[10]. The AODE satisfies the requirement of diversity by using ODEs with different super parents. However, due to their restricted structures, the accuracy of each ODE is not always high.

A very simple NBT that has no internal node is equivalent to an ODE, since the root node of the NBT corresponds to the super parent of the ODE and the conditional propability table of an NB at a leaf node of the NBT is equivalent to the conditional probability table of the ODE conditioned on the super parent value. The deeper the NBT grows, the more complex the NBT is. Therefore, an NBT can be considered as an extension of an ODE with higher representational flexibility than an ODE. However, the representational complexity of the NBT is so high that it could happen that an NBT with some feature on its root node has high accuracy whereas an NBT with other features has low accuracy.

Therefore, we propose a model averaging method for NBTs (shown in Fig. 2), each of which has a corresponding feature on its root node. Hereafter, we refer to our proposed method as an averaged naive Bayes trees (ANBT).

In this section, we describe the procedure for training the ANBT and for classification via the ANBT, as shown in Fig. 3.

3.1 Structure Learning

For learning the tree structure of each NBT, we utilize the same method as that used by Kohavi for an NBT; according to this method, we estimate the generalization accuracy by five-fold CV and select the feature of each internal node on the basis of the maximum and above-threshold improvement in generalization accuracy.

3.2 Parameter Estimation

Let NBT_j with a root feature A_j have B_j leaves and let $p_{jk}(k = 1, 2, \cdots, B_j)$ denote paths from the root to the leaves. We can estimate NB parameters at the leaf

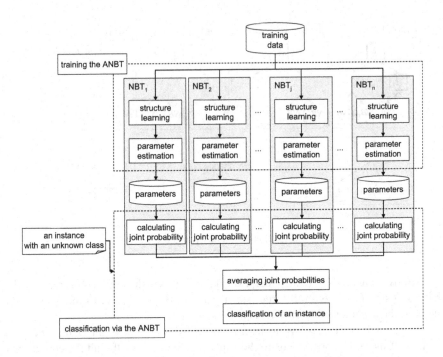

Fig. 3. Procedure for training the ANBT and classification via the ANBT

corresponding to the path p_{jk}, i.e., the probability $\hat{P}_{p_{jk}}(c)$ of a class c and the class-conditional probability $\hat{P}_{p_{jk}}(a_l|c)$ of a feature value a_l, as follows:

$$\hat{P}_{p_{jk}}(c) = \frac{N_{p_{jk},c} + \alpha_{p_{jk},c}}{\sum_c (N_{p_{jk},c} + \alpha_{p_{jk},c})} \tag{9}$$

$$\hat{P}_{p_{jk}}(a_l|c) = \frac{N_{p_{jk},c,a_l} + \alpha_{p_{jk},c,a_l}}{\sum_c (N_{p_{jk},c,a_l} + \alpha_{p_{jk},c,a_l})} \tag{10}$$

Here, $N_{p_{jk},c}$ is the frequency of a class c on the path p_{jk} and N_{p_{jk},c,a_l} is the frequency of a class c and a feature value a_l on the path p_{jk}. $\alpha_{p_{jk},c}$ and α_{p_{jk},c,a_l} are prior parameters corresponding to $N_{p_{jk},c}$ and N_{p_{jk},c,a_l}, respectively. The instantiation when $\alpha_{p_{jk},c} = \alpha_{p_{jk},c,a_l} = 1$ is called Laplace correction, and the instantiation when $\alpha_{p_{jk},c} = M/v_c$ and $\alpha_{p_{jk},c,a_l} = M/v_l$ is called M-estimation[11] or m-estimate[12], where v_c and v_l are the number of values of the class variable C and the feature variable A_l, respectively.

3.3 Calculation of Joint Probabilities

We calculate the joint probability of a class c and feature values a using NBT$_j$ parameters $\hat{\theta}_j \equiv \{\hat{P}_{p_{jk}}(c), \hat{P}_{p_{jk}}(a_l|c); c = c_1, \cdots, c_m, k = 1, \cdots, B_j, l = 1, \cdots, n\}$.

$$P_{\hat{\theta}_j}(c, \boldsymbol{a}) = \sum_{k=1}^{B_j} \delta(\boldsymbol{a}, p_{jk}) \hat{P}_{p_{jk}}(c) \prod_{l \neq j} \hat{P}_{p_{jk}}(a_l|c) \tag{11}$$

Here, $\delta(\boldsymbol{x}, p)$ is a function that takes 1 when the instance $\boldsymbol{x}$ belongs to the subset of the data that is assigned to the leaf of the path p, and 0 otherwise.

3.4   Averaging of Joint Probabilities

We average the joint probabilities of all NBTs.

$$\bar{P}(c, \boldsymbol{a}) = \frac{1}{n} \sum_{j=1}^{n} P_{\hat{\theta}_j}(c, \boldsymbol{a}) \tag{12}$$

When we assign a weight w_j to NBT$_j$ in averaging, the weighted-averaged joint probability is calculated as follows:

$$\bar{P}(c, \boldsymbol{a}) = \frac{1}{\sum_{j=1}^{n} w_j} \sum_{j=1}^{n} w_j P_{\hat{\theta}_j}(c, \boldsymbol{a}) \tag{13}$$

3.5   Classification of an Instance

When an instance $\boldsymbol{a}$ with an unknown class is given, we calculate the averaged joint probability for each class and decide a class $\hat{c}$ to the instance as follows:

$$\hat{c} = \operatorname*{argmax}_{c \in C} P(c|\boldsymbol{a}) \tag{14}$$

$$P(c|\boldsymbol{a}) \propto \bar{P}(c, \boldsymbol{a}) \tag{15}$$

Here, we assume that the posterior probability of a class $P(c|\boldsymbol{a})$ is proportional to the averaged joint probability $\bar{P}(c, \boldsymbol{a})$.

3.6   Handling of Continuous Features

ANBT can handle datasets with not only just discrete features but also continuous features, as follows:

- In structure learning, we decide the threshold of a continuous feature so as to maximize information gain in the same manner as that achieved by C4.5 algorithm.
- In parameter estimation, we utilize three techniques for handling continuous features:
  1. Discretization and Laplace estimation: discretize continuous features in the subsets of data at each leaf node and estimate multinomial parameters via Laplace estimation.
  2. Discretization and M-estimation: discretize continuous features in the subsets of data at each leaf node and estimate multinomial parameters via M-estimation.

Table 1. Parameter settings of ANBT

No.	Maximum depth	Parameter estimation	Weighting method
1)	2	Laplace correction	equally weighting
2)	2	M-estimation	equally weighting
3)	2	Gaussian parameter estimation	equally weighting
4)	2	Laplace correction	normalized mutual information (weightily)
5)	2	M-estimation	normalized mutual information (weightily)
6)	3	Laplace correction	equally weighting
7)	3	M-estimation	equally weighting
8)	3	Gaussian parameter estimation	equally weighting

   3. Gaussian parameter estimation: estimate Gaussian parameters, i.e., means and
      standard deviations instead of multinomial parameters, by assuming a Gaussian
      distribution on a continuous domain.
 – In calculating joint probabilities, we calculate probabilities differently for the three
   above-mentioned parameter estimation methods. Specifically, when we use the discretization and Laplace estimation method or the discretization and M-estimation
   method, we discretize input feature values and calculate the joint probabilities using multinomial parameters. On the other hand, when we use the Gaussian parameter estimation method, we calculate the joint probabilities using the values of the
   Gaussian probability density function.

4   Experimental Settings

We conducted experiments to compare the performances of our proposed ANBT
method and existing classifiers: NB, DT, NBT, random forests (RF)[13], AODE
(Laplace correction, M-estimation), weightily AODE (M-estimation) and Support Vector Machines (SVMs; linear kernel and quadratic kernel)[14]. RF is a classifier consisting of DTs trained using the bootstrap subsamples of training data to equip diversity.
RF classifies a newly observed instance by majority voting of the trained DTs. In order
to solve the quadratic programming problem of SVMs, we utilized sequential minimal
optimization[15,16], where the complexity parameter is set to 1.0.

   We tested some parameter settings of the ANBT, as shown in Table 1. Tested parameters are a combination of two maximum depths (two and three), three parameter
estimation methods (Laplace correction, M-estimation and Gaussian parameter estimation) and two weighting methods (equally weighting and weighting via normalized
mutual information, which we hereafter refer to as *weightily*). Here, tree depth is defined
as follows: a tree of maximum depth one is composed of the root node and the children
(equivalent to an ODE); a tree of maximum depth two is composed of the root node,
the children, and the grandchildren; a tree of maximum depth three is composed of the
root node, the children, the grandchildren, and the great-grandchildren. In other words,
an NBT of maximum depth $\mathcal{N}$ is $\mathcal{N}$-dependent. The purpose of limiting the tree depth
is to reduce space complexity. When we carry out parameter estimation via the M-
estimation method, M is set to 1.0. The ANBT with parameter estimation via Laplace

Table 2. Compared classifiers

Classifier	can handle discrete features	can handle continuous features (continuous-feature-compliant)
NB	Y	Y
DT	Y	Y
NBT	Y	Y
RF	Y	Y
AODE	Y	N
weightily AODE	Y	N
SVM	Y	Y
ANBT (equally weighting)	Y	Y
ANBT (weightily)	Y	N

correction or M-estimation discretizes continuous features before estimating NB parameters at each leaf node. When we use the Gaussian parameter estimation method for continuous features, we use Laplace correction for discrete features. Weighting via normalized mutual information is the same as weightily AODE. The weightily ANBT cannot handle continuous features, because normalized mutual information cannot be calculated for feature pairs in which either feature is continuous.

For comparative experiments, we utilized 36 public datasets in the classification task; these datasets were obtained from the University of California, Irvine, (UCI) Machine Learning Repository[17]; they are also published on the Weka[18] website. These datasets are the same as those used for experiments performed by Jiang et al.[8]. Further, we carried out the same preliminary processing as that by Jiang et al. did:

(a) Replace missing values: We replaced missing feature values with means or modes using the ReplaceMissingValues filter in Weka[18].
(b) Discretize continuous features: We applied 10-bin discretization to continuous features using the Discretize filter in Weka.
(c) Remove useless features: We considered features that do not vary at all or that vary too much (e.g., one value for one instance) as useless and removed such features using the RemoveUseless filter in Weka.

Table 2 shows whether the compared classifiers can handle continuous features. For testing each classifier, we prepared two preliminarily processed datasets as follows:

Pre-processed datasets I: These are datasets for testing classifiers that can handle continuous features, which we call *continuous-feature-compliant* classifiers, as well as classifiers that cannot handle continuous features, which we call *non-continuous-feature-compliant* classifiers. We applied preliminary data processes (a) - (c) to the original datasets.
Pre-processed datasets II: These are datasets for testing continuous-feature-compliant classifiers. We applied preliminary data processes other than process (b) to original datasets containing more than one continuous feature.

Table 3. Discription of datasets for tests

Dataset	#classes	#instances	Pre-processed dataset I		Pre-processed dataset II			
			#features discrete	#feature values discrete	#features total	continuous	discrete	#feature values discrete
1. anneal.ORIG	6	898	18	75	18	6	12	51
2. anneal	6	898	31	112	31	6	25	88
3. audiology	24	226	69	154	-	-	-	-
4. autos	7	205	22	107	25	15	10	60
5. balance-scale	3	625	4	8	4	4	0	0
6. breast-cancer	2	286	9	51	-	-	-	-
7. breast-w	2	699	9	29	9	9	0	0
8. horse-colic.ORIG	2	368	22	472	27	7	20	468
9. horse-colic	2	368	16	57	22	7	15	55
10. credit-rating	2	690	15	54	15	6	9	41
11. german_credit	2	1000	15	60	20	7	13	56
12. pima_diabetes	2	768	6	15	8	8	0	0
13. Glass	7	214	7	20	9	9	0	0
14. heart-c	5	303	11	27	13	6	7	19
15. heart-h	5	294	9	23	12	5	7	19
16. heart-statlog	2	270	9	18	13	13	0	0
17. hepatitis	2	155	16	33	19	6	13	26
18. hypothyroid	4	3772	25	63	27	6	21	45
19. ionosphere	2	351	33	144	33	33	0	0
20. iris	3	150	4	12	4	4	0	0
21. kr-vs-kp	2	3196	36	74	-	-	-	-
22. labor	2	57	11	27	16	8	8	21
23. letter	26	20000	15	154	16	16	0	0
24. lymphography	4	148	18	50	18	3	15	44
25. mushroom	2	8124	21	123	-	-	-	-
26. primary-tumor	22	339	17	37	-	-	-	-
27. segment	7	2310	18	171	18	18	0	0
28. sick	2	3772	25	56	27	6	21	45
29. sonar	2	208	21	42	60	60	0	0
30. soybean	19	683	35	100	-	-	-	-
31. splice	3	3190	60	287	-	-	-	-
32. vehicle	4	846	18	73	18	18	0	0
33. vote	2	435	16	32	-	-	-	-
34. vowel	11	990	13	62	13	10	3	19
35. waveform	3	5000	19	109	40	40	0	0
36. zoo	7	101	16	34	16	1	15	30

Table 3 shows the characteristics of 36 pre-processed datasets I and 28 pre-processed datasets II. Here, even though the SVMs can handle discrete features, we removed it from the lists of classifiers for comparative experiments using pre-processed datasets I.

We conducted 10-fold CV tests for comparing the classifier performances. We divided each dataset into 10 almost equal-sized blocks at random, and in each validation, one block was used for test data and the remaining were used for training classifiers. We tried 10 runs of 10-fold CV under different random seeds for randomizing data. Throughout all the tests, we measured the classification accuracy, i.e., the percentage of correctly classified instances, and compared mean accuracies of pairs of classifiers for the same tests by paired corrected re-sampled two-tailed t-tests[19] (significance level: 5.0 %). We used data mining software Weka[18] for implementing our proposed method and conducting experiments.

5   Experimental Results

Table 4 shows the experimental results of comparison of the ANBT with other classifiers. Each entry in row i and column j in Table 4 indicates #wins/#losses, where win (loss) indicates that the classifier in column j significantly outperforms (underperforms) that in row i in terms of classification accuracy.

As shown in Table 4, #wins of the ANBT against the NB exceeds #losses by 15 tests or more for any settings, #wins of the ANBT against the DT exceeds #losses by 12 tests or more, #wins of ANBT 2), 5), and 7) against the NBT exceeds #losses by 8 tests or more, and #wins of the ANBT against the RF exceeds #losses by 6 tests or more for any settings. This implies that the ANBT, particularly ANBTs 2), 5), and 7), outperform NB, DT, NBT, and RF. Table 4 shows that ANBTs 2), 5), and 7) outperform ANBTs 1), 4), and 6), respectively, in terms of more #wins and less #losses, where each compared pair is under the same condition except for in the parameter estimation method. This suggests that the ANBT performs better when it uses M-estimation than when it uses Laplace estimation. Among the various ANBT configurations, ANBT 7) performs particularly well and it achieves 7 wins for the AODE (M-estimation), which is a state-of-the-art classifier, and it achieves 3 wins for the weightily AODE, which is also a state-of-the-art classifier. ANBT 2) and ANBT 5), whose maximum depth is two and whose weights are equal and normalized mutual information, respectively, performs equivalently well in comparison tests for the AODE and weightily AODE and they

Table 4. #wins/#losses: Summary of mean accuracy comparisons in tests using 36 pre-processed datasets I

Classifier for comparison	ANBT					
	1)	2)	4)	5)	6)	7)
NB	16 / 0	19 / 0	15 / 0	16 / 0	16 / 1	17 / 1
DT	14 / 1	14 / 2	14 / 1	15 / 2	14 / 1	14 / 2
NBT	6 / 1	10 / 1	9 / 1	9 / 1	7 / 0	11 / 0
RF	9 / 2	11 / 1	8 / 2	8 / 1	11 / 2	12 / 1
AODE (Laplace estimation)	8 / 0	11 / 0	9 / 0	9 / 0	7 / 0	10 / 0
AODE (M-estimation)	1 / 2	6 / 0	3 / 2	6 / 0	3 / 2	7 / 0
weightily AODE	1 / 3	3 / 0	2 / 2	3 / 0	1 / 3	3 / 0

Table 5. #wins/#losses: Summary of mean accuracy comparisons in tests using 28 pre-processed datasets II

Classifier for comparison	ANBT					
	1)	2)	3)	6)	7)	8)
NB	16 / 1	15 / 1	16 / 1	16 / 1	15 / 1	16 / 1
DT	10 / 3	10 / 4	9 / 7	11 / 1	12 / 2	11 / 4
NBT	4 / 1	3 / 1	3 / 5	4 / 0	5 / 0	4 / 2
RF	3 / 5	2 / 3	2 / 7	3 / 2	2 / 1	3 / 4
SVM (linear kernel)	9 / 2	9 / 2	7 / 2	11 / 2	10 / 2	9 / 1
SVM (quadratic kernel)	7 / 6	6 / 4	5 / 9	8 / 4	9 / 4	7 / 7

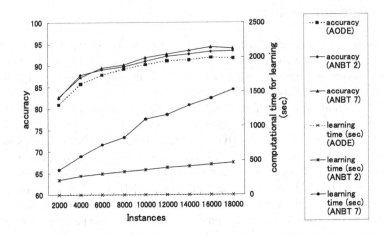

Fig. 4. Learning curve and the corresponding learning time for the letter dataset

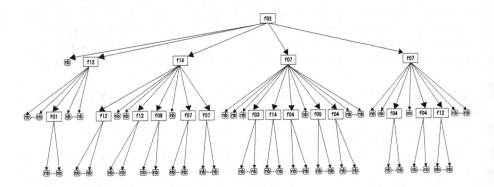

Fig. 5. An example of a resulting structure of an NBT of ANBT 7) trained with the letter dataset

achieve 6 wins for the AODE (M-estimation) and 3 wins for the the weightily AODE. This implies that weighting via normalized mutual information is not very effective in the case of the ANBT.

Table 5 shows the experimental results of comparison of the ANBT with continuous-feature-compliant classifiers. #wins of the ANBT against NB, DT and NBT exceeds #losses, similar to the results shown in Table 4; this suggests that the ANBT can treat continuous features better than NB, DT and NBT. However, #wins of the ANBT against RF does not exceed #losses in some cases; this suggests that the RF can treat continuous features better than the ANBT in some cases. Similar to Table 4, Table 5 suggests that the ANBT performs better when it uses M-estimation than when it uses Laplace estimation, as indicated by the comparison of ANBT 1) with ANBT 2) and that of ANBT 6) with ANBT 7). ANBT 3) and ANBT 8) underperform other ANBT configurations, and this suggests that the ANBT performs better when it discretizes continuous features and estimates multinomial parameters than when it uses Gaussian parameter estima-

Table 6. Detailed results: means and standard deviations of classification accuracy in tests with 36 pre-processed datasets I

Dataset	NB	DT	NBT	RF	AODE M-estimation	weightily AODE	ANBT 2)	ANBT 7)
1. anneal.ORIG	92.62±2.75	92.47±2.37	94.14±2.23	94.64±2.13	93.45±2.56	94.00±2.46	**94.42±2.42**	**94.65±2.66**
2. anneal	95.72±2.27	98.78±0.91	98.81±1.25	99.21±0.87	98.85±1.07	98.78±1.17	99.00±0.99	99.30±0.77
3. audiology	71.40±6.37	77.22±7.69	76.66±7.47	75.95±8.23	76.04±6.37	76.26±6.36	76.00±6.43	75.55±6.68
4. autos	72.30±10.31	76.39±9.55	80.20±9.08	86.32±7.54	84.93±7.32	86.49±7.57	89.49±6.94	89.69±6.77
5. balance-scale	71.08±4.29	69.32±3.89	70.83±3.91	69.92±3.98	69.29±3.85	69.43±3.97	69.55±3.90	69.43±3.92
6. breast-cancer	72.94±7.71	75.26±5.04	71.66±7.92	69.17±6.80	72.01±6.45	71.97±6.79	70.19±6.74	70.64±6.55
7. breast-w	97.25±1.79	94.65±2.51	97.07±1.89	95.19±2.64	96.98±1.84	97.01±1.89	96.97±1.82	96.84±2.02
8. horse-colic.ORIG	74.46±6.81	76.09±5.31	74.42±6.83	72.15±5.85	74.25±6.59	72.78±6.77	74.37±6.36	74.33±6.92
9. horse-colic	81.39±5.74	84.72±5.95	81.50±6.05	82.91±5.75	82.07±5.94	81.66±5.53	82.25±6.14	82.15±6.14
10. credit-rating	86.25±4.01	86.58±3.53	85.62±4.18	83.77±4.38	86.23±3.77	86.00±3.60	86.35±3.62	86.62±3.62
11. german_credit	75.43±3.84	72.17±3.49	74.85±4.23	72.46±3.98	76.01±3.92	76.06±3.91	75.43±3.77	75.37±3.62
12. pima_diabetes	77.85±4.67	77.34±4.91	77.27±4.74	76.29±4.43	78.20±4.40	78.03±4.44	78.19±4.45	78.10±4.47
13. Glass	74.39±7.95	75.23±9.46	78.16±8.67	75.50±8.51	78.35±8.03	78.12±7.94	79.84±8.30	79.56±8.27
14. heart-c	83.60±6.42	77.32±6.20	82.90±6.57	80.67±6.21	83.16±6.22	83.00±6.22	82.18±6.47	81.95±6.52
15. heart-h	84.46±5.92	80.96±6.90	83.64±6.51	81.63±6.62	84.84±5.99	85.52±5.97	84.09±6.14	84.15±6.25
16. heart-statlog	83.74±6.25	82.26±7.32	83.52±6.43	83.48±6.77	82.81±6.75	82.52±6.56	83.11±6.15	83.63±5.98
17. hepatitis	84.22±9.41	81.33±9.48	84.03±8.43	83.58±9.64	85.71±9.21	85.58±8.86	84.89±8.74	85.72±8.09
18. hypothyroid	98.48±0.59	99.28±0.42	99.04±0.53	99.17±0.50	98.97±0.46	99.31±0.37	99.36±0.40	99.49±0.38
19. ionosphere	90.77±4.76	89.49±5.12	91.57±4.73	92.68±4.03	93.68±4.06	93.68±4.07	93.67±3.84	93.76±4.04
20. iris	94.47±5.61	93.87±4.89	94.27±5.28	94.07±5.76	93.27±5.65	93.33±5.61	94.27±5.77	94.27±5.77
21. kr-vs-kp	87.79±1.91	99.44±0.37	97.81±2.05	98.87±0.61	91.19±1.58	94.18±1.25	**95.01±1.30**	**96.72±1.05**
22. labor	93.13±10.56	87.13±15.32	92.03±12.63	91.77±11.45	94.87±10.22	94.17±10.45	95.97±8.51	95.97±8.51
23. letter	74.00±0.88	78.75±0.78	84.74±0.85	90.09±0.60	91.46±0.63	91.71±0.58	**93.68±0.59**	**94.01±0.56**
24. lymphography	84.97±8.30	76.51±10.14	81.42±10.22	79.65±9.20	87.88±8.19	88.41±7.68	88.54±7.79	88.13±7.66
25. mushroom	95.52±0.78	100.00±0.00	100.00±0.00	100.00±0.00	99.96±0.06	99.98±0.04	100.00±0.00	100.00±0.00
26. primary-tumor	47.20±6.02	41.01±6.59	45.84±6.61	39.68±5.96	48.08±5.81	47.94±5.89	49.18±5.35	49.35±5.72
27. segment	91.67±1.69	95.23±1.37	94.67±1.38	96.24±1.18	96.90±1.04	96.86±1.05	97.22±1.00	97.24±0.95
28. sick	97.10±0.84	97.82±0.75	97.65±0.76	97.62±0.76	97.46±0.80	97.73±0.71	97.70±0.73	**97.89±0.74**
29. sonar	85.16±7.52	80.60±8.85	81.98±8.76	80.59±8.40	86.45±6.72	86.98±6.81	85.68±7.62	85.91±7.69
30. soybean	92.20±3.23	92.63±2.72	92.30±2.70	92.80±2.45	94.45±2.28	94.33±2.36	94.17±2.48	94.37±2.43
31. splice	95.42±1.14	94.17±1.28	95.42±1.14	90.09±1.78	96.34±0.95	96.36±0.95	96.07±1.13	96.10±1.14
32. vehicle	62.52±3.81	70.77±3.86	70.68±3.53	72.99±4.40	72.77±3.70	73.22±3.85	73.79±3.83	73.31±3.86
33. vote	90.21±3.95	96.27±2.79	94.78±3.32	95.95±2.83	94.53±3.17	94.46±3.17	95.67±2.84	95.88±2.88
34. vowel	65.23±4.53	79.54±4.01	78.08±4.19	88.98±3.01	87.39±3.12	84.19±3.43	**91.34±3.02**	**90.78±3.12**
35. waveform	80.72±1.50	76.36±1.77	82.84±1.81	81.00±1.72	82.46±1.50	86.51±1.50	86.55±1.49	86.45±1.43
36. zoo	93.98±7.14	92.61±7.33	93.56±6.96	96.25±5.42	96.15±5.44	97.42±4.82	96.83±5.07	96.43±5.39
Average	83.32	83.88	85.11	85.04	86.43	86.50	86.97	87.05

Table 7. Detailed results: means and standard deviations of classification accuracy in tests with 28 pre-processed datasets II

Dataset	NB	DT	NBT	RF	SVM linear kernel	SVM quadratic kernel	ANBT 2)	ANBT 7)
1. anneal.ORIG	64.03±4.95	94.14±2.32	93.93±2.28	95.48±1.89	87.44±2.97	90.13±2.77	**94.37±2.43**	**94.28±2.42**
2. anneal	86.50±3.39	98.57±1.04	98.50±1.28	99.43±0.72	97.46±1.60	99.52±0.68	98.83±1.19	99.10±0.93
4. autos	57.67±10.91	79.51±9.73	77.12±9.02	81.84±8.70	71.44±10.16	76.89±9.12	81.49±8.88	**82.72±8.41**
5. balance-scale	90.53±1.67	77.82±3.42	75.97±5.19	80.06±3.51	87.57±2.49	91.38±1.93	**76.93±4.43**	78.14±4.08
7. breast-w	96.12±2.16	94.89±2.49	96.31±2.29	96.05±2.17	96.77±2.03	96.35±2.21	97.01±1.86	96.98±1.89
8. horse-colic.ORIG	72.33±7.36	81.47±6.16	77.54±7.80	72.08±6.03	73.42±6.47	72.80±6.80	76.39±6.26	76.59±6.06
9. horse-colic	77.39±6.34	83.09±5.96	81.60±6.01	83.67±5.81	82.77±5.37	76.69±6.39	**82.80±5.93**	82.82±5.94
10. credit-rating	77.81±4.21	85.58±3.99	85.26±3.87	85.51±4.23	84.88±3.86	84.91±3.98	86.35±3.60	86.48±3.73
11. german_credit	75.16±3.48	71.25±3.17	74.27±4.22	73.75±3.47	75.09±3.42	69.28±4.31	**75.59±3.81**	**75.71±3.75**
12. pima_diabetes	75.75±5.32	74.49±5.27	75.24±5.23	74.46±4.65	76.80±4.54	77.32±4.46	75.68±4.60	75.58±4.63
13. Glass	49.45±9.50	67.63±9.31	69.90±8.60	75.98±8.64	57.36±8.77	62.49±8.80	**74.13±8.02**	74.40±7.95
14. heart-c	82.65±7.38	76.45±7.21	80.30±6.84	80.18±6.60	83.89±6.28	81.27±5.79	82.94±6.42	83.07±6.34
15. heart-h	82.83±6.05	78.04±7.08	82.15±6.64	80.59±6.25	82.84±6.47	82.97±6.19	84.36±5.68	83.92±5.96
16. heart-statlog	83.59±5.98	78.15±7.42	79.26±8.34	80.44±6.82	83.89±6.24	81.22±6.62	84.22±6.59	83.81±6.38
17. hepatitis	83.29±10.26	77.02±10.03	81.36±9.93	83.68±8.95	85.77±9.01	81.63±8.29	83.94±9.60	83.80±9.35
18. hypothyroid	95.29±0.74	99.58±0.34	99.56±0.35	99.28±0.42	93.58±0.45	93.61±0.57	**99.48±0.35**	**99.65±0.28**
19. ionosphere	82.17±6.14	89.74±4.38	89.95±4.50	92.88±4.41	88.07±5.32	90.69±4.68	91.60±4.28	92.85±4.20
20. iris	95.53±5.02	94.73±5.30	93.80±5.32	94.20±5.16	96.27±4.58	95.53±4.93	95.20±5.02	95.20±4.84
22. labor	93.87±10.63	81.23±13.68	90.70±12.18	89.40±12.31	93.50±9.63	92.63±12.07	92.77±11.62	92.93±11.60
23. letter	64.07±0.91	88.03±0.71	86.60±0.83	94.54±0.47	82.34±0.76	88.51±0.62	85.48±0.74	**89.27±0.64**
24. lymphography	83.13±8.89	75.84±11.05	80.80±8.85	80.74±8.88	86.48±7.68	83.45±8.30	87.11±8.27	86.99±8.74
27. segment	80.13±2.14	96.79±1.29	95.22±1.38	97.71±1.06	92.91±1.51	94.68±1.49	95.27±1.16	**97.12±1.17**
28. sick	92.89±1.37	98.81±0.56	97.99±0.85	98.47±0.59	93.87±0.12	95.27±0.63	**97.79±0.79**	98.20±0.70
29. sonar	67.71±8.66	73.61±9.34	77.07±9.65	81.02±7.57	76.60±8.27	83.80±7.84	78.10±9.49	78.91±9.25
32. vehicle	44.68±4.59	72.28±4.32	70.99±4.67	74.08±4.65	74.20±5.16	80.26±3.69	**72.42±3.12**	73.78±3.44
34. vowel	62.90±4.38	80.20±4.36	92.34±3.03	95.74±2.29	70.61±3.86	97.33±1.58	96.08±2.13	**95.72±2.22**
35. waveform	80.01±1.45	75.25±1.90	80.16±1.81	81.84±1.70	86.48±1.53	86.02±1.56	**82.66±1.42**	**82.83±1.50**
36. zoo	95.07±5.86	92.61±7.33	94.84±6.54	95.94±5.67	93.68±5.60	96.05±5.60	95.84±5.86	95.82±5.87
Average	78.31	83.46	84.95	86.39	84.14	85.81	86.60	87.02

tion, as Hsu et al.[20] showed that discretizing continuous features before estimating NB parameters is effective. Table 5 shows that ANBT 7), whose maximum depth is three, performs better than ANBT 2), whose maximum depth is two; this suggests that relaxing the maximum depth is effective in the case of a continuous-feature-compliant ANBT. All ANBTs except for ANBT 3) and ANBT 8) outperform the SVMs (linear kernel and quadratic kernel), which are state-of-the-art continuous-feature-compliant classifiers. Among the various ANBT configurations, ANBT 7) performs particularly well and achieves 10 wins and 2 losses for the SVM (linear kernel) and 9 wins and 4 losses for the SVM (quadratic kernel). Let us mention that SVMs could give higher accuracy by optimizing its parameters such as the complexity parameter.

Table 6 shows detailed results of means and standard deviations of classification accuracies in experiments for comparing ANBT 2) and ANBT 7) with other classifiers and Table 7 shows those in experiments for comparing ANBT 2) and ANBT 7) with continuous-feature-compliant classifiers. In Table 6, boldface numbers indicates significant improvement over the AODE (M-estimation). In Table 7, boldface numbers indicates significant improvement over the SVM (quadratic kernel) and underlined numbers indicates a significant decline. The bottom row of each table indicates the average of mean accuracies of each classifier and suggests that on an average, the ANBT, particularly ANBT 7), shows higher accuracy than other classifiers.

According to Table 6, the accuracy of the ANBT for the letter dataset, which has the largest number of instances and the largest number of classes among the 36 datasets, is significantly higher than that of the other classifiers. Fig. 4 shows the learning curve and the corresponding learning time for tests in which 2000 randomly selected instances of the letter dataset are used for test data in common. As shown in Fig. 4, the ANBT shows higher accuracy than the AODE (M-estimation), and it also shows a longer computational time for learning than the AODE (M-estimation). The time complexity of ANBT learning is the sum of time complexity of structure learning of NBTs and that of NB training at the NBT leaf nodes. It increases rapidly with the growth of NBTs, since larger NBTs have more internal and leaf nodes. The learning time of the ANBT is #attributes times the average learning time of NBTs.

Fig. 5 shows an example of a resulting structure of an NBT of ANBT 7) trained with the letter dataset. In this figure, a node with a label beginning with "f" represents a feature and that with the label of "nb" represents an NB classifier. The depth of paths ranges from one to three. From Fig. 4, we observe that a longer path results in better accuracy and longer learning time.

6   Conclusion

We have proposed a model averaging method for NBTs and presented experimental results of classification accuracy. The experimental results show that our ANBT outperforms the state-of-the-art classifiers AODE. These results suggest that the ANBT effectively takes advantage of the high accuracy and diversity of NBTs.

The highlights of experimental results of comparison tests can be summarized as follows: our ANBT 7) (maximum depth three, parameter estimation via M-estimation) outperforms the AODE (M-estimation) on classification accuracy for 7 datasets; further, ANBT 7) does not underperform the AODE (M-estimation) for any dataset.

In the future, we intend to determine the optimal parameter settings by conducting detailed experiments. Another future direction of this study is to reduce the computational complexity of the ANBT and to refine the method for selecting and pruning trees so as to improve the classification accuracy of the ANBT.

References

1. Friedman, N., Geiger, D., Goldszmidt, M.: Bayesian network classifiers. Machine learning 29(2), 131–163 (1997)
2. Chickering, D., Heckerman, D., Meek, C.: Large-sample learning of Bayesian networks is NP-hard. The Journal of Machine Learning Research 5, 1287–1330 (2004)
3. Hovold, J.: Naive Bayes spam filtering using word-position-based attributes. In: Proceedings of the Second Conference on Email and Anti-Spam, CEAS (2005)
4. Quinlan, J.: C4. 5: Programs for machine learning. Morgan Kaufmann, San Francisco (1993)
5. Kohavi, R.: Scaling up the accuracy of naive-Bayes classifiers: A decision-tree hybrid. In: Proceedings of the Second International Conference on Knowledge Discovery and Data Mining (KDD), pp. 202–207 (1996)
6. Sahami, M.: Learning limited dependence Bayesian classifiers. In: Proceedings of the Second International Conference on Knowledge Discovery and Data Mining (KDD), pp. 335–338 (1996)
7. Webb, G., Boughton, J., Wang, Z.: Not so naive Bayes: Aggregating one-dependence estimators. Machine Learning 58(1), 5–24 (2005)
8. Jiang, L., Zhang, H.: Weightily averaged one-dependence estimators. In: Yang, Q., Webb, G. (eds.) PRICAI 2006. LNCS (LNAI), vol. 4099, pp. 970–974. Springer, Heidelberg (2006)
9. Guyon, I., Elisseeff, A.: An Introduction to Variable and Feature Selection. The Journal of Machine Learning Research 3, 1157–1182 (2003)
10. Tang, E., Suganthan, P., Yao, X.: An analysis of diversity measures. Machine Learning 65, 247–271 (2006)
11. Cestnik, B., Bratko, I.: On estimating probabilities in tree pruning. In: Kodratoff, Y. (ed.) EWSL 1991. LNCS, vol. 482, pp. 138–150. Springer, Heidelberg (1991)
12. Mitchel, T.: Machine learning. McGraw Hill, New York (1997)
13. Breiman, L.: Random forests. Machine learning 45(1), 5–32 (2001)
14. Boser, B., Guyon, I., Vapnik, V.: A training algorithm for optimal margin classifiers. In: Proceedings of the fifth annual workshop on Computational learning theory (COLT), pp. 144–152 (1992)
15. Platt, J.: Sequential minimal optimization: A fast algorithm for training support vector machines. Advances in Kernel Methods-Support Vector Learning 208 (1999)
16. Keerthi, S., Shevade, S., Bhattacharyya, C., Murthy, K.: Improvements to Platt's SMO algorithm for SVM classifier design. Neural Computation 13(3), 637–649 (2001)
17. Asuncion, A., Newman, D.: UCI machine learning repository. University of California, Irvine, School of Information and Computer Sciences (2007)
18. Whitten, I., Frank, E.: Data Mining: Practical machine learning tools and techniques. Morgan Kaufmann, San Francisco (2005)
19. Nadeau, C., Bengio, Y.: Inference for the generalization error. Machine Learning 52(3), 239–281 (2003)
20. Hsu, C., Huang, H., Wong, T.: Why discretization works for naive Bayesian classifiers. In: Proceedings of the Seventeenth International Conference on Machine Learning (ICML), pp. 399–406 (2000)

Automatic Choice of Control Measurements

Gayle Leen[1], David R. Hardoon[2], and Samuel Kaski[1]

[1] Helsinki University of Technology
Department of Information and Computer Science
P.O. Box 5400, FIN-02015 TKK, Finland
{gleen@cis.hut.fi,samuel.kaski}@tkk.fi
[2] University College London
Dept. of Computer Science, Gower Street, London WC1E 6BT U.K.
D.Hardoon@cs.ucl.ac.uk

Abstract. In experimental design, a standard approach for distinguishing experimentally induced effects from unwanted effects is to design control measurements that differ only in terms of the former. However, in some cases, it may be problematic to design and measure controls specifically for an experiment. In this paper, we investigate the possibility of *learning to choose* suitable controls from a database of potential controls, which differ in their degree of relevance to the experiment. This approach is especially relevant in the field of bioinformatics where experimental studies are predominantly small-scale, while vast amounts of biological measurements are becoming increasingly available. We focus on finding controls for differential gene expression studies (case vs control) of various cancers. In this situation, the ideal control would be a healthy sample from the same tissue (the same mixture of cells as the tumor tissue), under the same conditions except for cancer-specific effects, which is almost impossible to obtain in practice. We formulate the problem of learning to choose the control in a Gaussian process classification framework, as a novel paired multitask learning problem. The similarities between the underlying set of classifiers are learned from the set of control tissue gene expression profiles.

1 Introduction

We approach the problem of learning to choose suitable control measurements for an experiment from a database of potential controls using a novel multi-task learning formulation. We begin by motivating the problem from a bioinformatics standpoint, and then formulate it in machine learning terms in Section 1.1.

Microarray technologies enable the simultaneous interrogation of the expression of thousands of genes, revealing the intricate workings of a cell on a molecular level. The ability to study the entire genomic profile in this way opens up many exciting research possibilities; biologists can characterise a cell in terms of its gene expression levels, and analyse how its profile varies between different conditions, leading to insights which could potentially benefit drug development,

Z.-H. Zhou and T. Washio (Eds.): ACML 2009, LNAI 5828, pp. 206–219, 2009.

disease diagnosis, functional genomics, and many other fields. Due to the potential of this research, vast amounts of gene expression measurements under different experimental conditions have been collected, and many public databases are available such as ArrayExpress [1] and the Gene Expression Omnibus [2].

A typical experimental set-up to investigate the effect of some factor, for instance a disease or drug treatment, is to compare each gene's expression level in the affected sample with a control sample. These differential gene expression studies can lead to identification of possible gene targets for further analysis, biomarkers for a disease etc. However, this procedure is prone to error; gene expression data is inherently noisy, due to factors such as measurement noise, patient-specific and laboratory-specific variation. Additionally, in general, these experiments only consider a small set of samples, since often there are only a few test cases (e.g. patients) available to the laboratory carrying out the analysis. The presence of these potential sources of noise makes it especially crucial to select a good set of control samples for a differential gene expression study.

However, designing controlled experiments may not be a straightforward task. Ideally, a large set of control samples would be measured by the same laboratory conducting the experimental study, but in practice, there is only a small control set (or none) available, which has to be augmented through selecting samples from public repositories of gene expression data. This task is problematic; in addition to the bias induced in samples due to laboratory and patient-specific effects, there is no established ontology for sample / tissue annotation, resulting in vague or missing labels, or terminology that is inconsistent between experiments. Furthermore, there may only be a very small number of the desired control samples available. One typical way to resolve this problem is to average over a large set of available samples, which are only partially related to the correct type of control sample. Obviously this approach would be suboptimal if there is a large number of unrelated samples, and a more sensible solution, which we address in this paper, would be to weight the pool of samples according to their relevance to the study.

There have been recent studies that propose a number of methodologies for gene expression analysis: clustering tissue and cell samples into a number of groups according to overlapping feature similarities [3,4,5], classifier methodology as an exploration technique to identify mislabeled and questionable tissue samples [6] and analysing the origin of tissue samples by explicitly modeling each tissue as a probabilistic sample from a population of related tissues [7]. However none of the current gene/tissue analysis studies, to the knowledge of the authors, explore the issue of learning how to identify suitable controls to affected samples when they cannot be specifically designed.

1.1   Control Sample Selection and Multitask Learning Approaches

This work proposes a novel approach to a frequently occurring and complex problem in experimental design for bioinformatics. We focus on the learning task of how to identify suitable controls for case samples, by using a novel paired multitask learning framework. We formulate the problem as follows: The suitable controls for each experiment form a group of controls. These groups will be considered as classes, and the task is to classify each case sample to one of these classes. In learning the classification, we need to use knowledge about the relationships between the case and the control samples. This pairing will in effect be transferred to new pairs.

Suppose that we have N_H control samples $\mathbf{Y} = \{\mathbf{y}_1, ...\mathbf{y}_{N_H}\}$, which we can classify into one of K control classes: $t_y \in \{1, ..., K\}$, so that we have a labeled data set $\mathcal{D}_Y = \{\mathbf{y}_n, t_{y,n}\}_{n=1}^{N_H}$. We also have N_C case samples $\mathbf{X} = \{\mathbf{x}_1, ...\mathbf{x}_{N_C}\}$, for which there are known mappings to control classes $\mathcal{D}_X = \{\mathbf{x}_n, t_{x,n}\}_{n=1}^{N_C}$. For a new case sample (and case type not contained in the training set), $\mathbf{x}_m$, we want to predict the control class, given the preexisting mappings $\mathcal{D}_X$ and $\mathcal{D}_Y$, guided by the relationships between the different control classes i.e.

$$p(t_m \mid \mathbf{x}_m, \mathcal{D}_X, \mathcal{D}_Y) \tag{1}$$

This is a multiclass (K classes) classification problem, which borrows statistical strength from $\mathcal{D}_Y$ about the relatedness of the classes. This represents the idea that if two control classes a and b are similar (found from the relationship between sets $\mathbf{Y}_a$ and $\mathbf{Y}_b$), then they are both likely to be used as control for the same case profile. In effect, $\mathbf{Y}$ augments the labeling for the control classes.

Our formulation of the control selection problem has resonance in several related subfields of machine learning, which address the issue of augmenting the data set for a learning problem with other partially related sources of information. The unifying concept is that the joint distribution of the inputs $\mathbf{x}$ and outputs t differs between the desired learning problem and the auxiliary learning problems; the existing approaches differ in the way that this shift is characterised. Transfer learning and multitask learning approaches [8,9,10] assume that information can be transferred from auxiliary, partially relevant tasks to the task(s) of interest, and generally assume the same input distributions $p(\mathbf{x})$ between tasks, with a task specific $p(t \mid \mathbf{x})$. Another family of approaches assumes that $p(t \mid \mathbf{x})$ remains unchanged between different tasks while the input domain $p(\mathbf{x})$ differs; they include learning under covariate shift [11,12] and domain adaptation [13]. In these terms, the novel problem that we address in this paper can be called *paired multitask learning*.

1.2   Paired Multitask Learning

In a traditional multitask learning scenario, there is a set of K related tasks[1] which we will here call *primary tasks*. For example, given a set of inputs and labels $\{x_n, t_n\}_{n=1}^N$, $t_n \in 1, ..., K$, eash task could be to classify the samples to one of the K classes, learned by finding $\{p(t_i \mid \mathbf{x}, \theta_i)\}_{i=1}^K$, where θ_i is the parameterisation for the ith classifier (see Figure 1a). Information could be shared among tasks, for instance through a shared parameter α (Figure 1b).

---

[1] In this paper we consider situations where all tasks have the same set of inputs and outputs.

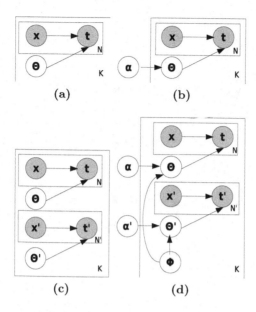

Fig. 1. Schematic illustration of statistical strength sharing in multitask learning scenarios. Learning a set of K tasks as in (a) amounts to finding different parameterisations $\theta_i, i = 1, ..., K$ for the tasks. If the tasks are assumed to be related, multitask learning approaches assume some shared structure across all K tasks through a common parameterisation via α (b). We consider the situation where there are K pairs of tasks (c), and propose the structure in (d) to share information between the tasks. There is shared structure within each task set's parameterisation θ, θ' through α, α' and across each of the K pairs through ϕ.

In our framework, we consider an additional level of dependencies: we have an auxiliary set of tasks $\{x'_n, t'_n\}_{n=1}^{N'}$, $t'_n \in 1, ..., K$, where $p(\mathbf{x}') \neq p(\mathbf{x})$ (Figure 1c). We transfer information about the *relatedness of the auxiliary set of tasks* to the set of tasks of interest, by adding one more level of parameterisation, Φ (Figure 1d). This is achieved by finding a corresponding set of classifiers $\{p(t'_i \mid \mathbf{x}, \theta'_i)\}_{i=1}^{K}$, coupled through $p(\theta'_1, ..., \theta'_K \mid \alpha')$. Information is shared between the two task sets, auxiliary and primary, through the shared parameterisation Φ which couples the pairs of corresponding tasks. The proposed model uses flexible assumptions about the shift between the auxiliary tasks and the primary tasks, since we assume different sets of conditional distributions, linked only through shared parameterisation Φ.

Our approach is remotely related to [9,10,14] in that the Gaussian process framework is used to capture inter-task similarity, and partially to the recent transfer learning approach [15] where a sample from any task is weighted to match the joint distribution of the target task $p(\mathbf{x}, t)$. The weights are derived from an input-output pair's probability of belonging to the target task, which is calculated from a multiclass classifier learned on the pool of samples. Whereas in

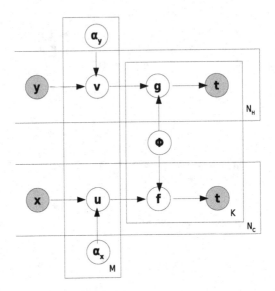

Fig. 2. Graphical model. The functions $\mathbf{f}_t$ and $\mathbf{g}_t$, which classify the control and case profiles $\mathbf{x}$ and $\mathbf{y}$, respectively, to the tth (of K) control classes, are related through parameters $\boldsymbol{\Phi}_t$.

[15] this is a prior step to the actual learning of tasks, in our work we learn inter-task information jointly with learning the tasks, through learning an auxiliary set of tasks.

The rest of this paper is organised as follows: In the following section, we discuss our proposed model for addressing the problem of selecting appropriate control samples with our modeling assumptions and model inference. We continue to give our experiments using the proposed model in Section 3 and finally, we give our concluding discussion in Section 4.

2   Gaussian Process Classification for Paired Multitask Learning

In this section we introduce a framework for paired multitask learning. There are K pairs of tasks, where the ith pair consists of learning to classify cancer profiles and control profiles to the ith control class. The graphical model is shown in Figure 2, and we next explain its structure.

We can classify the nth profile $\mathbf{y}_n$ from the control sample set into one of K classes, by learning the mappings to class labels; we assume that the probability of the labels depends on a set of K functions $\{f_1(\mathbf{y}_n), ..., f_K(\mathbf{y}_n)\}$ evaluated at

$\mathbf{y}_n$. Using a multinomial probit link function, we define the mapping between the function values and class labels as:

$$P(t_n = c \mid \{f_1(\mathbf{y}_n), ..., f_K(\mathbf{y}_n)\}) = E_{p(z)} \left(\prod_{i \neq c} \Phi(z + f_c(\mathbf{y}_n) - f_i(\mathbf{y}_n)) \right) \quad (2)$$

where $\Phi(z) = \int_{-\infty}^{z} \mathcal{N}(u \mid 0, 1) du$ is the cumulative normal density function. Similarly, we assume that an expression profile $\mathbf{x}_n$ from the case samples can be classified into the K classes, where the probability of the class labels also depend on a set of corresponding underlying functions $\{g_1(\mathbf{x}_n), ..., g_K(\mathbf{x}_n)\}$ in an analagous link function to (2). We denote the jth functions evaluated at the data points as $\mathbf{f}_j = [f_j(\mathbf{y}_1), ..., f_j(\mathbf{y}_{N_H})]^\top$ and $\mathbf{g}_j = [g_j(\mathbf{x}_1), ..., g_j(\mathbf{x}_{N_C})]^\top$, and across all K classes as $\mathbf{f} = [\mathbf{f}_1^\top, ..., \mathbf{f}_K^\top]^\top$ and $\mathbf{g} = [\mathbf{g}_1^\top, ..., \mathbf{g}_K^\top]^\top$.

In this work, we are interested in transferring information about the interrelatedness of the classes from one task (mapping control samples to tissue classes) to the main task (mapping cancer samples to tissue classes). For a standard multiclass Gaussian process classification task, the K functions are given Gaussian process priors, which are assumed to be uncorrelated across classes, i.e. $p(\mathbf{f}) = \mathcal{N}(\mathbf{f} \mid \mathbf{0}, \mathbf{K})$, where $\mathbf{K}$ is block diagonal in the class specific covariance functions $\mathbf{K}_1, ..., \mathbf{K}_K$. We take a different approach, and model each function as a linear combination of M latent functions where $M < K$; for the ith pair of functions this is:

$$\mathbf{f}_i = \sum_j \Phi_{i,j} \mathbf{u}_j, \qquad \mathbf{g}_i = \sum_j \Phi_{i,j} \mathbf{v}_j \quad (3)$$

where $\Phi_{i,j}$ is the weight of the jth latent function in the ith function, and $\mathbf{u}_j$ and $\mathbf{v}_j$ are the jth pair of latent functions. This formulation models dependencies between the functions for each multi-class classifier via $\Phi \in \Re^{K \times M}$, and this structure is shared between the two classification tasks. If we place Gaussian process priors over each latent function, $p(\mathbf{u}_j) = \mathcal{N}(\mathbf{u}_j \mid \mathbf{0}, \mathbf{K}_{u,j}), p(\mathbf{v}_j) = \mathcal{N}(\mathbf{v}_j \mid \mathbf{0}, \mathbf{K}_{v,j})$, then the distributions over $\mathbf{f}$ and $\mathbf{g}$ are:

$$p(\mathbf{f} \mid \Phi) = \int p(\mathbf{f} \mid \mathbf{u}, \Phi) p(\mathbf{u}) d\mathbf{u} = \mathcal{N}\left(\mathbf{f} \mid \mathbf{0}, (\Phi \otimes \mathbf{I}) \mathbf{K}_u (\Phi \otimes \mathbf{I})^\top\right) \quad (4)$$

$$p(\mathbf{g} \mid \Phi) = \int p(\mathbf{g} \mid \mathbf{v}, \Phi) p(\mathbf{v}) d\mathbf{v} = \mathcal{N}\left(\mathbf{g} \mid \mathbf{0}, (\Phi \otimes \mathbf{I}) \mathbf{K}_v (\Phi \otimes \mathbf{I})^\top\right) \quad (5)$$

where $\otimes$ denotes the Kronecker product, $\mathbf{K}_u$ and $\mathbf{K}_v$ are block diagonal in the class specific covariance functions $\mathbf{K}_{u,1}, ..., \mathbf{K}_{u,K}$ and $\mathbf{K}_{v,1}, ..., \mathbf{K}_{v,K}$ respectively. This prior captures correlations within the latent function sets; the cross covariance functions between $\mathbf{f}_i$ and $\mathbf{f}_j$ is given by $\sum_n \Phi_{i,n} \Phi_{j,n} \mathbf{K}_{u,n}$, and similarly for $\mathbf{g}_i$ and $\mathbf{g}_j$: $\sum_n \Phi_{i,n} \Phi_{j,n} \mathbf{K}_{v,n}$, and this relationship is shared across the two tasks via Φ. The model has similarities to the semiparametric latent factor model [14] in that statistical strength is shared across K GP's through a smaller set of M underlying functions. However in our approach, we use the learned relationship between one set of GP's on the control set to help train a set of GP's on the related set of case samples.

2.1   Inference in the Model

We use the data augmentation strategy as detailed in [16], by introducing auxiliary latent variables $\mathbf{f}'$ and $\mathbf{g}'$ in (2) for each classifier i.e. so that we can rewrite the probit link functions (linking $\{f_{n1}, ..., f_{nK}\}$ to $t_{x,n}$ in (2)) as:

$$P(t_{x,n} = c \mid \{f_{ni}\}_{i=1}^{K}) = \int P(t_{x,n} = c \mid \{f'_{ni}\}_{i=1}^{K}) \prod_{i=1}^{K} p(f'_{ni} \mid f_{ni}) df'_{ni}$$

$$= \int \delta(f'_{nc} > f'_{nk} \forall k \neq c) \prod_{i=1}^{K} \mathcal{N}(f'_{ni} \mid f_{ni}, 1) df'_{ni}$$

$$(6)$$

where we have denoted the ith function of $\mathbf{x}_n$, $f_i(\mathbf{x}_n)$, as f_{ni}. We similarly derive $P(t_{y,n} = c \mid \{g_{ni}\}_{i=1}^{K})$. To train the model we need to find the posterior distribution over $\Theta_x = \{\mathbf{f}', \mathbf{u}\}$, $\Theta_y = \{\mathbf{g}', \mathbf{v}\}$ and also the shared mixing matrix Φ, its hyperparameters ψ, and covariance function hyperparameters for both classifiers (which we will denote by α_x and α_y). The joint distribution over these quantities is given by

$$p(\mathbf{t}_x, \mathbf{t}_y, \Theta_x, \Theta_y, \Phi, \alpha_x, \alpha_y \mid \mathbf{X}, \mathbf{Y}, \psi) = \qquad (7)$$
$$p(\mathbf{t}_x, \Theta_x, \Phi, \alpha_x \mid \mathbf{X}) p(\mathbf{t}_y, \Theta_y, \Phi, \alpha_y \mid \mathbf{Y}) p(\Phi \mid \psi)$$

where

$$p(\mathbf{t}_x, \Theta_x, \Phi, \alpha_x \mid \mathbf{X}) = \qquad (8)$$
$$\prod_{n=1}^{N_C} \left[\sum_{i=1}^{K} \delta(f'_{ni} > f'_{nk} \forall k \neq i) \delta(t_{x,n} = i) \right] p(\mathbf{f}' \mid \mathbf{u}, \Phi) p(\mathbf{u} \mid \alpha_x, \mathbf{X})$$

and

$$p(\mathbf{t}_y, \Theta_y, \Phi, \alpha_y \mid \mathbf{Y}) = \qquad (9)$$
$$\prod_{n=1}^{N_H} \left[\sum_{i=1}^{K} \delta(g'_{ni} > g'_{nk} \forall k \neq i) \delta(t_{y,n} = i) \right] p(\mathbf{g}' \mid \mathbf{v}, \Phi) p(\mathbf{v} \mid \alpha_y, \mathbf{Y}).$$

We employ a variational approximation to the above, by finding an ensemble of approximating posterior distributions $Q(\Theta_x)Q(\Theta_y)Q(\Phi)Q(\alpha_x)Q(\alpha_y)$ to $p(\Theta_x, \Theta_y, \Phi, \alpha_x, \alpha_y \mid \mathbf{t}_x, \mathbf{t}_y, \mathbf{X}, \mathbf{Y})$ that maximise the lower bound on the marginal likelihood

$$\log p(\mathbf{t}_x, \mathbf{t}_y, \mid \mathbf{X}, \mathbf{Y}, \psi) \geq \qquad (10)$$
$$E_{Q(\Theta_x)Q(\Theta_y)Q(\Phi)Q(\alpha_x)Q(\alpha_y)} \{\log p(\mathbf{t}_x, \mathbf{t}_y, \Theta_x, \Theta_y, \Phi, \alpha_x, \alpha_y \mid \mathbf{X}, \mathbf{Y})\}$$
$$-E_{Q(\Theta_x)Q(\Theta_y)Q(\Phi)Q(\alpha_x)Q(\alpha_y)} \{\log Q(\Theta_x)Q(\Theta_y)Q(\Phi)Q(\alpha_x)Q(\alpha_y)\}.$$

The form of the $Q()'s$ are given below:

$$Q(\mathbf{u}) \propto \mathcal{N}(\tilde{\mathbf{f}}' \mid (\tilde{\Phi} \otimes \mathbf{I})\mathbf{u}, \mathbf{I})\mathcal{N}(\mathbf{u} \mid 0, \tilde{\mathbf{K}}) \tag{11}$$

$$Q(\mathbf{f}') \propto \mathcal{N}(\mathbf{f}' \mid (\tilde{\Phi} \otimes \mathbf{I})\tilde{\mathbf{u}}, \mathbf{I}) \prod_n [\delta(f'_{ni} > f'_{nk} \forall k \neq i)\delta(t_{x,n} = i)] \tag{12}$$

$$Q(\alpha_x) \propto \mathcal{N}(\tilde{\mathbf{u}} \mid 0, \mathbf{K})\mathcal{G}(\alpha_{x,i} \mid a_i, b_i) \tag{13}$$

where we use Gamma distributions over the hyperparameters of the covariance function a_i, $\{w_i, ..., w_{D_x}\}$, $\tilde{x}$ denotes the posterior mean of $Q(x)$, and similarly for $Q(\mathbf{v})$), $Q(\mathbf{g}')$), and $Q(\alpha_y)$). Finally,

$$Q(\Phi) \propto \mathcal{N}(\tilde{\mathbf{f}}' \mid (\Phi \otimes \mathbf{I})\tilde{\mathbf{u}}, \mathbf{I})\mathcal{N}(\tilde{\mathbf{g}}' \mid (\Phi \otimes \mathbf{I})\tilde{\mathbf{v}}, \mathbf{I}) \prod_{i=1}^{K} \mathcal{N}(\Phi_i \mid 0, \sigma_i \mathbf{I}) \tag{14}$$

where Φ_i denotes the ith row of Φ.

3 Experiments

In this section, we evaluate the model's performance on simulated data, and on a real world data set.

3.1 Experimental Details

For all the experiments in this paper we use the following: squared exponential covariance function $k(\mathbf{x}_i, \mathbf{x}_j) = a \exp -\frac{1}{2}(\mathbf{x}_i - \mathbf{x}_j)^\top \mathbf{W}(\mathbf{x}_i - \mathbf{x}_j)$ for all GP's, where a is a scale parameter, and $\mathbf{W}$ is a diagonal matrix with the inverse length scales $\{w_i, ..., w_D\}$ (D the dimension of the data) on the diagonal. We fix the noise level for each of the GP's to 1e-3, and use a distribution of $\mathcal{G}(1, 1)$ over the hyperparameters of the covariance functions. The optimization is sensitive to the initialisation; to initialise Φ, we first calculate a class similarity kernel between the means of each class in $\mathbf{Y}$, and then find the first M principal component vectors. We found that a linear kernel works best in practice.

3.2 Toy Data

We demonstrate the model's ability to generalise to unseen classes for a new data point $\mathbf{x}_n$, based on the relationship of the unseen class with the other classes learned from $\mathbf{Y}$ (where all classes are present), for an 8- class classification problem.

We generated a pair of data sets, each containing eight classes. The classes for each data set are shown in Figure 3 (a) and (b). For the two classification tasks, there is the same underlying structure to the set of class boundaries; they are a mixture of three latent functions, one for horizontal discrimination, one for

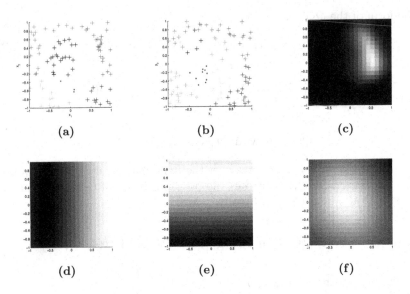

Fig. 3. Toy data experiment. Two sets of toy data: **X** in (a), and **Y** in (b) are from the same 8 classes (corresponding classes shown by the different coloured markers). The model learns an underlying set of functions for the first data set, as seen in (d) - (f), and can find a predictive distribution (c) over the missing class from the information learned from the auxiliary sets of tasks (b).

vertical discrimination, and one that discriminates between inside and outside of the circle. We remove the data points belonging to one of the classes for the first data set, and use these as test data. Figure 3 (c) shows the correctly inferred distribution for the missing class (green in (b)) evaluated over a grid over the input space. The bottom row of the figure shows the three inferred latent functions.

3.3   Cancer Profiles

Cancer is a complex disease, arising from genetic abnormalities which disrupt a cell's ordinary functioning, leading to uncontrolled cell proliferation. Identification of the mutated genes that drive the oncogenesis and gaining an understanding of cancer on a molecular basis is one of the key goals of cancer research; consequently there are ongoing coordinated efforts between clinicians, biologists and computer scientists to collect and analyse a large collection of genomic profiles over many different cancer types. While there is potentially a huge amount of useful information about cancer contained in such databases, its extraction presents many methodological challenges for bioinformaticians. Gene expression measurements are likely to contain bias due to factors such as patient-specific and laboratory-specific effects, and typically there are only a small number of samples available for each experimental condition. These factors make it problematic to select a set of pairs of control and normal tissue samples, such that the differential gene expression of the case samples is solely due to cancer-specific variation. However, we can exploit shared information between different sets of experiments: there are similarities (e.g. similar pathway activations) between different cancers, and similarities between normal tissue types.

We propose a solution to the problem of control measurement selection for a set of case profiles through using our model to learn the relationship between a pool of controls and cancer profiles, over a wide range of control classes. We show how the model can predict control classes for new cancer samples.

Data. Two publicly available gene expression data sets were taken from the NCBI's Gene Expression Omnibus [2]. The first data set is a series of cancer portraits (accession GSE2109, https://expo.intgen.org/geo/), consisting of 1911 clinically annotated gene expression profiles taken from 82 different tumor types in humans. The second data set is a set of 353 gene expression profiles taken over 65 different tissues in the normal human body (accession GSE3526, Neurocrine Biosciences, Inc.). Both data sets were preprocessed using RMA [17]. For each sample, we constructed a feature vector where each element represented the genes' activation in a known biological pathway, according to KEGG[2] gene sets from the Molecular Signatures Database (MSigDB) [18], resulting in a 200-dimensional vector of pathway activations. To calculate each pathway activation, we used the mean of the expression levels of the genes in each pathway. These feature vectors were used as inputs to the model. A visualization of the two data sets is given in Figure 4 (see caption for details).

Based on the annotations for both data sets, we manually classified the normal tissue samples into 35 control classes. For each class we used a maximum of 10 samples. Annotations for the tumor samples include the classification tissue. We then assigned the tumor samples to the control classes for cases where the mapping was evident. For each class we also allowed a maximum of 10 samples, if available, for the training set. Many classes contained few, or even no samples. The rest of samples (where the mapping was known) were assigned to a first test set ('Test Set 1'). Samples where the potential control class was ambiguous, e.g. samples where the classification tissue did not have an equivalent in the control set, or vague: *connective and soft tissue, ill defined*, were assigned to a second test set ('Test Set 2').

Results. We trained the model for different numbers of latent functions. Figure 5 shows the classification accuracy for the samples for each class in 'Test Set 1'. As a comparison method, we find control classes for each cancer sample by assigning it to the nearest control class using a K nearest neighbours classifier (optimal $K = 1$) trained over the control set. We picked the model with the highest predictive likelihood over the test set for the next set of experiments. Figure 6 visualizes the performance of the model on some samples selected from

---

[2] A manually curated database of gene pathways from http://www.genome.jp/kegg/

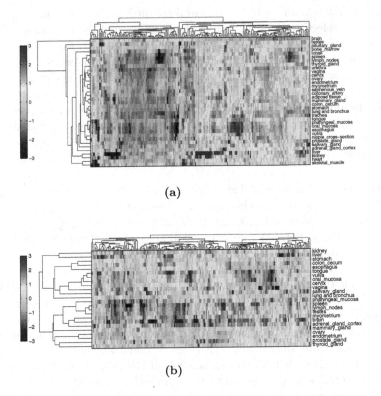

(a)

(b)

Fig. 4. (a): Heat map visualisation of **Y**, the set of control samples and (b): **X**, the set of tumor samples. Each row corresponds to the mean of the pathway activations (columns) in a class, and the data was clustered using standard hierarchical two dimensional clustering. This shows the similarities between different classes in each data set (note that the ordering of the columns differs between the figures).

'Test Set 2'. For each sample (rows), the predictive distribution over the control classes is visualized as a heat map. From the figure, we see that the model is able to make some sensible predictions for some of the tumor samples, when the control class is ambiguous.

The *gastroesophagal junction adenocarcinoma* samples (where the ideal control sample should be taken from the junction between the esophagus and the stomach) are mapped to *esophagus* and *stomach*. *Leiomyosarcoma* (4th row) and *uterine sarcoma* are both uterine cancers, and are mapped to *endometrium*, and most of the *colorectal adenocarcinoma* samples map to *colon cecum*. *Malignant melanoma* samples, a type of skin cancer, are mapped to *adipose tissue* and *lymph nodes*, which is a plausible prediction. Furthermore, the class *adipose tissue* was not in the training set, which verifies that the model was able to generalise to unseen classes. However, for cases such as the *bladder carcinomas* (rows 6 to 10) and the skin cancers (rows 11 to 15) where there does not appear to be a single appropriate control class, it is problematic to choose a suitable control.

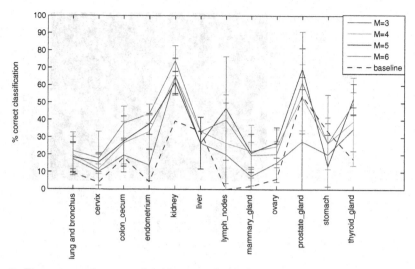

Fig. 5. Percentage of correct classifications for each control class in the test set, evaluated for different numbers of latent functions. The error bars correspond to ± 1 s.d. over 10 runs.

4   Discussion

In this paper, we highlighted the relevant, and frequently occurring, problem of control measurement selection in experimental design, primarily focusing on differential gene expression studies for analysing cancer. The inherently biased nature of gene expression measurements towards many factors (such as patient, laboratory and tissue-specific) can cause erroneous findings for differential gene expression studies, if the control measurements do not match the biases of the case samples. The possibility of error could be minimised by using a large set of appropriate control measurements, but these are seldom available in practice. However, there are large volumes of publicly available gene expression measurements, which could potentially be used as controls for the experiment.

We proposed a novel approach which can automatically find control measurements for a given cancer profile, from a pool of control samples, and a set of existing mappings to cancer samples. We train a model which jointly learns to classify the control and cancer samples into the K control classes, where the two sets of classifiers are constrained to be a mixture of underlying M functions where $M < K$. The functions can be different between the two sets, but the mixing matrix is the same. This approach can be viewed as a paired multitask learning problem, where the two sets of multiple tasks are the classification of control samples into control classes, and the classification of cancer samples into control classes. *Within* each task set, statistical strength is shared across the tasks by constraining the classifiers to be a mixture of underlying functions. Information about the relatedness of the tasks is shared *between* the two task sets, by constraining the mixing matrix to be the same.

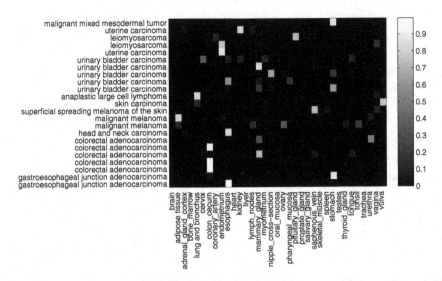

Fig. 6. Visualization of the probability distribution over the control classes (x axis) for some tumor samples (y axis) with unknown control classes

We found that the model was able to give reasonable performance on finding control measurements for a set of cancer portraits. For test data where the control class was known, the model mapped cancer profiles to one of 35 control classes with a better accuracy than a nearest neighbour classifier, trained over the control samples. However, in some cases (see Figure 6), particularly when a single appropriate control class did not exist, the model did not find appropriate control classes. This could be due to the way in which the model is constrained to be a mixture of underlying functions. Consequently, the model could be wasting its modelling power on accurately discriminating for a small number of classes, while the accuracy in predicting the other classes remains low. This is likely, given the spread of classification accuracies in Figure 5.

We feel that this is a promising approach to a difficult problem in bioinformatics, and a new innovative framework for constructing and solving structured multitask learning problems. We aim to further develop this framework by imposing alternative structural constraints between the two sets of tasks.

Acknowledgments. G.L. and S.K. belong to the Finnish Centre of Excellence on Adaptive Informatics Research. Funding: TEKES (grant no. 40101/07); PASCAL 2 Network of Excellence, ICT 216886 (partially to G.L. and S.K.).

References

1. Brazma, A., Parkinson, H., Sarkans, U., Shojatalab, M., Vilo, J., Abeygunawardena, N., Holloway, E., Kapushesky, M., Kemmeren, P., Garcia Lara, G., Oezcimen, A., Rocca-Serra, P., Sansone, S.-A.: ArrayExpress – A public repository for microarray gene expression data at the EBI. Nucleic Acids Research 31(1), 68–71 (2003)

2. Edgar, R., Domrachev, M., Lash, A.E.: Gene Expression Omnibus: NCBI gene expression and hybridization array data repository. Nucleic Acids Research 30(1), 207–210 (2002)
3. Knorr-Held, L., Best, N.G.: Shared component models for detecting joint and selective clustering of two diseases, Sonderforschungsbereich 386, Paper 183 (2000)
4. Yeung, K.Y., Haynor, D.R., Ruzzo, W.L.: Validating clustering for gene expression data. Bioinformatics 17(4), 309–318 (2001)
5. Ge, R., Ester, M., Gao, B.J., Hu, Z., Bhattacharya, B., Ben-Moshe, B.: Joint Cluster Analysis of Attribute Data and Relationship Data: The Connected k-Center Problem, Algorithms and Applications. ACM Transactions on Knowledge Discovery from Data 2(2), Article 7 (2008)
6. Furey, T.S., Cristianini, N., Duffy, N., Bednarski, D.W., Schummer, M., Haussler, D.: Support vector machine classification and validation of cancer tissue samples using microarray expression data. Bioinformatics 16(10), 906–914 (2000)
7. Gerber, G.K., Dowell, R.D., Jaakkola, T.S., Gifford, D.K.: Automated Discovery of Functional Generality of Human Gene Expression Programs. PLoS Comput. Biol. 3(8), e148 (2007), doi:10.1371/journal.pcbi.0030148
8. Zhang, J., Ghahramani, Z., Yang, Y.: Flexible Latent Variable Models for Multi-task Learning. Machine Learning 73(3), 221–242 (2008)
9. Bonilla, E.V., Chai, K.M.A., Williams, C.K.I.: Multi-task Gaussian Process Prediction. In: Neural Information Processing Systems (2008)
10. Yu, K., Tresp, V., Schwaighofer, A.: Learning Gaussian Processes from Multiple Tasks. In: 25th International Conference on Machine Learning (2008)
11. Storkey, A.J., Sugiyama, M.: Mixture Regression for Covariate Shift. In: Advances in Neural Information Processing Systems 19 (2007)
12. Sugiyama, M., Krauledat, M., Müller, K.-R.: Covariate shift adaptation by importance weighted cross validation. Journal of Machine Learning Research (8), 985–1005 (2007)
13. Daumé, H., Marcu, D.: Domain adaptation for statistical classifiers. Journal of Artifical Intelligence Research (26), 101–126 (2006)
14. Teh, Y.W., Seeger, M., Jordan, M.I.: Semiparametric latent factor models. In: Proceedings of the Eighth Conference on Artificial Intelligence and Statistics, AISTATS (2005)
15. Bickel, S., Bogojeska, J., Lengauer, T., Scheffer, T.: Multi-Task Learning for HIV Therapy Screening. In: 22nd International Conference on Machine Learning (2008)
16. Girolami, M., Rogers, S.: Variational Bayesian multinomial probit regression with Gaussian process priors. Neural Computation 18(8), 1790–1817 (2006)
17. Irizarry, R.A., Hobbs, B., Collin, F., Beazer-Barclay, Y.D., Antonellis, K.J., Scherf, U., Speed, T.P.: Exploration, Normalization, and Summaries of High Density Oligonucleotide Array Probe Level Data. Biostatistics 4(2), 249–264 (2003)
18. Subramanian, A., Tamayo, P., Mootha, V.K., Mukherjee, S., Ebert, B.L., Gillette, M.A., Paulovich, A., Pomeroy, S.L., Golub, T.R., Lander, E.S., Mesirov, J.P.: Gene set enrichment analysis: A knowledge-based approach for interpreting genome-wide expression profiles. Proceedings of the National Academy of Sciences of the United States of America 102(43), 15545–15550 (2005)

Coupled Metric Learning for Face Recognition with Degraded Images

Bo Li[1], Hong Chang[2,3], Shiguang Shan[2,3], and Xilin Chen[2,3]

[1] School of Computer Science and Technology, Harbin Institute of Technology,
Harbin, 150001, China
[2] Key Lab of Intelligent Information Processing, Chinese Academy of Sciences (CAS),
Beijing, 100190, China
[3] Institute of Computing Technology,
CAS, Beijing, 100190, China
{bli,hchang,sgshan,xlchen}@jdl.ac.cn

Abstract. Real-world face recognition systems are sometimes confronted with degraded face images, e.g., low-resolution, blurred, and noisy ones. Traditional two-step methods have limited performance, due to the disadvantageous issues of inconsistent targets between restoration and recognition, over-dependence on normal face images, and high computational complexity. To avoid these limitations, we propose a novel approach using coupled metric learning, without image restoration or any other preprocessing operations. Different from most previous work, our method takes into consideration both the recognition of the degraded test faces as well as the class-wise feature extraction of the normal faces in training set. We formulate the coupled metric learning as an optimization problem and solve it efficiently with a closed-form solution. This method can be generally applied to face recognition problems with various degrade images. Experimental results on various degraded face recognition problems show the effectiveness and efficiency of our proposed method.

1 Introduction

1.1 Degraded Face Recognition

During the past decades, face recognition has attracted a great deal of attention in the filed of pattern recognition and computer vision. Various face recognition algorithms have been proposed. An overview of popular methods is given in [1]. Unfortunately, the performance of face recognition systems will decrease in front of serious degraded test images, such as low-resolution images of only 12×12 pixels. Opposite to the *degraded images*, let us call the undegraded (e.g., clear, high-resolution and undamaged) ones as *normal images*.

Traditional methods for degraded face recognition usually take "two steps": image processing and recognition. Obviously, it is not a good choice to directly pass the degraded test images to the face recognition system enrolled with normal images, due to the sacrifice of important image details. A more commonly used way is to first restore the degraded images into normal ones. This kind of methods attempt to retrieve the loss

Z.-H. Zhou and T. Washio (Eds.): ACML 2009, LNAI 5828, pp. 220–233, 2009.

details of the degraded images. For this purpose, a lot of super-resolution and deblurring algorithms have been proposed, either for single image [2,3,4,5,6] or multi images [7,8,9]. Among these methods, learning-based super-resolution methods attract a lot of attention, such as [2,4,6,10]. More specific methods for face images have also been studied. Baker and Kanade [11] propose "hallucination" technique for face image super resolution. This learning-based method infers the high-resolution (HR) face images from an input low-resolution (LR) image using the face priors learned from training set. Liu et al. [12] propose a statistical modeling approach that integrates a global parametric model and a local nonparametric one.

However, there exist some limitations on the two-step methods. The main problem lies in that the preprocessing step only aims at minimizing the reconstruction error between restored images and the ground-truth, without any consideration of the subsequent recognition target. The restoration process with inconsistent target cannot always improve the performance of face recognition as much as we expected. Besides, the sophisticated super-resolution and deblurring algorithms used in this step are usually computational complex, which is not ideal for some real-time systems.

1.2 Recent Related Work

Recently, some algorithms without restoring the test image in the first step have been proposed, especially for low-resolution face recognition. Instead of reconstructing a HR image explicitly, the method proposed in [13] carries out the reconstruction in the Eigen-face space and only outputs the weights along the principal components for face recognition purpose. Jia and Gong [14] propose a tensor-based method to simultaneously super-resolve and recognize LR face images. Face images with different poses, expressions and illuminations are modeled in tensor space, where HR images in multi-modal can be estimated from the corresponding LR images in single-modal. This method is a tensor extension of the super-resolution problem for the single-modal LR face image. More recently, some researchers propose a new algorithm which could simultaneously carry out super-resolution and feature extraction for LR face recognition [15]. This method designs a new objective function which integrates the constraints for reconstruction and classification together. In face recognition tasks, the method requires to optimize the object function for each test image. Although the authors propose a speed up procedure, the computational complexity is still high, especially for real-world real-time applications.

Despite the recent progresses, most existing methods for degraded face recognition have limitations in the following aspects: (1) Inconsistent targets in restoration and recognition. Minimizing the reconstruction error in image restoration may not always guarantee good performance of the subsequent face recognition, even if the reconstruction is not explicitly performed [13,14]. (2) From empirical studies, normal images in the training (and gallery) set are not always good for recognition. On the contrary, high-frequency details and background noise in the normal images may decrease the recognition performance. (3) Computational complexity. Many super-resolution and deblurring methods, adopted in the preprocessing step, are time-consuming. Even the simultaneous method [15] is not efficient enough.

1.3   This Paper

To overcome the limitations stated above, we propose a new simple and elegant way to solve the problem of degraded face recognition. Different from previous methods, we address the problem through coupled metric learning which actually decides two transformations. One transformation maps degraded images to a new subspace, where higher recognition performance can be achieved; the other one maps normal images and class labels together to the same subspace for better class-wise feature representation. The coupled transformations are jointly determined by solving an optimization problem. The optimization procedure is fairly efficient with closed-form solutions.

Compared with some previous related methods, our work contributes in the following aspects. First, it is novel to learn coupled metrics, which takes into consideration both degraded face recognition and normal class-wise feature extraction. Second, it is very simple and efficient. Our method does not require any preprocessing (e.g., super-resolution or deblurring). The metric learning step is carried out in the off-line training phase. Third, it can be applied to face recognition with various degraded images as well as more general degraded object classification problems. Note that metric learning methods for classification has been studied by many researchers [16,17,18,19]. However, only a few methods try to learn coupled metrics and none of them has been successfully applied in the degraded face recognition problems mentioned above.

The rest of this paper is organized as follows. We propose the coupled metric learning method in Section 2, where problem formulation and optimization are presented in detail. Section 3 summarizes the algorithm for degraded face recognition and describes some implementation details. We then apply our method to several degraded face recognition problems in Section 4. Some discussions on nonlinear extension and other possible applications in given Section 5. Finally, Section 6 concludes the whole paper.

2   Coupled Metric Learning

2.1   Problem Formulation

Denote the feature vector of a degraded test image as $\mathbf{x}^p \in \mathcal{R}^m$. The training set consists of N normal facial images $\mathbf{y}_i \in \mathcal{R}^M$ and their degraded counter-parts $\mathbf{x}_i \in \mathcal{R}^m$, $i = 1, \ldots, N$. Suppose there are C classes. Let $\mathbf{t}_i$ denote the C-dimensional class indicator vector of $\mathbf{y}_i$ ($\mathbf{x}_i$), which can be seen as one column vector from a $C \times C$ identity matrix. The features of degraded and normal face images, $\mathbf{x}^p$, $\mathbf{x}_i$ and $\mathbf{y}_i (i = 1, \ldots, N)$, are usually constructed by concatenation of pixel values. m and M denote the dimensionalities of these feature vectors.

Let us represent the training set in the form of $\mathbf{X} = [\mathbf{x}_1, \ldots, \mathbf{x}_N]$ and $\mathbf{Y} = [\mathbf{y}_1, \ldots, \mathbf{y}_N]$, where $\mathbf{X}$ and $\mathbf{Y}$ are matrices with the sizes of $m \times N$ and $M \times N$, respectively. The labels of the training set can be represented using the class indicator matrix $\mathbf{T} = [\mathbf{t}_1, \ldots, \mathbf{t}_N]$. Actually, the degraded and normal face images form subspaces $\mathcal{X} \subset \mathcal{R}^m$ and $\mathcal{Y} \subset \mathcal{R}^M$. The label space $\mathcal{T}$ is spanned by $\{\mathbf{e}_1, \ldots, \mathbf{e}_C\}$, where C is the class number and $\mathbf{e}_i$ is a unit vector with only one non-zero value at the i-th entry. For example, $\mathbf{e}_1 = [1, 0, \ldots, 0]^T$ is a vector with size of $C \times 1$. Consequently, $\mathbf{t}_i = \mathbf{e}_j$ means that the i-th sample belongs to the j-th class.

Different from traditional methods, which recognize $\mathbf{x}^p$ by first reconstructing its normal counter-part, we address this challenging problem in the metric learning context.

2.2 Metric Learning in Coupled Spaces

The basic idea of our method is to learn coupled metrics which in fact map $\mathbf{X}$, $\mathbf{Y}$ and $\mathbf{T}$ to a joint new subspace $\mathcal{Z} \subset \mathcal{R}^d$, where the new distance measure is more ideal for face recognition.

On one hand, we try to learn a metric from degraded training images, $\mathbf{x}_i, i = 1, \ldots, N$, to get better class-separability. To this end, we define the linear transformation in matrix form as follows:

$$\widetilde{\mathbf{X}} = \mathbf{AX} = [\mathbf{Ax}_1, \ldots, \mathbf{Ax}_N], \tag{1}$$

where $\widetilde{\mathbf{X}}$ represent the new features in the d-dimensional transformed space $\mathcal{Z}$. In this way, the $d \times m$ matrix $\mathbf{A}$ maps the degraded features to a new space where face recognition is actually performed. We call the transformation matrix $\mathbf{A}$ *recognition matrix*.

On the other hand, we expect to extract better features from normal training images, $\mathbf{y}_i, i = 1, \ldots, N$, and preserve the class label information at the same time. More specifically, we define the linear transformation on $\mathbf{Y}$ and $\mathbf{T}$ as

$$\widetilde{\mathbf{Y}} = \mathbf{F}\hat{\mathbf{Y}} = \mathbf{F}[\mathbf{Y}^T \alpha \mathbf{T}^T]^T, \tag{2}$$

where $\mathbf{F}$ is the $d \times (M + C)$ transformation matrix. The transformation decided by $\mathbf{F}$ results in a new class-wise feature representation in the feature space $\mathcal{Z}$, we thus name it as *feature matrix*. For mathematical conciseness, we use $\hat{\mathbf{Y}} = [\mathbf{Y}^T \alpha \mathbf{T}^T]^T$ to denote the expanded normal training matrix in the following context. α is a scaling parameter. To estimate $\mathbf{A}$ and $\mathbf{F}$ defined above, we formulate the coupled metric learning as an optimization problem. The objective function to be minimized is defined as

$$J(\mathbf{A}, \mathbf{F}) = \|\widetilde{\mathbf{X}} - \widetilde{\mathbf{Y}}\| = \|\mathbf{AX} - \mathbf{F}\hat{\mathbf{Y}}\|, \tag{3}$$

with the constraints of $\widetilde{\mathbf{X}}\widetilde{\mathbf{X}}^T = \widetilde{\mathbf{Y}}\widetilde{\mathbf{Y}}^T = \mathbf{I}$. $\mathbf{I}$ denotes the identity matrix and $\| \cdot \|$ the Frobenius norm. Note that $\mathbf{X}$ and $\hat{\mathbf{Y}}$ are first centered by extracting the means from the vectors involved.

The optimization procedure is similar with that of Canonical correlation analysis (CCA) [20,21]. Let us consider the simplest case when we map $\mathbf{X}$ and $\hat{\mathbf{Y}}$ to 1-dimensional target space $\mathcal{Z}$. The object function becomes

$$J(\mathbf{a}, \mathbf{f}) = \|\mathbf{aX} - \mathbf{f}\hat{\mathbf{Y}}\|, \tag{4}$$

subject to $\|\mathbf{aX}\| = \|\mathbf{f}\hat{\mathbf{Y}}\| = 1$. Here, the recognition matrix $\mathbf{A}$ and the feature matrix $\mathbf{F}$ are reduced to row vectors $\mathbf{a}$ and $\mathbf{f}$ with sizes of m and $M + C$, respectively. Note that, in our procedure, d should be not larger than the minimum of m and $M + C$. Eqn. (4) is equal to the following maximization problem

$$\max_{\mathbf{a}, \mathbf{f}} \mathbf{a}\mathbf{X}\hat{\mathbf{Y}}^T\mathbf{f}^T, \text{ s. t. } \mathbf{a}\mathbf{X}\mathbf{X}^T\mathbf{a}^T = \mathbf{f}\hat{\mathbf{Y}}\hat{\mathbf{Y}}^T\mathbf{f}^T = 1. \tag{5}$$

The corresponding Lagrangian is

$$L(\mathbf{a}, \mathbf{f}, \lambda_1, \lambda_2) = \mathbf{a}\mathbf{X}\hat{\mathbf{Y}}^T\mathbf{f}^T - \lambda_1(\mathbf{a}\mathbf{X}\mathbf{X}^T\mathbf{a}^T - 1) - \lambda_2(\mathbf{f}\hat{\mathbf{Y}}\hat{\mathbf{Y}}^T\mathbf{f}^T - 1). \qquad (6)$$

Setting the derivatives of L w.r.t. $\mathbf{a}$ and $\mathbf{f}$ to 0's, we obtain

$$\mathbf{X}\hat{\mathbf{Y}}^T\mathbf{f}^T - \lambda_1\mathbf{X}\mathbf{X}^T\mathbf{a}^T = 0, \qquad (7)$$

$$\hat{\mathbf{Y}}\mathbf{X}^T\mathbf{a}^T - \lambda_2\hat{\mathbf{Y}}\hat{\mathbf{Y}}^T\mathbf{f}^T = 0. \qquad (8)$$

Subtracting $\mathbf{a}$ times (7) from $\mathbf{f}$ times (8) and considering the constrains, we can finally have $\lambda_1 = \lambda_2 = \lambda$,

$$\mathbf{X}\hat{\mathbf{Y}}^T(\hat{\mathbf{Y}}\hat{\mathbf{Y}}^T)^{-1}\hat{\mathbf{Y}}\mathbf{X}^T\mathbf{a}^T = \lambda^2\mathbf{X}\mathbf{X}^T\mathbf{a}^T, \qquad (9)$$

and

$$\mathbf{f} = \frac{1}{\lambda}\mathbf{a}\mathbf{X}\hat{\mathbf{Y}}^T(\hat{\mathbf{Y}}\hat{\mathbf{Y}}^T)^{-1}. \qquad (10)$$

Therefore, we can obtain the sequence of $\mathbf{a}$'s by solving a generalized eigendecomposition problem (Equation (9)) and then get $\mathbf{f}$'s from Equation (10). $\mathbf{A}$ is simply constructed by piling the first d largest eigenvectors of Equation (9) up. According to $\mathbf{A}$, $\mathbf{F}$ could be constructed. Note that, $\mathbf{X}\mathbf{X}^T$ and $\hat{\mathbf{Y}}\hat{\mathbf{Y}}^T$ is usually non-invertible. In this case, we carry out the regularization operations on them: $R(\mathbf{X}\mathbf{X}^T) = (\mathbf{X}\mathbf{X}^T + \kappa\mathbf{I})^{-1}$ and $R(\hat{\mathbf{Y}}\hat{\mathbf{Y}}^T) = (\hat{\mathbf{Y}}\hat{\mathbf{Y}}^T + \kappa\mathbf{I})^{-1}$, where κ is set to a small positive value (e.g., $\kappa = 10^{-6}$).

3   Degraded Face Recognition

Once we get the recognition matrix $\mathbf{A}$ and the feature matrix $\mathbf{F}$ through coupled metric learning, we may perform face recognition in the transformed new feature space. In this section, we present the overall algorithm and give some implementation details.

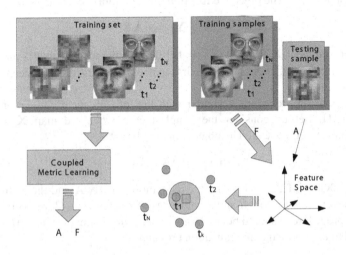

Fig. 1. An overview diagram for the algorithm

The overall algorithm is summarized in Figure 2. Note that the algorithm actually consists of two phases: an offline coupled metric learning phase and an online recognition phase. In the first phase, we learn the transformation matrices $\mathbf{A}$ and $\mathbf{F}$. In the second phase, degraded test image and the normal training images are jointly projected to a new feature space, using the learned transformations, respectively. The label of the test image is obtained using any distance-based classification method, such as k-nearest neighbor (kNN) classifier, support vector machines (SVMs), and so on. An overview diagram is shown in Figure 1 for better understanding the training and recognition procedure in our algorithm.

Input: Training degraded and normal face images and labels: $\mathbf{x}_i, \mathbf{y}_i, \mathbf{t}_i, i = 1, \ldots, N$
 a degraded test face image: $\mathbf{x}^p$
Offline Coupled metric learning:
 Learn coupled recognition matrix $\mathbf{A}$ and feature matrix $\mathbf{F}$
Online Recognition:
 Transform the test image $\mathbf{x}^p$ to $\widetilde{\mathbf{x}}^p$: $\widetilde{\mathbf{x}}^p = \mathbf{A}\mathbf{x}^p$
 Transform training images $\mathbf{y}_i$ to $\widetilde{\mathbf{y}}_i$ ($i = 1, \ldots, N$): $\widetilde{\mathbf{y}}_i = \mathbf{F}\mathbf{y}_i$
 Recognize $\widetilde{\mathbf{x}}^p$ from $\widetilde{\mathbf{y}}_i$
Output: Label $\mathbf{t}^p$ ($i = 1, \ldots, N$)

Fig. 2. Algorithm of degraded face recognition based on coupled metric learning

Note that our algorithm is ready to adopt the "probe-gallery" setting as well. Given a gallery set $(\mathbf{y}_i^g, \mathbf{t}_i^g), i = 1, \ldots, N_g$, with N_g being the number of normal face images in the set, we just need to transform these features to $\widetilde{\mathbf{y}}_i^g = \mathbf{F}\mathbf{y}_i^g$. The recognition is then performed on $\widetilde{\mathbf{x}}^p$ from $\widetilde{\mathbf{y}}_i^g$ ($i = 1, \ldots, N_g$).

There are only two parameters involved in our proposed method. It is easy to set them according to their physical meanings. In our experiments, the dimensionality of the transformed target space d and the scaling parameter α are simply determined using several rounds of cross-validations. The kNN classifiers are used in the final classification step for face recognition. The kNN classifiers are used in the final classification step for face recognition.

4 Experiments

In this section, we access the efficacy of our proposed method for some real-world problems. The experiments are generally performed on face images with various degradations.

4.1 Face Database and Degraded Images

We make use of the AR face database [22], which consists of over $3,200$ color frontal view faces of 135 individuals. For each individual, there are 26 images, recorded in two sessions separated by two weeks and with different illumination, expressions, and facial disguises. In our experiment, we select 14 images for each individual with only illumination and expression changes. The image set from the first session (the former 7 images) is considered as training set and the other (the latter 7 images) as testing set.

The original images are cropped into 72×72 pixels and aligned according to the positions of two eyes. The corresponding serious degraded images are obtained as following procedures. The blurred images are generated by convoluting the normal images with point spread function (PSF). The larger the diameter of PSF, the more blurring the processed images. In the experiments, we use a "disk" kernel with diameter of 13 pixels as PSF. To get low resolution images of size 12×12, we take the operations of blurring and then down-sampling (with down-sampling factor equal to 6). We synthesize partially occluded images by setting the pixel values inside a square occlusion area to 0. Different sizes of occlusion areas are used in our experiments. Figure 3 shows an example face image from AR database and its counter-parts in low-resolution, blurring, and partial occlusion (from left to right in the figure).

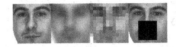

Fig. 3. A normal face image and the corresponding blurred one, low-resolution one, and occluded one

In our experiments, we normalize the columns of $\mathbf{X}$ and $\mathbf{Y}$ into the unit vectors. According to the direct sum kernels [23], we set a empirical value of $\alpha = 1$. Sometimes, this selection will affect the final recognition performance. The dimensionality settings of the features for different kinds of dagraded face recognition could be found in Table 1.

We compare our proposed method with some traditional baseline methods. More specifically, approaches of the following four classes are included in our comparative study: (1) directly use the degraded test images for recognition; (2) directly do recognition with the ground-truth test images without degradation; (3) restore the degraded test images using different classical algorithms and then do recognition (two-step or restoration-based method); (4) learn coupled metric for recognition (CML-based method). In these comparison, the enrollments are normal images. Besides, we center the features before hand in all experiments.

4.2 An Illustrative Example

First of all, we give a preliminary comparison between these methods for blurred face recognition. In this experiment, we use the wiener filter algorithm [24] to estimate the deblurred images. Figure 4 shows some face samples of the first five persons embedded in three-dimensional space, where markers with the same type belong to the same person. The solid markers denote the normal samples while the hollow denote the degraded. From Figure 4(a), we can see that the blurred samples from different persons are mixed with each other. Figure 4(b) shows the results using deblurred samples, where

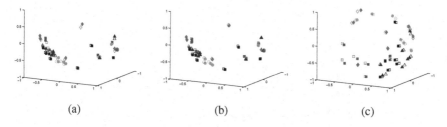

(a)   (b)   (c)

Fig. 4. Face samples of the first five persons embedded in 3-D space. (a) The blurred samples and the normal ones. (b) The deblurred samples and the normal ones. (c) The blurred samples and the normal ones after coupled metric learning. Here, the solid markers denote the normal samples while the hollow markers of the same type denote the degraded ones of the same person.

samples from the same person are closer, but samples from different persons are still difficult to discriminate. Finally, Figure 4(c) shows the results of coupled metric learning. We can see that the separability among different persons is increased. Intuitively, the situation in this subfigure will benefit classification. We also give the quantitative results of face recognition, as shown in Figure 5. Obviously, the performance of our method is significantly better than those of direct using degraded images and the restoration-based method. We can thus conclude that the restoration method could benefit recognition, while our proposed method is more outstanding for this purpose. Note that for visualization clarity, we select 35 samples from 5 persons in Figure 4(a)-(c), while the recognition rates in Figure 5 are computed using all samples.

Table 1. Dimensionality settings for the experiments of degraded face recognition on AR

Blurred Faces	Low-resolution Faces	Partially Occluded Faces
PCA: 945	PCA: 150	PCA: 250
LDA: 134	LDA: 134	LDA: 134
CML: 945	CML: 144	CML: 250

4.3 Face Recognition with Degraded Images

We further perform experiments on face recognition with degraded images, in order to demonstrate the effectiveness of the proposed algorithm and verify the discussions in previous sections. We compare different methods on face recognition tasks with degradation types including low resolution, blurring and partial occlusion, respectively.

On Blurred Faces. Firstly, let us test the coupled metric learning on the problem of face recognition with blurred test images. The blurred test images are generated from the normal test ones by convoluting with a "disk" PSF with diameter of 13 pixels. We apply two restoration-based methods by adopting Wiener filter [24] and Lucy-Richardson algorithm [25,26] in the deblurring step and compare these methods with ours.

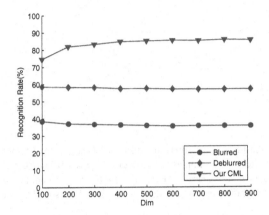

Fig. 5. The recognition rate curves for the illustrative experiments shown in Figure 4

As for the two traditional methods, we adopt two conventional face features, i.e., Eigen-faces (with PCA) [27] and Fisher-faces (with LDA) [28]. As for our method, we simply use pixel values as features. In the implementation, 945 Eigen-faces and 134 Fisher-faces are used for different methods. Notice that Fisher-faces are no more than 134 since the data set we used has 135 subjects. For our method, we use 945 Eigen-faces as the test features and original image intensities as the enrolled features.

The curves of cumulative recognition rates for different methods are shown in Figure 6. We compared our method with the different restoration and feature extraction methods which include: Normal-PCA (normal test images), Normal-LDA , Blurred-PCA (blurred test images), Blurred-LDA, Wiener-PCA (blurred test images restored by Wiener filter [24]), Wiener-LDA, Lucy-PCA (blurred test images restored by Lucy-Richardson algorithm [25,26]), Lucy-LDA, and Our CML (coupled metric learning method).

Our coupled metric learning method gets the best performance than all two-steps restoration-based methods. It archives the rank-1 recognition rate of 87.1%, even better than 85.6% of high-resolution LDA. From the blurred example shown in Figure 3, we can see that the blurring is very serious. Even though, our method still gets more satisfactory recognition results than normal situation. The reason may lie in that our metric learning method can find a feature space where the normal images, class information and the blurred ones are perfectly fused. Actually, multi-scale feature fusion is an effective scheme to improve the performance of face recognition systems [15].

On Low-Resolution Faces. Then, we test our method on the problem of face recognition with low-resolution test images. In this experiment, 72×72 high-resolution face images and the corresponding 12×12 low-resolution ones are used as normal and degraded images, respectively. We compare our method with two restoration-based methods using "spline" interpolation and learning-based super-resolution [4].

As the previous experiment, we also adopt 150 Eigen-faces and 134 Fisher-faces in the algorithms involved in comparison. As for our method, we simply use pixel values as features for both the LR test images and the normal enrolled ones.

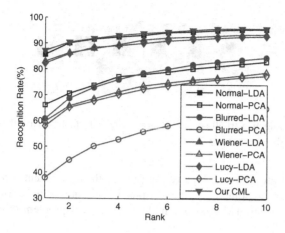

Fig. 6. Cumulative recognition results on blurred face recognition with different restoration and feature extraction methods

We compute the accumulative recognition rates for different methods that as shown in Figure 7. In the figure, the methods involved in our comparative study include Normal-LDA (HR test images), Normal-PCA, Spline-LDA (LR test images interpolatd by spline), Spline-PCA, LSR-LDA (LR test images super-resolved by learning-based method [4]), LSR-PCA, and Our CML. As we can see from Figure 7, our CML-based method outperforms most other methods and only a little worse than LDA with real high-resolution images. It achieves the best recognition rate of 86.3% for rank-1. The best rank-1 performance runner-up is 85.6%, given by high-resolution LDA. The performance of two other methods, learning-based method LDA and spline interpolation LDA, reach 82.0% and 74.4%, respectively.

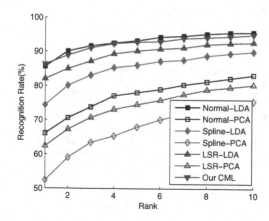

Fig. 7. Cumulative recognition results on LR face recognition with different restoration and feature extraction methods

Fig. 8. Examples of synthetic partially occluded face images (the first row) and the corresponding restoration by HNE (the second row)

On Partially Occluded Faces. Finally, we present the experiments on partially occluded face recognition. Sometimes, the test face images are occluded by other object such as dark glasses. Similarly, our method can solve this problem through coupled metric learning.

We test our method on a synthetic set of partially occluded face images. We generate the occluded face images by setting the pixel intensity in a square area to zero. The sizes of the occlusion areas range from 20×20 to 45×45, with each expansion leading to 5-pixel larger in the area edges. Figure 8 shows some examples in the synthetic set and corresponding restored ones by holistic neighbor embedding (HNE) [29]. We compare our method with a two-step method which first restores the occluded images by HNE. The number of neighbors involved in the reconstruction is 7. As for our method, the partially occluded samples are represented by the first 250-dimensional features from PCA.

From the results show in Figure 9, the two-step method get very good performance, since the face images have strong priors and the un-occluded parts are accurate (see Figure 8). On the other hand, our method gets even better result than HNE-based methods and is comparable with the Normal-LDA method using the ground-truth images without occlusion.

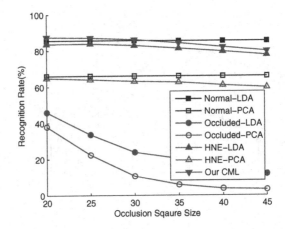

Fig. 9. Recognition rates on occluded face recognition with different restoration and feature extraction methods. The horizontal axis represents different sizes of occlusion squares.

5 Discussions

5.1 Real Dagraded Face Image

In Experiments, the synthesized dagraded face images are a little different from the real-world dagraded face image. For example, the low-resolution face images are noisy ones usually. Obviously, the performance of the original proposed method based on coupled metric learning will reduce. In this case, we can use the features which are nonsensitive for noise, such as Gabor wavelets, instead of the original intensity feature.

5.2 Nonlinear Extension with Kernels

The coupled metric learning method proposed in Section 2.2 can be extended from global linear forms to kernel version, since better feature representation and class separability is usually achieved through nonlinear mappings.

5.3 Other Possible Applications

It is worth noting that besides face recognition with degraded images, our proposed method can also be used to restore the degraded faces, as some super-resolution and de-blurring methods do. As for super-resolution problems, the LR face to be reconstructed is mapped to a new space (learned by training HR and LR images), where we can do the reconstruction similar to the methods in [4] or [10]. Making good use of the relationship between LR and labeled HR images through coupled metric learning, our approach is more promising than some related super-resolution methods. Due to the space limitation, the experiments on face restoration with coupled metric learning are beyond the scope of this paper.

6 Concluding Remarks

This paper proposes a novel approach using coupled metric learning to address the problem of degraded face recognition. Different from some previous work, this method considers both the recognition of the degraded test faces and class-wise feature extraction of normal faces. The coupled metric learning is formulated as an optimization problem and can be solved efficiently with a closed-form solution. Experiments on various degraded face recognition tasks show that our method can get very satisfactory results.

In our future work, we will continue to study degraded face recognition problems along the metric learning direction. A possible extension of the current work is to study an new objective function which can put different weights on the coupled transformations. This is reasonable as one set of features may contain richer information than the other. However, it is not very straightforward if we still expect clean and closed-form solutions. Other possible research topics include investigating the relationship between reconstruction and recognition, as well as studying the break points and performance bounds for specific degraded face recognition.

232   B. Li et al.

Acknowledgement

This work is partially supported by National Natural Science Foundation of China under contract Nos.60803084, 60833013, and 60832004; 100 Talents Program of CAS, and National Basic Research Program of China (973 Program) under contract 2009CB320902.

References

1. Zhao, W., Chellappa, R., Rosenfeld, A., Phillips, P.: Face recognition: a literature survey. ACM Computing Surveys, 399–458 (2003)
2. Baker, S., Kanade, T.: Limits on super-resolution and how to break them. IEEE Transactions on Pattern Analysis and Machine Intelligence 24(9), 1167–1183 (2002)
3. Biemond, J., Lagendijk, R., Mersereau, R.: Iterative methods for image deblurring. Proceedings of the IEEE 78(5), 856–883 (1990)
4. Chang, H., Yeung, D., Xiong, Y.: Super-resolution through neighbor embedding. In: Proceedings of the IEEE Computer Society Conference on Computer Vision and Pattern Recognition, vol. 1, pp. 275–282 (2004)
5. Fergus, R., Singh, B., Hertzmann, A., Roweis, S., Freeman, W.: Removing camera shake from a single photograph. ACM Transactions on Graphics 25(3), 784–794 (2006)
6. Freeman, B., Pasztor, E., Carmichael, O.: Learning low-level vision. International Journal of Computer Vision 40(1), 25–47 (2000)
7. Capel, D., Zisserman, A.: Super-resolution from multiple views using learnt image models. In: Proceedings of the IEEE Computer Society Conference on Computer Vision and Pattern Recognition, pp. 627–634 (2001)
8. Chen, J., Yuan, L., Tang, C., Quan, L.: Robust dual motion deblurring. In: Proceedings of the IEEE Computer Society Conference on Computer Vision and Pattern Recognition, pp. 1–8 (2008)
9. Elad, M., Feuer, A.: Super-resolution reconstruction of image sequences. IEEE Transactions on Pattern Analysis and Machine Intelligence 21(9), 817–834 (1999)
10. Yang, J., Wright, J., Huang, T., Ma, Y.: Image super-resolution as sparse representation of raw patches. In: Proceedings of the IEEE Computer Society Conference on Computer Vision and Pattern Recognition (2008)
11. Baker, S., Kanade, T.: Hallucinating faces. Automatic Face and Gesture Recognition (2000)
12. Liu, C., Shum, H., Zhang, C.: A two-step approach to hallucinating faces: global parametric model and local nonparametric model. In: Proceedings of the IEEE Computer Society Conference on Computer Vision and Pattern Recognition, pp. 192–198 (2001)
13. Gunturk, B., Batur, A., Altunbasak, Y., Hayes, M., Mersereau, R.: Eigenface-domain super-resolution for face recognition. IEEE Transactions on Image Processing 12(5), 597–606 (2003)
14. Jia, K., Gong, S.: Multi-modal tensor face for simultaneous super-resolution. In: Proceedings of the Tenth IEEE International Conference on Computer Vision, pp. 1683–1690 (2005)
15. Hennings-Yeomans, P., Baker, S., Kumar, B.: Simultaneous super-resolution and feature extraction for recognition of low-resolution faces. In: Proceedings of the IEEE Computer Society Conference on Computer Vision and Pattern Recognition, pp. 1–8 (2008)
16. Domeniconi, C., Peng, J., Gunopulos, D.: Locally adaptive metric nearest-neighbor classification. IEEE Transactions on Pattern Analysis and Machine Intelligence 24(9), 1281–1285 (2002)

17. Frome, A., Singer, Y., Sha, F., Malik, J.: Learning globally-consistent local distance functions for shape based image retrieval and classification. In: Proceedings of the Eleventh IEEE International Conference on Computer Vision (2007)
18. Goldberger, J., Roweis, S., Hinton, G., Salakhutdinov, R.: Neighbourhood component analysis. In: Advances in Neural Information Processing Systems 17, pp. 513–520 (2005)
19. Weinberger, K., Blitzer, J., Saul, L.: Distance metric learning for large margin nearest neighbor classification. In: Weiss, Y., Schölkopf, B., Platt, J. (eds.) Advances in Neural Information Processing Systems 18. MIT Press, Cambridge (2006)
20. Hotelling, H.: Relations between two sets of variates. Biometrika 28, 312–377 (1936)
21. Hardoon, D., Szedmak, S., Shawe-Taylor, J.: Canonical correlation analysis: an overview with application to learning methods. Neural Computation 16, 2639–2664 (2004)
22. Martinez, A., Benavente, R.: The AR face database. Technical report, CVC Tech. Report No. 24 (1998)
23. Zhang, Z.: Learning metrics via discriminant kernels and multidimensional scaling: toward expected euclidean representation. In: Proceedings of the Twentieth International Conference on Machine Learning, Washington, DC, USA (2003)
24. Gonzalez, R.C., Woods, R.E.: Digital Image Processing. Prentice Hall, Englewood Cliffs (2002)
25. Lucy, L.B.: An iterative technique for the rectification of observed distributions. Astronomical Journal 79(6), 745–754 (1974)
26. Richardson, W.: Bayesian-based iterative method of image restoration. Journal of the Optical Society of America 62(1), 55–59 (1972)
27. Turk, M., Pentland, A.: Eigenfaces for recognition. Journal of Cognitive Neuroscience 3(1), 71–86 (1991)
28. Belhumeur, P., Hespanha, J., Kriegman, D.: Eigenfaces vs. Fisherfaces: recognition using class specific linear projection. IEEE Transactions on Pattern Analysis and Machine Intelligence 19(7), 711–720 (1997)
29. Park, J., Oh, Y., Ahn, S., Lee, S.: Glasses removal from facial image using recursive error compensation. IEEE transactions on pattern analysis and machine intelligence 27(5), 805–811 (2005)

Cost-Sensitive Boosting: Fitting an Additive Asymmetric Logistic Regression Model*

Qiu-jie Li, Yao-bin Mao, Zhi-quan Wang, and Wen-bo Xiang

School of Automation, Nanjing University of Sci. & Tech.,
Nanjing, P.R. China, 210094
{liqiujie_1,maoyaobin}@163.com

Abstract. Conventional machine learning algorithms like boosting tend to equally treat misclassification errors that are not adequate to process certain cost-sensitive classification problems such as object detection. Although many cost-sensitive extensions of boosting by directly modifying the weighting strategy of correspond original algorithms have been proposed and reported, they are heuristic in nature and only proved effective by empirical results but lack sound theoretical analysis. This paper develops a framework from a statistical insight that can embody almost all existing cost-sensitive boosting algorithms: fitting an additive asymmetric logistic regression model by stage-wise optimization of certain criterions. Four cost-sensitive versions of boosting algorithms are derived, namely CSDA, CSRA, CSGA and CSLB which respectively correspond to Discrete AdaBoost, Real AdaBoost, Gentle AdaBoost and LogitBoost. Experimental results on the application of face detection have shown the effectiveness of the proposed learning framework in the reduction of the cumulative misclassification cost.

Keywords: Boosting, Cost-Sensitive Boosting, Logistic Regression.

1 Introduction

Minimum classification error is generally used as a criterion in many classifier designs that take advantage of machine learning techniques. In such criterion, the costs of two kinds of classification errors, false positive and false negative for a typical two-class problem are treated equally. However, in real applications such as fraud detection, medical diagnosis and various problems in business decision-making, different types of misclassifications are of different costs (more often than not, low false negative is preferred), thus should be treated differently. For example, in fraud detection systems, undetected frauds with high transaction amounts are obviously more costly. In medical diagnosis, the errors committed in diagnosing someone as healthy when one has a life-threatening disease is usually considered to be far more serious (higher costs) than the opposite type of error of diagnosing someone as ill when one is in fact healthy. In object detection based on computer vision, achieving extremely high detection rates is more important than achieving low false alarms. One way to obtain an asymmetric training

* This work was partially supported by Open Fund of Province Level Key Laboratory for Colleges and Universities in Jiangsu Province under Grant No.KJS0802.

Z.-H. Zhou and T. Washio (Eds.): ACML 2009, LNAI 5828, pp. 234–247, 2009.

algorithm is directly extending a symmetric one to its cost-sensitive version. Here we take boosting as a typical example. Since its generation, boosting as a family of general and effective methods for improving the accuracy of any given learning algorithm has become one class of the most popular algorithms in machine learning. The first practical boosting algorithm introduced by Freund and Schapire [1] is AdaBoost (also called Discrete AdaBoost). By sequentially applying to re-weighted versions of the training data, a weak classifier is boosted to a strong one through a weighted majority vote of many weak classifiers. Then Schapire and Singer [2] extended it to a generalized version called Real AdaBoost in which real-value prediction instead of binary-value is used. Many theories were proposed to try to explain the tremendous success achieved by boosting, including generalization error analysis [2], game theory [3], VC theory [4] and much beyond. One promising interpretation from a statistical view was given by Friedman *et al.* [5]. They proved that boosting is an approximation to additive logistic regression model and developed two more direct approximations, Gentle AdaBoost and LogitBoost along this thought.

The above-mentioned boosting algorithms are cost-insensitive that are not sufficiently effective for data-imbalanced applications. Various cost-sensitive extensions have been proposed in the literature. Although several kinds of improvements exist, most of the cost-sensitive boosting algorithms are derived from Discrete AdaBoost, for instance, AdaCost [6], CSB0, CSB1, CSB2 [7], AdaC1, AdaC2, AdaC3 [8] and asymmetric boosting [9]. Viola and Jones [10] recently presented a cost-sensitive generalization of Real AdaBoost and successfully applied it to face detection. However, to the best of our humble knowledge, no discussion on cost-sensitive extensions of Gentle AdaBoost and LogitBoost are reported though these two algorithms have been successfully applied in many important fields [11,12]. Another insufficiency is that most of the current cost-sensitive boosting algorithms directly modify the weighting strategy of their original correspondings which achieve successful empirically results but lack sound theoretical analysis.

This paper develops a framework from a statistical insight that can embody almost all existing cost-sensitive boosting algorithms: fitting an additive asymmetric logistic regression model by stage-wise optimization of certain criterions. The main contributions of the paper are on three aspects: (1) A framework for constructing cost-sensitive boosting algorithms is developed from a view point of statistics. Cost-sensitive boosting is considered as fitting an additive asymmetric logistic regression model. A cost-sensitive boosting is obtained through stage-wisely optimizing a target criterion. Four cost-sensitive boosting algorithms using diverse criterions and optimization methods are studied, namely CSDA, CSRA, CSGA and CSLB that respectively correspond to Discrete AdaBoost, Real AdaBoost, Gentle AdaBoost and LogitBoost. (2) Based upon the developed framework, the effectiveness of the cost-sensitive versions of Discrete AdaBoost and Real AdaBoost proposed in [6,7,8,10] is interpreted theoretically, which achieve cost-sensitive by directly modifying the weighting strategy. (3) To the best of our humble knowledge, it is the first time for the other two boosting algorithms, Gentle AdaBoost and LogitBoost to be extended to their asymmetric versions. Finally, experimental results on the application of face detection have shown the effectiveness of the proposed learning framework in the reduction of the cumulative misclassification cost.

2 Bayes Optimal Classification

We begin with Bayes optimal classification to construct cost-sensitive boosting algorithms. A cost matrix C is defined in formula (1) for a two-class problem,

$$C = \begin{pmatrix} c_{++} & c_{+-} \\ c_{-+} & c_{--} \end{pmatrix} \tag{1}$$

where c_{-+} is the cost of predicting an actual $+1$ (positive) label to -1 (negative). Similar meanings for c_{+-}, c_{++} and c_{--}. Assume y is the prediction output, given a sample x, the optimal prediction should minimize the expected loss according to the Bayesian decision theory. For example, if and only if the expected loss of predicting $+1$ is not greater than that of predicting -1, the prediction output takes $+1$, mathematically described as follows:

$$c_{++}P(y = +1|x) + c_{+-}P(y = -1|x) \le c_{-+}P(y = +1|x) + c_{--}P(y = -1|x). \tag{2}$$

Let $p(x)$ denote the posterior probabilities $P(y = +1|x)$ and write (2) in half log-ratio form, we obtain a discriminant function $F(x)$, where

$$F(x) = \frac{1}{2} \log \frac{(c_{-+} - c_{++})p(x)}{(c_{+-} - c_{--})(1 - p(x))} = \frac{1}{2} \log \frac{p(x)}{1 - p(x)} + \frac{1}{2} \log \frac{c_{-+} - c_{++}}{c_{+-} - c_{--}}. \tag{3}$$

The prediction for x is given by $\mathbf{sign}(F(x))$. Note that the decision threshold is influenced by cost matrix C which is problem dependent. The cost of a certain type of error may be conditional on the circumstances. For example, in detection of fraud, the cost of missing a particular case of fraud will depend on the amount of money involved in that particular case. In this paper, only constant cost is considered.

3 Boosting

Boosting works by sequentially applying a single classification algorithm to reweighted versions of the training data and taking a weighted majority voting mechanism to aggregate sequence of classifiers thus produced. Boosting has been proved to be a general and effective method for improving the accuracy of a given learning algorithm in many applications. Lots of theories are proposed to explain why this simple strategy results in dramatic improvements in performance. Schapire and Singer [2] demonstrated that boosting is a process which greedily minimizes the upper bound of training errors. The exponential form loss, $e^{-yF(x)}$ used in boosting is motivated as a differentiable upper bound to the misclassification error (no loss to a correct prediction and a unit loss to any misclassification, i.e. 0-1 loss)

$$e^{-yF(x)} \ge \begin{cases} 1, & \text{if } yF(x) < 0, \\ 0, & \text{if } yF(x) \ge 0. \end{cases} \tag{4}$$

However, the bound is too loose to use in practice. A reasonable explanation of the mechanism of boosting was given by Friedman *et al.* [5] in terms of statistical principles. They showed that for two-class problem both Discrete and Real AdaBoost algorithms fit an additive logistic regression model $F(x)$, where

$$F(x) = \frac{1}{2} \log \frac{p(x)}{1 - p(x)} = \sum_{i=1}^{M} f_i(x), \tag{5}$$

specially $F(x) = \sum_{i=1}^{M} c_i f_i(x)$ for Discrete AdaBoost. It is obvious that $F(x)$ is the discriminant function in 0-1 loss case of (3). Both AdaBoost algorithms estimate $F(x)$ by stage-wisely minimizing an exponential criterion as described by (6) whose minimum is at (5)

$$J(F(x)) = E[e^{-yF(x)}]. \tag{6}$$

Given an imperfect estimate $F_{m-1}(x)$, the stage-wise minimization algorithm seeks an improved estimation $F_m(x) = F_{m-1}(x) + c_m f_m(x)$ or $F_m(x) = F_{m-1}(x) + f_m(x)$ in each operation round. More direct approximations also can be achieved, which lead to the development of Gentle AdaBoost and LogitBoost [5]. Gentle AdaBoost minimizes the criterion $E[e^{-yF(x)}]$ by Newton updates while LogitBoost takes a likelihood-based procedure to estimate $F(x)$. According to (5), the binomial probabilities are derived as

$$p(x) = \frac{1}{1 + e^{-2F(x)}}, \tag{7}$$

therefore, the binomial log-likelihood is

$$l(y, p(x)) = -\log(1 + e^{-2yF(x)}). \tag{8}$$

Then Newton steps are used to maximize the Bernoulli likelihood

$$El[F(x)] = -E[\log(1 + e^{-2yF(x)})]. \tag{9}$$

Table 1 has briefly summarized the four boosting algorithms mentioned above. For further details, reader is suggested to refer to [5].

Table 1. Summarization of four boosting algorithms. Boosting can be viewed as approximations to an additive symmetric logistic regression model and different boosting algorithms use diverse criterions and optimization methods.

Name	$F(x)$	$f_i(x)$	Criterion	Optimization method
DA	$F(x) = \sum_{i=1}^{M} c_i f_i(x)$	$f_i(x) \in \{-1, 1\}, c_i > 0$	$E[e^{-yF(x)}]$	Newton-like steps
RA	$F(x) = \sum_{i=1}^{M} f_i(x)$	$f_i(x) \in R$	$E[e^{-yF(x)}]$	calculus of stationary points
GA	$F(x) = \sum_{i=1}^{M} f_i(x)$	$f_i(x) \in R$	$E[e^{-yF(x)}]$	Newton steps
LB	$F(x) = \sum_{i=1}^{M} f_i(x)$	$f_i(x) \in R$	$-E[\log(1 + e^{-2yF(x)})]$	Newton steps

4   Cost-Sensitive Boosting

The original boosting algorithms use 0-1 loss that is not appropriate to asymmetric classification problems. Derived from Discrete AdaBoost, various cost-sensitive boosting
extensions have been proposed. Most of such extensions directly modify the weighting
strategies of the original boosting algorithms which, despite of being demonstrated effective by experiments, are heuristic and lack of theoretic analysis. Here, we succinctly
construct a set of cost-sensitive boosting algorithms in the framework of fitting an additive asymmetric logistic regression model, which builds on the work of Friedman *et
al.* [5] and can embody almost all existing cost-sensitive boosting algorithms.

4.1   Cost-Sensitive Criterions

Since the original boosting algorithms only use 0-1 loss, as described in section 3, they
are cost-insensitive. To make the algorithm cost-sensitive, an asymmetric loss should
be introduced. Here a 0-r loss is used, where the cost of the false negative is r times
larger than that of the false positive. Then, the discriminant function $F(x)$ is turned to
an asymmetric logistic regression model as shown in (10)

$$F(x) = \frac{1}{2} \log \frac{p(x)r}{1 - p(x)} = \frac{1}{2} \log \frac{p(x)}{1 - p(x)} + \frac{1}{2} \log r. \tag{10}$$

It shows that the decision threshold is $\frac{1}{2} \log r$ smaller than that of (5), which makes more
positive samples be classified correctly in the expense of false positive increase. In this
paper, instead of directly decreasing the decision threshold, we construct cost-sensitive
versions of boosting to process the asymmetric classification problems. We view cost-
sensitive boosting as approximations to an additive asymmetric regression model by
optimizing certain cost-sensitive criterions which should be deliberately designed.

Two points are considered for the design of cost-sensitive criterion: (1) the optimization solution should be in the form of an asymmetric logistic regression, and (2) the
constructed cost-sensitive boosting algorithms should be fully compatible with their
cost-insensitive correspondings. We use

$$J(F(x)) = E[r^{y^*} e^{-yF(x)}] \tag{11}$$

as the cost-sensitive criterion, where $y^* = \frac{y+1}{2} \in \{0, 1\}$. It is obvious that the first
condition is satisfied, and we will show that the second condition is also satisfied in
section 4.2. Moreover, similar to the upper bound of the 0-1 loss, $e^{-yF(x)}$, $r^{y^*} e^{-yF(x)}$
can be motivated as the upper bound of the 0-r loss:

$$r^{y^*} e^{-yF(x)} \geq \begin{cases} r, & \text{if } yF(x) < 0, y = 1, \\ 1, & \text{if } yF(x) < 0, y = -1, \\ 0, & \text{if } yF(x) \geq 0. \end{cases} \tag{12}$$

Criterion (11) can be directly applied to obtain CSDA, CSRA and CSGA. While for
CSLB, since the criterion is a log-likelihood, certain process is needed. Consider the
binomial probabilities $p(x)$ according to (10), where

$$p(x) = \frac{1}{1 + re^{-2F(x)}}. \tag{13}$$

Then, the binomial log-likelihood is

$$l(y, p(x)) = -\log(1 + r^y e^{-2yF(x)}).$$ (14)

So the cost-sensitive criterion for CSLB is

$$El[F(x)] = -E[\log(1 + r^y e^{-2yF(x)})].$$ (15)

4.2 CSDA, CSRA, CSGA and CSLB

Four cost-sensitive boosting algorithms are deduced in detail in this subsection following the framework of stage-wise fitting the additive asymmetric logistic regression model.

For Real AdaBoost and Gentle AdaBoost, given $F(x) = \sum_{i=1}^{M} f_i(x)$, the criterion is

$$J(F(x)) = E[r^{y^*} e^{-yF(x)}] = E[r^{y^*} e^{-y \sum_{i=1}^{M} f_i(x)}] = E\left[\prod_{i=1}^{M} r^{\frac{y^*}{M}} e^{-yf_i(x)}\right].$$ (16)

Note that unlike decreasing decision threshold to perform cost-sensitive predictions in (10), the asymmetric cost r^{y^*} is assigned in every iteration to adjust the sequentially produced classifiers. Let $k = r^{\frac{1}{M}}$, $F_0(x) = 0$, $F_m(x) = \sum_{i=1}^{m} f_i(x)$, $m = 1, 2, ..., M$, the criterion of the m-th estimate $F_m(x)$ is

$$J(F_m(x)) = E[k^{my^*} e^{-yF_m(x)}]$$ (17)

which also holds for Discrete AdaBoost.

CSDA. *CSDA fits an additive asymmetric logistic regression model via Newton-like updates for minimizing* $J(F(x)) = E[r^{y^*} e^{-yF(x)}]$.

The procedure begins with $F_0(x) = 0$ and $w_0(x) = k^{y^*}$. For the m-th round of iteration, we have $F_{m-1}(x)$ and seek a better estimate $F_m(x) = F_{m-1}(x) + c_m f_m(x)$ by the following three steps.

Step 1: Fix c_m (and x), seek the optimal $f_m(x)$. Expand $J(F_m + c_m f_m)$ to second order about $f_m(x) = 0$,

$$
\begin{aligned}
J(F_{m-1} + c_m f_m) &= E[k^{my^*} e^{-yF_{m-1} + c_m f_m} | x] \\
&\approx E\left[k^{my^*} e^{-yF_{m-1}} (1 - yc_m f_m + \frac{c_m^2 y^2 f_m^2}{2}) | x\right] \\
&= E\left[k^{my^*} e^{-yF_{m-1}} (1 - yc_m f_m + \frac{c_m^2}{2}) | x\right].
\end{aligned}
$$ (18)

Then

$$
\begin{aligned}
f_m(x) &= \arg\min_f J(F_{m-1} + c_m f_m) \\
&= \arg\min_f E_{w_{m-1}}\left[1 - yc_m f_m + \frac{c_m^2}{2} | x\right] \\
&= \arg\max_f E_{w_{m-1}}[yf_m | x]
\end{aligned}
$$ (19)

where the notation $E_w[.|x]$ refers to a weighted conditional expectation

$$E_w[g(x,y)|x] = \frac{E[w(x,y)g(x,y)|x]}{E[w(x,y)|x]} \tag{20}$$

and $w_{m-1} = w_{m-1}(x,y) = k^{my^*}e^{-yF_{m-1}(x)}$. Note that $c_m \geq 0$, the solution is

$$f_m(x) = \begin{cases} 1, & \text{if } E_{w_{m-1}}[y|x] > 0, \\ -1, & \text{otherwise.} \end{cases} \tag{21}$$

Step 2: Given $f_m(x)$, determine c_m by directly minimizing $J(F_{m-1} + c_m f_m)$:

$$c_m = \arg\min_f E_{w_{m-1}}[e^{-c_m y f_m(x)}] = \frac{1}{2}\log\frac{1-e_m}{e_m} \tag{22}$$

where $e_m = E_{w_{m-1}}[1_{[y \neq f_m(x)]}]$.

Step 3: Update the weights for the $(m+1)$-th iteration:

$$w_m = k^{(m+1)y^*}e^{-yF_m} = w_{m-1}k^{y^*}e^{-yc_m f_m}. \tag{23}$$

CSRA. *CSRA builds an additive asymmetric logistic regression model by stage-wise and approximate optimization of* $J(F(x)) = E[r^{y^*}e^{-yF(x)}]$.

The procedure begins with $F_0(x) = 0$ and $w_0(x) = k^{y^*}$. For the m-th round, given $F_{m-1}(x)$ we seek an improved estimate $F_m(x) = F_{m-1}(x) + f_m(x)$.

Step 1: Minimize $J(F_{m-1}(x) + f_m(x))$ at each x.

$$\begin{aligned} J(F_{m-1} + f_m) &= E[k^{my^*}e^{-yF_{m-1}+f_m}] \\ &= e^{-f_m}E[k^{my^*}e^{-yF_{m-1}}1_{[y=1]}|x] \\ &\quad + e^{f_m}E[k^{my^*}e^{-yF_{m-1}}1_{[y=-1]}|x]. \end{aligned} \tag{24}$$

Dividing through by $E[k^{my^*}e^{-yF_{m-1}}|x]$ and setting the derivative $f_m(x)$ to zero we get

$$f_m(x) = \frac{1}{2}\log\frac{E_{w_{m-1}}[1_{[y=1]}|x]}{E_{w_{m-1}}[1_{[y=-1]}|x]} = \frac{1}{2}\log\frac{P_{w_{m-1}}(y=1|x)}{P_{w_{m-1}}(y=-1|x)}, \tag{25}$$

where $w_{m-1} = w_{m-1}(x,y) = k^{my^*}e^{-yF_{m-1}}$.

Step 2: Update the weights for the $(m+1)$-th iteration:

$$w_m = k^{(m+1)y^*}e^{-yF_m} = w_{m-1}k^{y^*}e^{-yf_m}. \tag{26}$$

CSGA. *CSGA fits an additive asymmetric logistic regression model using Newton steps for minimizing* $J(F(x)) = E[r^{y^*}e^{-yF(x)}]$.

The procedure begins with $F_0(x) = 0$ and $w_0(x) = k^{y^*}$. Like Real AdaBoost, for the m-th round, we have $F_{m-1}(x)$ and seek an improved estimate $F_m(x) = F_{m-1}(x) + f_m(x)$ on condition of $J(F_{m-1}(x) + f_m(x))$.

Step 1:

$$\frac{\partial J(F_{m-1} + f_m)}{\partial f_m}\bigg|_{f_m=0} = -E[k^{my^*}e^{-yF_{m-1}}y|x], \tag{27}$$

$$\frac{\partial^2 J(F_{m-1} + f_m)}{\partial f_m^2}\bigg|_{f_m=0} = E[k^{my^*}e^{-yF_{m-1}}|x]. \tag{28}$$

Hence the Newton update $f_m(x)$ is

$$f_m(x) = \frac{E[k^{my^*}e^{-yF_{m-1}}y|x]}{E[k^{my^*}e^{-yF_{m-1}}|x]} = E_{w_{m-1}}[y|x], \tag{29}$$

where $w_{m-1} = w_{m-1}(x, y) = k^{my^*}e^{-yF_{m-1}}$.

Step 2: Update the weights for the $(m + 1)$-th iteration:

$$w_m = k^{(m+1)y^*}e^{-yF_m} = w_{m-1}k^{y^*}e^{-yf_m}. \tag{30}$$

CSLB. *CSLB uses Newton steps for fitting an additive asymmetric logistic model by maximum likelihood* $El[F(x)] = -E[\log(1 + r^y e^{-2yF(x)})]$.
Consider the posterior probabilities $p(x)$,

$$p(x) = \frac{1}{1 + re^{-2F(x)}} = \frac{1}{1 + re^{-2\sum_{i=1}^{m} f_i(x)}} = \frac{1}{1 + \prod_{i=1}^{m} k^y e^{-2f_i(x)}}. \tag{31}$$

The procedure begins with $F_0(x) = 0$ and $p_0(x) = \frac{1}{1+k^y}$. For the m-th round,

Step 1: Compute the probabilities $p_{m-1}(x)$ through

$$p_{m-1}(x) = \frac{1}{1 + k^{(m-1)y}e^{-2yF_{m-1}(x)}}. \tag{32}$$

Step 2:

$$\frac{\partial E[l(F_{m-1} + f_m)]}{\partial f_m}\bigg|_{f_m=0} = 2E[y^* - p_{m-1}|x], \tag{33}$$

$$\frac{\partial^2 E[l(F_{m-1} + f_m)]}{\partial f_m^2}\bigg|_{f_m=0} = 2E[p_{m-1}(1 - p_{m-1})|x]. \tag{34}$$

The Newton update is then performed

$$f_m(x) = \frac{1}{2}\frac{E[y^* - p_{m-1}|x]}{E[p_{m-1}(1 - p_{m-1})|x]} = \frac{1}{2}E_{w_m}\left[\frac{y^* - p_{m-1}}{p_{m-1}(1 - p_{m-1})}\right], \tag{35}$$

where $w_m = w_m(x) = p_{m-1}(x)(1 - p_{m-1}(x))$.
Table 2 summarizes the proposed four cost-sensitive boosting algorithms. The cost-insensitive boosting procedures can be achieved when $r = 1$ ($k = 1$).

Table 2. Summarization of four cost-sensitive boosting algorithms. Cost-sensitive boosting can be viewed as approximations to an additive asymmetric logistic regression model by optimizing certain cost-sensitive criterions. The cost-sensitive versions can be easily implemented with little modification (shown in bold) to the original boostings. For CSDA, CSRA and CSGA, the weight-update step is modified to make the next iteration put more focus on those misclassified positive samples. For CSLB, the probability update step is modified to decrease the posterior probabilities of positives so that the target of the learning algorithm will give a higher priority to classify positive samples correctly.

Name	Criterion	Modification
CSDA	$E[\mathbf{r}^{y^*}e^{-yF(x)}]$	$w_0(x) = \mathbf{k}^{y^*}$, $w_m = w_{m-1}\mathbf{k}^{y^*}e^{-yc_mf_m}$
CSRA	$E[\mathbf{r}^{y^*}e^{-yF(x)}]$	$w_0(x) = \mathbf{k}^{y^*}$, $w_m = w_{m-1}\mathbf{k}^{y^*}e^{-yf_m}$
CSGA	$E[\mathbf{r}^{y^*}e^{-yF(x)}]$	$w_0(x) = \mathbf{k}^{y^*}$, $w_m = w_{m-1}\mathbf{k}^{y^*}e^{-yf_m}$
CSLB	$-E[\log(1 + \mathbf{r}^{y}e^{-2yF(x)})]$	$p_0(x) = 1/(1+\mathbf{k}^{y})$, $p_m = 1/(1+\mathbf{k}^{\mathbf{my}}e^{-2yF_m})$

4.3   Comparison with Other Cost-Sensitive Boosting Algorithms

The proposed cost-sensitive boosting algorithms are constructed from a statistical insight and most of the previous cost-sensitive boosting algorithms can be embodied in this framework. It is shown that the only difference in implementation between CSDA and CSRA and their original correspondings is whether the asymmetric cost factor k^{y^*} is used in the weight-update step. Here take CSDA as an example, for the $(m+1)$-th round, the weighs are

$$w_m = w_{m-1} \cdot \begin{cases} ke^{-c_m|f_m|}, & \text{if } y = +1, f_m \geq 0, \\ ke^{c_m|f_m|}, & \text{if } y = +1, f_m < 0, \\ e^{c_m|f_m|}, & \text{if } y = -1, f_m > 0, \\ e^{-c_m|f_m|}, & \text{if } y = -1, f_m \leq 0, \end{cases} \tag{36}$$

where false negative receives more weight increment than false positive does, meanwhile, true positive loses less weights than true negative when $k > 1$. This modification strategy is consistent with what most heuristic cost-sensitive boosting algorithms have. Among those algorithms, AdaC2 (the most effective one in AdaCX algorithms proposed in [8]), CSB2 [7], Asymmetric AdaBoost [10] and certain calculations in Ada-Cost [6] use just the same modification as what we do in this paper. Our framework has given these algorithms a solid demonstration in principle. Masnadi-Shirazi and Vasconcelos [9] have proposed a cost-sensitive extension of Discrete AdaBoost namely asymmetric boosting, which is based on a different criterion with ours to fit an asymmetric logistic model. They defined a cost of C_1 for false positives and C_2 for misses, the asymmetric logistic transform has the following form:

$$F(x) = \frac{1}{C_1 + C_2} \log \frac{p(x)C_1}{(1-p(x))C_2}. \tag{37}$$

The corresponding cost-sensitive criterion is defined as (38):

$$E\left[1_{[y=1]}e^{-yC_1F(x)} + 1_{[y=-1]}e^{-yC_2F(x)}\right]. \tag{38}$$

Following Schapire and Singer [2], the cost-sensitive algorithm is deduced with tedious steps and has an extremely complex form.

We also extended Gentle AdaBoost and LogitBoost to their cost-sensitive versions for the first time to the best of our humble knowledge. For CSGA, the similar modification as CSDA and CSRA is used in the weight-update step to increase the weight of those misclassified positive samples more hardly. While for CSLB, the intrinsic mechanism is very different. The modification is in the probability update step, the posterior probabilities of all positives have been decreased at each round, therefore the target of learning algorithm will put more focus on classifying positive samples correctly.

5   Experiments

Considering the similar implementation, many previous papers have demonstrated the effectiveness of the proposed CSDA and CSRA algorithms in the asymmetric classification problems such as credit card fraud detection, medical diagnosis, face detection and so on [6,7,8,10]. In this paper we performed the experiments in the domain of face detection which is intrinsically a cost-sensitive classification problem. Achieving extremely high detection rates rather than low errors is a target addressed by good face detectors. Experiments follow the general form presented in Viola and Jones [13]. A training and a test set consist of fix-sized faces and non-faces were acquired (24×24 pixels for each picture). Both the training and the test sets contain 1000 face samples

Table 3. Cumulative misclassification cost for CSDA, CSRA, CSGA and CSLB with different c, k. It can be clearly seen that cumulative misclassification cost is significantly decreased when $k > 1$. We also observe that cost-sensitive boosting algorithms achieve low false negative rate with increased false positive rate that can be reduced using cascade form.

Name	k	FN_{rate}	FP_{rate}	$c = 1$	$c = 5$	$c = 10$
	1.00	7.60%	**4.40%**	**0.120**	0.424	0.804
CSDA	1.02	5.20%	7.20%	0.124	0.332	0.592
	1.04	**4.60%**	8.30%	0.129	**0.313**	**0.543**
	1.00	5.50%	**3.35%**	0.089	0.309	0.584
CSRA	1.02	4.40%	4.35%	**0.088**	0.264	0.484
	1.04	**3.70%**	5.85%	0.096	**0.244**	**0.429**
	1.00	5.60%	**2.65%**	0.083	0.307	0.587
CSGA	1.02	4.10%	4.05%	**0.082**	0.246	0.451
	1.04	**3.40%**	5.65%	0.091	**0.227**	**0.397**
	1.00	7.10%	**3.35%**	0.105	0.389	0.744
CSLB	1.02	5.10%	5.05%	**0.102**	0.306	0.561
	1.04	**3.50%**	7.45%	0.110	**0.250**	**0.425**

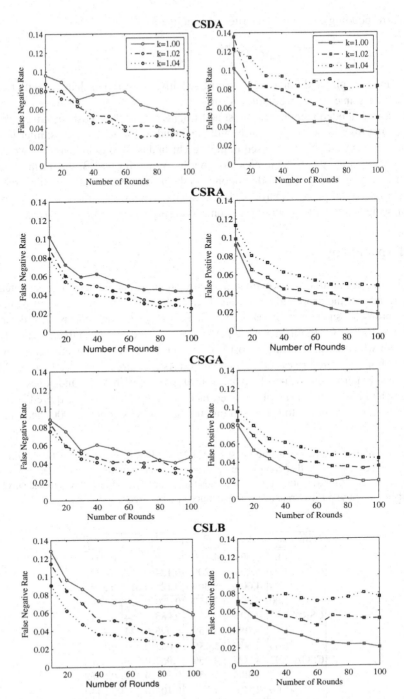

Fig. 1. FN_{rate}, FP_{rate} for CSDA, CSRA, CSGA and CSLB, as a function of the iteration number. Notice that lower false negative rate is achieved when $k > 1$, with tolerated higher false positive rate.

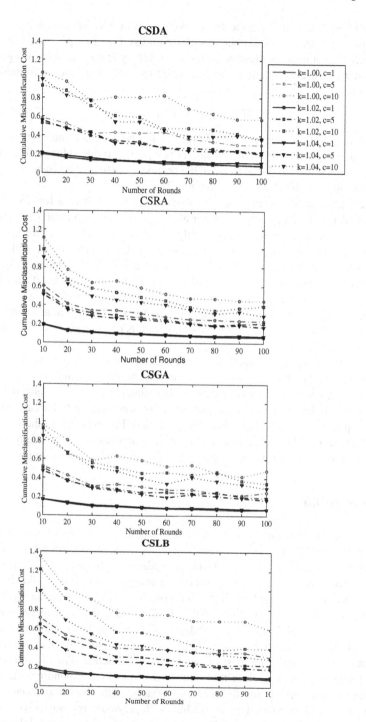

Fig. 2. Cumulative misclassification cost for CSDA, CSRA, CSGA and CSLB, as a function of the iteration number. It is clear that cumulative misclassification cost is reduced more significantly as the cost factor c increases.

and 2000 non-face samples respectively. Four types of harr-like features are used and the feature set is quite large (about 134,736 features). Boosting provides feature selection where the set of weak classifier is the set of binary features, *i.e.* boosting stumps.

Cumulative misclassification cost is defined for evaluating the performance of all algorithms,

$$CMC = c \times FN_{rate} + FP_{rate} \tag{39}$$

where FN_{rate} is false negative rate, FP_{rate} is false positive rate and c is the cost factor. Notice that c is different from the parameter $r(k)$, c is introduced for performance evaluation while $r(k)$ is introduced to make the original boosting cost-sensitive. It becomes cumulative misclassification error when $c = 1$.

Table 3 presents cumulative misclassification cost on test data for CSDA, CSRA, CSGA, CSLB with different c and k for all 50 rounds. We can clearly see that cumulative misclassification cost is significantly decreased when $k > 1$. We also observe that cost-sensitive boosting algorithms achieve low false negative rate with increased false positive rate that can be reduced using cascade form as reported in [13]. Figure 1 and 2 show false negative rate, false positive rate, cumulative misclassification cost for CSDA, CSRA, CSGA and CSLB with different k, as a function of the iteration number. The conclusions obtained from Table 3 hold true for almost the whole iterative process and it is clear that cumulative misclassification cost is reduced more significantly as the cost factor c increases.

Note that in the practical application, a face detector is usually trained to a cascade form to improve the detection speed and decrease the false alarm rate. The only added parameter k introduced by the cost-sensitive boosting can be preset related to the predefined minimum detection rate and maximum false acceptance rate for each stage [14]. In other applications k can have different definition. For example, in the classification problems with imbalanced data such as medical diagnosis, k can be determined based on the proportion of positive and negative training examples in the training set.

6 Conclusions

This paper has investigated cost-sensitive boosting algorithms for asymmetric classification problems. We developed a framework for cost-sensitive extensions of boosting algorithms from a statistical view: fitting an additive asymmetric logistical regression model. A diversity of cost-sensitive criteria are designed for estimating the discriminant function, all in asymmetric logistical form. Four cost-sensitive boosting algorithms are deduced in detail, namely CSDA, CSRA, CSGA and CSLB that respectively correspond to Discrete AdaBoost, Real AdaBoost, Gentle AdaBoost and LogitBoost, specially, it's the first time for Gentle AdaBoost and LogitBoost to be extended to their cost-sensitive versions to the best of our humble knowledge. The difference between the cost-sensitive boosting algorithms and their original correspondings only exists in the update steps, making our algorithms compatible with the original ones in implementation. We also compared our algorithms with previous proposals, showing that our framework can embody most of them and give these algorithms a solid demonstration in principle. Empirical evaluations have shown significant reductions in the cumulative misclassification cost of the cost-sensitive boostings over those of the cost-insensitive ones have achieved at no expense of computational payload increase.

References

1. Freund, Y., Schapire, R.E.: A Decision-Theoretic Generalization of on-Line Learning and an Application on Boosting. Journal of Computer and System Sciences 55(1), 119–139 (1997)
2. Schapire, R.E., Singer, Y.: Improved Boosting Algorithms Using Confidence-Rated Predictions. Journal of Machine Learning 37(3), 297–336 (1999)
3. Freund, Y., Schapire, R.E.: Game Theory, on-Line Prediction and Boosting. In: Proceedings of the Ninth Annual Conference on Computational Learning Theory, pp. 325–332 (1996)
4. Schapire, R.E., Freund, Y., Bartlett, P., Lee, W.: Boosting the Margin: a New Explanation for the Effectiveness of Voting Methods. The Annals of Statistics 26(5), 1651–1686 (1998)
5. Friedman, J., Hastie, T., Tibshirani, R.: Additive Logistic Regression: a Statistical View of Boosting. The Annals of Statistics 28(2), 337–407 (2000)
6. Fan, W., Stolfo, S.J., Zhang, J., Chan, P.K.: Adacost: Misclassification Cost-Sensitive Boosting. In: ICML, pp. 97–105 (1999)
7. Ting, K.M.: A Comparative Study of Cost-Sensitive Boosting Algorithms. In: ICML, pp. 983–990 (2000)
8. Sun, Y., Kamel, M.S., Wong, A.K.C., Wang, Y.: Cost-Sensitive Boosting for Classification of Imbalanced Data. Pattern Recognition 40(12), 3358–3378 (2007)
9. Masnadi-Shirazi, H., Vasconcelos, N.: Asymmetric Boosting. In: ICML, pp. 609–619 (2007)
10. Viola, P., Jones, M.: Fast and Robust Classification Using Asymmetric AdaBoost and a Detector Cascade. In: Neural Information Processing Systems, pp. 1311–1318 (2001)
11. Lienhart, R., Kuranov, A., Pisarevsky, V.: Empirical Analysis of Detection Cascades of Boosted Classifiers for Rapid Object Detection. In: Michaelis, B., Krell, G. (eds.) DAGM 2003. LNCS, vol. 2781, pp. 297–304. Springer, Heidelberg (2003)
12. Tuzel, O., Porikli, F., Meer, P.: Human Detection via Classification on Riemannian Manifolds. In: CVPR, pp. 1–8 (2007)
13. Viola, P., Jones, M.: Rapid Object Detection Using a Boosted Cascade of Simple Features. In: CVPR (2001)
14. Pham, M.T., Hoang, V.D., Cham, T.J.: Detection with Multi-Exit Asymmetric Boosting. In: CVPR (2008)

On Compressibility and Acceleration of Orthogonal NMF for POMDP Compression

Xin Li, William K. Cheung, and Jiming Liu

Department of Computer Science, Hong Kong Baptist University
Hong Kong

Abstract. State space compression is one of the recently proposed approaches for improving POMDP's tractability. Despite its initial success, it still carries two intrinsic limitations. First, not all POMDP problems can be compressed equally well. Also, the cost of computing the compressed space itself may become significant as the size of the problem is scaled up. In this paper, we address the two issues with respect to an orthogonal non-negative matrix factorization recently proposed for POMDP compression. In particular, we first propose an eigenvalue analysis to evaluate the compressibility of a POMDP and determine an effective range for the dimension reduction. Also, we incorporate the interior-point gradient acceleration into the orthogonal NMF and derive an accelerated version to minimize the compression overhead. The validity of the eigenvalue analysis has been evaluated empirically. Also, the proposed accelerated orthogonal NMF has been demonstrated to be effective in speeding up the policy computation for a set of robot navigation related problems.

1 Introduction

Partially observable Markov decision process (POMDP) is a mathematical framework commonly adopted for agent planning in a stochastic environment with the state information only partially observed. It has been widely applied to a number of areas including robotics [1], assistive systems [2], medical decision support [3], etc. Solving a POMDP problem is to find the optimal policy for an agent to maximize its long-term reward, e.g., to navigate efficiently to reach a given goal. Due to the partially observable assumption, the underlying policy is typically defined as a mapping between the probability distribution over the finite set of the true states (so called *belief state*) and the action to be performed. The belief states form a continuous space (henceforth called the *belief space*).

Computing POMDP's optimal policy is challenging. The exact solution for POMDP problem based on a value iteration technique was first proposed in [4]. In [5], it has been shown that the complexity for obtaining the optimal policy with t steps ahead is $O(\zeta_{t-1}^{|Z|})$ where Z is the set of the possible observations and ζ_i is the space complexity of the value function at the ith iteration. Due to the intractability of POMDP problems, a number of *approximation* methods have been proposed in literature. The value-directed compression (VDC) [6] engenders

Z.-H. Zhou and T. Washio (Eds.): ACML 2009, LNAI 5828, pp. 248–262, 2009.

a size-reduced POMDP to compute the policy. Belief compression [7] reduces
the POMDP problem to a grid-like MDP problem by exploring belief space's
sparsity via exponential principal component analysis (EPCA). An orthogonal
nonnegative matrix factorization (O-NMF) was proposed in [8], where the belief
compression and VDC approaches were integrated to achieve a higher belief
space compression ratio. Among the POMDP compression methods, there exist
at least two fundamental research issues to be addressed. One is whether a
POMDP problem could be effectively compressed or not. Another is that the
overhead incurred for the compression could itself be a time-consumption task.

In this paper, we base on the O-NMF based compression [8] and propose an
eigenvalue analysis for estimating to what extent a particular POMDP problem
can be compressed to be detailed in Section 4. We argue that the eigenvalue
distribution can reflect the sparsity of the belief space. Then, we propose in
Section 5 an accelerated O-NMF using interior-point gradient acceleration for
reducing the underlying compression overhead. We have carried out the eigenvalue analysis to a number of POMDP problems of different types and found
that those related to robot navigation could be more effectively compressed. This
finding is later on validated via simulation. In addition, we have evaluated the
newly derived accelerated O-NMF and showed that it can significantly reduce
the compression overhead and thus the whole policy computation process. All
the experimental results can be found in Section 6. The conclusion and future
work are presented in Section 7.

2   POMDP Basics

A POMDP model is mathematically characterized by a tuple $< \mathcal{S}, \mathcal{A}, \mathcal{Z}, T,$
$O, R >$, which contains a finite set of true states $\mathcal{S}$, a finite set of agent's ac-
tions $\mathcal{A}$, state transition probabilities $T : \mathcal{S} \times \mathcal{A} \rightarrow \Pi(\mathcal{S})$ where $\Pi(\mathcal{S})$ is the
set of distributions over $\mathcal{S}$, a reward function which depends on the action and
the state $R : \mathcal{S} \times \mathcal{A} \rightarrow \mathcal{R}$, a finite set of observations $\mathcal{Z}$ and the corresponding
set of observation probabilities $O : \mathcal{S} \times \mathcal{A} \rightarrow \Pi(\mathcal{Z})$ where $\Pi(\mathcal{Z})$ is the set of
distributions over $\mathcal{Z}$. Solving POMDP problems typically requires the use of
the concept of belief state which is defined as a probability mass function over
the true states, given as $b = (b(s_1), b(s_2), ...b(s_{|\mathcal{S}|}))$, where $s_i \in \mathcal{S}, b(s_i) \geq 0$,
and $\sum_{s_i \in \mathcal{S}} b(s_i) = 1$. Given a belief state b_t, the new belief state in the next
simulation step is defined based on the Bayesian Updating rule as:

$$
\begin{aligned}
b_{t+1}(s_j) &= P(s_j | z, a, b_t) \\
&= \frac{O(s_j, a, z) \sum_{s_i \in \mathcal{S}} T(s_i, a, s_j) b_t(s_i)}{\sum_{s_j \in \mathcal{S}} O(s_j, a, z) \sum_{s_i \in \mathcal{S}} T(s_i, a, s_j) b_t(s_i)}
\end{aligned}
\tag{1}
$$

where $a \in \mathcal{A}$ and $z \in \mathcal{Z}$. The reward function of the belief state b_j can be computed as $\rho(b_j, a) = \sum_{s_i \in \mathcal{S}} b_j(s_i) R(s_i, a)$. The transition function over the belief
states becomes $\tau(b_i, a, b_j) = p(b_j | b_i, a) = \sum_{z \in \mathcal{Z}} P(z | a, b_i)$, where z subjects to
holding Eq.(1). To compute the optimal policy $\pi : \mathcal{R}^{|\mathcal{S}|} \rightarrow \mathcal{A}$ iteratively, a value
function is typically involved in each iteration, given as:

$$V(b_i) = max_a[\rho(b_i, a) + \gamma \sum_{b_j} \tau(b_i, a, b_j)V(b_j)] \qquad (2)$$

$$\pi^*(b_i) = argmax_a[\rho(b_i, a) + \gamma \sum_{b_j} \tau(b_i, a, b_j)V(b_j)] \qquad (3)$$

where γ is the discounting factor. The value functions are essentially a set of $|\mathcal{S}|$-dimensional hyper-planes (also called α vectors), each associated with an action. Eq.(3) can be then rewritten as $V(b) = \max_{\alpha_i} \alpha_i^T b$. The optimal policy is defined as the "envelop" of these hyper-planes which is piecewise linear and convex (PWLC). A number of existing algorithms make use of some computational shortcuts based on the PWLC property for solving POMDP problems [5].

3   A Review of POMDP Compression

In this section, we present three recently proposed methodologies for POMDP compression, namely *belief compression, value-directed compression* and *orthogonal NMF-based compression*. The first one is based on EPCA and the second one is based on Krylov subspace decomposition, while the last one integrates the strengths of the former two approaches via an orthogonal NMF.

3.1   Belief Compression

Belief compression [7] is a recently proposed approach which projects the high-dimensional belief space to a low-dimensional one using EPCA. The principle behind is to eliminate the redundancy in computing the optimal policy for the belief space. Compared with conventional principal components analysis (PCA), EPCA is more effective in the dimension reduction due to its adoption of non-linear projection. Also, it can ensure all the elements of the compressed belief states and their reconstructions to be positive as each belief state should itself be a probability distribution.

However, using EPCA results in loss of the PWLC property of the POMDP value function. Thus, conventional policy computation techniques cannot be used for the compressed problem. While an auxiliary MDP can be created using the sampled belief states as its true states so that the policy of the original POMDP can be approximated, the quality of the resulting policy becomes depending on not only the effectiveness of the belief compression, but also the accuracy the policy computation approximation.

3.2   Value-Directed Compression

Value-directed compression (VDC) [6] uses a linear projection based on the minimal *Krylov* subspace to reduce the dimensionality of the belief space. The reward and state transition functions of the compressed POMDP are derived in such a way that the value functions for the belief states defined before and after the compression take the same value throughout the iterations. Thus the quality of the computed policies before and after the compression should remain identical in principle.

Compared with the EPCA-based belief compression approach, VDC does not explore the distribution characteristics of the belief space as what the belief compression approach does but computes a sub-space which is invariant to the compression projection matrix. The POMDP compressed via VDC is still a POMDP problem with a reduced dimension and all the existing algorithms for policy computation are thus applicable. However, computing the *Krylov* sub-space is time-consuming as a large number of linear programming problems are to be solved and yet a high compression ratio cannot be guaranteed.

3.3   Orthogonal NMF-Based Compression

Orthogonal NMF-based compression (O-NMF) [8] explores the belief space's sparsity and keeps the problem value-directed. In addition, the use of O-NMF makes the problem solving better posed, where the O-NMF assumes a special orthogonality constraint given as

$$V = WH, \quad s.t. \quad WW^T = I. \tag{4}$$

To compute W and H, the following updating rules are proposed in [8]:

$$W_{ik} \leftarrow W_{ik} \sqrt{\frac{(VH^T)_{ik}}{(WHH^T + VH^TW^TW - WHH^TW^TW)_{ik}}} \tag{5}$$

$$H_{kj} \leftarrow H_{kj} \frac{(W^TV)_{kj}}{(W^TWH)_{kj}}. \tag{6}$$

which minimize $\|V - WH\|_F^2$ as well as $\|WW^T - I\|_F^2$ using the Lagrange Multiplier method. The convergence of the updating rules have been proved in [8]. With the aid of O-NMF, the closed-form solutions for the low-dimensional generalized transition function $\widetilde{G}^{<a,z>}$ and low-dimensional reward function $\widetilde{R}$ are given as

$$\widetilde{G}^{<a,z>} = FG^{<a,z>}F^T \tag{7}$$

$$\widetilde{R} = FR. \tag{8}$$

The high-dimensional α vectors can be computed as $\alpha = \widetilde{\alpha}F$, where $G^{<a,z>}$ denote the $|S| \times |S|$ matrix with each of its elements representing the state transition probability given a specified pair of action a and observation z, that is $G(s_j, s_i)^{<a,z>} = P(s_j|s_i, a, z) = O(s_j, a, z)T(s_i, a, s_j)$.

4   An Eigenvalue Analysis on POMDP's Compressibility

All the compression methods introduced in Section 3 have shown their effectiveness and efficiency on some POMDP problems but not all. Without exhaustively testing different reduced dimensions, one could hardly tell the best dimension for a POMDP problem to be compressed. In this section, we study POMDP's compressibility and estimate to what extent a particular POMDP problem can be compressed. We propose an eigenvalue analysis to be performed on the generalized transition function G to estimate the effective reduced dimension for both G and F so that the overall compression can be as "lossless" as possible. The following two Lemmas show how the compressibility of POMDP associates with the intrinsic property of G. The proofs can be found in the Appendix.

Lemma 1. Given $G^{<a,z>}$ of dimensions $n \times n$, there exists an $n \times m$ matrix F, where $F^T F = I$ and m is no less than the rank of $G^{<a,z>}$ such that a lossless value-directed compression can be resulted, i.e. $G^{<a,z>} F = F \widetilde{G}^{<a,z>}$.

Lemma 2. Given $G^{<a,z>}$ of dimensions $n \times n$, there exists an $n \times m$ matrix F which induces a lossless value-directed compression $G^{<a,z>} F = F \widetilde{G}^{<a,z>}$ for all of the action and observation pairs $< a, z >$, where $F^T F = I$ and m is no less than the $\text{rank}(\sum_{a,z} G^{<a,z>})$.

According to Lemma 2, we define $rank(\sum_{a,z} G^{<a,z>})$ to be the estimated reduced dimension for the lossless compression setting. Unfortunately for some problems, the value of $rank(\sum_{a,z} G^{<a,z>})$ is quite close to the problem's original dimension. For example, $rank(\sum_{a,z} G^{<a,z>})$ are 88, 604, 1069 for Hallway2 ($dimension = 92$), Hall68by9[1] ($dimension = 612$), Hall68way16 ($dimension = 1088$) respectively. This means that lossless compression is hard to achieve. Inspired by PCA, we sorted the eigenvalues of $\sum_{a,z} G^{<a,z>}$ for Hallway2, Tag [9], Hall68by9, Hall68by16, RockSample4by4 and RockSample5by5 and plotted them in Figure 1.[2] We observed that Hallway2, Hall68by9, and Hall68by16 have sharp drops at the dimensions 30, 200, and 250 respectively. This hints that those sharp drops could be considered as the compressibility limits of the problems. In Section 6, we will show that the corresponding dimensions in fact are quite close to the effective compression ratio found empirically. In addition, a careful comparison among the sub-figures in Figure 1 indicates that the Hall68byn^2 problems of higher resolutions can be more effectively compressed. Also, as shown in Figure 1, the sum of the eigenvalues of those truncated dimensions is about 10 percent of the overall sum of all the eigenvalues for Hallway2, Hall68by9, and Hall68by16 problems. However, for the Tag, RockSample4by4 and RockSample5by5 problems, we need to set the reduced dimension to be no less than 620, 200 and 640 respectively to guarantee a less than 10 percent "loss". Thus, this implies that the Tag, RockSample4by4 and RockSample5by5 problems may not be able to benefit too much from the compression approach.

---

[1] $Hall68by n^2$ problem is described in detail in Section 6. For other benchmark problems, one can refer to http://www.pomdp.org/pomdp/examples/index.shtml for more detailed description.

[2] For a better observation of these sorted eigenvalue, we scaled the eigenvalue matrix's non-diagonal elements via $E(i,j) = 1/2(E(i-1,j) + E(i,j+1))$.

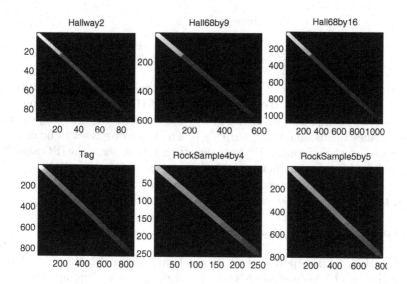

Fig. 1. An illustration of the sorted eigenvalues for Hallway2, Hall68by9, Hall68by16, Tag, Rocksample4by4, Rocksample5by5 problems

5   On Acceleration of Orthogonal NMF

As mentioned in Section 1, reducing the compression overhead is another fundamental issue to be addressed for POMDP compression. In this section, we studied the use of the interior-point gradient method to speed up the convergence of O-NMF [8]. With the overhead of O-NMF further minimized, we show that large-scale robot navigation problems ($\sim$ 1000 states) could be solved in a considerable reduced time.

5.1   Interior-Point Gradient (IPG) Algorithm

The interior-point gradient algorithm was proposed in [10] to solve the large-scale totally nonnegative least squares problem given as

$$\min_{x_i \geq 0} f(x) \equiv ||Ax - b||^2 \tag{9}$$

where $b_i > 0$, A is a $m \times n$ matrix with $m \geq n$ and $A_{i,j} \geq 0$. x can be iteratively solved via the proposed updating rule

$$x^{k+1} = x^k + \alpha_k p^k \tag{10}$$

where p^k is a scaled negative gradient direction and α_k is the step-size parameter for each iteration. More specifically, they are defined as

$$q^k = A^T b - A^T A x^k$$
$$p^k = x^k./(A^T A x^K) \circ q$$

$$\alpha_k = min(\tau_k \hat{\alpha}_k, \alpha_k^*)$$
$$\hat{\alpha}_k = \max \alpha \; : \; x^k + \alpha p^k \geq 0$$
$$\alpha_k^* = (p^k)^T q./(p^k)^T A^T A p^k$$

where the symbols "$\circ$" and "$./$" denote the component-wise multiplication and division respectively, $\hat{\alpha}_k$ is the exact minimizer of the objective function in the searching direction p, and α_k^* is the largest step-size in the searching direction which satisfies the inequality $x^k + \alpha p^k \geq 0$. The value of τ is set to be close to 1 such that α_k can be guaranteed to be no less than 1. In general, the IPG algorithm is a scaled gradient descent method with the guarantee of keeping all the iterations positive by appropriately choosing the step-size. Its convergence and efficiency in solving the totally nonnegative problems have been shown in [10].

5.2   IPG Acceleration for O-NMF

In this section, we incorporate the IPG acceleration into O-NMF [8] to derive a new version of O-NMF called O-NMF-S and its corresponding updating rules. Since another orthogonal NMF with the constraint $W^T W = I$ [11] was used in O-NMF as the initialization of W and H, O-NMF-S includes two parts: the acceleration of the initialization step and the acceleration of Eqs.(5) and (6). The following Lemmas verify the effectiveness of the speedup of the corresponding derived W-updating rules.

Lemma 3 (Initialization Acceleration). By setting

$$\alpha_k = \min(\tau_k \hat{\alpha}_k, \alpha_k^*)$$
$$\hat{\alpha}_k = \max(\alpha : W_{i.}^T + \alpha p \geq 0)$$
$$\alpha_k^* = \frac{p^T q}{p^T (HH^T + W^T WHV^T (W^T)^{-1} - W^T WHH^T)p} \qquad (11)$$
$$p = W^T \circ \sqrt{HV^T./(HH^T W^T + W^T WHV^T - W^T WHH^T W^T)} - W^T \qquad (12)$$
$$q = HV^T - HH^T W^T - W^T WHV^T + W^T WHH^T W^T. \qquad (13)$$

it can be guaranteed that $\alpha_k^* \geq 1$ and $\alpha_k \geq 1$.

Lemma 4 (W – updating Rule Acceleration). By setting

$$\alpha_k = \min(\tau_k \hat{\alpha}_k, \alpha^*)$$
$$\hat{\alpha}_k = \max(\alpha : W_{i.}^T + \alpha p \geq 0)$$
$$\alpha_k^* = \frac{p^T q}{p^T (HV^T W)p}$$
$$p = W^T \circ \sqrt{(HV^T)./(HV^T WW^T)} - W^T$$
$$q = HV^T - HV^T WW^T$$

it can be guaranteed that $\hat{\alpha}_k > 1$, $\alpha_k^* \geq 1$ and $\alpha_k \geq 1$.

The proofs for Lemma 3 and 4 can be found in the Appendix. For the acceleration of H's iteration, we simply adopt the acceleration rule over the conventional NMF [10] since the constraint $W^T W = I$ is irrelevant to H-updating. In regard to the convergence of the two versions of W-updating rules, it can be easily proved by substituting $A^T A$ appeared in [10] with $HH^T + W^T W H V^T (W^T)^{-1} - W^T W H H^T$ and $H V^T W W^T$ respectively. Algorithm 1 illustrates the detailed updating steps for O-NMF-S.

Algorithm 1. An Accelerated O-NMF (O-NMF-S)

1: **Input:** data V, reduced dimension l
2: **Output:** W, H
3: /* initialize W and H */
4: Set W and H randomly.
5: **for** $i = 1$ **to** $iterNum$ **do**
6:   $q = HV^T - HV^T WW^T$, $p = W^T \circ \sqrt{(HV^T)./(HV^T WW^T)} - W^T$
7:   $\alpha = \min(\frac{p^T q}{p^T (HV^T W)p}, \tau \max(\hat{\alpha} : W_{i.}^T + \hat{\alpha}p \geq 0))$
8:   $W^T = W^T + \alpha \circ p$
9:   $q = W^T V - W^T WH$, $p = (H./W^T WH) \circ q$
10: $\alpha = \min(\frac{p^T q}{p^T W^T Wp}, \tau \max\{\hat{\alpha} : H_{.j} + \hat{\alpha}p \geq 0\})$
11: $H = H + \alpha \circ p$
12: **end for**
13: /* update W, H using the speedup updating rule for O-NMF */
14: **while** $\delta > \epsilon$ **do**
15:   $\delta = \|V - WH\|_F^2$, $\lambda = VH^T W^T - WHH^T W^T$
16:   set -ve λ_{ik} to 0
17:   $q = HV^T - HH^T W^T - W^T \lambda^T$, $p = W^T \circ \sqrt{(HV^T)./(HH^T W^T + W^T \lambda^T)} - W^T$
18:   $\alpha = \min(\frac{p^T q}{p^T (HH^T + W^T W H V^T (W^T)^{-1} - W^T W H H^T)p}, \tau \max(\hat{\alpha} : W_{i.}^T + \hat{\alpha}p \geq 0))$
19:   $W^T = W^T + \alpha \circ p$
20:   $q = W^T V - W^T WH$, $p = (H./W^T WH) \circ q$
21:   $\alpha = \min(\frac{p^T q}{p^T W^T Wp}, \tau \max\{\hat{\alpha} : H_{.j} + \hat{\alpha}p \geq 0\})$
22:   $H = H + \alpha \circ p$
23: **end while**
24: **return**

---

6 Performance Evaluation

To evaluate the performance of O-NMF-S regarding its scalability, we created a scalable synthetic navigation problem. With the problem created, we can easily generate robot navigation problems at different resolutions for the particularly designed maze. The number of states can range from 68 to ~ 1000 states or even more.

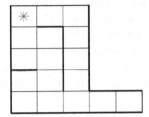

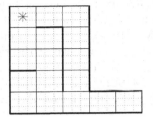

Fig. 2. Navigation environment of Hall68by1 problem

Fig. 3. Navigation environment of Hall68by9 problem

6.1 $Hall68byn^2$ - A Robot Navigation Problem with Different Grid Resolutions

Navigating robots in a stochastic environment is typically modelled using POMDP. The problem includes partitioning a geographical region into smaller grids as states and specifying the transition probabilities among those grids. In general, a model with a higher grid resolution can lead to a smoother navigation behavior of the robot. However, the complexity of the POMDP problem increases as the grid resolution increases. We designed a robot navigation problem called Hall68byn^2 whose grid resolution can be increased by changing the value of n. It defines problems with different numbers of true states depending on the grid resolution. The objective is to find a specific goal as shown in Figure 2. It is equivalent to a POMDP model with $68n^2$ states (4 possible orientations in each of the 17 rooms, n is an positive integer indicating the resolution per room), 5 actions and 17 observations. Figure 3 shows the maze of Hall68by9 where $n = 3$. One can see that each state in the Hall68by1 (Figure 2) is splitted into 9 sub-states in Hall68by9, and thus the complexity of the problem is "scaled up". Note that, for the scaled-up problem, we assume that the observation associated with each sub-state is the same as that of the corresponding state before splitting. Reaching one of the goal states will yield a +1 reward and then the next state will be set randomly to a non-goal state to restart the navigation task. Thus, the initial belief is $b_0 = (\frac{1}{64n^2}, ..., \frac{1}{64n^2}, 0.0, 0.0, ..., 0.0, 0.0, \frac{1}{64n^2}, ..., \frac{1}{64n^2})^T$. The zeros are corresponding to $4n^2$ goal states. The discount factor was set to be 0.95 in our experiments. We used this set of problems to simulate POMDP problems of different sizes. Also, we believe that the proposed compression approach is especially scalable for this type of problems as indicated by the eigenvalue analysis result shown in Figure 1.

6.2 Effectiveness of O-NMF-S for Dimension Reduction

Figures 4 and 5 show the change of the orthogonality and the least squared residue over time measured for different speedup versions of the orthogonal NMF proposed in [11]. In particular, we implemented the original version [11], versions with W accelerated, H accelerated, and WH accelerated. We observed that the

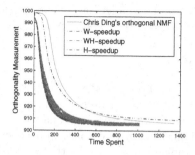

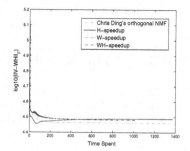

Fig. 4. Orthogonality comparison

Fig. 5. The least squared residues comparison

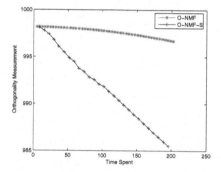

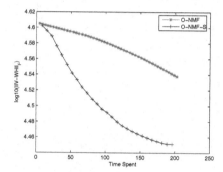

Fig. 6. Comparison on the orthogonality between O-NMF-S and O-NMF

Fig. 7. Comparison on the least squared residue between O-NMF-S and O-NMF

orthogonality curves dropped dramatically for the versions with W-updating acceleration involved (labeled as W-speedup and WH-speedup in Figiure 4) when compared with the original version [11]. This indicates the effectiveness of the accelerated version used for the initialization step. Figure 5 shows that WH-speedup version has a converged squared residue a bit bigger than that of the original version but still comparable. This hints that IPG acceleration tends to speed up convergence for better satisfaction of the orthogonal construction at the expense of the reconstruction accuracy. To summarize, given a fixed number of iterations or allowed time, the accelerated initialization step for O-NMF could achieve a better orthogonality as well as a comparable least square residue.

Figures 6 and 7 [3] show a comparison between O-NMF-S and O-NMF[8] again in terms of orthogonality derivation and reconstruction least squared residue. We observe that, given a fixed period of time, O-NMF-S achieves significantly

---

[3] The number of iteration steps of O-NMF-S is smaller than that of O-NMF. It is considerable because each iteration step of O-NMF-S related to more computations of matrix multiplications than that of O-NMF, thus will spend more time.

improvement in achieving good orthogonality and a smaller least squared residue. This further confirms the effectiveness of our newly proposed O-NMF-S.

6.3   POMDP Compression by O-NMF-S

To compute the policies after the O-NMF-S compression, Perseus [12] was used in our experiments. Perseus is a randomized point-based value iteration method which is known to be effective and yet efficient in computing high quality policies given a POMDP. Figures 8 and 9 show the performance comparison between O-NMF-S and O-NMF for the Hall68by9 problem. It is obvious that O-NMF-S achieves a much better average reward at a less computational cost over all the reduced dimensions.

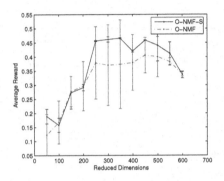 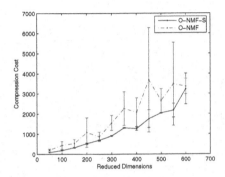

Fig. 8. Comparison on policy performance between O-NMF-S and O-NMF   **Fig. 9.** Comparison on compression efficiency between O-NMF-S and O-NMF

Table 1. Performance comparison for Hall68byn^2

(*T_p* - policy computing time; *T_c* - compression time)

PROBLEM (STATES/ACTIONS/OBS.)	REWARD	TIME(SEC.) $T_p + T_c$	REDUCED DIM.
HallwayBY9(*500 samples*) (612s 5a 17o)			
PERSEUS	0.3889± 0.0285	3000	*N/A*
PERSEUS	0.4576± 0.0406	5399	*N/A*
PERSEUS+$ONMF$	0.4355± 0.0215	500+967	250
PERSEUS+$ONMF$-S	0.4403± 0.0182	500+688	250
HallwayBY16(*500 samples*) (1088s 5a 21o)			
PERSEUS	0.2542± 0.0216	5000	*N/A*
PERSEUS	0.3079± 0.0082	20000	*N/A*
PERSEUS+O-NMF	0.3895± 0.0221	1000+2477	300
PERSEUS+O-NMF-S	0.3918 ± 0.0207	1000+1883	300

Table 1 tabulates the results obtained in terms of (a) running time (with breakdown of the time needed by O-NMF-S, O-NMF, and Perseus) and (b) the policy's average award. Hall68by9 and Hall68by16 have been tested. The reduced dimension settings were picked according to the sharp drops identified according to Figure 1. According to Table 1, it can be observed that O-NMF-S can result in policies of quality almost the same as that without it at a reasonable reduced dimension for both Hall68by9 (from 612 to 250) and Hall68by16(from 1088 to 300). Also, the computational speedup brought by the O-NMF-S was quite significant when compared with Perseus.

7 Conclusion and Future Work

In this paper, we proposed (1) an eigenvalue analysis to estimate the compression limit of a POMDP problem, and (2) an accelerated version of O-NMF based on the interior point gradient acceleration technique to minimize the POMDP compression overhead cost. We validated the correctness of the eigenvalue analysis by applying it to POMDP problems of different types. We found that among the problems tested, the compression approach is more effective for robot navigation related problems. Also, the speedup achieved via O-NMF-S has been empirically found to be significant for some POMDP problems with more than 1000 states.

This work can be further extended at least in the following two directions: (1) The eigenvalue analysis shows that for most of the problems, there exist specific pairs of $< a, z >$ where the corresponding values of $rank(G^{<a,z>})$ are far less than the number of states. Thus, schemes for decomposing POMDP according to the properties of $< a, z >$ pairs worth future research efforts. (2) It is also interesting to investigate how to effectively interleave the policy computation step based on our approach and the policy execution to achieve online learning under the partial observation scenario.

Acknowledgments

This work is partially supported by HKBU Faculty Research Grant (FRG/08-09/124).

References

1. Simmons, R., Koenig, S.: Probabilistic robot navigation in partially observable environments. In: Proceedings of the International Joint Conference on Artificial Intelligence (IJCAI 1995), July 1995, pp. 1080–1087 (1995)
2. Hoey, J., Poupart, P., Boutilier, C., Mihailidis, A.: POMDP models for assistive technology. Technical report, Intelligent Assistive Technology and Systems Lab (IATSL), University of Toronto (2005)
3. Hauskrecht, M., Fraser, H.: Planning treatment of ischemic heart disease with partially observable markov decision processes. Artificial Intelligence in Medicine 18, 221–244 (2000)

4. Smallwood, R.D., Sondik, E.J.: The optimal control of partially observable Markov processes over a finite horizon. Operations Research 21, 1071–1088 (1973)
5. Cassandra, A.R.: Exact and approximate algorithms for partially observable Markov decision processes. PhD thesis, Brown University, U.S.A (1998)
6. Poupart, P., Boutilier, C.: Value-directed compression of POMDPs. In: Proceedings of Advances in Neural Information Processing Systems 15, pp. 1547–1554. MIT Press, Cambridge (2003)
7. Roy, N., Gordon, G., Thrun, S.: Finding approximate POMDP solutions through belief compression. Journal of Artificial Intelligence Research 23, 1–40 (2005)
8. Li, X., Cheung, W.K., Liu, J.: Improving POMDP's tractability via belief compression and clustering. IEEE Transactions on Systems, Man and Cybernetics, Part B (to appear, 2009)
9. Pineau, J., Roy, N., Thrun, S.: A hierarchical approach to POMDP planning and execution. In: Proceedings of Workshop on Hierarchy and Memory in Reinforcement Learning (ICML) (June 2001)
10. Merritt, M., Zhang, Y.: An interior-point gradient method for large-scale totally nonnegative least squares problems. Technical Report TR04-08, Department of Computational and Applied Mathematics, Rice University, Houston, Texas, U.S.A (May 2004)
11. Ding, C., Li, T., Peng, W., Park, H.: Orthogonal nonnegative matrix t-factorizations for clustering. In: Proceedings of the 12th ACM SIGKDD International Conference on Knowledge Discovery and Data Mining, Philadelphia, PA, USA, pp. 126–135 (2006)
12. Spaan, M.T.J., Vlassis, N.: Perseus: Randomized point-based value iteration for POMDPs. Journal of Artificial Intelligence Research 24, 195–220 (2005)
13. Gonzalez, E.F., Zhang, Y.: Accelerating the lee-seung algorithm for non-negative matrix factorization. Technical Report TR05-02, Department of Computational and Applied Mathematics, Rice University, Houston, Texas, U.S.A (March 2005)

Appendix

The Proof of Lemma 1

With the constraint $F^T F = I$, $G^{<a,z>} F = F \widetilde{G}^{<a,z>}$ can be rewritten as

$$\widetilde{G}^{<a,z>} = F G^{<a,z>} F^T. \tag{14}$$

As G is an $n \times n$ square matrix, an eigen-decomposition on G would be

$$G^{<a,z>} = P \Lambda_G P^{-1}$$

where P is formed from the corresponding eigenvectors of $G^{<a,z>}$, and Λ_G is the diagonal matrix formed from the eigenvalues of $G^{<a,z>}$. Then Eq.(14) can be rewritten as

$$\widetilde{G}^{<a,z>} = (FP) \Lambda_G (FP)^{-1}. \tag{15}$$

Thus, $G^{<a,z>}$ and $\widetilde{G}^{<a,z>}$ are isomorphic and share the same group of bases. Thus, the necessary condition for Eq.(15) to hold is $m \geq rank(G^{<a,z>})$. □

The Proof of Lemma 2

According to Lemma 1, by setting $m = \max\limits_{a,z}(rank(G^{<a,z>}))$, the following equations will hold, given as

$$\widetilde{G}^{<a_1,z_1>}F = F\widetilde{G}^{<a_1,z_1>} \tag{16}$$

$$\cdots \cdots \cdots$$

$$\widetilde{G}^{<a_{|A|},z_{|O|}>}F = F\widetilde{G}^{<a_{|A|},z_{|O|}>}. \tag{17}$$

Obviously, $\sum\limits_{a,z} G^{<a,z>}F = F\sum\limits_{a,z}\widetilde{G}^{<a,z>}$ is the necessary condition for satisfying Eq.(16), up to Eq.(17). Again, from Lemma 1, $\sum\limits_{a,z} G^{<a,z>}$ and $\sum\limits_{a,z}\widetilde{G}^{<a,z>}$ are isomorphic, thus $m \geq rank(\sum\limits_{a,z} G^{<a,z>})$. $\qquad\square$

The Proof of Lemma 3

The proof relies on one proposition described in [13] and two easily proved common rules which are leveraged in the following derivations. The proposition is

$$\eta = \frac{(e-u)^T Diag(x \circ Bx)(e-u)}{(e-u)^T Diag(x)(B)Diag(x)(e-u)} \geq 1$$

where $B_{i,j} \geq 0$, $x_i \geq 0$, $Diag(x)$ is the diagonal matrix with its leading diagonal as the vector x, e is the vector of all ones, and u is any vector $u \in \mathbb{R}^n$. Again the symbol "$\circ$" denotes the component-wise multiplication. The two rules are given as

$$(v_1 \circ v_2)(v_3 \circ v_4) = v_1 Diag(v_2^T \circ v_3)v_4 \tag{18}$$

$$(v_1 \circ v_2)M(v_3 \circ v_4) = v_1 Diag(v_2^T)M Diag(v_3)v_4 \tag{19}$$

where v_1, v_2, v_3, and v_4 are n-dimensional row vectors, and M is an $n \times n$ matrix. Now, let us consider the case

$$W^T + p = W^T \circ \sqrt{HV^T./(HH^TW^T + W^TWHV^T - W^TWHH^TW^T)}.$$

Since $HH^TW^T + W^TVH^TW^T - W^TWHH^TW^T$ is forced to be larger than zero [8], $W^T + p > 0$. It is not difficult to show that $\hat{\alpha}_k \geq 1$. Next, we are to prove $\alpha_k^* \geq 1$. By substituting Eq.(13) to rewrite α^*, substituting Eq.(12) into Eq.(11), as well as setting

$$B = HH^T + W^TWHV^T(W^T)^{-1} - W^TWHH^T,$$

we get

$$\begin{aligned}
\alpha_k^* &= \frac{(W^T \circ \sqrt{HV^T./BW^T} - W^T)^T(HV^T - BW^T)}{(W^T \circ \sqrt{HV^T./BW^T} - W^T)^T(B)(W^T \circ \sqrt{HV^T./BW^T} - W^T)} \\
&= \frac{((e - \sqrt{HV^T./BW^T})^T \circ W)(BW^T \circ (e - HV^T./BW^T))}{((e - \sqrt{HV^T./BW^T})^T \circ W)(B)(W^T \circ (e - \sqrt{HV^T./BW^T}))}.
\end{aligned} \tag{20}$$

For the numerator of Eq.(20), we leverage Eq.(18) by taking $(e-\sqrt{HV^T./BW^T})^T, W, BW^T, (e-HV^T./BW^T)$ as v_1, v_2, v_3, v_4 respectively. Then α_k^* becomes

$$\alpha_k^* = \frac{((e-\sqrt{HV^T./BW^T})^T)Diag(W^T \circ BW^T)((e-HV^T./BW^T))}{((e-\sqrt{HV^T./BW^T})^T \circ W)(B)(W^T \circ (e-\sqrt{HV^T./BW^T})}. \quad (21)$$

For the denominator of Eq.(21), we leverage Eq.(19) by taking $(e-\sqrt{HV^T./BW^T})^T, W, B, W^T, (e-\sqrt{HV^T./BW^T})$ as v_1, v_2, M, v_3, v_4 respectively, then α_k^* becomes

$$\alpha_k^* = \frac{((e-\sqrt{HV^T./BW^T})^T)Diag(W^T \circ BW^T)((e-HV^T./BW^T))}{(e-\sqrt{HV^T./BW^T})^T Diag(W^T)(B)Diag(W^T)(e-\sqrt{HV^T./BW^T})}.$$

Furthermore, with the substitution of $x = W^T$ and

$$u = \sqrt{HV^T./(HH^TW^T + W^TWHV^T - W^TWHH^TW^T)},$$

α_k^* becomes

$$\alpha_k^* = \frac{(e-u)^T Diag(x \circ Bx)(e-u.^2)}{(e-u)^T Diag(x)(B)Diag(x)(e-u)}. \quad (22)$$

Note that the square in Eq.(22) is taken component-wise. Thus α_k^* becomes

$$\alpha_k^* = \frac{(e-u)^T Diag(x \circ Bx)((e+u) \circ (e-u))}{(e-u)^T Diag(x)(B)Diag(x)(e-u)}$$
$$= \frac{(e-u)^T Diag(x \circ Bx)Diag(e+u)(e-u)}{(e-u)^T Diag(x)(B)Diag(x)(e-u)}. \quad (23)$$

Since $u \geq 0$, we obtain

$$\alpha_k^* \geq \frac{(e-u)^T Diag(x \circ Bx)(e-u)}{(e-u)^T Diag(x)(B)Diag(x)(e-u)}.$$

Thus $\alpha_k^* \geq \eta \geq 1$. Consequentially, we have $\alpha_k \geq 1$. $\qquad\square$

The Proof of Lemma 4
Similar to the proof of Lemma 3, $W^T + p = W^T \circ \sqrt{(HV^T)./(HV^TWW^T)} > 0$ which gives $\hat{\alpha}_k > 1$. To prove $\alpha_k^* \geq 1$, we rewrite α_k^* to be the exact form as Eq.(23) by substituting x, B, u with $W^T, HV^TWW^T, \sqrt{(HV^T)./(HV^TWW^T)}$ respectively. Again, $\alpha_k^* \geq \eta \geq 1$, and thus $\alpha_k \geq 1$. $\qquad\square$

Building a Decision Cluster Forest Model
to Classify High Dimensional Data
with Multi-classes

Yan Li and Edward Hung

Department of Computing, The Hong Kong Polytechnic University
Hung Hom, Hong Kong
{csyanli,csehung}@comp.polyu.edu.hk

Abstract. In this paper, a decision cluster forest classification model is proposed for high dimensional data with multiple classes. A decision cluster forest (DCF) consists of a set of decision cluster trees, in which the leaves of each tree are clusters labeled with the same class that determines the class of new objects falling in the clusters. By recursively calling a variable weighting k-means algorithm, a decision cluster tree can be generated from a subset of the training data that contains the objects in the same class. The set of m decision cluster trees grown from the subsets of m classes constitute the decision cluster forest. Anderson-Darling test is used to determine the stopping condition of tree growing. A DCF classification (DCFC) model is selected from all leaves of the m decision cluster trees in the forest. A series of experiments on both synthetic and real data sets have shown that the DCFC model performed better in accuracy and scalability than the single decision cluster tree method and the methods of k-NN, decision tree and SVM. This new model is particularly suitable for large, high dimensional data with many classes.

Keywords: Clustering, classification, W-k-means, forest.

1   Introduction

One challenge in data mining is classification of high dimensional data with multiple classes [1]. This kind of data may occur in application fields such as text mining, multimedia mining and bio-informatics. To solve this problem, the ADCC method was proposed in [2], which builds a classification model from high dimensional data. Given a training data set, the ADCC algorithm recursively calls the variable weighting k-means algorithm (W-k-means) [3] to generate a decision cluster tree. Each node with its dominant class forms a decision cluster. ADCC uses the leaves of the tree as the classification model, and leaves are labeled with their dominant classes to determine the classes of new objects falling in the clusters. Experiment results have shown that this decision cluster classification model was effective and efficient in classifying high dimensional data [2].

Z.-H. Zhou and T. Washio (Eds.): ACML 2009, LNAI 5828, pp. 263–277, 2009.

One shortcoming of this ADCC method is that the algorithm generates some weak decision clusters in which no single class dominates. Existence of weak clusters in the model can affect classification performance of the model. It has been shown that classification accuracy could be improved after weak decision clusters were removed from the model [4]. Weak clusters occur because objects of different classes are mixed in the clustering process to generate decision clusters. If we assume that objects in the same class have their own cluster distributions, we can separate objects of different classes according to the object class labels and generate decision clusters from objects in each class. Then, we combine the decision clusters of different classes to form the decision cluster classification model. In this way, weak decision clusters can be avoided.

In this paper, we propose a Decision Cluster Forest method to build a decision cluster classification model from high dimensional data with multiple classes. Instead of building a single decision cluster tree from the entire training data, we build a set of cluster trees from subsets of the training data set to form a decision cluster forest. Each tree in the forest is built from the subset of objects in the same class. The proposition for this method is that the objects in the same class tend to have their own spatial distributions in the data space. Therefore, decision clusters of objects in the same class are found. The decision clusters in the same tree have the same dominant class. In this way, no weak cluster is created in such decision cluster tree. A decision cluster model can be selected from any subset of leaf decision clusters from the decision cluster forest so the model is called a decision cluster forest classification model (DCFC).

The decision cluster forest method has advantages of classifying data with multiple classes because the DCFC model is guaranteed to contain decision clusters in all classes. In other multi-class classification methods, such as decision trees, the information of small classes is often under represented in the model. The error-correcting output codes method (ECOC) was designed to solve multi-class learning problem by learning multiple binary classification models and matching the classification results with the designed codeword to correct misclassifications [5]. This method is more like an ensemble method but the challenge is on the design of code word. In contrast, the DCFC model is a more intuitive and direct multi-class classification method and easy to use.

In growing a decision cluster tree from a subset of objects in the same class, we adopt the W-k-means algorithm to reduce the effect of noisy attributes in high dimensional data. We also grow a binary decision cluster tree and use Anderson Darling Test [6,7] as a stopping criterion in tree growing. We have conducted a series of experiments on both synthetic and real data sets to demonstrate the efficiency and accuracy of the DCFC method. Compared with other classification methods, including ADCC, k-NN, J48 (a decision tree algorithm) and SMO (one of SVM algorithms), our experimental results have shown that the DCFC method has performed better than those methods in classification accuracy on large high dimensional sparse data sets. Thus the results demonstrate that the DCFC method is more suitable for large, high dimensional data with many classes.

Clustering methods have been explored to solve classification problems [8,9]. An early example of using the k-means clustering algorithm to build a cluster tree classification model can go back to early 80's [10]. In this work, a binary cluster tree was built by interactively executing the k-means clustering algorithm. At each node, a further partition was determined by the percentage of the dominant class in the cluster node. However, only small numeric data could be classified and every time only two sub-clusters are formed. In 2000, Huang et al. proposed a new interactive approach to build a decision cluster classification model [11]. In this approach, the k-prototypes clustering algorithm was used to partition the training data, and a visual cluster validation method [12] was adopted to verify the partitioning result at each cluster node. The above two interactive methods are not adequate for high dimensional data with noise because the clustering algorithms used are not able to handle noisy attributes and it is too time consuming to involve human judgment. The concept clustering tree which is a decision tree where each node as well as each leaf corresponds to a cluster is proposed by Blockeel et al. [13]. The nodes of decision cluster tree proposed in our method are also clusters, but the process of the tree construction including partition method, stopping criteria and the number of sub-clusters are totally different.

The rest of this paper is organized as follows. In Section 2, we briefly review the decision cluster classification model and the method to generate a decision cluster tree. In Section 3, we introduce the decision cluster forest classification model and the algorithm for model building. In Section 4, experimental results and comparisons are reported. In Section 5, we conclude this paper.

2   Decision Cluster Classification Model

In this section, we briefly review the decision cluster classification model and the method to build a model by growing a decision cluster tree with a k-means clustering algorithm [2,11].

The decision cluster classification model follows the proposition that given a training data with a mixture density, objects in the same class tend to be spatially close in the data space [11]. Let $X = x_1, x_2, ..., x_n$ be a training data set of n classified objects, each described by m attributes and labeled in one of m classes. A decision cluster classification model can be simply generated in the following steps:

1. Use a clustering algorithm to cluster X into k clusters without using the class label information,
2. For each cluster, calculate the class distribution and label it with the most frequent class as the dominant class. A cluster labeled with a dominant class is a decision cluster,
3. The set of decision clusters constitutes a decision cluster classification model or a DCC model,

4. When a new object is submitted to the DCC model, the distances between the object and the centers of decision clusters are computed and the dominant class of the decision cluster with the shortest distance to the object is assigned to the object.

In principle, any subset of decision clusters forms a DCC model. However, the model performance on classification accuracy depends on the purity of decision clusters, i.e., high percentage of the dominant class in clusters.

For large complex data with varying distribution densities, we can use a clustering algorithm to generate a decision cluster tree as shown in Figure 1. The root X, which is the training data set, is partitioned into 3 clusters C_1^0, C_2^0, C_3^0. Here, the superscript indicates the level of the node from which the clusters are generated and the subscript is the cluster number in this level. Clusters C_1^0 and C_3^0 are further partitioned into 2 and 3 sub-clusters respectively, which form level 2 of the cluster tree. Subsequently, two clusters C_2^1 and C_4^1 are further partitioned into three and two sub clusters respectively, which form level 3 of the cluster tree. We define a node partition as the clustering of the node. For example, the clustering of root consists of 3 clusters C_1^0, C_2^0, C_3^0. A decision cluster tree can be represented as a sequence of nested clusterings, for instance

$$X(C_1^0(C_1^1, C_2^1(C_1^2, C_2^2, C_3^2)), C_2^0, C_3^0(C_3^1, C_4^1(C_4^2, C_5^2), C_5^1))$$

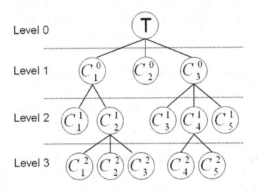

Fig. 1. Example of a cluster tree

Given a training data X and a clustering algorithm $\mathbf{f}$, a cluster tree can be generated automatically by recursively calling $\mathbf{f}$ to partition the nodes, starting from the root node. In this process the following decisions have to be made:

- Which clustering algorithm should be used for generating a clustering at each node?
- How many clusters should be generated at each node?
- Where to stop at each path of the tree?

In [2], the authors have proposed the ADCC algorithm to automatically generate a decision cluster tree from high dimensional data. This ADCC algorithm uses $W\text{-}k$-means [3] to generate the clustering at each node. Because $W\text{-}k$-means is able to reduce the impact of noisy attributes by assigning smaller weights to them in clustering, it implicitly performs attribute selection in the clustering process. As such, W-k-means is more suitable for high dimensional data with noisy attributes. At each node, the number of clusters is equal to the number of classes in the subset of the data in the node. Multiple termination conditions are used to stop growing of the tree at a node, including the size, class purity and Anderson-Darling test [6].

After a decision cluster tree is created, the leaves of the tree are extracted as the decision cluster model. Experimental results have shown that the decision cluster model outperformed other classification methods in high dimensional data classification [2]. However, we have also observed that not all leaf decision clusters had high percentage in the dominant classes. These weak decision clusters could affect the performance of the decision cluster model. In the view of data distribution, these weak clusters often occur at the boundaries of other decision clusters with strong dominant classes. By removing the weak decision clusters from the model, the classification accuracy could be improved slightly [4]. In the next section, we propose a decision cluster forest approach to tackle this problem.

3   Decision Cluster Forest

3.1   Decision Cluster Forest (DCF)

A Decision Cluster Forest (DCF) consists of a set of decision cluster trees. Each tree grows from a subset of the training data that contains objects in the same class. If the training data X has m classes, the decision cluster forest will have m decision cluster trees. Given a decision cluster forest, a Decision Cluster Forest Classification (DCFC) model can be built by simply extracting the leaves of decision cluster trees.

All decision clusters in a DCFC model from the decision cluster forest have strong dominant classes with 100 percentage distribution. This approach follows the proposition that in a large high dimensional data set with multiple classes, the objects in each class tend to occupy a spatial region with its own mixture density distribution. Therefore, the mixture densities can be discovered in a clustering process from the set of objects in the same class.

Let X be a training data set of n objects in m classes. We divide X into m subsets $(X_1, X_2, \ldots, X_k)$, each with objects in the same class. For each subset, we use a clustering algorithm to build a decision cluster tree in which all nodes have the same dominant class. The decision cluster forest consists of m decision cluster trees with nodes of m different dominant classes. Figure 2 illustrates a decision cluster forest with different decision cluster trees.

Given a decision cluster forest, we can select any subset of decision clusters from multiple trees to build a DCFC model. Similar to the single decision cluster tree method, the performance of a DCFC model on classification accuracy also depends on the quality of the decision cluster trees and the selection of decision clusters to be included in the model. Therefore, the following two processes are crucial: (1) generation of a set of decision cluster trees and (2) selection of a subset of the decision clusters from the decision cluster forest for the classification model.

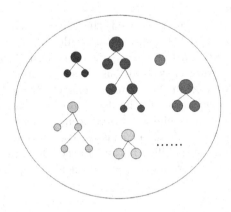

Fig. 2. Distribution of decision cluster trees in a decision cluster forest

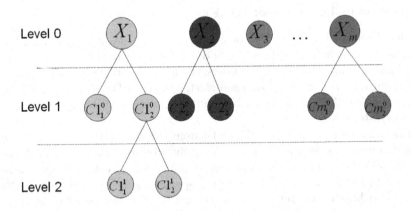

Fig. 3. Generation of decision cluster trees for a decision cluster forest

Given a subset X_i from the training data set X, a decision cluster tree can be built with a process similar to that used in building a single decision cluster tree. However, since all objects in the data set are in the same class, class distribution calculation is not necessary. When executing W-k-means on each inner node, the parameter k (the number of clusters should be divided) of W-k-means is unknown. To avoid decreasing the clustering quality due to a too large k value, we set parameter k to 2 and generate a hierarchical binary cluster tree. That is, we perform a binary partition to result in a binary tree at each inner node. All decision clusters are assigned the same dominant class. A decision cluster tree with the same dominant class is shown in Figure 3. Multiple decision cluster trees are generated from different subsets of training data with different dominant classes marked in different colors.

Take the left tree as an example. This binary decision cluster tree is generated from the data subset X_1 shown in the root. The root X_1 is first partitioned into 2 clusters $C1_1^0, C1_2^0$. Here, the cluster index $C1$ indicates the clusters are generated from the data of class 1. Therefore, all clusters have the same dominant class 1. The superscript indicates the level of the node from which the clusters are generated and the subscript is the cluster number in this level. Cluster $C1_2^0$ is further partitioned into 2 sub-clusters $C1_1^1, C1_2^1$, which form level 2 of this decision cluster tree. This binary partition process continues until the stopping criteria are satisfied at the two leaves. Other decision clusters trees marked in different colors are generated from different subsets of the training data in a similar way.

In a real world training data set, the distribution of classes is often unbalanced. Some classes have more objects than others. Therefore, the depths of decision cluster trees are different. In the extreme case such as X_3, a decision cluster tree may have only the root because the number of objects is too small. In this case, we treat X_3 as one decision cluster in the model.

3.2   DCF Classification (DCFC) Model

The decision cluster forest in Figure 3 can be represented in a sequence of nested clusterings as follows:

$$\{X1(C1_1^0, C1_2^0(C1_1^1, C1_2^1)); X2(C2_1^0, C2_2^0); ...; Xm(Cm_1^0(Cm_1^1, Cm_2^1), Cm_2^0)\}$$

In this sequence, the decision clusters with the same dominant class are grouped in the top level of X_i for $1 \le i \le m$. Any subset of disjoint decision clusters from the top levels downward can make a DCFC model. There are many ways to select classification models from a decision cluster forest. In this work, we select the leaf nodes of each decision cluster tree in the decision cluster forest to build the DCFC model. Each decision cluster in the model is represented by its center and dominant class.

The DCFC model classifies new objects as follows:

1. Compute the distances between a new object and the centers of the decision clusters with a distance function;
2. Assign to the object the dominant class of the decision cluster with the shortest distance to the object

3.3   DCFC Algorithm

Table 1 shows the DCFC algorithm to automatically build a decision cluster forest and select a leaf-based DCFC model. The input to the algorithm is a training data set with m classes and the output from the algorithm is a DCFC model. The algorithm first divides the input data X into m subsets. For each subset, it calls the W-k-means algorithm to generate a decision cluster tree. At each node, it calls Stop-Test() function to test the stopping criteria to determine whether to call W-k-means to partition the node into two children nodes or turn the node to a leaf node. After all decision cluster trees are generated, a decision cluster forest is obtained. Finally, the set of all leaf decision clusters is returned as the DCFC model.

Table 1. DCFC Algorithm

Input: A training data set X (with m classes).

Output: A classification model $DCFC\text{-}model$.

Begin

1. partition X into m training data subsets,
    $X = X_1, X_2, ..., X_m$.(Section 3.1)

2. for each training data subset X_i, $i = 1...m$;

3. initialize a decision cluster tree DCT_i with root $\{X_i\}$;

4. sign the root as *internal node*;

5. for each internal node X in DCT_i

6.   if $(Stop\text{-}Test(X))$ sign X as *leaf node*;

7.   run W-k-means on X to produce two new sub-clusters;

8.   sign the two new sub-clusters as *internal node* and add them into DCT_i;

9. end for

10. end for

11. include all leaf nodes from all m trees into the classification model $DCFC\text{-}model$
    and represent each node by its center and its dominant class;

12. return $DCFC\text{-}model$;

End

The $Stopping\text{-}Test(X)$ algorithm is given in Table 2. Since all data objects belong to the same class, the class label information is no longer useful in the tree generation process. In this algorithm, only two stopping criteria are considered: the number of objects in the data subset and the Anderson Darling (AD) Test [6,7]. If a node cluster is too small, it need not be further partitioned. If objects in a cluster follow a normal distribution, it is a good cluster and does not need a further partition. Anderson Darling (AD) Test is a powerful statistical method to determine whether a sample comes from a specified distribution or not. If the testing result satisfies the AD Test criteria, this cluster node is treated as a leaf node. Otherwise, it is further partitioned into sub-clusters.

Table 2. Algorithm of $Stopping\text{-}Test(C)$

Input: the node C which contains n objects.

Output: a Boolean value $Stop$ which is either

(i) TRUE which means "stop" or (ii) FALSE, otherwise.

Remarks:

δ: the threshold of the number of data in C (e.g., 10).

Begin

1. $Stop$ = FALSE;

2.   if ($n < \delta$ OR $AD\text{-}Test(C)$)

3. $Stop$ = TRUE and label C with the dominant class label;

4.   return $Stop$;

End

---

4   Experiments

In this section, we present the experiments we have conducted on both synthetic and real-world data sets. The experimental results demonstrate that DCFC outperforms some existing classification algorithms, including original k-NN, decision tree (J48), SVM (SMO) and ADCC [2] in terms of speed, scalability and classification accuracy. Weka[1] implementations of J48 and SMO were used in our comparisons. DCFC algorithm and ADCC were implemented in java.

All experiments were conducted on an Intel(R) Xeon(R), 1.60 GHz computer with 8GB memory. We compared the accuracy and execution time of these classification algorithms. In our experiments, our setting is similar to [2] where default parameters were used for J48 and SMO in Weka. For J48, the minimal number of instances per leaf is set as 2. The SMOs were trained with a linear kernel where the complexity parameter C was set as 1.0. For k-NN, the number of neighbors, k, equals to 1. For ADCC, we use the same parameter settings as in [2]. DCFC needs one parameter δ presented in $Stopping\text{-}Test(C)$, which is presented in Section 3.3.

---

[1] http://www.cs.waikato.ac.nz/~ml/weka/

In *Stopping-Test(C)* algorithm (see table 2), we first consider the size of the current node and then consider the result of *AD-TEST(C)* when we determine whether the current node C should be further divided. Since it takes longer for *AD-TEST(C)* than judging the size, we want to filter the obvious condition under which the node need not be further divided. Thus, the size threshold δ can be set to values so that the further clustering of C should be stopped obviously when the data size in node C is less than δ. δ depends on the size of the root node and is always set to 5% of the number of samples in the root node. If the size of a node is smaller than 5% of the size of the root, we consider this node as a leaf automatically.

4.1   Experiment on Text Data

We compared DCFC with other three well-known classification algorithms and ADCC on some high dimensional real data sets. These data sets are taken from the UCI machine learning data repository[2].

Table 3 lists eight data sets built from the Twenty Newsgroups data. The original text data was first preprocessed to strip the news messages from the special tags and the email headers and eliminate the stem words and stop words. The dimension (word) in each document was weighted by the Term Frequency (TF). We preprocessed these data sets by deleting some dimensions with smallest TF values. The resulting text data contain several thousand frequent words. Different data sets have different cluster properties. Some of them have semantically similar classes (e.g. T_2), whereas others contain semantically different classes (e.g. T_1). Some of them have overlapping words (dimensions) (e.g. T_5), while some of them contain the unbalanced number of documents in each class (e.g. T_8).

Five classification algorithms, DCFC, ADCC, k-NN, J48 and SMO, were tested on these text data sets. Each classification algorithm was run 10 times on each data set using different 30% of data as testing test (70% as training data). The deviation of results of each algorithm is not large, so we report their average. Figure 4 shows the classification accuracy and execution time on these text data sets. We can see that DCFC achieves higher accuracy than ADCC, k-NN and J48 in most cases. DCFC and ADCC are comparably faster than other algorithms. DCFC achieves higher or close accuracy to SMO but much faster than that.

4.2   Experiment on Other Real Data

Table 4 lists the other two real data sets which are also taken from UCI machine learning repository.

Table 5 lists the results of execution time and classification accuracy generated by five classification methods from these two real data sets. These two data sets are comparably simple and low dimensional. DCFC works as well as ADCC which is more suitable for large, high dimensional data.

---

[2] http://archive.ics.uci.edu/ml/datasets.html

Table 3. Text data sets generated from the 20-Newsgroups data

Data sets	classes	Dimension	size
T_1	alt.attheism	3939	200
	Comp.graphics		200
T_2	talk.politics.mideast	5795	200
	Talk.politics.misc		200
T_3	comp.sys.imb.pc.hardware	2558	200
	Comp.sys.mac.hardware		200
T_4	alt.attheism	5856	300
	talk.religion.misc		300
T_5	rec.autos	3154	400
	Rec.motocycles		400
T_6	rec.autos	3979	200
	Rec.motocycles		400
T_7	comp.graphics	7924	200
	rec.sport.baseball		200
	sci.space		200
	talk.politics.mideast		200
T_8	comp.graphics	8549	300
	rec.sport.baseball		200
	sci.space		100
	talk.politics.mideast		50

Table 4. Other real data sets from UCI

Data set name	Instances	Dimensions	Classes	Training	testing
Waveform	5000	40	3	3500	1500
Reuters	9980	337	10	6986	2994

4.3 Scalability

We conducted experiments on synthetic data sets to demonstrate and compare the scalability between our new DCFC algorithm and other classification algorithms. The results show that our method is more scalable when the number of samples and dimensions increase.

We adopted the same two groups of synthetic data sets with different numbers of dimensions and samples (shown in Table 6) as in Ref. [2]. These data sets were used to compare the scalability of ADCC. Each data set contains three

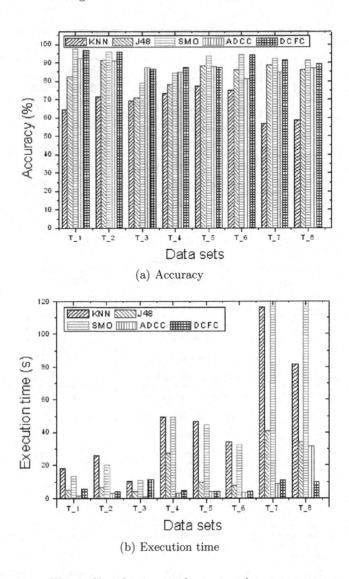

(a) Accuracy

(b) Execution time

Fig. 4. Classification results on text data sets

clusters randomly generated with normal distributions. The class labels are randomly arranged. In each run, we used 70% of data as the training data and the remaining 30% as the testing data. We recorded the total execution time and compared the performance of DCFC with original k-NN, J48 and ADCC with different number of dimensions and samples respectively.

Figure 5 shows the execution time against the number of dimensions and Figure 6 shows the execution time against the number of samples. Because the curve of the result of k-NN ranges too large to demonstrate the curves of other

Table 5. Classification results of data sets in Table 4 by five classification methods

Data sets	Waveform		Reuters	
Metrics	Time(s)	Acc.(%)	Time(s)	Acc.(%)
DCFC	2.5031	84.4	81.922	68
ADCC	2.516	83.8	74.952	68.97
k-NN	15.688	70.8	485.625	57.95
J48	1.2	73.26	35.23	65.29
SMO	1.94	85	392.91	65.86

Table 6. Two groups of synthetic data sets (each having three classes)

Data sets	Dimensions	Instances	Data sets	Dimensions	Instances
A1	5	5,000	B1	4	3,000
A2	20	5,000	B2	4	9,000
A3	50	5,000	B3	4	15,000
A4	100	5,000	B4	4	30,000
A5	200	5,000	B5	4	45,000
A6	300	5,000	B6	4	60,000
A7	400	5,000	B7	4	75,000
A8	500	5,000	B8	4	90,000

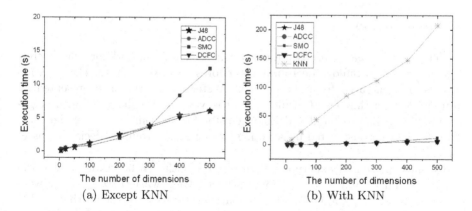

(a) Except KNN   (b) With KNN

Fig. 5. Execution time vs dimension number

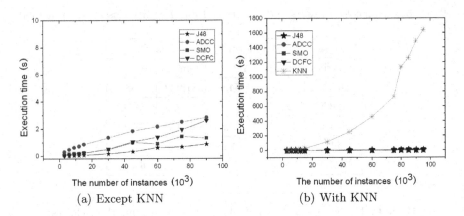

(a) Except KNN   (b) With KNN

Fig. 6. Execution time vs data size

algorithms clearly, we presented the results in two sub-figures. Sub-figure (a) is the comparison results of DCFC, ADCC, SMO and J48, and Sub-figure (b) is the comparison results adding k-NN. We can see that the execution time increased linearly for DCFC, ADCC, SMO and J48 whereas the execution time for k-NN increased rapidly when the number of instances approached 90000. Although the execution time for k-NN increased linearly as the number of dimensions increased, the increase was much faster than DCFC, SMO, ADCC and J48. Meanwhile, the execution time for DCFC, ADCC, SMO and decision tree (J48) are comparable to each other and scalable on this group of data sets. SMO grows faster over dimensions than DCFC, ADCC and J48.

5   Conclusion

In this paper, we have proposed a new classification framework for using a clustering method to build a decision cluster forest. Decision cluster classification models are generated from the decision cluster forest. We have presented an automatic algorithm DCFC which uses the variable weighting k-means algorithm W-k-means to build a decision cluster forest from a training data set. The classifier is constructed by selecting all leaf nodes from the decision cluster forest.

We have presented experimental results on both synthetic and real world data sets and compared the performance of DCFC with those of other well-known classification methods. The comparison results have shown that DCFC has advantages in classifying large, high dimensional data with multiple classes. In our future work, we will conduct experiments to compare the DCFC model with the ECOC multi-class classification method.

Acknowledgement

This work has been partially supported by grants PolyU 5174/07E from Hong Kong RGC, 1-ZV5P, G-U524 and A-PA5S from PolyU.

References

1. Piatetsky-Shapiro, G., Djeraba, C., Getoor, L., Grossman, R., Feldman, R., Zaki, M.: What are the grand challenges for data mining? In: KDD 2006 panel report. SIGKDD Explorations, vol. 8, pp. 70–77 (2006)
2. Li, Y., Hung, E., Chung, K., Huang, J.: Building a decision cluster classification model by a variable weighting k-means method. In: Wobcke, W., Zhang, M. (eds.) AI 2008. LNCS (LNAI), vol. 5360, pp. 337–347. Springer, Heidelberg (2008)
3. Huang, J., Ng, M., Rong, H., Li, Z.: Automated variable weighting in k-means type clustering. IEEE Transactions on Pattern Analysis and Machine Intelligence 27, 657–668 (2005)
4. Jing, L., Huang, J., Ng, M.K., Rong, H.: A feature weighting approach to building classification models by interactive clustering. In: Torra, V., Narukawa, Y. (eds.) MDAI 2004. LNCS (LNAI), vol. 3131, pp. 284–294. Springer, Heidelberg (2004)
5. Dietterich, T., Bakiri, G.: Solving multiclass learning problems via error-correcting output codes. Journal of Artificial Intelligence Research 2, 263–286 (1995)
6. Anderson, T.W., Darling, D.A.: Asymptotic theory of certain "goodness-of-fit" criteria based on stochastic processes. The Annals of Mathematical Statistics 23, 193–212 (1952)
7. Stephens, M.A.: Edf statistics for goodness of fit and some comparisons. Journal of the American Statistical Association 69, 730–737 (1974)
8. Kyriakopoulou, A., Kalamboukis, T.: Text classification using clustering. In: ECML-PKDD Discovery Challenge Workshop Proceedings (2006)
9. Zhang, B., Srihari, S.N.: Fast k-nearest neighbor classification using cluster-based trees. IEEE Transactions on Pattern Analysis and Machine Intelligence 26, 525–528 (2004)
10. Mui, J., Fu, K.: Automated classification of nucleated blood cells using a binary tree classifier. IEEE Transactions on Pattern Analysis and Machine Intelligence 2, 429–443 (1980)
11. Huang, Z., Ng, M., Lin, T., Cheung, D.: An interactive approach to building classification models by clustering and cluster validation. In: Leung, K.-S., Chan, L., Meng, H. (eds.) IDEAL 2000. LNCS, vol. 1983, pp. 23–28. Springer, Heidelberg (2000)
12. Huang, Z., Lin, T.: A visual method of cluster validation with fastmap. In: Terano, T., Chen, A.L.P. (eds.) PAKDD 2000. LNCS, vol. 1805, pp. 153–164. Springer, Heidelberg (2000)
13. Blockeel, H., Raedt, L., Ramong, J.: Top-down induction of clustering trees. In: Proceedings of the 15th International Conference on Machine Learning, pp. 55–63 (1998)

Query Selection via Weighted Entropy in Graph-Based Semi-supervised Classification

Krikamol Muandet[1], Sanparith Marukatat[2], and Cholwich Nattee[1]

[1] School of Information, Computer and Communication Technology
Sirindhorn International Institute of Technology
Thammasat University, 131 Tiwanont Rd. Bangkadi
Pathum Thani, Thailand 12000
[2] National Electronic and Computer Technology Center
National Science and Technology Development Agency
111 Thailand Science Park
Pathum Thani, Thailand 12120

Abstract. There has recently been a large effort in using unlabeled data in conjunction with labeled data in machine learning. Semi-supervised learning and active learning are two well-known techniques that exploit the unlabeled data in the learning process. In this work, the active learning is used to query a label for an unlabeled data on top of a semi-supervised classifier. This work focuses on the query selection criterion. The proposed criterion selects the example for which the label change results in the largest pertubation of other examples' label. Experimental results show the effectiveness of the proposed query selection criterion in comparison to existing techniques.

Keywords: Graph-Based Semi-supervised Learning, Active Learning.

1 Introduction

There has recently been a large effort in using unlabeled data in conjunction with labeled data in machine learning. Indeed, in supervised learning, a huge amount of labeled data, e.g., manual annotation of legitimate and spam emails, is needed to train an accurate classifier. Labeled examples are however expensive and difficult to obtain since they require experienced annotators. Unlabeled instances, on the other hand, are easy to collect. Although they are usually useless in traditional supervised learning, unlabeled data may contain relevant information that can produce considerable improvement in learning accuracy. This is why several works [1,2,3,4] focus on how to exploit both labeled and unlabeled data to train the classifier. Semi-supervised learning and active learning are two well-known settings that have been considered to incorporate both kinds of data in learning process.

Semi-supervised learning has drawn large attention in the past decade due to its significance in many real-world problems. With few labeled examples, the goal of this learning method is then to create as accurate as or even better classifier than what we would obtain from traditional supervised learning. Many

Z.-H. Zhou and T. Washio (Eds.): ACML 2009, LNAI 5828, pp. 278–292, 2009.
© Springer-Verlag Berlin Heidelberg 2009

successful techniques for semi-supervised learning have been proposed in various frameworks, e.g., generative models [5,6], Gaussian processes [2,7,8], and information regularizations [9,10]. Amongst these tecniques, semi-supervised learning on graph has received the most attention, as shown by the number of works on this subject including graph mincut [11,12], learning with local and global consistency [13], and manifold regularizations [14,15], for example. Interested readers should consult [16] for an extensive literature review of semi-supervised learning.

The goal of active learning coincides with that of semi-supervised learning, which aims at reducing the use of labeled data. The interesting aspect of active learning is that the learning system actively queries the instance whose label will be assigned by human annotators. As a result, the number of labeled examples required is usually less than what would actually be when learned using normal supervised learning. Active learning has been studied for many real-world problems such as text classification [17,18]. Further information on active learning can be obtained from [19].

In this work, we propose the novel algorithm based on the combination of active learning and semi-supervised classification. More precisely, the system actively queries an instance from a pool of unlabeled instances. The human annotator will give the true label of this example. Then labeled examples along with the rest of unlabeled instances are used in standard semi-supervised classification. Due to its capability of evaluating the model successively, this learning framework becomes more effective than traditional semi-supervised learning processes. In this work we introduce a novel evaluative measure called *weighted entropy* that is used as a criterion for active query selection on top of graph-based semi-supervised classification.

This paper is organized as follows. Section 2 briefly reviews the related works. Next, we discuss the problem of graph-based semi-supervised learning in Sec. 3. The query selection algorithm based on the weighted entropy is then presented in Sec. 4. Experimental results are shown in Sec. 5, followed by conclusions in Sec. 6.

2   Related Works

All active learning techniques focus on how to find the optimal query, which generally involves evaluating the informativeness of unlabeled examples. There have been many proposed querying strategies that work well in practice. Random sampling [20] is the simplest querying strategy. This strategy is often used for preliminary testing of learning algorithms in active learning since it is very easy to implement. However, random method is not very effective in practice. For example, if the dataset contains much less number of positive examples than negative examples, then it is likely that a random sampling will select less positive examples than negative examples. As a result, Lewis and Gale [17] proposed a querying framework called uncertainty sampling. In this framework, the algorithm asks an annotator to label those examples whose class membership is the most uncertain. This strategy significantly reduces the amount of training data that would have to be manually labeled to achieve a desired level of accuracy. However, uncertainty sampling is prone to query outliers, which may not be "representative" of other examples in the distribution [21,22].

Due to this problem, the estimated error reduction framework has been proposed. It prefers the queries that minimize the expected future error of the learning system. Roy and McCallum applied this framework in text classification using Naive Bayes [21]. Zhu et al. [22] combined it with a semi-supervised classification. Both techniques result in an improvement over both random and uncertainty sampling. However, the estimated error reduction may be the most expensive query selection framework for many model classes since it requires estimating the expected future error over the set of unlabeled data for each query. This usually means that a new model must be incrementally trained for each possible query, leading to an increase in computational cost for some models, such as a logistic regression models and neural networks. Fortunately, this is not the case for graph-based semi-supervised classification with Gaussian random field models [22] for which the incremental training technique is exact and efficient. Therefore, combining active learning with semi-supervised classification guarantees that this approach will be fairly practical.

3 Graph-Based Semi-supervised Classification

Graph-based semi-supervised classification methods utilize a weighted graph whose nodes represent labeled and unlabeled data points. The weighted edges reflect the similarities between nodes. Most existing algorithms are based on the assumption that the labeling is smooth on the graph, i.e., similar data points tend to have similar labels. This is called the *smoothness assumption* [3].

Let $\mathcal{X} = \{x_1, \ldots, x_l, x_{l+1}, \ldots, x_n\} \subset \mathbb{R}^m$ be a set of n data points. The first l data points are labeled as $y = [y_1, y_2, \ldots, y_l]^{\mathrm{T}}$ with $y_i \in Y = \{0, 1\}$. The rest of $u = n - l$ data points are initially unlabeled. Let L and U denote sets of labeled and unlabeled examples, respectively. The goal of semi-supervised learning is to utilize labeled data together with unlabeled data to construct a classifier.

The graph is represented by a matrix $W = [w_{ij}]_{n \times n}$. The non-negative edge weight w_{ij} between node i, j is computed as

$$w_{ij} = \exp(-\sum_{d=1}^{m} (x_{i,d} - x_{j,d})^2 / \sigma_d^2) \ , \tag{1}$$

where $x_{i,d}$ is the d-th component of x_i and σ_d is the bandwidth hyperparameters for the dimension d.

In general, graph-based semi-supervised classification technique searches for a real-valued function f on graph and then assigns labels based on f. Given the weight matrix W and the real-valued function f, the inconsistency on graph can be define according to the smoothness assumption as

$$E(f) = \sum_{i,j=1}^{n} w_{ij}(f(i) - f(j))^2 \, , \tag{2}$$

where $f(i)$ and $f(j)$ are the function values evaluated on node i and j, respectively. For a labeled example $1 \le i \le l$, $f(i)$ is considered fixed to y_i. The inconsistency term (2) can also be written in quadratic form $f^{T}\Delta f$, where $f = [f(1), .., f(n)]^{T}$ and Δ is known as the combinatorial graph Laplacian matrix defined as $\Delta = D - W$, where the matrix D is a diagonal matrix whose entries $D_{ii} = \sum_{j=1}^{n} w_{ij}$ is the degree of node i. Minimization of $E(f)$ forces f to take values y_i on labeled data points and varies smoothly on unlabeled data points in accordance with the weight matrix W. It is not difficult to show that the optimal function is given by:

$$f_U = (D_{UU} - W_{UU})^{-1}W_{UL}f_L = -\Delta_{UU}^{-1}\Delta_{UL}f_L \, , \tag{3}$$

where f_U is a vector of function values evaluated on all unlabeled data points, $f_L = [y_1, y_2, \ldots, y_l]^{T}$, and $f = [f_L; f_U]$. All related matrices are defined as follows:

$$W = \begin{pmatrix} W_{LL} & W_{LU} \\ W_{UL} & W_{UU} \end{pmatrix}, \Delta = \begin{pmatrix} \Delta_{LL} & \Delta_{LU} \\ \Delta_{UL} & \Delta_{UU} \end{pmatrix} \, .$$

The functional (3) is called the *soft-label* function since its values do not directly specify the class membership of unlabeled examples, but can be interpreted as a probability of being in each class. The most obvious method to transform soft-label to hard-label is by thresholding, e.g., classify x_i, $l \le i \le n$, as being in class 1 if $f(i) > 0.5$, and in class 0 otherwise. This method generally works well when the classes are well-separated. However, this is generally not the case in many practical applications. In such cases, using simple thresholding may result in an unbalanced classification.

Class Mass Normalization (CMN) is another method to transform the soft-label to hard-label [3]. The class distribution of the data is adjusted to match the class priors, that can be obtained from the labeled examples. For example, if the prior class proportion of class 1 and 0 is p and $1 - p$, respectively, then an unlabeled examples x_k is classified as class 1 iff $p \cdot (f(k)/\sum_i f(i)) > (1 - p) \cdot ((1 - f(k))/\sum_i(1 - f(i)))$. This method works well when we have sufficient labeled examples to determine the class prior that accurately represents the true class distribution.

4   Weighted Entropy

We propose a novel technique to perform active learning with graph-based semi-supervised learning. This active learning technique selects queries from the unlabeled data by considering their influence on other available examples. Each unlabeled example is evaluated by looking at the characteristics of the overall soft-label function when its label is altered. That is, we prefer the example that if its label is changed, will result in (1) the large change in soft-label values of other unlabeled examples that can probably alter their class membership and (2) the large change in soft-label values of other unlabeled examples that rarely changes the class membership of other examples, but increases the confidence of the labeling function. We propose *weighted entropy* as an evaluation function which can simultaneously take both criteria into account. The criterion function is derived from the harmonic energy minimization function with the Gaussian random field model [3].

4.1 Problem Formulation

In graph-based setting, we need to know a soft-label function when different labels are assigned to unlabeled node k to evaluate its impact to other unlabeled examples. Thus it is necessary to define an efficient update of soft-label function after knowing one more label. Following the derivation in [22], we add one new node with value f_0 to the graph. The new node is connected to unlabeled node k with weight w_0. Thus, as $w_0 \to \infty$, we effectively assign label f_0 to node k.

Note that the harmonic energy minimization function is $f_U = -\Delta_{UU}^{-1}\Delta_{UL}f_L = (D_{UU}-W_{UU})^{-1}W_{UL}f_L$. After adding the new node to the graph, let the matrices $\tilde{D}_{UU}, \tilde{W}_{UL}$, and $\tilde{f}_L$ be the updated versions of D_{UU}, W_{UL}, and f_L, respectively. Since the new node is a labeled node in the graph, the labeling function can be updated as

$$
\begin{aligned}
\tilde{f}_U &= (\tilde{D}_{UU} - W_{UU})^{-1}\tilde{W}_{UL}\tilde{f}_L \\
&= (w_0 e_k e_k^{\mathrm{T}} + D_{UU} - W_{UU})^{-1}(w_0 f_0 e_k + W_{UL}f_L) \\
&= (w_0 e_k e_k^{\mathrm{T}} + \Delta_{UU})^{-1}(w_0 f_0 e_k + W_{UL}f_L) ,
\end{aligned}
\tag{4}
$$

where e_k is a column vector with 1 in position k and 0 elsewhere. Using matrix inversion lemma, we obtain

$$
\begin{aligned}
(w_0 e_k e_k^{\mathrm{T}} + \Delta_{UU})^{-1} &= \Delta_{UU}^{-1} - \frac{\Delta_{UU}^{-1}(\sqrt{w_0}e_k)(\sqrt{w_0}e_k)^{\mathrm{T}}\Delta_{UU}^{-1}}{1 + (\sqrt{w_0}e_k)^{\mathrm{T}}\Delta_{UU}^{-1}(\sqrt{w_0}e_k)} \\
&= \mathcal{L} - \frac{w_0\mathcal{L}_{|k}\mathcal{L}}{1 + w_0\mathcal{L}_{kk}}
\end{aligned}
\tag{5}
$$

where we use $\mathcal{L}$ to denote Δ_{UU}^{-1} and $\mathcal{L}_{|k}$ is a square matrix with $\mathcal{L}$'s k-th column and 0 elsewhere. With some calculations, we derive (5) into

$$
\tilde{f}_U = f_U + \frac{w_0 f_0 - w_0 f(k)}{1 + w_0\mathcal{L}_{kk}}\mathcal{L}_{.k}
\tag{6}
$$

where $f(k)$ is the soft-label of unlabeled node k, and $\mathcal{L}_{.k}$ is the k-th column vector in $\mathcal{L}$. To force the label at node k to be f_0, we let $w_0 \to \infty$ to obtain

$$
\tilde{f}_U = f_U + \frac{f_0 - f(k)}{\mathcal{L}_{kk}}\mathcal{L}_{.k} .
\tag{7}
$$

Consequently, we can now formulate the equation to compute the soft-label function after knowing the label of a particular example. The functional (7) can be used to compute the possible soft-label functions after assigning different labels to example x_k. For binary classification in which $Y = \{0, 1\}$, we obtain

$$f_U^{(x_k,y)} = f_U + (y - f(k))\frac{(\Delta_{UU}^{-1})_{\cdot k}}{(\Delta_{UU}^{-1})_{kk}} \quad , \tag{8}$$

where $f_U^{(x_k,y)}$ is the soft-label function after we assign label $y \in Y$ to unlabeled examples x_k.

To derive the evaluation function, we start by defining the uncertainty $I(f_{U,i})$ of the soft-label value at node i. The total uncertainty at node i when we assign different labels to example x_k is $\sum_{y \in Y} I(f_{U,i}^{(x_k,y)})$. This work uses entropy function to measure this uncertainty, i.e., $I(f_{U,i}) = -f_{U,i} \log_2 f_{U,i}$. By weighting each uncertianty term with the probability of x_k belonging to each class, we can write the evaluation function as

$$\mathcal{C}(k) = \sum_{i=1}^{u} \sum_{y=0,1} p(y_k = y|L) I(f_{U,i}^{(x_k,y)})$$

$$= \sum_{i=1}^{u} \sum_{y=0,1} -p(y_k = y|L) \left[f_{U,i}^{(x_k,y)} \log_2 f_{U,i}^{(x_k,y)} \right] \quad . \tag{9}$$

This quantity is called the *weighted entropy* measure. Given a set of label data L, $p(y_k = y|L)$ is the true label distribution at example x_k. This term makes $\mathcal{C}(k)$ not computable, however it can still be estimated by the soft-label value, i.e. , $p(y_k = 1|L) \approx f_{U,k}$ and $p(y_k = 0|L) \approx 1 - f_{U,k}$. As a result, the *estimated weighted entropy* is defined as

$$\hat{\mathcal{C}}(k) = \sum_{i=1}^{u} -f_{U,k} \left[f_{U,i}^{(x_k,1)} \log_2 f_{U,i}^{(x_k,1)} \right] - (1 - f_{U,k}) \left[f_{U,i}^{(x_k,0)} \log_2 f_{U,i}^{(x_k,0)} \right] \quad . \tag{10}$$

The weighted entropy measures how much the selected query affects the labels of other unlabeled examples when its own label is changed. Note that the soft-label function determines the probability of examples being in positive class, i.e., $y_i = 0$ if $f(i) \le 0.5$ and $y_i = 1$ otherwise.

4.2   Analysis of Weighted Entropy

In this work, the query selection is based on the influence when its label is changed. Given a particular example x_k, the behavior of weighted entropy can be analyzed as follows:

1. The value of weighted entropy is minimized when the expected soft-label values $f_U^{(x_k,0)}$ and $f_U^{(x_k,1)}$ have large difference, e.g., $f_U^{(x_k,1)}$ is close to 1

and $f_U^{(x_k,0)}$ is close to 0, or vice versa. This means that the example x_k has a significant effect on the soft-label values of other unlabeled examples. Furthermore, such effect will likely cause the alternation of class membership of most examples. As a result, this suggests that we should know the true label of x_k as early as possible.

2. The value of weighted entropy is also minimized when both $f_U^{(x_k,1)}$ and $f_U^{(x_k,0)}$ become closer to the boundary value of the soft-label function. This means that even if changing the label of x_k does not alter labels of other examples but if it results in more confident labeling function, then x_k will be considered as an important example by weighted entropy measure. In this case, we can see that the formulation of estimated weighted entropy (10) is similar to the expected estimated risk defined in [22]. The expected estimated risk is computed by summing a weighted estimated risks of $f_U^{(x_k,y)}$ using $\min(f_U^{(x_k,y)}, 1 - f_U^{(x_k,y)})$. In contrast, by using the weighted entropy, the proposed method focuses on the behavior of soft-label values of unlabeled data when the label of queried example is changed. Therefore, this method can effectively exploit necessary information needed to evaluated unlabeled data.

It is worth mentioning that we expect these two cases to happen in different periods during the query selection process. It is obvious that the queries corresponding to the first case will likely become the most preferable candidates at the beginning of the process, when labels are prone to change. After some times when labels become more resistant to change, the most preferable candidates will fall into the second case, which tends to improve the confidence of the classifier. Therefore, selecting the queries based on the proposed criterion assures that examples that have negative effects on the performance as a result of learning process will be discovered as early as possible. Consequently, the query selection will become safer for the subsequent learning process. The final query selection criterion (11) called *Minimum Weighted Entropy* (MinWE) for query selection is shown below.

Minimum Weighted Entropy
For a set of unlabeled examples U, select an example x_k resulting in

$$k = \arg\min_{k'} \; \hat{C}(k') \; , \tag{11}$$

where $\hat{C}(k')$ is the estimated weighted entropy of $x_{k'}$.

Algorithm 1 summarizes the active query selection using minimum weighted entropy. Note that in each iteration we need to update the inverse graph Laplacian with the row/column for x_k removed, whose computational complexity is $\mathcal{O}(u^3)$ in general. To avoid the computational cost of the matrix inversion, the matrix inversion lemma is applied to compute this matrix from Δ_{UU} and Δ_{UU}^{-1} (see appendix B of [22] for the derivation). As a result, the overall time complexity of the proposed algorithm is $\mathcal{O}(u^2)$.

Algorithm 1. Active Query Selection with Minimum Weighted Entropy

---
1: Choose first r examples randomly.
2: **while** Need more queries **do**
3:     Update the inverse graph Laplacian Δ_{UU}^{-1}.
4:     Compute $\hat{C}(k)$ using (10) for all $x_k \in U$.
5:     Query x_k according to (11).
6:     Receive the answer y_k.
7:     Add (x_k, y_k) into L and remove x_k from U.
8: **end while**

---

5   Experimental Results

5.1   Experimental Setup

In our experiments, Random, MaxUncertain, MinRisk, and MinWE query selection methods are evaluated on different datasets, namely, handwritten digits dataset[1] and benchmark datasets for semi-supervised learning[2]. All experiments are performed in the transductive setting, i.e., test data coincides with training data used in the learning process. In each experiment, the first two labeled examples are chosen by random selection. Since random initialization can consequentially influence the results, each experiment is repeated 10 times and the average accuracy as well as its ± 1 standard deviation are reported.

In Random method, the next instance is queried randomly from a set of available unlabeled instances. In contrast, the MaxUncertain method queries an unlabeled instance whose soft-label value is closest to 0.5, meaning that the instance possesses the highest uncertainty on its true label. The MinRisk method selects an instance that minimizes the expected estimated risk defined in [22]. Similarly, the MinWE method queries an instance that minimizes the estimated weighted entropy proposed in this work. For comparison, the classification accuracy attained by each method is evaluated on the unlabeled instances at each iteration after adding the new query to the set of labeled examples.

5.2   Overview of Results

Figure 1 shows the accuracy and weighted entropy values on handwritten digits dataset as increasing number of labeled examples are acquired using different query selection methods. The task is to classify the digit "0" against the digit "8". The digits are 16×16 grid, with pixel values ranging from 0 to 255. Thus each digit is represented by a 256-dimensional vector. The dataset consists of 1,080 images, with 540 images randomly selected from each class. All data points and their pairwise similarities are represented by a fully connected graph with weights $w_{ij} = \exp(-d_{ij}^2/0.25)$, where d_{ij} is the Euclidean distance between x_i and x_j.

---

[1] http://www-i6.informatik.rwth-aachen.de/~keysers/usps.html
[2] http://www.kyb.tuebingen.mpg.de/ssl-book/benchmarks.html

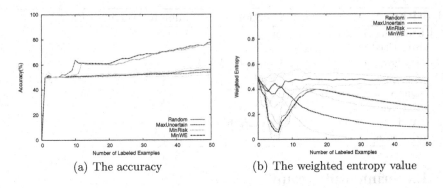

(a) The accuracy   (b) The weighted entropy value

Fig. 1. (a) Accuracy and (b) the weighted entropy values on the handwritten digits dataset("0" versus "8" classification problem) of four different query selection methods

As illustrated in Fig.1(a), `Random` and `MaxUncertain` methods achieve approximately the same level of accuracy, which increases very slowly as the system acquires more labeled examples. In contrast, the accuracy attained by `MinRisk` and `MinWE` methods rises very rapidly. Although they lead to roughly the same results in the later stage of the learning process, the `MinWE` method discloses the informative queries earlier than the `MinRisk` method as depicted at the 10th query in Fig.1(a).

Fig.1(b) shows the weighted entropy value versus the number of labeled examples. The `MinRisk` and `MinWE` methods generate the similar trend of weighted entropy values, which tend to decrease with the increasing number of labeled examples. Though exhibiting the similar criteria, the `MinWE` method produces a slighly lower values of weighted entropy than the `MinRisk` method. Another interesting point in this experiment is the weighted entropy values of `MaxUncertain` method that decrease substantially. According to Fig.3, this case usually occurs when the values of soft-label are very close to either 0 or 1 because the majority of queries come from the same class. This degenerate case of `MaxUncertain` method primarily leads to low accuracy.

Note that an advantage of `MinWE` method over `MinRisk` method is the ability to initially identify important queries that have significant impact on the soft-label value of other unlabeled examples if their labels are changed. Thus it follows immediately that imposing the true labels on these examples lessens the label alteration in the following stages. The `MinRisk` method choose queries based on how much they will improve the expected accuracy. It does not take into account the fact that this improvement can change some of influential examples' labels that probably degrades the true accuracy.

Figure 2 supports our previous claim on the advantages of `MinWE` method. The percentage of label changes in a set of unlabeled examples of handwritten digits dataset is calculated at each iteration after the next query is obtained. Therefore, this figure nearly reflects the reliability of each query selection method, i.e., the reliable method rarely alters the labels of examples compared to unreliable

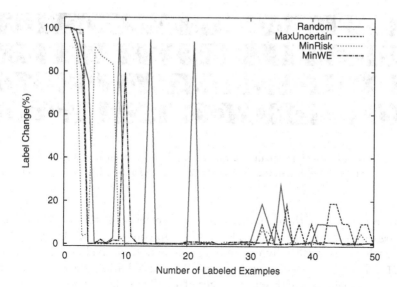

Fig. 2. The percentage of labels on unlabeled examples that change as the increasing number of labeled examples is obtained in handwritten digits dataset

ones. Intuitively, the `Random` method is unreliable because the labels frequently change throughout the learning process. Nonetheless, the `MinRisk` and `MinWE` methods are more reliable since the variation of labels occurs essentially in the initial stages of the learning process and then becomes less likely to occur in the subsequent stages. Moreover, although `MinRisk` method substantially lessens the generalization error and the the alteration of labels in the beginning, the correctness of labeling function at some data points may not be guaranteed. As a result, obtaining more label examples likely changes labels of some examples as illustrated in Fig. 2 when there are approximately 8, 32, and 48 labeled examples. Note that, on the other hand, the labels of most influential queries are assigned in the initial stages by `MinWE` method. Therefore, the variation of labels in the subsequent stages is minimized, leading to more reliable labeling function.

Another issue we want to address in this work is the meaningfulness of the queries selected by each methods. It is worthwhile to mention because it also affects the performance of the classifier. If the selected queries are ambiguous, it is with high chance that the assigned labels will be incorrect and knowing their labels will not provide any useful information. This problem may not be recognized in easy tasks such as handwritten recognition or face detection, for example, but it becomes more realistic when we handle the complicated problems, in which it is not convenient to visualize the data. Figure 3 illustrates the 20 most frequently queried instances across all trials. Each row shows the most frequent queries, sorted by the number of times they are selected.

Figure 4 shows the results of the experiments on benchmark datasets for semi-supervised learning. The original benchmark consists of eight datasets, but in this

Fig. 3. Top twenty most frequent queries of the handwritten digits obtained during the learning process using `Random`, `MaxUncertain`, `MinRisk`, and `MinWE` methods

work we use only six of them to assess the query selection methods[3]. The datasets are categorized into two groups. The first group consists of `g241c`, `g241n`, and `Digit1`, which are artificially created. The second group includes `USPS`, `COIL2`, and `BCI`, all of which are derived from real data. In addition, the classes of `USPS` dataset are imbalanced with relative sizes of 1:4. Thus the performance on these data sets is the indication of the performance in the real applications. See [23] for the detail of each dataset. In this experiment, we construct a weighted k nearest neighbors graph with weights $w_{ij} = \exp(-d_{ij}^2/2\sigma)$ if it is an edge between x_i and x_j, and 0 otherwise. For all datasets, $k = 5$ and σ is fixed as the median of the pairwise distance between adjacent nodes on the graph.

Figure 4 confirms an effectiveness of the proposed query selection method. As can be seen in figure, the `MinWE` method outperforms other query selection methods and is slighly better than `MinRisk` method in almost all datasets, except `Digit1` dataset. Other than the superior experimental results on artificial datasets such as `g241c` and `g241n`, `MinWE` method also achieve the highest accuracy in `USPS`, `COIL2`, and `BCI` datasets that are obtained from the real data. This is therefore the indication of expected classification performance of `MinWE` method in practice.

In almost all datasets, the `MaxUncertain` query selection method possesses the lowest accuracy among all other methods. In some datasets, e.g., `USPS`, labeling the most uncertain unlabeled example may harm the accuracy as illustrated in Fig. 4(b). This underlines the fact that in semi-supervised learning, more labeled examples does not mean higher accuracy. Intuitively, the uncertainty on class membership of a particular instance does not adequately indicate the informativeness of the instance. The evaluation based on this measure suggests only that the instance is hard to classify, but does not imply how, after knowing its label, the instance will influence the rest of unlabeled instances. Consequently,

---

[3] The `g241c`, `g241n`, `Digit1`, `USPS`, `COIL2`, and `BCI` datasets are used extensively to assess the performance of several semi-supervised learning techniques, whereas few techniques utilize `Text` and `SecStr` datasets, which have special characteristics. Therefore, this work considers only those six datasets for the convenience of the experiments and the reliability of the results.

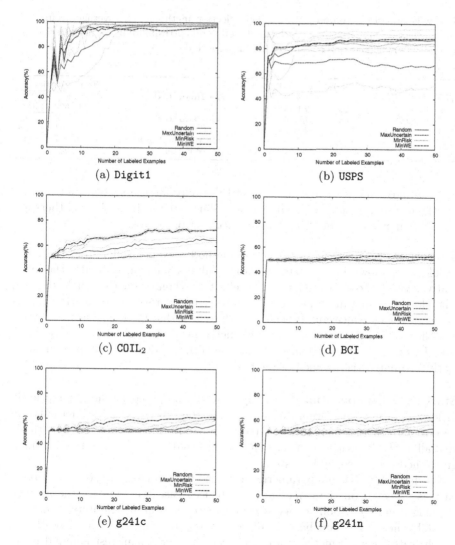

Fig. 4. The accuracy of different query selection methods on benchmark dataset for semi-supervised learning. The selected datasets includes (a) `Digit1`, (b) `USPS`, (c) `COIL`$_2$, (d) `BCI`, (e) `g241c`, and (f) `g241n`. The figure presents an average accuracy and its ±1 standard deviation.

in practice, we recommend to use `MaxUncertain` only for preliminary testing, especially in the difficult problems, for the following reasons:

1. The uncertainty on class membership does not suggest, neither directly nor indirectly, the influence of the queried instance on the labels of other instances. Thus providing its label may not considerably improve, but probably diminish the predictive accuracy.

Table 1. Statistical comparison of query selection methods. The average accuracy of query selection methods on each data set (left) and corresponding results of Friedman and Wilcoxon signed ranks tests (right) are reported.

	Random	MaxUncertain	MinRisk	MinWE		Test Statistics			
Digit1	88.49	93.04	93.57	92.72	**Friedman Test**	N	Chi-Square	df	Sig.
USPS	83.71	68.31	81.48	86.23		6	11.600	3	.009
COIL2	59.06	51.68	65.96	70.44	**Wilcoxon Test**			Z	Sig.
BCI	50.34	49.91	51.72	52.71	MinWE-Random			-2.201	.028
g241c	51.67	50.22	52.30	59.48	MinWE-MaxUncertain			-1.992	.046
g241n	51.41	50.92	52.01	59.38	MinWE-MinRisk			-1.992	.046

2. The most uncertain instance is hard not only for the system, but also for the human annotators to categorize as illustrated in Fig.3. Hence, there is a high chance that the given label is incorrect or useless.

Fortunately, the MinWE query selection method does not suffer from these two problems as shown by the experimental results. That is, it evaluates the informativeness of those instances by the effect they make on the whole dataset. Therefore, labeling more instances guarantee to improve the predictive performance of the classification. It is also worth to note that the MinWE achieve better classification results than the MinRisk method even though its evaluative measure does not directly take into account the accuracy of classification, compared to the estimated risk defined in the MinRisk.

Statistical Comparison. To justify that the proposed method significantly yields an improve performance over the existing query selection methods, we need to perform a proper statistical test over all datasets with the null hypothesis that all methods perform equally well. As suggested by [24] we used the Friedman and Wilcoxon signed ranks tests.

Friedman test is a nonparametric test used to compare three or more observations repeated on the same subjects. In this case, we compare the average learning accuracy of four query selection methods over six datasets to inspect the difference in medians between different methods. Under the null hypothesis, which states that all methods are equivalent, the Friedman test is found to be significant $\chi^2(3, N = 6) = 11.6$ and $p < .01$, as shown in Table 1. This merely indicates the differences in learning accuracy among the four methods.

Next, after obtaining a significant Friedman test, the follow-up tests need to be conducted to evaluate comparisons between pairs of query selection methods. As indicated earlier, we use the Wilcoxon signed ranks test. Since we are only interested in whether the proposed method yields an improved performance over the existing ones, the pairwise comparisons between the MinWE method and the others, namely, Random, MaxUncertain, and MinRisk are conducted. The null hypothesis for each comparison states that there is no difference in learning accuracy between two methods, whereas alternative hypothesis states that the first method, i.e., MinRisk, gives higher accuracy than the second one. According to the result of Wilcoxon test in Table 1, all three comparisons are significant at the .05 alpha level, leading us to reject the null hypothesis and conclude that the MinWE method significantly outperforms Random, MaxUncertain, and MinRisk methods.

6 Conclusions

This paper proposes a new query selection criterion for active learning. The proposed criterion, called minimum weighted entropy, selects the example for which the label change results in the largest pertubation of other examples' label. It relies on a graph-based semi-supervised learning technique to efficiently compute the weighted entropy for the selection process. Experimental results show the advantage of the proposed selection criterion over the existing criterion on several datasets.

Acknowledgments. This work is supported in part by the Young Scientist and Technologist Programme or YSTP (SIIT-NSTDA:S1Y48/F-002) of the National Science and Technology Development Agency, Thailand.

References

1. Krishnapuram, B., Williams, D., Xue, Y., Hartemink, A.J., Carin, L., Figueiredo, M.A.T.: On semi-supervised classification. In: NIPS (2004)
2. Zhu, X., Ghahramani, Z.: Semi-supervised learning: From gaussian fields to gaussian processes. Technical report, School of CS, CMU (2003)
3. Zhu, X., Ghahramani, Z., Lafferty, J.: Semi-supervised learning using gaussian fields and harmonic functions. In: ICML, pp. 912–919 (2003)
4. Minton, S., Knoblock, C.A.: Active + semi-supervised learning = robust multi-view learning. In: Proceedings of ICML 2002, 19th International Conference on Machine Learning, pp. 435–442 (2002)
5. Nigam, K., Mccallum, A., Thrun, S., Mitchell, T.: Text classification from labeled and unlabeled documents using em. Machine Learning, 103–134 (1999)
6. Baluja, S.: Probabilistic modeling for face orientation discrimination: learning from labeled and unlabeled data. In: Proceedings of the 1998 conference on Advances in neural information processing systems II, pp. 854–860. MIT Press, Cambridge (1999)
7. Lawrence, N.D., Jordan, M.I.: Semi-supervised learning via gaussian processes. In: Saul, L.K., Weiss, Y., Bottou, L. (eds.) Advances in Neural Information Processing Systems 17, pp. 753–760. MIT Press, Cambridge (2005)
8. Chu, W., Sindhwani, V., Ghahramani, Z., Keerthi, S.S.: Relational learning with gaussian processes. In: Schölkopf, B., Platt, J., Hoffman, T. (eds.) Advances in Neural Information Processing Systems 19, pp. 289–296. MIT Press, Cambridge (2007)
9. Szummer, M., Jaakkola, T.: Information regularization with partially labeled data. In: Advances in Neural Information Processing Systems 15. MIT Press, Cambridge (2003)
10. Tommi, A.C., Jaakkola, T.: On information regularization. In: Proceedings of the 19th UAI (2003)

11. Blum, A., Lafferty, J., Rwebangira, M.R., Reddy, R.: Semi-supervised learning using randomized mincuts. In: ICML 2004: Proceedings of the twenty-first international conference on Machine learning, p. 13. ACM, New York (2004)
12. Blum, A., Chawla, S.: Learning from labeled and unlabeled data using graph mincuts. In: ICML 2001: Proceedings of the Eighteenth International Conference on Machine Learning, pp. 19–26. Morgan Kaufmann Publishers Inc., San Francisco (2001)
13. Zhou, D., Bousquet, O., Lal, T.N., Weston, J., Schölkopf, B.: Learning with local and global consistency. In: Advances in Neural Information Processing Systems 16, pp. 321–328. MIT Press, Cambridge (2003)
14. Belkin, M., Matveeva, I., Niyogi, P.: Regularization and semi-supervised learning on large graphs. In: Shawe-Taylor, J., Singer, Y. (eds.) COLT 2004. LNCS (LNAI), vol. 3120, pp. 624–638. Springer, Heidelberg (2004)
15. Belkin, M., Niyogi, P., Sindhwani, V.: Manifold regularization: A geometric framework for learning from labeled and unlabeled examples. Journal of Machine Learning Research 7, 2399–2434 (2006)
16. Zhu, X.: Semi-supervised learning literature survey. Technical Report 1530, Computer Sciences, University of Wisconsin-Madison (2005), http://www.cs.wisc.edu/~jerryzhu/pub/ssl_survey.pdf
17. Lewis, D.D., Gale, W.A.: A sequential algorithm for training text classifiers. In: SIGIR 1994: Proceedings of the 17th annual international ACM SIGIR conference on Research and development in information retrieval, pp. 3–12. Springer-Verlag New York, Inc, New York (1994)
18. Mccallum, A.K.: Employing em in pool-based active learning for text classification. In: Proceedings of the 15th International Conference on Machine Learning, pp. 350–358. Morgan Kaufmann, San Francisco (1998)
19. Settles, B.: Active learning literature survey. Computer Sciences Technical Report 1648, University of Wisconsin–Madison (2009)
20. Cochran, W.G.: Sampling Techniques. John Wiley and Sons, Chichester (1977)
21. Roy, N., Mccallum, A.: Toward optimal active learning through sampling estimation of error reduction. In: Proc. 18th International Conf. on Machine Learning, pp. 441–448. Morgan Kaufmann, San Francisco (2001)
22. Zhu, X., Lafferty, J., Ghahramani, Z.: Combining active learning and semi-supervised learning using gaussian fields and harmonic functions. In: ICML 2003 workshop on The Continuum from Labeled to Unlabeled Data in Machine Learning and Data Mining, pp. 58–65 (2003)
23. Chapelle, O., Schölkopf, B., Zien, A. (eds.): Semi-Supervised Learning. MIT Press, Cambridge (2006)
24. Demšar, J.: Statistical comparisons of classifiers over multiple data sets. J. Mach. Learn. Res. 7, 1–30 (2006)

Learning Algorithms for Domain Adaptation

Manas A. Pathak and Eric H. Nyberg

Language Technologies Institute
School of Computer Science
Carnegie Mellon University
Pittsburgh, PA 15213, USA
{manasp,ehn}@cs.cmu.edu

Abstract. A fundamental assumption for any machine learning task is to have training and test data instances drawn from the same distribution while having a sufficiently large number of training instances. In many practical settings, this ideal assumption is invalidated as the labeled training instances are scarce and there is a high cost associated with labeling them. On the other hand, we might have access to plenty of labeled data from a different domain, which can provide useful information for the present domain. In this paper, we discuss adaptive learning techniques to address this specific problem: learning with little training data from the same distribution along with a large pool of data from a different distribution. An underlying theme of our work is to identify situations when the auxiliary data is likely to help in training with the primary data. We propose two algorithms for the domain adaptation task: dataset reweighting and subset selection. We present theoretical analysis of behavior of the algorithms based on the concept of domain similarity, which we use to formulate error bounds for our algorithms. We also present an experimental evaluation of our techniques on data from a real world question answering system.

1 Introduction

In machine learning tasks, it is assumed that the labeled *primary training data* is similar to the test data in order to expect good accuracy on the test data. Further, it is important to have training data of a sizable amount in order to build a reliable model for classification. In practice, while it is usually time-consuming and expensive to acquire labeled instances which are similar to the test data, we might often have plenty of labeled data from an *auxiliary source* which is somewhat different from our test data. Despite this overall difference between the datasets, there may be parts of the auxiliary data that are similar and thus useful. More specifically, here we explore algorithms based on the assumption that the auxiliary data distribution is a mixture between the primary data distribution and a different distribution. In NLP tasks such as named-entity recognition, parsing, text classification, etc. we usually have plenty of data from standard text corpora (e.g. Penn Treebank (1)) but little data for specialized genres of text we might be interested in processing. In user-centric tasks such as spam detection, handwriting/voice recognition, etc., there is usually little labeled data for each individual user who is using the system while there is a large amount of labeled data for all users combined.

Z.-H. Zhou and T. Washio (Eds.): ACML 2009, LNAI 5828, pp. 293–307, 2009.

A domain adaptation learning setting involves a *primary domain* $\mathcal{D}_P$ and an *auxiliary domain* $\mathcal{D}_A$. We mostly consider the problem of learning with two domains, although it is possible to generalize to handle multiple domains (2). We denote the datasets sampled from primary and auxiliary domains as $(x_1^p, y_1^p), ...(x_{N_P}^p, y_{N_P}^p)$ and $(x_1^a, y_1^a), ...(x_{N_A}^a, y_{N_A}^a)$, where N_P and N_A are the sample sizes of the primary and auxiliary datasets respectively. A learning algorithm takes labeled training instances sampled from the two domains and estimates a labeling function for the primary domain. The test dataset for evaluating the learning algorithm is sampled from the primary domain. The central issue in such a learning setting is that we have few primary training instances and relatively large number of auxiliary training instances. The labeled primary instances can be thought of having a high *cost* associated with them as opposed to auxiliary instances, which have a lower cost. If we had large number of primary instances to start with, we could perform learning well enough with standard supervised learning approaches and would not need the auxiliary data. As it is difficult to learn a good labeling function with a small primary training dataset, we aim to make use of auxiliary data for learning. In fact, identifying situations when using auxiliary data with primary data helps in training is a fundamental question we are trying to answer in this paper.

2   Related Work

A major area of domain adaptation research is extending conventional supervised learning algorithms to handle data from multiple domains. Wu and Dietterich (3) proposed an extension to support vector machines called C-SVM for training with inadequate primary data and low quality auxiliary data in an image classification task. This is done by formulating the optimization problem with two separate loss functions and slack variables for each dataset. While they reported a noticeable improvement using auxiliary data, they did not present a quantitative study over varying auxiliary data. Liao et al. (4) improved on this work using a logistic regression based approach "M-Logit", with additional parameters for each auxiliary instance for controlling their contributions in the training process. However, in both of these works, there was limited analysis about identifying parameters which control the contribution of auxiliary data. There were some heuristics presented in (4) based on the size of primary and auxiliary data, but we found them to be unstable in practice. Also, there was no study of the underlying conditions of primary and auxiliary data when the domain adaptation is likely to provide an improvement. These two works were a major influence early on in our research. We modeled our domain adaptation algorithms (see section 3) with ideas drawn from these works. There have been several empirical studies of domain adaptation (5; 6; 7). Jiang and Zhai (7) suggest purely empirical algorithms which in part inspire our theoretical analysis. We hope that our analysis can provide deeper understanding of the workings of these algorithms.

While there have been a large number of investigations into domain adaptation from an experimental perspective, theoretical studies have been limited in number. Only recently, Ben-David et al. (8), presented the analysis of structure learning for domain adaptation in specific natural language problems. This was extended Blitzer et al. (9)

who theoretically formulated the problem in terms of a domain similarity metric (10) and provided an elegant theory based on VC dimensions for a simple domain adaptation classifier. The work by Blitzer et al. is a major inspiration for this work. Our goal was to develop a similar theory for other domain adaptation algorithms to gain a better understanding about their behavior.

3   Domain Adaptation Algorithms

In this section, we propose various different learning algorithms for the domain adaptation task. Although the strategies are applicable generally, we present them in the framework of a logistic regression classifier for concreteness and simplicity. We model the class posterior probabilities with a sigmoid function $\sigma(s) = \frac{1}{1+e^{-s}}$. We assume the training set instances x_i with labels $y_i \in \{-1, 1\}$ to be i.i.d. The data log-likelihood $\ell(w)$ is equal to

$$\ell(w) = \sum_i \log \sigma(y_i w^T x_i). \tag{1}$$

The classification algorithm involves maximizing the data log-likelihood with respect to w and using the maximum likelihood value of $\hat{w}$ to classify the test instances using the sigmoid function $\sigma(\hat{w}^T x_i)$. The data log-likelihood ℓ is convex and can be optimized by gradient ascent.

3.1   Baseline

We treat the simple combination of primary and auxiliary data as a baseline domain adaptation technique. The data log-likelihood of the combined primary and auxiliary training set becomes

$$\ell(w) = \sum_i^{N_P} \log \sigma(y_i^p w^T x_i^p) + \sum_i^{N_A} \log \sigma(y_i^a w^T x_i^a) \tag{2}$$

With $N_A > N_P$, the effect of the auxiliary dataset in the training typically overwhelms that of the primary dataset. When the model is applied to the primary test set, the classification accuracy will be low compared to a model trained with adequate primary data. We use this intuition to develop algorithms which take the differences in domains into account.

3.2   Auxiliary Dataset Reweighting

In this algorithm, we decrease the importance of auxiliary dataset in training by a predefined amount. This will cause the learned model be more aligned towards primary domain distribution while retaining the generalization properties provided by large auxiliary instances. In context of the logistic regression framework defined above, we reduce the contribution of auxiliary instances in training by multiplying its label by a parameter $\alpha \in [0, 1]$.

$$\ell(w, \alpha) = \sum_{i}^{N_P} \log \sigma(y_i^p w^T x_i^p) + \sum_{i}^{N_A} \log \sigma(y_i^a w^T x_i^a \cdot \alpha) \tag{3}$$

When α is close to 0, $\log \sigma(y_i^a w^T x_i^a \cdot \alpha)$ will have a constant value and the auxiliary training instances (x_i^a, y_i^a) are discounted from training. The importance of auxiliary dataset in the training increases with α and when $\alpha = 1$, we get our baseline algorithm.

Even though this algorithm is simplistic, it provides insights into learning with two domains. The algorithm performs quite well in practice when there is little difference between primary and auxiliary datasets. Identifying $\hat{\alpha}$ which minimizes the primary test error raises some interesting questions.

3.3   A1: Auxiliary Subset Selection

While the auxiliary dataset reweighting algorithm works usually well in practice, it makes a simplistic assumption of treating all auxiliary instances equally. This can result in problems when we have a multi-modal auxiliary distribution and only part of it matches the primary data. One strategy is to directly select the auxiliary instances which have lower primary risk. As the information of primary distribution is not available, we approximate this by the primary empirical risk and identify subsets of auxiliary dataset which minimize it. As there are 2^{N_A} possible subsets of instances which can be selected, evaluating the primary empirical risk for each one of them is intractable. Even if we restrict ourselves to subset of a fixed predefined size k, we are left with $\binom{N_A}{k}$ selections which is still very large as N_A is large for most practical situations.

We propose a greedy algorithm called "A1"[1] for selecting the auxiliary subsets in an efficient way. Let S denote the size k set of auxiliary dataset to be selected. We start with initializing $S^{(0)}$ to k randomly selected auxiliary instances. We use the complete primary data and auxiliary data which is in $S^{(t)}$ for training the model $w^{(t+1)}$.

$$\hat{w}^{(t+1)} = \underset{w}{\operatorname{argmax}} \sum_{i}^{N_P} \log \sigma(y_i^p w^T x_i^p) + \sum_{(x_i^a, y_i^a) \in S^{(t)}} \log \sigma(y_i^a w^T x_i^a) \tag{4}$$

We apply this model $w^{(t+1)}$ to the auxiliary dataset and calculate the error between predicted value and auxiliary label for each auxiliary instance. The top k auxiliary instances minimizing this error are assigned to $S^{(t+1)}$.

$$S^{(t+1)} = \underset{top\ 'k'\ i's}{\operatorname{argmin}} |y_i^a - 2\sigma(w^T x_i^a) + 1| \tag{5}$$

As the auxiliary labels are in $\{-1, 1\}$, we scale the predicted values given by the sigmoid function which are in $[0, 1]$ to that range. We repeat steps 4 and 5 until the selection into S does not change. In practice we observe that this algorithm converges within a few iterations.

As was the case with α in dataset reweighting, the value of selection parameter plays an important role in the algorithm. If k is too small, it is similar to training on the

---

[1] The choice of name is non-mnemonic.

primary data which is small by itself. If k is too large, it is similar to optimizing the risk function on the combined primary and auxiliary data. In this way, the optimal value of k would be related to how similar the two domains are.

3.4   Soft A1 Algorithm

In the original A1 algorithm, we select k auxiliary instances for training and disregard the rest. One possible improvement is to discount the auxiliary instances partially. Instead of a set S, we consider an auxiliary weight vector $z \in [0,1]^{N_A}$. The value of z_i indicates how much (x_i^a, y_i^a) is discounted in training. In order to avoid the learned model to discount all of auxiliary data, we fix $\sum z_i = k$. The parameter k has similar properties as in the original A1 algorithm.

We start with a randomly initialized $z^{(0)}$, summing to k and compute the learned model $w^{(t+1)}$.

$$\hat{w}^{(t+1)} = \underset{w}{\text{argmax}} \sum_i^{N_P} \log \sigma(y_i^p w^T x_i^p) + \sum_i^{N_A} \log \sigma(y_i^a w^T x_i^a \cdot z_i^{(t)}) \tag{6}$$

We apply this model $w^{(t+1)}$ to the auxiliary dataset and calculate the error between predicted value and auxiliary label for each auxiliary instance. We assign $z_i^{(t+1)}$ the following value indicating how close was the predicted value to the true auxiliary label.

$$z_i^{(t+1)} = 1 - |y_i^a - 2\sigma(w^T x_i^a) + 1| \tag{7}$$

In order to maintain the $\sum z_i = k$ constraint, we assign the z_i values starting in the decending order. Once the cumulative sum becomes k, we set the remaining $z_i = 0$ as shown below.

```
sum = 0;
for (j in 1 to Na) {
      sum = sum + jth largest value in z;
      if (sum >  k) break;
}
set all values < jth largest value in z to 0;
```

3.5   Asymptotic Time Complexity Analysis

A unit step of the gradient ascent optimization in the logistic regression algorithm is computation of the log-likelihood function ℓ and its gradient. This involves the dot product $w^T x_i$ and summing over all instances x_i. In terms of the number of instances N and features d, the time complexity of this step is $O(Nd)$. The gradient ascent optimization terminates only after convergence in either ℓ or w, hence its asymptotic time complexity is data dependent. The time complexity of the unit step in the baseline procedure is $O((N_P + N_A) d)$. As the auxiliary dataset reweighting algorithm involves multiplying the auxiliary instance term in the log-likelihood computation by a fixed parameter α, the time complexity of its unit step will also be $O((N_P + N_A) d)$. The unit step of computation in the A1 algorithm is executing logistic regression algorithm over k auxiliary instances are used along with N_P primary instances till convergence, which in turn will have a unit step of time complexity $O((N_P + k)\, d)$. Hence, it is much more expensive to execute the A1 algorithm than baseline or auxiliary dataset reweighting algorithm. For the soft A1 algorithm, the unit step of computation is the logistic regression algorithm with N_P primary instances and all N_A auxiliary instances multiplied by the z vector. This in turn will have a unit step time complexity same as the baseline of $O((N_P + N_A)\, d)$. Hence, the soft A1 algorithm is more expensive to compute than the original A1 algorithm. To summarize, the order of asymptotic time complexities of the algorithms will be: baseline = auxiliary dataset reweighting $<<$ A1 $<$ Soft A1.

4 Theoretical Analysis

4.1 Domain Similarity

The central issue with domain adaptation is that we have limited primary data and auxiliary data is abundant but different from primary data. We aim to quantify the magnitude of the difference in this section. There are many standard measures of differences between probability distributions like KL-divergence and ℓ_P distance. However, as we do not have the knowledge of the exact distributions of the two domains beforehand, we need to estimate these measures from finite primary and auxiliary training datasets which will always have an associated error. To avoid the inconsistency in estimating the distance between domains from finite data samples, we consider a distance metric defined on hypothesis classes called $d_{\mathcal{H}}$ distance, which was originally proposed by (11) and (10). It has been recently used by (8; 9) as a foundation for investigating theoretical properties of domain adaptation.

Let $\mathcal{H}$ be a hypothesis class having a finite VC dimension of a given instance space $\mathcal{X}$. Let $\mathcal{A}_{\mathcal{H}} \subseteq 2^{\mathcal{X}}$ be the set of subsets of $\mathcal{X}$ such that $\mathcal{A}_{\mathcal{H}} = \{a | \exists\, h \in \mathcal{H}\ s.t.\ a = supp\, h\}$. Intuitively, $\mathcal{A}_{\mathcal{H}}$ contains sets of those $x \in \mathcal{X}$ which are labeled positively by some $h \in \mathcal{H}$. We calculate the absolute difference in probability of each subset $a \in \mathcal{A}_{\mathcal{H}}$, belonging to both the domain distributions $\mathcal{D}_P$ and $\mathcal{D}_A$. The $d_{\mathcal{H}}$ distance is the maximum difference in probability across all such subsets. It indicates that given a set of instances which are both classified in the same way, how different are the chances that would they have been generated from the two distributions.

Definition 1. *Let* $\mathcal{A}_{\mathcal{H}} \subseteq 2^{\mathcal{X}}, \{x | x \in \mathcal{X}, h \in \mathcal{H}, h(x) = 1\} \in \mathcal{A}_{\mathcal{H}}$. $d_{\mathcal{H}}$ *distance between two distributions* $\mathcal{D}_P$ *and* $\mathcal{D}_A$ *is defined as*

$$d_{\mathcal{H}}(\mathcal{D}_P, \mathcal{D}_A) = \sup_{a \in \mathcal{A}_{\mathcal{H}}} |\mathcal{D}_P(a) - \mathcal{D}_A(a)|. \qquad (8)$$

In this original formulation, the $d_{\mathcal{H}}$ distance is not directly helpful when understanding the behavior of classifiers. Given the hypothesis space $\mathcal{H}$, we construct a symmetric difference set $\mathcal{H}\Delta\mathcal{H} = \{h(x)\ xor\ h'(x) | h, h' \in \mathcal{H}\}$. For any two hypotheses $h, h' \in \mathcal{H}$ which disagree in their labels for some $x \in \mathcal{X}$ ($h(x) = 1, h'(x) = 0$ or $h(x) = 0, h'(x) = 1$), there exists a hypothesis in $\mathcal{H}\Delta\mathcal{H}$ which labels x as 1. Similarly, if ($h(x) = h'(x) = 0, h(x) = h'(x) = 1$), there exists a hypothesis in $\mathcal{H}\Delta\mathcal{H}$ which labels x as 0. As before, the support set $\mathcal{A}_{\mathcal{H}\Delta\mathcal{H}}$ contains all $x \in \mathcal{X}$ such that $h(x) \neq h'(x)$ for some two hypothesis $h, h' \in \mathcal{H}$. However, computing $d_{\mathcal{H}\Delta\mathcal{H}}$ is NP-hard even for hypothesis spaces with finite VC dimension (10). Instead, we approximate this by training a linear classifier to discriminate between the primary and auxiliary domains. We evaluate the classifier over held-out data from the two domains and use the accuracy of classification as an empirical estimate $\hat{d}_{\mathcal{H}\Delta\mathcal{H}} \in [0, 1]$. For domains that are fully separable by a linear classifier, we will have $\hat{d}_{\mathcal{H}\Delta\mathcal{H}} = 1$. When both domains are completely indistinguishable, we have $\hat{d}_{\mathcal{H}\Delta\mathcal{H}} = 0$.

4.2   Learning Bounds

We begin by establishing a bound between primary and auxiliary risk functions in terms of the distance metric. The following useful theorem follows from using $\mathcal{A}_{\mathcal{H}\Delta\mathcal{H}}$ as the basis set for $d_{\mathcal{H}\Delta\mathcal{H}}$ distance. The proof of all theorems is given in the Appendix.

Theorem 1. *Let h, h' be any two hypothesis belonging to the hypothesis space $\mathcal{H}$. Given the two domains $\mathcal{D}_P$ and $\mathcal{D}_A$, under the 0-1 loss function L_{01} and the corresponding risk functions R_P and R_A defined over the two domains,*

$$|R_A(h(x), h'(x)) - R_P(h(x), h'(x))| \le d_{\mathcal{H}\Delta\mathcal{H}}(\mathcal{D}_P, \mathcal{D}_A) \quad (9)$$

In section 3.1, we discussed the baseline domain adaptation algorithm which involves training with combined primary and auxiliary data. Theoretically, this is equivalent to minimizing the sum of primary and auxiliary risk together. The baseline hypothesis h^* is given by

$$h^* = \underset{h \in \mathcal{H}}{\operatorname{argmin}} \; R_P(h) + R_A(h). \quad (10)$$

We denote the risk of the baseline hypothesis by $\lambda = R_P(h^*) + R_A(h^*)$. This quantity is useful in getting an understanding of the domains. If λ is large, we cannot expect to do well on the while minimizing the auxiliary risk.

In both the auxiliary dataset reweighting and A1 algorithms, we reduce the importance of auxiliary dataset in training by introducing the α and k parameters. As both the parameters have the same effect of increasing the contribution of auxiliary data while varying between 0 and 1, we refer to both of them by α. The situation is slightly different in the A1 algorithm as size of the effective auxiliary dataset changes, but the theory generally holds in principle. We are working on developing this theory using additional information about the algorithm. The learning procedure minimizes the following combined true and empirical risk functions which we call true and empirical α risks.

$$R_\alpha(h) = R_P(h) + \alpha R_A(h), \hat{R}_\alpha(h) = \hat{R}_P(h) + \alpha \hat{R}_A(h) \quad (11)$$

We denote the hypothesis minimizing the empirical α risk $\hat{R}_\alpha(h)$ by $\hat{h}_\alpha$. As was the case before, when $\alpha = 0, R_\alpha(h)$ is equivalent to the primary risk and when $\alpha = 1, R_\alpha(h)$ is equivalent to the baseline risk. We now present a concentration of measure analysis to establish uniform convergence of empirical and true α risk. For clarity, let us denote the total number of instances $N_P + N_A$ by N and β as the fraction of primary instances $\frac{N_P}{N}$. The fraction of auxiliary instances will be $1 - \beta$. Using a similar argument and linearity of expectations, we show that $\hat{R}_\alpha$ is unbiased in the following lemma.

Lemma 1. *For a given hypothesis $h \in \mathcal{H}$, $\hat{R}_\alpha(h)$ is an unbiased estimator of $R_\alpha(h)$.*

$$\mathbb{E}\hat{R}_\alpha(h) = R_\alpha(h).$$

We use this result with the Hoeffding's inequality to establish the following error bound between the true and estimated α risk.

Theorem 2. *Let $\mathcal{H}$ be a hypothesis space with $VC(\mathcal{H}) = C$. Let N denote the total training instances $N_P + N_A$. Let β denote the fraction of primary instances $\frac{N_P}{N}$. Then, with probability $1 - \delta$, for every $h \in \mathcal{H}$,*

$$|\hat{R}_\alpha(h) - R_\alpha(h)| \leq \sqrt{\left(\frac{1}{\beta} + \frac{\alpha^2}{1 - \beta}\right) \frac{C \log(2N/C) - \log \delta}{2N}}.$$

We use theorem 2 to formulate the following bound between the hypothesis $\hat{h}_\alpha$ which minimizes the α risk and the hypothesis h_P^* which minimizes the true primary risk.

Theorem 3. *Let $\mathcal{H}$ be a hypothesis space with $VC(\mathcal{H}) = C$. Let N denote the total training instances $N_P + N_A$. Let β denote the fraction of primary instances $\frac{N_P}{N}$. Let the true primary risk minimizer be $h_P^* = \underset{h \in \mathcal{H}}{\operatorname{argmin}}\ R_P(h)$ and the empirical α risk minimizer be $\hat{h}_\alpha = \underset{h \in \mathcal{H}}{\operatorname{argmin}}\ \hat{R}_\alpha(h)$. Then, with probability $1 - \delta$,*

$$R_P(\hat{h}_\alpha) < R_P(h_P^*) + \frac{2}{1+\alpha}\sqrt{\left(\frac{1}{\beta} + \frac{\alpha^2}{1 - \beta}\right) \frac{C \log(2N/C) - \log \delta}{2N}} + \alpha[\lambda + d_{\mathcal{H}\Delta\mathcal{H}}(\mathcal{D}_P, \mathcal{D}_A)]$$

The auxiliary dataset reweighting and A1 algorithms are equivalent to training on the primary dataset when $\alpha = 0$. If we substitute $\alpha = 0$ in the bound given by Theorem 3, we have the original uniform convergence bound on training with primary data.

h_P^* is the best hypothesis we can have as it will have the minimum true primary risk. This hypothesis will have the minimum possible error on the on the primary test data. As the auxiliary dataset reweighting algorithm minimizes the empirical α risk, the classifier generated by this algorithm will satisfy the above theorem. The key result of Theorem 3 is that we have a bound on how much excess error $\hat{h}_\alpha^*$ can possibly have on the primary test dataset compared to the best possible classifier. It is important to note that this bound contains two terms: $\frac{\alpha^2}{1-\beta}$ can be thought of excess error caused due to limited primary data ($1 - \beta$ factor). In the second term, α interacts with $d_{\mathcal{H}\Delta\mathcal{H}}$ which can be thought of excess error caused due to the difference between primary and auxiliary distribution. Hence, the parameter α effectively tries to make a trade-off between these two factors.

Theorem 3 gives us an upper bound on the excess primary risk of the hypothesis minimizing empirical α risk as a function of α, say, $f(\alpha)$. We denote the α independent expression inside the square root by c_1. To find the value of α minimizing the excess primary risk, we have:

$$\frac{\partial}{\partial \alpha} f(\alpha) = \frac{-2}{(1+\alpha)^2} \sqrt{\left(\frac{1}{\beta} + \frac{\alpha^2}{1-\beta}\right) c_1} + \frac{1}{1+\alpha} \left[\left(\frac{1}{\beta} + \frac{\alpha^2}{1-\beta}\right) c_1\right]^{-1/2} \frac{2\alpha c_1}{1-\beta}$$
$$+ \lambda + d_{\mathcal{H} \Delta \mathcal{H}}(\mathcal{D}_P, \mathcal{D}_A) = 0. \tag{12}$$

Due to the nature of the equation, there is no simple closed form solution for α. For a given setting, as we do not know the value of true baseline risk λ beforehand, this would only give us a theoretical expression for α. We can get an empirical estimate using an estimate of λ from the primary and auxiliary datasets.

5   Experimental Results

We report experiences with our question answering (QA) system. For each question, our QA system produces answer candidates which are scored with 296 feature scores which are used to produce a ranked list of answers. We use *question error*, given by the percent of questions for which an incorrect answer appears ranked first as a metric to evaluate the ranking quality. We reduce the ranking of answer candidates to a classification problem. In training, manually vetted answer keys are used to label each question-answer pair as correct or incorrect. For testing, the probability of an answer being correct is used to rank the answers per question. For the primary dataset we used factoid questions from the Text REtrieval Conference (TREC) QA track (12). TREC 8-10, which consist of 1,200 questions with 9,485 question-answer pairs, was used for training while TREC 11, which consists of 500 questions and 4,742 question-answer pairs, was used for test. As an auxiliary dataset we made use of a question set used for internal development which covers a broad range of general reference knowledge topics such as history, geography, arts and entertainment. This auxiliary QA set consists of 2,500 questions and 9,579 question-answer pairs.

We first evaluate the auxiliary data reweighting algorithm on this dataset. The primary test errors for different values of α are shown in Figure 1(a). We compare the classifiers trained on primary datasets of two different sizes $N_P = 1897$ (solid) and $N_P = 4742$ (dashed) both with full auxiliary data $N_A = 9579$. The error for the larger primary dataset is lower than the error for the smaller dataset. The error when only using the primary data $\alpha = 0$ is lower than the error when using the primary and auxiliary data together $\alpha = 1$. This can be attributed to the qualitative differences in the two datasets. We observe that the minimum test error for the algorithm in both cases is observed around $\alpha = 0.4$ and $\alpha = 0.2$ for indicating that we require less contribution from auxiliary data for training as our primary data increases.

We evaluate both the original and soft versions of A1 algorithm on this dataset. We compare the primary test question error for the classifiers trained on primary datasets of two different sizes $N_P = 1897$ shown in Figure 1(b) and $N_P = 4742$ shown in Figure 1(c), both with full auxiliary data $N_A = 9579$. We see that the test error is clearly lower when we use larger primary dataset. Similar to the auxiliary dataset reweighting, the test error increases with increasing k up to a point and then starts increasing. This means

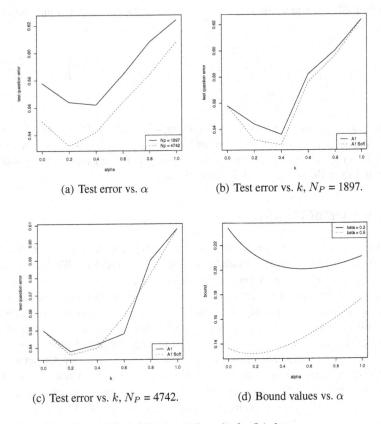

(a) Test error vs. α

(b) Test error vs. k, $N_P = 1897$.

(c) Test error vs. k, $N_P = 4742$.

(d) Bound values vs. α

Fig. 1. Experimental results for QA data

that as we go on adding auxiliary instances in training, the algorithm is able to identify parts of the auxiliary data which are useful for training. Also, we see that the soft A1 algorithm performs a lot better than the original A1 algorithm. This can be attributed to the relative flexibility of the algorithm in choosing the auxiliary instances. For the larger primary dataset with $N_P = 4742$, we see that the test error for the A1 algorithm increases beyond a point after adding additional auxiliary instances. This means that the primary data is sufficient for training and adding auxiliary instances introduces noise. However, the soft A1 algorithm does not perform much better than the original A1 algorithm, but it still has the minimum error at $k = 0.2$. Again, we observe that the value of k which has the minimum error is lower for larger value of N_P.

It is seen that the performance of both the algorithms is varies with the choice of α and k parameters and the changing contribution of the auxiliary data in training. As we have seen before, there is no simple closed form solution for the α parameter minimizing the theoretical excess risk given by Theorem 3. However, we can choose the parameter values empirically by performing cross-validation using different samples of primary data and auxiliary data, and evaluating on remainder of the primary data.

We evaluate the bound given in Theorem 3 for this dataset. For the sake of comparing the primary datasets of two sizes, we assign $\lambda + d_{\mathcal{H}\Delta\mathcal{H}}$ a constant value 0.1. β takes the values 0.2 and 0.5. Figure 1(d) shows the error bounds for this setup. We see that the experimental results are very similar in shape to the bound values. For both the curves, the minimum value $\hat{\alpha}$ is smaller for $\beta = 0.5$ as compared to $\beta = 0.2$ which is confirmant with our theory. Note that the values of the bound are relative to the true primary risk and not be directly compared to test error values. We see that the experimental results are very similar in shape to the bound values. For both the curves, the minimum value $\hat{\alpha}$ is smaller for $\beta = 0.5$ as compared to $\beta = 0.2$ which is confirmant with our theory.

Table 1. Primary test errors and execution times (in seconds) for QA data

(a) $N_P = 1897$.

Algorithm	Parameter	Test Error	Time
Primary		0.588	32.52
Baseline		0.624	153.87
Reweighting	$\alpha = 0.4$	0.562	161.07
A1	$k = 0.4$	0.536	325.15
Soft A1	$k = 0.4$	**0.528**	721.38
M-Logit		0.554	448.62

(b) $N_P = 4742$.

Algorithm	Parameter	Test Error	Time
Primary		0.550	53.67
Baseline		0.608	260.75
Reweighting	$\alpha = 0.2$	**0.532**	274.03
A1	$k = 0.2$	0.538	629.87
Soft A1	$k = 0.2$	0.536	1132.23
M-Logit		0.546	627.52

In Tables 1(a) and 1(b), we report the comparative performance of each algorithm including primary test errors and execution times[2] for the two primary dataset samples of size $N_P = 1897$ and $N_P = 4742$. We used our implementation of M-Logit based on (4) as a state of the art method for comparison. The errors were always found to be lower when we train with the larger primary dataset. For $N_P = 1897$, which is 20% of the primary data, the soft A1 algorithm achieves a substantial 15.39% improvement over the baseline and 10.20% improvement over training with primary data. This error is even lower than the baseline approach with $N_P = 4742$ which is 50% of the primary data. We have a 13.16% improvement over the baseline and 4% improvement with training with primary data alone. For $N_P = 1897$, the M-Logit algorithm performs better than baseline and dataset reweighting algorithms, while for $N_P = 4742$, the reweighting algorithm performs better than M-Logit. In both cases, the A1 and soft A1 algorithm have lower primary test errors than M-Logit. The A1 and soft A1 algorithms take much longer to execute than the baseline and reweighting algorithms, which in turn are much slower than training with primary data.

6   Conclusion

In this paper, we investigated the problem of domain adaptation which is learning with little training data from the same distribution along with large amount of data from a

---

[2] Execution times are reported for the experiments conducted in the R programming environment running over a 64-bit GNU/Linux machine with dual 2 GHz processors and 3 GB RAM.

different distribution. We introduced our domain adaptation algorithms in the logistic regression classifier framework. The auxiliary dataset reweighting algorithm modifies the contribution of auxiliary data in training with the α parameter. The A1 algorithm efficiently selects the instances from auxiliary data which are likely to minimize error in training with primary data. We also discussed a soft variant of A1 algorithm which partially discounts auxiliary instances from training by a fixed amount. We explored the concept of domain similarity with the hypothesis class based $d_{\mathcal{H}\Delta\mathcal{H}}$ distance metric and used it to develop a theoretical framework for analyzing domain adaptation methods. We presented an error bound for the auxiliary dataset reweighting algorithm which indicated a tradeoff between the domain similarity and size of the primary training data. We presented an experimental analysis of our algorithms over data from a question answering system. The experimental results were found to be closely aligned with our theory.

A few directions for making further enhancements in this research include performing a wider evaluation with other datasets and along with comparisons other techniques. It would be interesting to compare our approach to other paradigms like semi-supervised learning. As we alluded before, we are currently working towards establishing tighter error bounds for the A1 algorithm. We think there is also some real potential for improvement by exploring the problem with other complexity classes like Rademacher complexity. In all of the work here, we considered only simple cost functions where all auxiliary instances and features are equal in training. Extending our approaches to handle more complex cases would also be an interesting direction.

Acknowledgements

We would like to thank the anonymous reviewers for their insightful comments.

References

[1] Marcus, M.P., Santorini, B., Marcinkiewicz, M.A.: Building a large annotated corpus of english: The penn treebank. Computational Linguistics 19(2), 313–330 (1994)
[2] Crammer, K., Kearns, M.J., Wortman, J.: Learning from multiple sources. In: Advances in Neural Information Processing Systems 19, pp. 321–328 (2006)
[3] Wu, P., Dietterich, T.: Improving svm accuracy by training on auxiliary data sources. In: Proceedings of the 21st International Conference on Machine Learning, pp. 871–878 (2004)
[4] Liao, X., Xue, Y., Carin, L.: Logistic regression with an auxiliary data source. In: Proceedings of the 22nd International Conference on Machine Learning, pp. 505–512 (2005)
[5] Chelba, C., Acero, A.: Adaptation of maximum entropy capitalizer: Little data can help a lot. In: Proceedings of EMNLP 2004, pp. 285–292 (2004)
[6] Bickel, S., Bruckner, M., Scheffer, T.: Discrimminative learning for differing training and test distributions. In: Proceedings of the 24th International Conference on Machine Learning, pp. 81–88 (2007)
[7] Jiang, J., Zhai, C.: Instance weighting for domain adaptation in nlp. In: Proceedings of the 45th Annual Meeting of the Association of Computational Linguistics, pp. 264–271 (2007)
[8] Ben-david, S., Blitzer, J., Crammer, K., Pereira, F.: Analysis of representations for domain adaptation. In: NIPS (2006)

[9] Blitzer, J., Crammer, K., Kulesza, A., Pereira, F., Wortman, J.: Learning bounds for domain adaptation. In: Advances in Neural Information Processing Systems 20, pp. 129–136 (2007)

[10] Kifer, D., Ben-David, S., Gehrke, J.: Detecting change in data streams. In: Proceedings of the Thirtieth international conference on very large databases, pp. 180–191 (2004)

[11] Devroye, L., Gyorfi, L., Lugosi, G.: A probabilistic theory of pattern recognition, pp. 271–272. Springer, Heidelberg (1996)

[12] Voorhees, E.M., Harman, D.: Overview of the eighth text retrieval conference (trec-8). In: Proceedings of the Eighth Text REtrieval Conference, TREC-8 (1999)

Appendix

– Proof of Theorem 1.

$$\left| R_A(h(x), h'(x)) - R_P(h(x), h'(x)) \right| = \left| \mathbb{E}_{x \sim A} L_{01}(h(x), h'(x)) - \mathbb{E}_{x \sim P} L_{01}(h(x), h'(x)) \right|$$

$$= \left| \sum_{x \sim A} L_{01}(h(x), h'(x)) \mathcal{D}_A(x) - \sum_{x \sim P} L_{01}(h(x), h'(x)) \mathcal{D}_P(x) \right|$$

$$= \left| \sum_{x \in A} L_{01}(h(x), h'(x)) \mathcal{D}_A(x) + \sum_{x \notin A} L_{01}(h(x), h'(x)) \mathcal{D}_A(x) \right.$$

$$\left. + \sum_{x \in A} L_{01}(h(x), h'(x)) \mathcal{D}_P(x) + \sum_{x \notin A} L_{01}(h(x), h'(x)) \mathcal{D}_P(x) \right| \quad \text{(for some } A \in \mathcal{A}_{\mathcal{H} \Delta \mathcal{H}})$$

$$= \left| \sum_{x \in A} \mathcal{D}_A(x) - \sum_{x \in A} \mathcal{D}_P(x) \right| = |\mathcal{D}_A(A) - \mathcal{D}_P(A)| \le \sup_{A \in \mathcal{A}_{\mathcal{H} \Delta \mathcal{H}}} |\mathcal{D}_A(A) - \mathcal{D}_P(A)| = d_{\mathcal{H} \Delta \mathcal{H}}(\mathcal{D}_P, \mathcal{D}_A)$$

$\square$

– Proof of Lemma 1. We define the random variables $Z_1, ... Z_{N_P}$ to take the values $\frac{1}{\beta}|y - h(x)| \in \left[0, \frac{1}{\beta}\right]$ corresponding to the primary instances $x_1, ..., x_{N_P} \in \mathcal{X}_P \sim \mathcal{D}_P$. Similarly, define the random variables $Z_{N_P+1}, ..., Z_N$ to take the values $\frac{1}{1-\beta}|y - h(x)| \in \left[0, \frac{1}{1-\beta}\right]$ corresponding to the auxiliary instances $x_1, ..., x_{N_A} \in \mathcal{X}_A \sim \mathcal{D}_A$. From the definition of α risk, we have

$$\hat{R}_\alpha(h) = \hat{R}_P(h) + \alpha \hat{R}_A(h) = \frac{1}{N_P} \sum_{x \in \mathcal{X}_P} |y - h(x)| + \frac{\alpha}{N_A} \sum_{x \in \mathcal{X}_A} |y - h(x)|$$

$$= \frac{1}{N} \left[\frac{N}{N_P} \sum_{x \in \mathcal{X}_P} |y - h(x)| + \frac{N\alpha}{N_A} \sum_{x \in \mathcal{X}_A} |y - h(x)| \right] = \frac{1}{N} \left[\frac{1}{\beta} \sum_{x \in \mathcal{X}_P} |y - h(x)| + \frac{\alpha}{1-\beta} \sum_{x \in \mathcal{X}_A} |y - h(x)| \right]$$

$$= \frac{1}{N} \left[\sum_{i=1}^{N_P} Z_i + \sum_{i=N_P+1}^{N_A} Z_i \right] = \frac{1}{N} \sum_{i=1}^{N} Z_i = \bar{Z}.$$

$$\mathbb{E}\hat{R}_\alpha(h) = \mathbb{E}\left[\hat{R}_P(h) + \alpha \hat{R}_A(h) \right] = \mathbb{E}\left[\frac{1}{N_P} \sum_{x \in \mathcal{X}_P} |y - h(x)| + \frac{\alpha}{N_A} \sum_{x \in \mathcal{X}_A} |y - h(x)| \right]$$

$$= \frac{1}{N_P} \sum_{x \in \mathcal{X}_P} \mathbb{E}|y - h(x)| + \frac{\alpha}{N_A} \sum_{x \in \mathcal{X}_A} \mathbb{E}|y - h(x)| = \frac{1}{N_P} \sum_{x \in \mathcal{X}_P} R_P(h) + \frac{\alpha}{N_A} \sum_{x \in \mathcal{X}_A} R_A(h)$$

$$= R_P(h) + \alpha R_A(h) = R_\alpha(h).$$

□

- Proof of Theorem 2.

$$\mathbb{P}\left[|\hat{R}_\alpha(h) - R_\alpha(h)| > t\right] = \mathbb{P}\left[|\bar{Z} - \mathbb{E}\bar{Z}| > t\right] \leq 2exp\left[\frac{-2N^2t^2}{\sum_{i=1}^{N}(\max(Z_i) - \min(Z_i))^2}\right]$$

$$= 2exp\left[\frac{-2N^2t^2}{\sum_{i=1}^{N_P} \frac{1}{\beta^2} + \sum_{i=1}^{N_P} \frac{\alpha^2}{(1-\beta)^2}}\right] = 2exp\left[\frac{-2N^2t^2}{\frac{N_P}{\beta^2} + \frac{N_A\alpha^2}{(1-\beta)^2}}\right]$$

$$= 2exp\left[\frac{-2N^2t^2}{\frac{N}{\beta} + \frac{N\alpha^2}{1-\beta}}\right] = 2exp\left[\frac{-2Nt^2}{\frac{1}{\beta} + \frac{\alpha^2}{1-\beta}}\right].$$

The above result holds true for a single $h \in \mathcal{H}$. To generalize for the whole hypothesis class, we proceed with the standard uniform bound argument with VC dimensions as our growth function. Let $VC(\mathcal{H}) = C$.

$$\forall h \in \mathcal{H}, \mathbb{P}\left[|\hat{R}_\alpha(h) - R_\alpha(h)| > t\right] \leq \left(\frac{2N}{C}\right)^C exp\left[\frac{-2Nt^2}{\frac{1}{\beta} + \frac{\alpha^2}{1-\beta}}\right].$$

It should be noted that as $N \to \infty$, $\hat{R}_\alpha(h) \to R_\alpha(h)$ in probability. Hence, $\hat{R}_\alpha(h)$ is a consistent estimator of $R_\alpha(h)$.
We equate the RHS by δ and solve for t.

$$t = \sqrt{\left(\frac{1}{\beta} + \frac{\alpha^2}{1-\beta}\right)\frac{C\log(2N/C) - \log\delta}{2N}}.$$

With this, we get the following relationship.

$$\forall h \in \mathcal{H}, \mathbb{P}\left[|\hat{R}_\alpha(h) - R_\alpha(h)| > \sqrt{\left(\frac{1}{\beta} + \frac{\alpha^2}{1-\beta}\right)\frac{C\log(2N/C) - \log\delta}{2N}}\right] = \delta.$$

□

- Proof of Theorem 3. For any $h \in \mathcal{H}$,

$$|R_\alpha(h) - R_P(h)| = \alpha|R_A(h)|$$

$$= \alpha\left[|R_A(h) - R_A(h, h^*)| + |R_A(h, h^*) - R_P(h, h^*)| + |R_P(h, h^*) - R_P(h)| + R_P(h)\right]$$

$$\leq \alpha\left[|R_A(h^*)| + |R_A(h, h^*) - R_P(h, h^*)| + |R_P(h^*)| + R_P(h)\right] \quad \text{(triangle inequality)}$$

$$\leq \alpha\left[(R_A(h^*) + R_P(h^*)) + d_{\mathcal{H}\Delta\mathcal{H}}(\mathcal{D}_P, \mathcal{D}_A) + R_P(h)\right] \quad \text{(theorem 1)}$$

$$= \alpha\left[\lambda + d_{\mathcal{H}\Delta\mathcal{H}}(\mathcal{D}_P, \mathcal{D}_A) + R_P(h)\right].$$

$$\Rightarrow R_P(h) \leq R_\alpha(h) - \alpha[\lambda + d_{\mathcal{H}\Delta\mathcal{H}}(\mathcal{D}_P, \mathcal{D}_A) + R_P(h)]$$

$$(1 + \alpha)R_P(h) \leq R_\alpha(h) - \alpha[\lambda + d_{\mathcal{H}\Delta\mathcal{H}}(\mathcal{D}_P, \mathcal{D}_A)]$$

$$R_P(h) \leq \frac{1}{1+\alpha}R_\alpha(h) - \frac{\alpha}{1+\alpha}[\lambda + d_{\mathcal{H}\Delta\mathcal{H}}(\mathcal{D}_P, \mathcal{D}_A)] \tag{13}$$

We apply this inequality for the empirical α risk minimizer $\hat{h}_\alpha$.

$$R_P(\hat{h}_\alpha) \leq \frac{1}{1+\alpha} R_\alpha(\hat{h}_\alpha) - \frac{\alpha}{1+\alpha}[\lambda + d_{\mathcal{H}\Delta\mathcal{H}}(\mathcal{D}_P, \mathcal{D}_A)]$$

$$\leq \frac{1}{1+\alpha} \hat{R}_\alpha(\hat{h}_\alpha) + \frac{1}{1+\alpha} \sqrt{\left(\frac{1}{\beta} + \frac{\alpha^2}{1-\beta}\right) \frac{C\log(2N/C) - \log\delta}{2N}}$$
$$- \frac{\alpha}{1+\alpha}[\lambda + d_{\mathcal{H}\Delta\mathcal{H}}(\mathcal{D}_P, \mathcal{D}_A)] \quad \text{(theorem 2)}$$

$$\leq \frac{1}{1+\alpha} \hat{R}_\alpha(h_P^*) + \frac{1}{1+\alpha} \sqrt{\left(\frac{1}{\beta} + \frac{\alpha^2}{1-\beta}\right) \frac{C\log(2N/C) - \log\delta}{2N}}$$
$$- \frac{\alpha}{1+\alpha}[\lambda + d_{\mathcal{H}\Delta\mathcal{H}}(\mathcal{D}_P, \mathcal{D}_A)] \quad (\hat{h} = \operatorname*{argmin}_{h\in\mathcal{H}} \hat{R}_\alpha(h))$$

$$\leq \frac{1}{1+\alpha} R_\alpha(h_P^*) + \frac{2}{1+\alpha} \sqrt{\left(\frac{1}{\beta} + \frac{\alpha^2}{1-\beta}\right) \frac{C\log(2N/C) - \log\delta}{2N}}$$
$$- \frac{\alpha}{1+\alpha}[\lambda + d_{\mathcal{H}\Delta\mathcal{H}}(\mathcal{D}_P, \mathcal{D}_A)] \quad \text{(theorem 2)}$$

$$\leq R_P(h_P^*) + \frac{2}{1+\alpha} \sqrt{\left(\frac{1}{\beta} + \frac{\alpha^2}{1-\beta}\right) \frac{C\log(2N/C) - \log\delta}{2N}}$$
$$+ \alpha[\lambda + d_{\mathcal{H}\Delta\mathcal{H}}(\mathcal{D}_P, \mathcal{D}_A)].$$

$\square$

Mining Multi-label Concept-Drifting Data Streams Using Dynamic Classifier Ensemble*

Wei Qu, Yang Zhang, Junping Zhu, and Qiang Qiu

College of Information Engineering, Northwest A&F University
Yangling, Shaanxi Province, P.R. China, 712100
{lex,zhangyang,junpinzhu,qiuqiang}@nwsuaf.edu.cn

Abstract. The problem of mining single-label data streams has been extensively studied in recent years. However, not enough attention has been paid to the problem of mining multi-label data streams. In this paper, we propose an improved binary relevance method to take advantage of dependence information among class labels, and propose a dynamic classifier ensemble approach for classifying multi-label concept-drifting data streams. The weighted majority voting strategy is used in our classification algorithm. Our empirical study on both synthetic data set and real-life data set shows that the proposed dynamic classifier ensemble with improved binary relevance approach outperforms dynamic classifier ensemble with binary relevance algorithm, and static classifier ensemble with binary relevance algorithm.

Keywords: Multi-label; Data Stream; Concept Drift; Binary Relevance; Dynamic Classifier Ensemble.

1 Introduction

Modem organization generate tremendous amount of data by real-time production systems at unprecedented rates, which is known as data streams [1]. Other than the data volume, the underlying processes generating the data changes during the time, sometimes radically [2]. These changes can induce more or less changes in target concept, and is known as concept drift [3].

For the traditional classification tasks, each example is assigned to a single label l from a set of disjoint class labels L [5], while for multi-label classification tasks, each example can be assigned to a set of class labels $Y \subseteq L$. The problem of mining single-label data streams has been extensively studied in recent years. However, not enough attention has been paid to the problem of mining multi-label data streams, and such data streams are common in the real world. For example, a user is interested in the articles of certain topics over time, which may belong to more than one text category. Similarly, in medical diagnosis, a patient may suffer from multiple illnesses at the same time. During the medical treatments at different periods, some of them are cured while others still remain.

---

* This work is supported by Young Cadreman Supporting Program of Northwest A&F University (01140301). Corresponding author: Zhang Yang.

Z.-H. Zhou and T. Washio (Eds.): ACML 2009, LNAI 5828, pp. 308–321, 2009.

In this paper, an dynamic classifier ensemble approach is proposed to tackle multi-label data streams with concept drifting. We partition the incoming data stream into chunks, and then we learn an improved binary relevance classifier from each chunk. For classifying a test example, the weight of each base classifier in the ensemble is set by the performance of the classifier on the neighbors of the test example, which are found in most up-to-date chunk by k-nearest neighbor algorithm. Our experiment on both synthetic data set and real-life data set shows that the proposed approach has better performance than the comparing algorithms.

The rest of the paper is organized as follows. Section 2 reviews the related works. Section 3 presents our improved binary relevance methods. Section 4 outlines our algorithm for mining multi-label concept-drifting data streams. Section 5 gives our experiment result, and section 6 concludes this paper and gives our future work.

2   Related Works

To the best of our knowledge, the only work on multi-label data stream classification is our previous work [4], while a lot of works on classification of multi-label data sets and classification of single-label data streams could be found in the literature.

Multi-label Classification. Multi-label classification methods could be categorized into two groups [5]: problem transformation methods, and algorithm adaptation methods. The first group of methods transforms the original multi-label classification problems into one or more binary classification problems [6]. The second group of methods extends traditional algorithms to cope with multi-label classification problems directly [6], including decision trees [7], boosting [8], probabilistic methods [9], neural networks and support vector machines [10,11,12,13], lazy methods [14,15] and associative methods [16]. Compared with problem transformation methods, the algorithm adaptation methods is always time-consuming, which makes it not suitable for mining multi-label data streams, as data streams are always characterized by large volume of data.

Data Stream Classification. Incremental or online data stream approaches, such as CVFDT [2], are always time and space consuming. Ensemble learning is among the most popular and effective approaches to handle concept drift, in which a set of concept descriptions built over different time intervals is maintained, predictions of which are combined using a form of voting, or the most relevant description is selected [17]. Many ensemble based classification approaches have been proposed by the research community [1,18,19,20,21,22]. Two initial papers on classifying data streams by ensemble methods combine the results of base classifiers by static majority voting [18] and static weighted voting [1]. And later, some dynamic methods [19,20,21,22] were proposed. In dynamic integration, each base classifier receive a weight proportional to its local accuracy instead of the global classification accuracy. It is concluded in [22] that dynamic methods perform better than static methods.

Multi-label Data Streams. As a multi-label example can belongs to more than one category simultaneously, single-label data streams could be looked as a particular type of multi-label data streams. Therefore, any single-label data streams can be tackled by multi-label data stream classification approaches. As we mentioned before, multi-label data streams are common in the real-life applications. In our previous work [4], we transform each data chunk from the stream into $|L|$ single-label data chunks by binary relevance method, and an ensemble of classifiers are trained from the transformed chunks, and the classifiers in the ensemble are weighted based on their expected classification accuracy on the test data under the time-evolving environment. Although binary relevance method is effective, it does not take the dependency among labels into account. In this paper, we use the classified label vector as a new feature to help classify the other related labels. In our previous work, we adopt static ensemble method, in this paper, we employ dynamic integrated ensemble method to mining concept drifting multi-label data streams.

3   Binary Relevance

Here, we introduce some notations that are used in the rest of the paper. Let's write (x, y) for an example in the training data set, with $x = \{x_1, x_2, \cdots, x_{|F|}\}^T$ being a $|F|$-dimensional feature vector, and $y = \{y_1, y_2, \cdots, y_{|L|}\}$ being a $|L|$-dimensional label vector.

3.1   Binary Relevance

Binary Relevance (BR) learning is a widely used problem transformation method [5]. For each example, if original example contains label l, ($l \in L$), then it is labeled as l in the generated single label data set D_l, and as $\neg l$ otherwise [5]. By this way, BR method transforms the original multi-label data set into $|L|$ single-label data sets $D_1, D_2, ..., D_{|L|}$. For each data set D_l, a binary classifier is trained, and the final labels for each testing example can be determined by combining the classification results of all these $|L|$ binary classifiers.

3.2   Improved Binary Relevance

In [13], Godbole *et al.* proposed the SVM-HF method, which could be considered as an extension to BR [6]. Suppose that $i \in L$ and $j \in L$ are two labels, Godbole *et al.* assume that if i is a good indicator of j, then we can gain better performance on j with the information provided by i [13].

Firstly, for each label l ($l \in L$), a binary classifier B_l is trained on $D_l = \{(x, y_l)\}$, and thus we get the the BR classifier $B = \{B_1, B_2, \cdots, B_{|L|}\}$. Then, for each label l, B_l is used to classify example (x, y_l) in data set D_l, and the output

v_l is used as an additional feature, which is added into the original data set. In this way, the original data set is transformed into $D' = \{(x, v_1, v_2, ..., v_{|L|}, y)\}$.

Learning algorithm. In this paper, we made two modification to the SVM-HF method proposed in [13]. We will refer the modified method as the improved BR method. When learning for label l. 1) We set the weights of additional features, which are unrelated to l, to 0, as they provide no useful information for learning label l, and furthermore, they may induce noise information. Here, any certain feature selection algorithm, such as mutual information, and information gain, can be used to determine whether a label is related to l. 2) For a related additional feature i, we set its weight to the training accuracy of B_i on D_i. If B_i has poor classification performance, then there may exist a lot of noise in the classification result of B_i on D_i, which means that the additional feature introduced by label i may not be very helpful for learning l, and therefore it should be assigned with low weight. The detailed algorithm is showed in Algorithm 1.

Algorithm 1. The learning algorithm for improved BR method

Input:
　　　　D: the original data set;
　　　　L: the set of labels;
Output:
　　　　B: the BR classifier;
　　　　IB: the improved BR classifier;
1: transform D into $|L|$ single-label data sets, $\{D_1, D_2, \cdots, D_{|L|}\}$;
2: **for** each $l \in L$ **do**
3:　　train a binary classifier B_l from D_l;
4:　　$V_l = B_l.\text{Classify}(D_l)$;
5:　　add V_l as a new feature to D;
6:　　set the weight of feature V_l in D to the training accuracy of B_l;
7: **end for**
8: $B = \{B_1, B_2, \cdots, B_{|L|}\}$;
9: transform D into $|L|$ single-label data sets, $\{D'_1, D'_2, \cdots, D'_{|L|}\}$;
10: **for** $l \in L$ **do**
11:　　set the weights of new added features which are unrelated to l to 0;
12:　　train a binary classifier IB_l from D'_l;
13: **end for**
14: $IB = \{IB_1, IB_2, \cdots, IB_{|L|}\}$;
15: **return** B, IB;

Classification algorithm. In the testing phase, for a given test example t, firstly, t is classified by B, and the label vector outputted by B is added into t. Then, the final label for t is determined by combining the classification results from all the binary classifiers in IB.

4 Mining Multi-label Data Streams by Dynamic Classifier Ensemble

The incoming multi-label data stream could be partitioned into sequential chunks with the same chunk size, $S_1, S_2, \cdots, S_n$, with S_n being the most up-to-date chunk. We train a binary relevance classifier B for each S_i ($i \leq n$), then each chunk S_i is transformed into a data chunk S_i' with $|L|$ additional features, which are the outputs of B on S_i. On each transformed data chunk S_i', we train an improved BR classifier IB.

In data stream scenario, it is impossible and unnecessary to keep and use all the base classifiers, so we only keep K recently trained classifiers, discarding the old classifiers, as classifiers trained on recent data chunks may have better ability to represent the current concept. Algorithm 2 gives an outline of the dynamic classifier ensemble approach.

Algorithm 2. The learning algorithm for classifying multi-label data streams

---

Input:
  S_n: the most up-to-date chunk;
  EB: the ensemble of previously trained BR classifiers;
  EIB: the ensemble of previously trained improved BR classifiers;
  K: the maximal capacity of the classifier ensemble;
Output:
  EB: the updated ensemble of BR classifiers;
  EIB: the updated ensemble of improved BR classifiers;
 1: Train a BR classifier, B, and an improved BR classifier, IB, on S_n;
 2: **if** $|EB| == K$ **then**
 3:    Remove the oldest classifier in EB;
 4:    Remove the oldest classifier in EIB;
 5: **end if**
 6: $EB=EB \bigcup \{B\}$;
 7: $EIB=EIB \bigcup \{IB\}$;
 8: **return** EB, EIB;

In static weighted voting ensemble approach, the weight of each classifier in the ensemble are derived from the most up-to-date chunk, S_n. However, as different testing examples are associated with different classification difficulties, if we use global setting in the ensemble to predict a certain example, the weights may not be appropriate for this example. So it is better to weight the classifier dynamically in the testing phase than in the training phase [22]. Our approach is motivated by this idea, given a test example t, some training neighbors of t are found by k-nearest neighbor method. We derive the weights of the classifiers based on the performance of the classifiers on the the training neighbors in S_n. Here, we argue S_n has the most similar concept with t.

Algorithm 3. The classification algorithm for classifying multi-label data streams

Input:

　　t: the testing example;

　　S_n: the most up-to-date chunk;

　　S'_n: the transformed most up-to-date chunk;

　　EB: the ensemble of previously trained BR classifiers;

　　EIB: the ensemble of previously trained improved BR classifiers;

　　L: the set of labels;

　　q: the number of neighbors;

Output:

　　V': the predicted label vector for t;

1: $Neighbor_t = \text{FindNeighbors}(S_n, t, q)$;

2: **for** $EB_i \in EB$ **do**

3:　　$Y_i = EB_i.\text{Classify}(Neighbor_t)$;

4:　　$Z = \text{GetTrueLabels}(Neighbor_t)$;

5:　　$Weight_i = \text{Evaluate}(Y_i, Z)$;

6:　　$BY_i = IB_i.\text{Classify}(t)$;

7: **end for**

8: $V = \text{CombineLabels}(BY, Weight, L)$;

9: add V as new features to t

10: $Neighbor'_t = \text{FindNeighbors}(S'_n, t, q)$;

11: **for** $EIB_i \in EIB$ **do**

12:　　$Y'_i = EIB_i.\text{Classify}(Neighbor'_t)$;

13:　　$Z' = \text{GetTrueLabels}(Neighbor'_t)$;

14:　　$Weight'_i = \text{Evaluate}(Y'_i, Z')$;

15:　　$IBY_i = EIB_i.\text{Classify}(t)$;

16: **end for**

17: $V' = \text{CombineLabels}(IBY, Weight', L)$;

18: **return** V';

The classification algorithm for classifying multi-label data streams is listed in algorithm 3. In step 1 and 10, the function $FindNeighbors()$ is used to find the neighbors for t by k-nearest neighbor method with Euclidean distance. We get the weight for each classifier in step 3-5 and 12-14, the $Evaluate()$ function in step 5 and 14 can be any multi-label evaluation metrics. In sept 17, the final label vector for t is combined with the outputs of EIB using algorithm 4.

5   Experiments

In this section, we report our experiment results on both synthetic data set and real-life data set. The algorithms are implemented in Java with help of WEKA[1] and Mulan[2] software package.

---

[1] http://www.cs.waikato.ac.nz/ml/weka/

[2] http://mlkd.csd.auth.gr/multilabel.html

Algorithm 4. Class label combination algorithm by weighted voting

---

Input:

    IBY: the predicted class labels by classifiers in EIB;

    $Weight$: the weights of classifiers in EIB;

    L: the set of labels;

Output:

    V: the combined label vector;

1: **for** each $l \in L$ **do**

2:    $neg=pos=0$;

3:    **for** $IBY_i \in IBY$ **do**

4:      **if** $IBY_{i_l} == NEG$ **then**

5:        $neg = neg + Weight_i$;

6:      **else**

7:        $pos = pos + Weight_i$;

8:      **end if**

9:    **end for**

10:   **if** $neg > pos$ **then**

11:     $V_l = NEG$;

12:   **else**

13:     $V_l = POS$;

14:   **end if**

15: **end for**

16: **return** V;

5.1 Synthetic Data

We create synthetic multi-label data streams with drifting concepts based on moving hyperplanes [1,2]. A hyperplane in d-dimensional space is denoted by [1,2]:

$$\sum_{i=1}^{d} a_i x_i = a_0 \tag{1}$$

Thus, $|L|$ hyperplanes in d-dimensional space are denoted by:

$$\begin{cases} \sum_{i=1}^{d} a_{1_i} x_i = a_{1_0} \\ \sum_{i=1}^{d} a_{2_i} x_i = a_{2_0} \\ \quad \cdots \\ \sum_{i=1}^{d} a_{|L|_i} x_i = a_{|L|_0} \end{cases} \tag{2}$$

For label n, examples satisfying $\sum_{i=1}^{d} a_{n_i} x_i \geq a_{n_0}$, $(1 \leq n \leq |L|)$, are labeled as positive, and examples satisfying $\sum_{i=1}^{d} a_{n_i} x_i < a_{n_0}$ as negative. Examples are generated randomly, which distributed uniformly in multi-dimensional space $[0,1]^d$. Weights a_{n_i}, $(1 \leq i \leq d)$, are initialize by random values in range of $[0,1]$. We set $a_{n_0} = \frac{1}{2} \sum_{i=1}^{d} a_{n_i}$, $(1 \leq n \leq |L|)$, so that each hyperplane cuts the multi-dimensional space into two parts of the same volume. Thus, for each label, roughly half of the examples are positive, and the other half are negative. Noise is introduced by switching the labels of $p\%$ of the training examples randomly.

There is no dependency among labels in the data streams generated in the above way. However, multi-label classification tasks are always characterized by dependency among different labels [6]. In this subsection, we set $|L| = 4$, and introduce dependency into multi-label data streams in the following way:

$$\begin{cases} \sum_{i=1}^{d} a_{1_i} x_i + \sum_{i=1}^{f} a_{1_{(d+i)}} x_i = a_{1_0} \\ \sum_{i=1}^{d} a_{1_i} x_i + \sum_{i=1}^{f} a_{2_{(d+i)}} x_i = a_{2_0} \\ \sum_{i=1}^{d} a_{3_i} x_i + \sum_{i=1}^{f} a_{3_{(d+i)}} x_i = a_{3_0} \\ \sum_{i=1}^{d} \frac{1}{2}(a_{1_i} + a_{3_i}) x_i + \sum_{i=1}^{f} a_{4_{(d+i)}} x_i = a_{4_0} \end{cases} \quad (3)$$

For label 1, we set $a_{1_{(d+1)}}, a_{1_{(d+2)}}, \cdots, a_{1_{(d+f)}}$ to 0, thus it implies $\sum_{i=1}^{d} a_{1_i} x_i + \sum_{i=1}^{f} a_{1_{(d+i)}} x_i = \sum_{i=1}^{d} a_{1_i} x_i$. For label 2, we set $a_{2_{(d+1)}}, a_{2_{(d+2)}}, \cdots, a_{2_{(d+f)}}$ close to 0, so, we get $\sum_{i=1}^{d} a_{1_i} x_i + \sum_{i=1}^{f} a_{1_{(d+i)}} x_i \approx \sum_{i=1}^{d} a_{1_i} x_i + \sum_{i=1}^{f} a_{2_{(d+i)}} x_i$. Thus, if we know the classification result of label 1, we can use this information to help to classify label 2. As the feature space of label 1 is actually d-dimensional, and the feature space of label 2 is $(d+f)$-dimensional, it is harder to learn the concept for label 2 than label 1. The learning task becomes even harder when there is not enough training examples in each data chunk under data stream scenario. However, as label 1 and label 2 have similar concept, we can build the classifier for class label 2 with the help of the classification result on label 1. The term $\sum_{i=1}^{d} \frac{1}{2}(a_{1_i} + a_{3_i}) x_i$ in the formula of label 4 implies that the concept of label 4 depends on 2 concepts, say, label 1, and label 3.

In our experiments, we set $d = 50$, $f = 100$. Parameter f should have a large value, so that to learn from $(d + f)$-dimension feature space is much harder than to learn from d-dimension space. And for the same reason, d should not have a small value. We set $p = 5$ following [1], and concept drifting is simulated by a series of parameters. We write *chunkSize* for the count of examples in each data chunk from the training stream; k for the total number of dimensions whose weights are involved in concept drifting; $t \in R$ for the magnitude of the change (every N examples) for weights $a_{n_1}, a_{n_2}, \cdots, a_{n_k}$, $(1 \leq n \leq |L|)$; and $s_j \in \{-1, 1\}$, $(1 \leq j \leq k)$, for the direction of change for weights a_{n_j}. Weights change continuously, i.e., a_{n_j} is adjusted by $s_j \cdot t/N$ after each example is generated. Furthermore, there is a possibility of 10% that the change would reverse direction after every N examples are generated, that is to say, s_j is replaced by $-s_j$ with possibility of 10%. Also, each time the weights are updated, we recomputed $a_{n_0} = \frac{1}{2} \sum_{i=1}^{d} a_{n_i}$, so as to keep the class distribution of positive and negative samples.

In our experiments, we set $k = 10$. Parameter k should not have a very small value, since in that case, concept drift is not obvious, and hence the ensemble size does not matter. On the other hand, for a very large value of k, the concept drift is hard to capture. We set $t = 0.1$, $N = 1000$ following [1]. Information gain algorithm is used to select the additional features which are related to the label. And the number of neighbors, q, is set to 10% of the number of examples in the most up-to-date chunk, S_n. For each of the experiment setting, ten trails of experiments are conducted, and the averaged result is reported here. Furthermore, t-Test is used to illustrate the significance of performance difference between our proposed algorithm and comparing algorithms.

In the synthetic data set, the number of positive labels are nearly equal to that of negative labels, and some examples are labeled as negative in all the labels, so accuracy [13], precision [13], recall [13] and F_1 [23] are not suit for measuring performance of this data set. Hence, we use hamming loss [8] only, and use it as evaluation matric in Algorithm 3.

Here, we write DB_K, DI_K, and SI_K for the classification result for dynamic classifier ensemble with BR algorithm, dynamic classifier ensemble with improved BR algorithm, and static classifier ensemble with improved BR algorithm, respectively. Here, K represents the maximal capacity of the classifier ensemble. In SI_K, the weight of the base classifier is derived by evaluating it on the most up-to-date chunk. The classifier with the lowest weight is discarded when the ensemble is full.

Experiment with *chunkSize*. In this group of experiment, we set $K = 6$, and Naive Bayes algorithm is used as base classifier. Figure 1 shows *chunkSize* could affects hamming loss of the classifiers. Both large and small chunk size can drive up hamming loss. The ensemble can neither capture the concept drifts occurring inside the large data chunk, nor learn a good classifier with not enough training examples in small data chunk [1]. It is shown that DB algorithm gets less hamming loss with large *chunkSize* than DI and SI does, as it is hard to learn the concept of label 2 and label 4 with small data chunk. It is also shown that DI always outperforms DB, which may suggest the strong ability of DI to use the classification result of label 1 and 3 to learn label 2 and label 4.

Experiment with K. In this group of experiment, we set *chunkSize* $= 1000$, and Naive Bayes algorithm is used as base classifier. Figure 2 shows when maximal capacity of the ensemble increases, due to the increasing diversity inside the ensemble, the hamming loss drops significantly [1]. It is obviously that DB, DI, and SI will get a minimal hamming loss value when K is large enough.

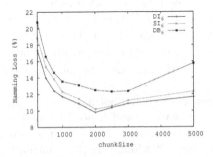

Fig. 1. Experiment with *chunkSize*

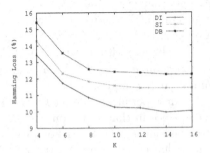

Fig. 2. Experiment with K

Table 1. Hamming Loss(%) of Dynamic Ensemble with Naive Bayes Algorithm

chunkSize	DB_1	DI_1	DB_2	DI_2	DB_4	DI_4	DB_6	DI_6	DB_8	DI_8
250	29.65	**26.45**	29.24	**26.23**	23.55	**20.22**	20.76	**17.32**	18.82	**15.55**
500	24.32	**21.36**	24.22	**21.28**	18.95	**16.12**	16.57	**14.00**	15.07	**12.66**
750	21.74	**19.13**	21.43	**19.06**	16.69	**14.43**	14.63	**12.43**	13.41	**11.34**
1000	19.86	**17.56**	19.72	**17.42**	15.42	**13.45**	13.56	**11.73**	12.57	**10.84**
t-Test	+		+		+		+		+	

Table 2. Hamming Loss(%) of Dynamic Ensemble with C4.5 Algorithm

chunkSize	DB_1	DI_1	DB_2	DI_2	DB_4	DI_4	DB_6	DI_6	DB_8	DI_8
250	35.79	**34.16**	35.12	**33.32**	29.80	**27.34**	27.02	**23.91**	25.23	**21.72**
500	33.20	**31.70**	32.95	**30.98**	27.23	**24.27**	24.64	**21.45**	22.97	**19.76**
750	31.80	**30.24**	31.19	**29.71**	25.83	**23.40**	23.11	**20.39**	21.64	**18.83**
1000	31.16	**29.60**	30.41	**28.94**	25.02	**23.03**	22.59	**20.29**	21.35	**18.71**
t-Test	+		+		+		+		+	

Table 3. Hamming Loss(%) of Dynamic Ensemble with SVM Algorithm

chunkSize	DB_1	DI_1	DB_2	DI_2	DB_4	DI_4	DB_6	DI_6	DB_8	DI_8
250	41.45	**19.13**	38.48	**18.33**	31.41	**16.55**	27.87	**16.43**	25.58	**16.50**
500	23.25	**16.07**	21.39	**15.48**	16.54	**12.45**	14.35	**11.37**	13.29	**10.83**
750	18.48	**13.72**	17.72	**13.24**	13.78	**10.89**	12.25	**10.07**	11.38	**9.59**
1000	16.54	**12.64**	16.08	**12.33**	12.78	**10.49**	11.28	**9.84**	10.54	**9.84**
t-Test	+		+		+		+		+	

DI **algorithm vs.** *DB* **algorithm.** In this group of experiment, we compare the classification performance of *DI* algorithm with *DB* algorithm. Table 1, 2, and 3 gives the classification result measured in hamming loss for *DI* and *DR* with Naive Bayes, C4.5, and SVM as base classifier, respectively. Again, as shown in table 1, 2, and 3, *chunkSize* and K affect hamming loss significantly. And from t-Test, it is shown that with different base classifiers, say, Naive Bayes, C4.5, and SVM, *DI* always outperforms *DB* significantly, which may suggest the strong ability of *DI* to utilize the dependency among different labels.

Dynamic ensemble vs. static ensemble. In this group of experiment, we compare the classification performance of static classifier ensemble with dynamic classifier ensemble. Here, we use improved BR algorithm. Table 4, 5, and 6 gives hamming loss result with Naive Bayes, C4.5, and SVM as base classifier, respectively. From t-Test, it could be concluded that, dynamic classifier ensemble offers better performance than static classifier ensemble with improved BR algorithm.

Table 4. Hamming Loss(%) of Ensemble Improved BR with Naive Bayes Algorithm

chunkSize	SI_1	DI_1	SI_2	DI_2	SI_4	DI_4	SI_6	DI_6	SI_8	DI_8
250	27.48	**26.45**	27.29	**26.23**	21.53	**20.22**	18.85	**17.32**	17.01	**15.55**
500	22.78	**21.36**	23.25	**21.28**	17.16	**16.12**	15.34	**14.00**	14.38	**12.66**
750	20.65	**19.13**	20.33	**19.06**	16.08	**14.43**	13.86	**12.43**	12.44	**11.34**
1000	18.97	**17.56**	18.85	**17.42**	14.30	**13.45**	12.32	**11.73**	11.82	**10.84**
t-Test	+		+		+		+		+	

Table 5. Hamming Loss(%) of Ensemble Improved BR with C4.5 Algorithm

chunkSize	SI_1	DI_1	SI_2	DI_2	SI_4	DI_4	SI_6	DI_6	SI_8	DI_8
250	35.56	**34.16**	34.66	**33.32**	28.76	**27.70**	24.33	**23.91**	21.99	**21.72**
500	32.78	**31.70**	31.98	**30.98**	25.59	**23.72**	21.25	**21.45**	19.56	19.76
750	31.38	**30.24**	30.86	**29.71**	25.03	**23.05**	21.05	**20.39**	18.84	**18.83**
1000	30.35	**29.60**	29.02	**28.94**	24.34	**22.72**	20.42	**20.29**	**18.65**	18.71
t-Test	+		+		+		+			

Table 6. Hamming Loss(%) of Ensemble Improved BR with SVM Algorithm

chunkSize	SI_1	DI_1	SI_2	DI_2	SI_4	DI_4	SI_6	DI_6	SI_8	DI_8
250	21.12	**19.13**	19.02	**18.33**	16.25	**16.55**	16.97	**16.43**	16.95	**16.50**
500	17.54	**16.07**	15.92	**15.48**	12.64	**12.45**	11.35	**11.37**	11.35	**10.83**
750	14.30	**13.72**	13.88	**13.24**	10.81	**10.89**	10.29	**10.07**	9.84	**9.59**
1000	12.65	**12.64**	12.64	**12.33**	10.43	**10.12**	10.01	**9.84**	10.22	**9.84**
t-Test	+		+		+		+		+	

5.2 Real-Life Data

In this subsection, we report our experiment result on RCV1-v2[3] text corpus. which is a popular multi-label text data set, and many evaluations have been documented [6] on this corpus. There are 4 main categories in this data set, namely, CCAT, ECAT, GCAT, and MCAT.

As we have shown the ability of the proposed algorithms to cope with gradual concept drift in the previous subsection, here, we preprocess the RCV1-v2 text corpus to simulate sudden concept drift. The scenario involves a user that is interested in different categories over time [24], which is shown in table 7. For example, in period 1, the user is interested in category CCAT and ECAT, and uninterested in GCAT and MCAT; in period 2, he loses his interest in CCAT and begin to show interests in GCAT. By this way, we simulate sudden concept drift.

---

[3] http://www.jmlr.org/papers/volume5/lewis04a/lewis04a.pdf

Table 7. The User's Interests Over Time

Period	Interested categories	Uninterested categories
1	CCAT,ECAT	GCAT,MCAT
2	ECAT,GCAT	CCAT,MCAT
3	GCAT,MCAT	CCAT,ECAT
4	ECAT,MCAT	GCAT,CCAT
5	CCAT,MCAT	GCAT,ECAT
6	GCAT,CCAT	MCAT,ECAT
7	CCAT,ECAT	GCAT,MCAT
...	...	...

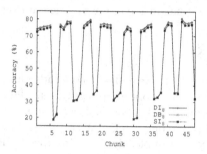

Fig. 3. Accuracy

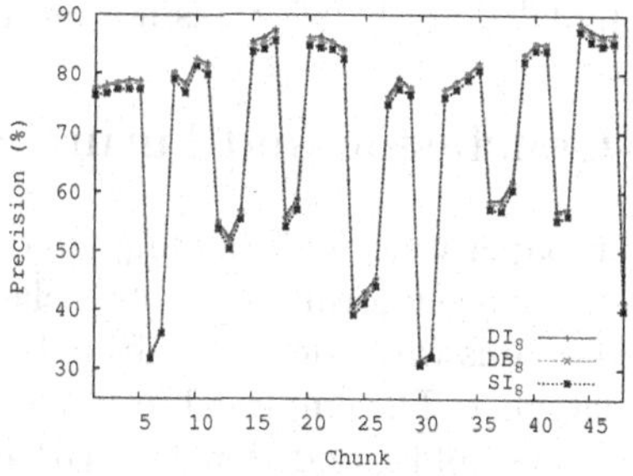

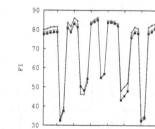

Fig. 4. Precision

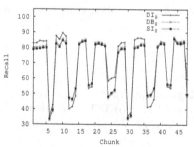

Fig. 5. Recall

Fig. 6. F_1

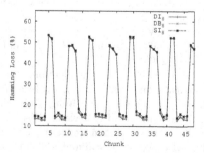

Fig. 7. Hamming Loss

For each of the experiments, ten trails of experiments are conducted, and the averaged classification performance is reported. We randomly select 49000 examples in each trial. Sudden concept drift occurs after every 6 data chunks, and 49 data chunks are generated with the data from the original text corpus. We set $chunkSize = 1000$, $K = 8$, and Naive Bayes is used as base classifier. We use five different evaluation metrics in algorithm 3 in the five experiments, say, accuracy, precision, recall, F_1, and hamming loss. Figures 3, 4, 5, 6 and 7 show the classification result measured in accuracy, precision, recall, F_1, and hamming loss, respectively. The results show that all the three algorithms, say, DI, DB, and SI, are able to track concept drift, with DI algorithm have slightly better result than DB and SI. It could be observed that the performance of DI is very close to that of DB, this may due to the reason that the dependency among labels in RCV1-v2 is not as strong as that inside our synthetic data set.

6   Conclusion and Future Work

In this paper we propose an improved binary relevance method to take advantage of dependence information among class labels, and propose to classify multi-label data streams with concept drifting by dynamic classifier ensemble with improved BR classifier. The empirical results showed that, comparing with static weighted voting ensemble, and dynamic BR classifier ensemble, dynamic improved BR classifier ensemble method offers better predictive performance. In our future work, we plan to tackle with the multi-label data streams with large label space, since BR classifier will build a binary classifier for each label, it will consume a lot of memory space.

References

1. Wang, H., Fan, W., Yu, P.S., Han, J.: Mining concept-drifting data streams using ensemble classifiers. In: Proceeding of the 9th ACM SIGKDD International Conference on Knowledge Discovery and Data Mining, pp. 226–235. ACM Press, New York (2003)
2. Hulten, G., Spencer, L., Domingos, P.: Mining Time-Changing Data Streams. In: ACM SIGKDD, pp. 97–106 (2001)
3. Widmer, G., Kubat, M.: Learning in the presence of concept drift and hidden contexts. Machine Learning 23(1), 69–101 (1996)
4. Qu, W., Zhang, Y., Zhu, J., Wang, Y.: Mining concept-drifting multi-label data streams using ensemble classifiers. In: Fuzzy Systems and Knowledge Discovery, Tianjin, China (to appear, 2009)
5. Tsoumakas, G., Katakis, I.: Multi-label classification: An overview. International Journal of Data Warehousing and Mining 3(3), 1–13 (2007)
6. Tsoumakas, G., Katakis, I., Vlahavas, I.: Mining Multi-label Data. In: Maimon, O., Rokach, L. (eds.) Data Mining and Knowledge Discovery Handbook, 2nd edn. Springer, Heidelberg (2009), http://mlkd.csd.auth.gr/multilabel.html
7. Clare, A., King, R.: Knowledge discovery in multi-label phenotype data. In: Siebes, A., De Raedt, L. (eds.) PKDD 2001. LNCS (LNAI), vol. 2168, pp. 42–53. Springer, Heidelberg (2001)

8. Schapire, R.E., Singer, Y.: Boostexter: a boosting-based system for text categorization. Machine Learning, 135–168 (2000)
9. McCallum, A.: Multi-label text classification with a mixture model trained by em. In: Proceedings of the AAAI 1999 Workshop on Text Learning, pp. 681–687 (1999)
10. Crammer, K., Singer, Y.: A family of additive online algorithms for category ranking. Journal of Machine Learning Research, 1025–1058 (2003)
11. Elisseeff, A., Weston, J.: A kernel method for multi-labeled classification. In: Advances in Neural Information Processing Systems, pp. 681–687 (2002)
12. Zhang, M.L., Zhou, Z.H.: Multi-label neural networks with applications to functional genomics and text categorization. IEEE Transactions on Knowledge and Data Engineering, 1338–1351 (2006)
13. Godbole, S., Sarawagi, S.: Discriminative methods for multi-labeled classification. In: Dai, H., Srikant, R., Zhang, C. (eds.) PAKDD 2004. LNCS (LNAI), vol. 3056, pp. 22–30. Springer, Heidelberg (2004)
14. Zhang, M.L., Zhou, Z.H.: A k-nearest neighbor based algorithm for multi-label classification. In: Proceedings of the 1st IEEE International Conference on Granular Computing, pp. 718–721 (2005)
15. Brinker, K., Hullermeier, E.: Case-based multilabel ranking. In: Proceedings of the 20th International Conference on Artificial Intelligence (IJCAI 2007), Hyderabad, India, pp. 702–707 (2007)
16. Thabtah, F., Cowling, P., Peng, Y.: Mmac: A new multi-class, multi-label associative classification approach. In: Proceedings of the 4th IEEE International Conference on Data Mining, ICDM 2004, pp. 217–224 (2004)
17. Tsymbal, A.: The problem of concept drift: definitions and related work. Technical Report TCD-CS-2004-15, Department of Computer Science, Trinity College Dublin, Ireland (2004)
18. Street, W., Kim, Y.: A streaming ensemble algorithm (SEA) for large scale classification. In: KDD 2001, 7th International Conference on Knowledge Discovery and Data Mining, San Francisco, CA, August 2001, pp. 377–382 (2001)
19. Zhu, X., Wu, X., Yang, Y.: Dynamic classifier selection for effective mining from noisy data streams. In: Proceedings of the 4th international conference on Data Mining (ICDM 2004), pp. 305–312 (2004)
20. Kolter, J.Z., Maloof, M.A.: Dynamic weighted majority: a new ensemble method for tracking concept drift. In: ICDM 2003, 3rd International Conference on Data Mining, pp. 123–130 (2003)
21. Zhang, Y., Jin, X.: An automatic construction and organization strategy for ensemble learning on data streams. ACM SIGMOD Record 35(3), 28–33 (2006)
22. Tsymbal, A., Pechenizkiy, M., Cunningham, P.: Puuronen. S.: Handling local concept drift with dynamic integration of classifiers: domain of antibiotic resistance in nosocomial infections. In: Proceedings of 19th International Symposium on Computer-Based Medical Systems (CBMS 2006), pp. 679–684 (2006)
23. Gao, S., Wu, W., Lee, C.-H., Chua, T.-S.: A maximal figure-of-merit approach to text categorization. In: SIGIR 2003, pp. 174–181 (2003)
24. Katakis, I., Tsoumakas, G., Vlahavas, I.: Dynamic Feature Space and Incremental Feature Selection for the Classification of Textual Data Streams. In: ECML/PKDD 2006 International Workshop on Knowledge Discovery from Data Streams, Berlin, Germany, pp. 107–116 (2006)

Learning Continuous-Time Information Diffusion Model for Social Behavioral Data Analysis

Kazumi Saito[1], Masahiro Kimura[2], Kouzou Ohara[3], and Hiroshi Motoda[4]

[1] School of Administration and Informatics, University of Shizuoka
52-1 Yada, Suruga-ku, Shizuoka 422-8526, Japan
k-saito@u-shizuoka-ken.ac.jp
[2] Department of Electronics and Informatics, Ryukoku University
Otsu 520-2194, Japan
kimura@rins.ryukoku.ac.jp
[3] Department of Integrated Information Technology, Aoyama Gakuin Univesity
Kanagawa 229-8558, Japan
ohara@it.aoyama.ac.jp
[4] Institute of Scientific and Industrial Research, Osaka University
8-1 Mihogaoka, Ibaraki, Osaka 567-0047, Japan
motoda@ar.sanken.osaka-u.ac.jp

Abstract. We address the problem of estimating the parameters for a continuous time delay independent cascade (CTIC) model, a more realistic model for information diffusion in complex social network, from the observed information diffusion data. For this purpose we formulate the rigorous likelihood to obtain the observed data and propose an iterative method to obtain the parameters (time-delay and diffusion) by maximizing this likelihood. We apply this method first to the problem of ranking influential nodes using the network structure taken from two real world web datasets and show that the proposed method can predict the high ranked influential nodes much more accurately than the well studied conventional four heuristic methods, and second to the problem of evaluating how different topics propagate in different ways using a real world blog data and show that there are indeed differences in the propagation speed among different topics.

1 Introduction

The rise of the Internet and the World Wide Web accelerates the creation of various large-scale social networks, and considerable attention has been brought to social networks as an important medium for the spread of information [1,2,3,4,5]. Innovation, topics and even malicious rumors can propagate through social networks in the form of so-called "word-of-mouth" communications. This forms a virtual society forming various kinds of communities. Just like a real world society, some community grows rapidly and some other shrinks. Likewise, some information propagates quickly and some other only slowly. Good things remain and bad things diminish as if there is a natural selection. The social network offers a nice platform to study a mechanism of society dynamics and behavior of humans, each as a member of the society. In this paper, we address the problem of how information diffuses through the social network, in

Z.-H. Zhou and T. Washio (Eds.): ACML 2009, LNAI 5828, pp. 322–337, 2009.

particular how different topics propagate differently by inducing a diffusion model that can handle continuous time delay.

There are several models that simulate information diffusion through a network. A widely-used model is the *independent cascade (IC)*, a fundamental probabilistic model of information diffusion [6,7], which can be regarded as the so-called *susceptible/infected/recovered (SIR) model* for the spread of a disease [2]. This model has been used to solve such problems as the *influence maximization problem* which is to find a limited number of nodes that are influential for the spread of information [7,8] and the *influence minimization problem* which is to suppress the spread of undesirable information by blocking a limited number of links [9]. The IC model requires the parameters that represent diffusion probabilities through links to be specified in advance. Since the true values of the parameters are not available in practice, this poses yet another problem of estimating them from the observed data [10].

One of the drawbacks of the IC model is that it cannot handle time-delays for information propagation, and we need a model to explicitly represent time delay. Gruhl et al. is the first to extend the IC model to include the time-delay [3]. Their model now has the parameters that represent time-delays through links as well as the parameters that represent diffusion probabilities through links. They presented a method for estimating the parameter values from the observed data using an EM-like algorithm, and experimentally showed its effectiveness using sparse Erdös-Renyi networks. However, it is not clear what they are optimizing in deriving the update formulas of the parameter values. Further, they treated the time as a discrete variable, which means that it is assumed that information propagate in a synchronized way in a sense that each node can be activated only at a specific time. In reality, time flows continuously and thus information, too, propagates on this continuous time axis. For any node, information must be able to be received at any time from other nodes and must be allowed to propagate to yet other nodes at any other time, both in an asynchronous way. Thus, for a realistic behavior analyses of information diffusion, we need to adopt a model that explicitly represents continuous time delay.

In this paper, we deal with an information diffusion model that incorporates continuous time delay based on the IC model (referred to as CTIC model), and propose a novel method for estimating the values of the parameters in the model from a set of information diffusion results that are observed as time-sequences of infected (active) nodes. What makes this problem difficult is that incorporating time-delay makes the time-sequence observation data structural. There is no way of knowing from the data which node activated which other node that comes later in the sequence. We introduce an objective function that rigorously represents the likelihood of obtaining such observed data sequences under the CTIC model on a given network, and derive an iterative algorithm by which the objective function is maximized. First we test the convergence performance of the proposed method by applying it to the problem of ranking influential nodes using the network structure taken from two real world web datasets and show that the parameters converge to the correct values by the iterative procedure and can predict the high ranked influential nodes much more accurately than the well studied conventional four heuristic methods. Second we apply the method to the problem of behavioral analysis of topic propagation, i.e., evaluating how different topics propagate in different ways, using a real world blog data and show that there are indeed differences in the propagation speed among different topics.

2   Information Diffusion Model and Learning Problem

We first define the IC model according to [7], and then introduce the continuous-time IC model. After that, we formulate our learning problem.

We mathematically model the spread of information through a directed network $G = (V, E)$ without self-links, where V and E ($\subset V \times V$) stands for the sets of all the nodes and links, respectively. We call nodes *active* if they have been influenced with the information. In the model, it is assumed that nodes can switch their states only from inactive to active, but not from active to inactive. Given an initial set S of active nodes, we assume that the nodes in S have first become active at an initial time, and all the other nodes are inactive at that time.

In this paper, node u is called a *child node* of node v if $(v, u) \in E$, and node u is called a *parent node* of node v if $(u, v) \in E$. For each node $v \in V$, let $F(v)$ and $B(v)$ denote the set of child nodes of v and the set of parent nodes of v, respectively,

$$F(v) = \{w \in V; (v, w) \in E\}, \quad B(v) = \{u \in V; (u, v) \in E\}.$$

2.1   Independent Cascade Model

Let us describe the definition of the IC model. In this model, for each link (u, v), we specify a real value $\lambda_{u,v}$ with $0 < \lambda_{u,v} < 1$ in advance. Here $\lambda_{u,v}$ is referred to as the *diffusion probability* through link (u, v).

The diffusion process unfolds in discrete time-steps $t \geq 0$, and proceeds from a given initial active set S in the following way. When a node u becomes active at time-step t, it is given a single chance to activate each currently inactive child node v, and succeeds with probability $\lambda_{u,v}$. If u succeeds, then v will become active at time-step $t + 1$. If multiple parent nodes of v become active at time-step t, then their activation attempts are sequenced in an arbitrary order, but all performed at time-step t. Whether or not u succeeds, it cannot make any further attempts to activate v in subsequent rounds. The process terminates if no more activations are possible.

2.2   Continuous-Time Independent Cascade Model

Next, we extend the IC model so as to allow continuous-time delays, and refer to the extended model as the *continuous-time independent cascade (CTIC) model*.

In the CTIC model, for each link $(u, v) \in E$, we specify real values $r_{u,v}$ and $\kappa_{u,v}$ with $r_{u,v} > 0$ and $0 < \kappa_{u,v} < 1$ in advance. We refer to $r_{u,v}$ and $\kappa_{u,v}$ as the *time-delay parameter* and the *diffusion parameter* through link (u, v), respectively.

The diffusion process unfolds in continuous-time t, and proceeds from a given initial active set S in the following way. Suppose that a node u becomes active at time t. Then, node u is given a single chance to activate each currently inactive child node v. We choose a delay-time δ from the exponential distribution with parameter $r_{u,v}$. If node v

is not active before time $t + \delta$, then node u attempts to activate node v, and succeeds with probability $\kappa_{u,v}$. If u succeeds, then v will become active at time $t + \delta$. Under the continuous time framework, it is unlikely that multiple parent nodes of v attempt to activate v for the activation at time $t + \delta$. But if they do, their activation attempts are sequenced in an arbitrary order. Whether or not u succeeds, it cannot make any further attempts to activate v in subsequent rounds. The process terminates if no more activations are possible.

For an initial active set S, let $\varphi(S)$ denote the number of active nodes at the end of the random process for the CTIC model. Note that $\varphi(S)$ is a random variable. Let $\sigma(S)$ denote the expected value of $\varphi(S)$. We call $\sigma(S)$ the *influence degree* of S for the CTIC model.

2.3 Learning Problem

For the CTIC model on network G, we define the time-delay parameter vector r and the diffusion parameter vector κ by

$$r = (r_{u,v})_{(u,v)\in E}, \quad \kappa = (\kappa_{u,v})_{(u,v)\in E}.$$

In practice, the true values of r and κ are not available. Thus, we must estimate them from past information diffusion histories observed as sets of active nodes.

We consider an observed data set of M independent information diffusion results,

$$\mathcal{D}_M = \{D_m;\ m = 1, \cdots, M\}.$$

Here, each D_m is a time-sequence of active nodes in the mth information diffusion result,

$$D_m = \langle D_m(t);\ t \in \mathcal{T}_m \rangle, \quad \mathcal{T}_m = \langle t_m, \cdots, T_m \rangle,$$

where $D_m(t)$ is the set of all the nodes that have first become active at time t, and $\mathcal{T}_m$ is the observation-time list; t_m is the observed initial time and T_m is the observed final time. We assume that for any active node v in the mth information diffusion result, there exits some $t \in \mathcal{T}_m$ such that $v \in D_m(t)$. Let $t_{m,v}$ denote the time at which node v becomes active in the mth information diffusion result, i.e., $v \in D_m(t_{m,v})$. For any $t \in \mathcal{T}_m$, we set

$$C_m(t) = \bigcup_{\tau \in \mathcal{T}_m \cap \{s;\ s < t\}} D_m(\tau)$$

Note that $C_m(t)$ is the set of active nodes before time t in the mth information diffusion result. We also interpret D_m as referring to the set of all the active nodes in the mth information diffusion result for convenience sake. In this paper, we consider the problem of estimating the values of r and κ from $\mathcal{D}_M$.

3 Proposed Method

We explain how we estimate the values of r and κ from $\mathcal{D}_M$. Here, we limit ourselves to outline the derivations of the proposed method due to the lack of space. We also briefly mention how we do behavioral analysis with the method.

3.1   Likelihood Function

For the learning problem described above, we strictly derive the likelihood function $\mathcal{L}(r, \kappa; \mathcal{D}_M)$ with respect to r and κ to use as our objective function.

First, we consider any node $v \in D_m$ with $t_{m,v} > 0$ for the mth information diffusion result. Let $\mathcal{A}_{m,u,v}$ denote the probability density that a node $u \in B(v) \cap C_m(t_{m,v})$ activates the node v at time $t_{m,v}$, that is,

$$\mathcal{A}_{m,u,v} = \kappa_{u,v} r_{u,v} \exp(-r_{u,v}(t_{m,v} - t_{m,u})). \tag{1}$$

Let $\mathcal{B}_{m,u,v}$ denote the probability that the node v is not activated from a node $u \in B(v) \cap C_m(t_{m,v})$ within the time-period $[t_{m,u}, t_{m,v}]$, that is,

$$\mathcal{B}_{m,u,v} = 1 - \kappa_{u,v} \int_{t_{m,u}}^{t_{m,v}} r_{u,v} \exp(-r_{u,v}(t - t_{m,u})) dt$$

$$= \kappa_{u,v} \exp(-r_{u,v}(t_{m,v} - t_{m,u})) + (1 - \kappa_{u,v}). \tag{2}$$

If there exist multiple active parents for the node v, i.e., $\eta = |B(v) \cap C_m(t_{m,v})| > 1$, we need to consider possibilities that each parent node succeeds in activating v at time $t_{m,v}$. However, in case of the continuous time delay model, we can ignore simultaneous activations by multiple active parents due to the continuous property. Thus, the probability density that the node v is activated at time $t_{m,v}$, denoted by $h_{m,v}$, can be expressed as

$$h_{m,v} = \sum_{u \in B(v) \cap C_m(t_{m,v})} \mathcal{A}_{m,u,v} \left(\prod_{x \in B(v) \cap C_m(t_{m,v}) \setminus \{u\}} \mathcal{B}_{m,x,v} \right).$$

$$= \prod_{x \in B(v) \cap C_m(t_{m,v})} \mathcal{B}_{m,x,v} \sum_{u \in B(v) \cap C_m(t_{m,v})} \mathcal{A}_{m,u,v} (\mathcal{B}_{m,u,v})^{-1}. \tag{3}$$

Note that we are not able to know which node u actually activated the node v. This can be regarded as a hidden structure.

Next, for the mth information diffusion result, we consider any link $(v, w) \in E$ such that $v \in C_m(T_m)$ and $w \notin D_m$. Let $g_{m,v,w}$ denote the probability that the node w is not activated by the node v within the observed time period $[t_m, T_m]$. We can easily derive the following equation:

$$g_{m,v,w} = \kappa_{v,w} \exp(-r_{v,w}(T_m - t_{m,v})) + (1 - \kappa_{v,w}). \tag{4}$$

Here we can naturally assume that each information diffusion process finished sufficiently earlier than the observed final time, i.e., $T_m \gg \max\{t; D_m(t) \neq \emptyset\}$. Thus, as $T_m \to \infty$ in equation (4), we assume

$$g_{m,v,w} = 1 - \kappa_{v,w}. \tag{5}$$

Therefore, by using equations (3), (5), and the independence properties, we can define the likelihood function $\mathcal{L}(r, \kappa; \mathcal{D}_M)$ with respect to r and κ by

$$\mathcal{L}(r, \kappa; \mathcal{D}_M) = \prod_{m=1}^{M} \left(\prod_{t \in T_m} \prod_{v \in D_m(t)} h_{m,v} \prod_{v \in D_m} \prod_{w \in F(v) \setminus D_m} g_{m,v,w} \right). \tag{6}$$

Here, we retained the product with respect to $v \in D_m(t)$ for completeness, but in practice there is only one v in $D_m(t)$.

In this paper, we focus on the above situation (i.e., equation (5)) for simplicity, but we can easily modify our method to cope with the general one (i.e., equation (4)). Thus, our problem is to obtain the values of r and κ, which maximize equation (6). For this estimation problem, we derive a method based on an iterative algorithm in order to stably obtain its solution.

3.2 Estimation Method

We describe our estimation method. Let $\bar{r} = (\bar{r}_{u,v})$ and $\bar{\kappa} = (\bar{\kappa}_{u,v})$ be the current estimates of r and κ, respectively. For each $v \in D_m$ and $u \in B(v) \cap C_m(t_{m,v})$, we define $\alpha_{m,u,v}$ by

$$\alpha_{m,u,v} = \mathcal{A}_{m,u,v}(\mathcal{B}_{m,u,v})^{-1} / \sum_{x \in B(v) \cap C_m(t_{m,v})} \mathcal{A}_{m,x,v}(\mathcal{B}_{m,x,v})^{-1}. \tag{7}$$

Let $\bar{\mathcal{A}}_{m,u,v}$, $\bar{\mathcal{B}}_{m,u,v}$, $\bar{h}_{m,v}$, and $\bar{\alpha}_{m,u,v}$ denote the values of $\mathcal{A}_{m,u,v}$, $\mathcal{B}_{m,u,v}$, $h_{m,v}$, and $\alpha_{m,u,v}$ calculated by using $\bar{r}$ and $\bar{\kappa}$, respectively.

From equations (3), (5), (6), we can transform our objective function $\mathcal{L}(r, \kappa; \mathcal{D}_M)$ as follows:

$$\log \mathcal{L}(r, \kappa; \mathcal{D}_M) = Q(r, \kappa; \bar{r}, \bar{\kappa}) - H(r, \kappa; \bar{r}, \bar{\kappa}), \tag{8}$$

where $Q(r, \kappa; \bar{r}, \bar{\kappa})$ is defined by

$$Q(r, \kappa; \bar{r}, \bar{\kappa}) = \sum_{m=1}^{M} \left(\sum_{t \in T_m} \sum_{v \in D_m(t)} Q_{m,v} + \sum_{v \in D_m} \sum_{w \in F(v) \setminus D_m} \log(1 - \kappa_{v,w}) \right),$$

$$Q_{m,v} = \sum_{u \in B(v) \cap C_m(t_{m,v})} \log(\mathcal{B}_{m,u,v}) + \sum_{u \in B(v) \cap C_m(t_{m,v})} \bar{\alpha}_{m,u,v} \log \left(\mathcal{A}_{m,u,v}(\mathcal{B}_{m,u,v})^{-1} \right) \tag{9}$$

and $H(r, \kappa; \bar{r}, \bar{\kappa})$ is defined by

$$H(r, \kappa; \bar{r}, \bar{\kappa}) = \sum_{m=1}^{M} \sum_{t \in T_m} \sum_{v \in D_m(t)} \sum_{u \in B(v) \cap C_m(t_{m,v})} \bar{\alpha}_{m,u,v} \log \alpha_{m,u,v}. \tag{10}$$

Since $H(r, \kappa; \bar{r}, \bar{\kappa})$ is maximized at $r = \bar{r}$ and $\kappa = \bar{\kappa}$ from equation (10), we can increase the value of $\mathcal{L}(r, \kappa; \mathcal{D}_M)$ by maximizing $Q(r, \kappa; \bar{r}, \bar{\kappa})$ (see equation (8)). Note here that although $\log \mathcal{A}_{m,u,v}$ is a linear combination of $\log \kappa_{u,v}$, $\log r_{u,v}$, and $r_{u,v}$, $\log \mathcal{B}_{m,u,v}$ cannot be written as such a linear combination (see equations (1), (2)). In order to cope with this problem of $\log \mathcal{B}_{m,u,v}$, we transform $\log \mathcal{B}_{m,u,v}$ in the same way as above, and define $\beta_{m,u,v}$ by

$$\beta_{m,u,v} = \kappa_{u,v} \exp(-r_{u,v}(t_{m,v} - t_{m,u})) / \mathcal{B}_{m,u,v}$$

Finally, as the solution which maximizes $Q(r, \kappa; \bar{r}, \bar{\kappa})$, we obtain the following update formulas of our estimation method:

$$r_{u,v} = \frac{\sum_{m \in M_{u,v}^+} \bar{\alpha}_{m,u,v}}{\sum_{m \in M_{u,v}^+} (\bar{\alpha}_{m,u,v} + (1 - \bar{\alpha}_{m,u,v})\bar{\beta}_{m,u,v})(t_{m,v} - t_{m,u})},$$

$$\kappa_{u,v} = \frac{1}{|\mathcal{M}_{u,v}^+| + |\mathcal{M}_{u,v}^-|} \sum_{m \in \mathcal{M}_{u,v}^+} (\bar{\alpha}_{m,u,v} + (1 - \bar{\alpha}_{m,u,v})\bar{\beta}_{m,u,v}),$$

where $\mathcal{M}_{u,v}^+$ and $\mathcal{M}_{u,v}^-$ are defined by

$$\mathcal{M}_{u,v}^+ = \{m \in \{1, \cdots, M\};\ u, v \in D_m,\ v \in F(u),\ t_{m,u} < t_{m,v}\},$$
$$\mathcal{M}_{u,v}^- = \{m \in \{1, \cdots, M\};\ u \in D_m,\ v \notin D_m,\ v \in F(u)\}.$$

Note that we can regard our estimation method as a kind of the EM algorithm. It should be noted here that each time iteration proceeds the value of the likelihood function never decreases and the iterative algorithm is guaranteed to converge.

3.3   Behavioral Analysis

Thus far, we assumed that the parameters (time-delay and diffusion) can vary with respect to links but remain the same irrespective of the topic of information diffused, following Gruhl et al. [3]. However, they may be sensitive to the topic.

Our method can cope with this by assigning m to a topic, and placing a constraint that the parameters depends only on topics but not on links throughout the network G, that is $r_{m,u,v} = r_m$ and $\kappa_{m,u,v} = \kappa_m$ for any link $(u, v) \in E$. This constraint is required because, without this, we have only one piece of observation for each (m, u, v) and there is no way to learn the parameters. Noting that we can naturally assume that people behave quite similarly for the same topic, this constraint should be acceptable. Under this setting, we can easily obtain the parameter update formulas. Using each pair of the estimated parameters, (r_m, κ_m), we can analyze the behavior of people with respect to the topics of information, by simply plotting (r_m, κ_m) as a point of 2-dimensional space.

3.4   Simple Case Analysis

Here, we analyze a few basic properties of the proposed method under simple settings. Assume that a node v became active at time t after receiving certain information. We denote the active parent nodes of v by $u_1, \cdots, u_N$. First, we consider a simple case that diffusion parameter κ is 1 for all links, time-delay parameter r is a constant and the same for all links, and the activation times of $u_1, \cdots, u_N$ are all zeros. Then, as is given in equation (3), the probability density that the node v is activated at time t by one of the parent nodes, can be expressed as follows

$$h_v = \sum_{n=1}^{N} r \exp(-rt) \left(1 - \int_0^t r \exp(-r\tau)d\tau\right)^{N-1} = Nr \exp(-Nrt).$$

Similarly, for the case that the parent nodes $u_1, \cdots, u_N$ became active at times $t_1, \cdots t_N$ $(< t)$, respectively, we easily obtain the following probability.

$$h_v = Nr \exp\left(-Nr\left(t - \frac{1}{N}\sum_{n=1}^{N} t_n\right)\right).$$

The maximum likelihood is attained by maximizing $\log h_v$ with respect to r, and the average delay time is obtained as follows:

$$r^{-1} = N\left(t - \frac{1}{N}\sum_{n=1}^{N} t_n\right).$$

We can see that this estimation is N times larger than the simple average of time differences. In other words, the information diffuses more quickly when there exist multiple active parents, i.e., r^{-1}/N, and this fact matches our intuition. Thus simple statistics such as the average delay time may fail to provide the intrinsic property of information diffusion phenomena, and this suggests that an adequate information diffusion model is vital.

Next, we consider another simple case that the diffusion parameter κ and the time-delay parameter r are both uniform and constant for all links, and the activation times of $u_1, \cdots, u_N$ are all zeros. Here both parameters are variables. Then the probability density that the node v is activated at time t can be expressed as follows

$$h_v = N\kappa r \exp(-rt)(\kappa \exp(-rt) + (1 - \kappa))^{N-1}.$$

Now, we consider maximizing $f(\kappa, r) = \log h_v$ with respect to κ and r. The first- and second-order derivatives of $f(\kappa, r)$ with respect to κ are given by

$$\frac{\partial f(\kappa, r)}{\partial \kappa} = \frac{1}{\kappa} + (N - 1)\frac{\exp(-rt) - 1}{\kappa \exp(-rt) + (1 - \kappa)}$$

$$\frac{\partial^2 f(\kappa, r)}{\partial \kappa \partial \kappa} = -\frac{1}{\kappa^2} - (N - 1)\left(\frac{\exp(-rt) - 1}{\kappa \exp(-rt) + (1 - \kappa)}\right)^2.$$

Since the above second-order derivative is negative definite for a given parameter r, we note that there exists a unique global solution to κ. The corresponding derivatives with respect to r are given by

$$\frac{\partial f(\kappa, r)}{\partial r} = \frac{1}{r} - t - (N - 1)\frac{t\kappa \exp(-rt)}{\kappa \exp(-rt) + (1 - \kappa)}$$

$$\frac{\partial^2 f(\kappa, r)}{\partial r \partial r} = -\frac{1}{r^2} + (N - 1)\frac{t^2 \kappa(1 - \kappa)\exp(-rt)}{(\kappa \exp(-rt) + (1 - \kappa))^2}.$$

Unfortunately, we cannot guarantee that the above second-order derivative is negative definite. However, most likely, this value is negative when $r \ll 1$, and can be positive when $r \gg 1$ in which case the shape of the objective function can be complex. We can speculate that the convergence is better for a smaller value of r. Later, in our experiments, we empirically evaluate this point by using the method described in 3.1 and 3.2 with $r = 2$ and $r = 0.5$, which are in the range that is widely explored by many existing studies. Clearly, we need to perform further theoretical and empirical studies because we are simultaneously estimating both diffusion and time-delay parameters, κ and r. However, the experiments show that our method is stable for the range of parameters we used, indicating that the likelihood function has favorable mathematical properties.

4   Experiments with Artificial Data

We evaluated the effectiveness of the proposed learning method using the topologies of two large real network data. First, we evaluated how accurately it can estimate the parameters of the CTIC model from $\mathcal{D}_M$. Next, we considered applying our learning method to the problem of extracting influential nodes, and evaluated how well our learned model can predict the high ranked influential nodes with respect to influence degree $\sigma(v)$, $(v \in V)$ for the true CTIC model.

4.1   Experimental Settings

In our experiments, we employed two datasets of large real networks used in [9], which exhibit many of the key features of social networks. The first one is a trackback network of Japanese blogs. The network data was collected by tracing the trackbacks from one blog in the site goo^1 in May, 2005. We refer to this network data as the blog network. The blog network was a strongly-connected bidirectional network, where a link created by a trackback was regarded as a bidirectional link since blog authors establish mutual communications by putting trackbacks on each other's blogs. The blog network had 12,047 nodes and 79,920 directed links. The second one is a network of people that was derived from the "list of people" within Japanese Wikipedia. We refer to this network data as the Wikipedia network. The Wikipedia network was also a strongly-connected bidirectional network, and had 9,481 nodes and 245,044 directed links.

Here, we assumed the simplest case where $r_{u,v}$ and $\kappa_{u,v}$ are uniform throughout the network G, that is, $r_{u,v} = r$, $\kappa_{u,v} = \kappa$ for any link $(u, v) \in E$. One reason behind this assumption is that we can make fair comparison with the existing heuristics that are solely based on network structure (see 4.2). Another reason is that there is no need to acquire observation sequence data that at least pass through every link once. This drastically reduces the amount of data to learn the parameters. Then, our task is to estimate the values of r and κ. According to [7], we set the value of κ relatively small. In particular, we set the value of κ to a value smaller than $1/\bar{d}$, where $\bar{d}$ is the mean out-degree of a network. Since the values of $\bar{d}$ were about 6.63 and 25.85 for the blog and the Wikipedia networks, respectively, the corresponding values of $1/\bar{d}$ were about 0.15 and 0.03. Thus, as for the true value of the diffusion parameter κ, we decided to set $\kappa = 0.1$ for the blog network and $\kappa = 0.01$ for the Wikipedia network. As for the true value of the time-delay parameter r, we decided to investigate two cases: one with a relatively high value $r = 2$ (a short time-delay case) and the other with a relatively low value $r = 0.5$ (a long time-delay case) in both networks. We used the training data $\mathcal{D}_M$ in the learning stage, which is constructed by generating each D_m from a randomly selected initial active node $D_m(0)$ using the true CTIC model. T_m was chosen to be effectively ∞.

We note that the influence degree $\sigma(v)$ of a node v is invariant with respect to the values of the delay-parameter r. In fact, the effect of r is to delay the timings when nodes become active, that is, parameter $r_{u,v}$ only controls how soon or late node v actually becomes active when node u activates node v. Therefore, nodes that can be activated

---

^{1} http://blog.goo.ne.jp/

are in indeed activated eventually after a sufficiently long time has elapsed, which is the case here, i.e. $T_m = \infty$. Thus, we can evaluate the $\sigma(v)$ of the CTIC model by the influence degree of v for the corresponding IC model. We estimated the influence degrees $\{\sigma(v); v \in V\}$ using the method of [8] with the parameter value $10,000$, where the parameter represents the number of bond percolation processes (we do not describe the method here due to the page limit). The average value and the standard deviation of the influence degrees was 87.5 and 131 for the blog network, and 8.14 and 18.4 for the Wikipedia network.

4.2   Comparison Methods

We compared the predicted result of the high ranked influential nodes for the true CTIC model by the proposed method with four heuristics widely used in social network analysis.

The first three of these heuristics are "degree centrality", "closeness centrality", and "betweenness centrality". These are commonly used as influence measure in sociology [11], where the out-degree of node v is defined as the number of links going out from v, the closeness of node v is defined as the reciprocal of the average distance between v and other nodes in the network, and the betweenness of node v is defined as the total number of shortest paths between pairs of nodes that pass through v. The fourth is "authoritativeness" obtained by the "PageRank" method [12]. We considered this measure since this is a well known method for identifying authoritative or influential pages in a hyperlink network of web pages. This method has a parameter ε; when we view it as a model of a random web surfer, ε corresponds to the probability with which a surfer jumps to a page picked uniformly at random [13]. In our experiments, we used a typical setting of $\varepsilon = 0.15$.

4.3   Experimental Results

First, we examined the parameter estimation accuracy by the proposed method. Let r_0 and κ_0 be the true values of the parameters r and κ, respectively, and let $\hat{r}$ and $\hat{\kappa}$ be the values of r and κ estimated by the proposed method, respectively. We evaluated the learning performance in terms of the error rates,

$$\mathcal{E}_r = \frac{|r_0 - \hat{r}|}{r_0}, \quad \mathcal{E}_\kappa = \frac{|\kappa_0 - \hat{\kappa}|}{\kappa_0}.$$

Table 1 shows the average values of $\mathcal{E}_r$ and $\mathcal{E}_\kappa$ for different numbers of training samples, M. For each M we repeated the same experiment 5 times independently, and for each experiment we tried 5 different initial values of the parameters that are randomly drawn from $[0,1]$ with uniform distribution. The convergence criterion is

$$|\kappa^{(n)} - \kappa^{(n+1)}| + |r^{(n)} - r^{(n+1)}| < 10^{-12},$$

where the superscript (n) indicates the value for the nth iteration. Our algorithm converged at around 40 iterations for the blog data and 70 iterations for the Wikipedia data.

Table 1. Learning performance by the proposed method

Blog network (r = 2)			Wikipedia network (r = 2)			Blog network (r = 0.5)			Wikipedia network (r = 0.5)		
M	$\mathcal{E}_r$	$\mathcal{E}_\kappa$	M	$\mathcal{E}_r$	$\mathcal{E}_\kappa$	M	$\mathcal{E}_r$	$\mathcal{E}_\kappa$	M	$\mathcal{E}_r$	$\mathcal{E}_\kappa$
20	0.013	0.015	20	0.036	0.034	20	0.011	0.012	20	0.026	0.028
40	0.010	0.010	40	0.024	0.016	40	0.010	0.007	40	0.021	0.023
60	0.008	0.008	60	0.013	0.015	60	0.009	0.005	60	0.018	0.021
80	0.007	0.007	80	0.012	0.013	80	0.004	0.004	80	0.014	0.012
100	0.005	0.005	100	0.006	0.011	100	0.004	0.004	100	0.007	0.006

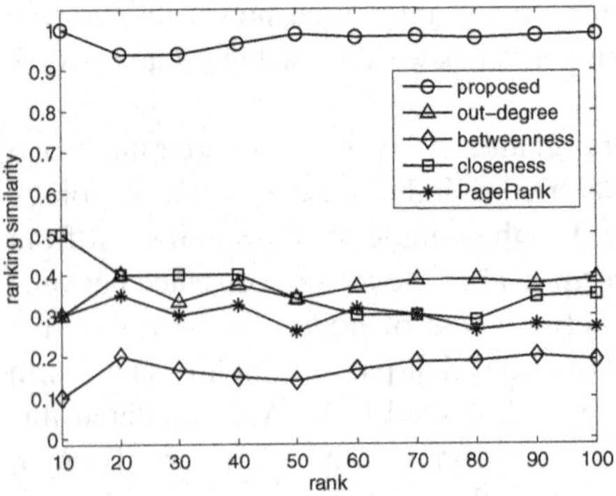

(a) blog network ($r = 2$) 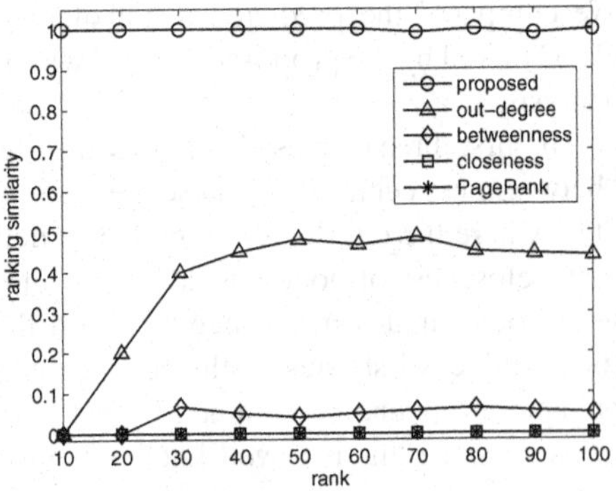   (b) Wikipedia network ($r = 2$)

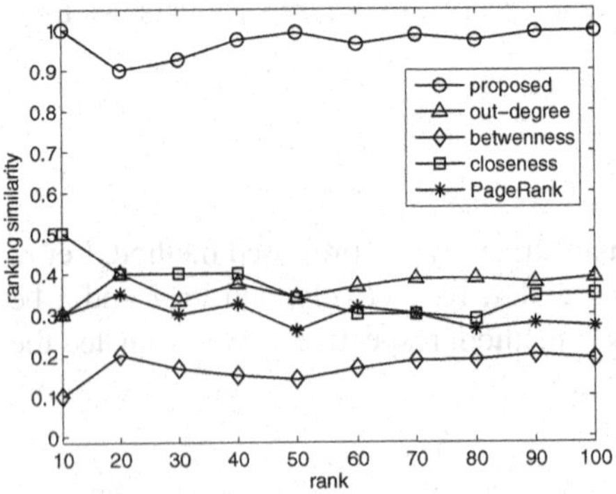

(c) blog network ($r = 0.5$) 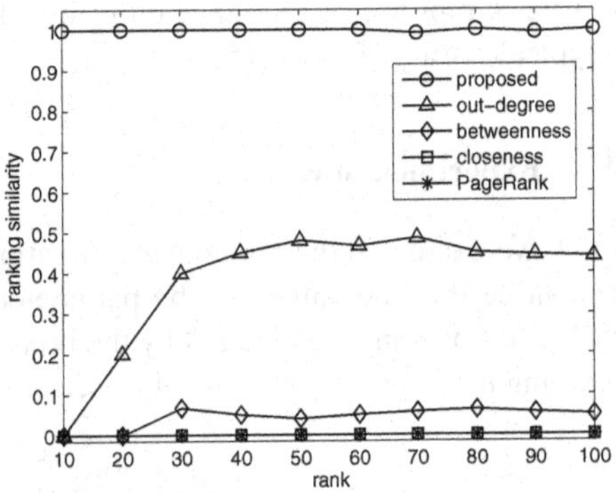   (d) Wikipedia network ($r = 0.5$)

Fig. 1. Performance comparison in extracting influential nodes

Further, it is observed, as predicted by the simple case analysis in 3.4, that the convergence was faster for a smaller value of r. The converged values are close to the true values when there is a reasonable amount of training data. The results demonstrate the effectiveness of the proposed method.

Next, we compared the proposed method with the out-degree, the betweenness, the closeness, and the PageRank methods in terms of the capability of ranking the influential nodes. For any positive integer k ($\leq |V|$), let $L_0(k)$ be the true set of top k nodes, and let $L(k)$ be the set of top k nodes for a given ranking method. We evaluated the performance of the ranking method by the *ranking similarity* $F(k)$ at rank k, where $F(k)$ is defined by

$$F(k) = \frac{|L_0(k) \cap L(k)|}{k}.$$

We focused on ranking similarities only at high ranks since we are interested in extracting influential nodes. Figures 1a and 1c show the results for the blog network, and Figures 1b and 1d show the results for the Wikipedia network, where the true value of r is $r = 2$ and $r = 0.5$ for Figures 1a and 1b, and Figures 1c and 1d, respectively. In these figures, circles, triangles, diamonds, squares, and asterisks indicate ranking similarity $F(k)$ as a function of rank k for the proposed, the out-degree, the betweenness, the closeness, and the PageRank methods, respectively. For the proposed method, we plotted the average value of $F(k)$ at k for 5 experimental results stated earlier in the case of $M = 100$. The proposed method gives far better results than the other heuristic based methods for the both networks, demonstrating the effectiveness of our proposed learning method.

5 Behavioral Analysis of Real World Blog Data

We applied our method to behavioral analysis using a real world blog data based on the method described in 3.3 and investigated how each topic spreads throughout the network.

5.1 Experimental Settings

The network we used is a real blogroll network in which bloggers are connected to each other. We note that when there is a blogroll link from blogger y to another blogger x, this means that y is a reader of the blog of x. Thus, we can assume that topics propagate from blogger x to blogger y. According to [14], we suppose that a topic is represented as a URL which can be tracked down from blog to blog. We used the database of a blog-hosting service in Japan called *Doblog*[2]. The database is constructed by all the Doblog data from October 2003 to June 2005, and contains $52, 525$ bloggers and $115, 552$ blogroll links.

We identified all the URLs mentioned in blog posts in the Doblog database, and constructed the following list for each URL from all the blog posts that contain the URL:

$$\langle (v_1, t_1), \cdots, (v_k, t_k) \rangle, \quad (t_1 < \cdots < t_k),$$

where v_i is a blogger who mentioned the URL in her/his blog post published at time t_i. By taking into account the blogroll relations for the list, we estimated such paths that the URL might propagate through the blogroll network. We extracted $7, 356$ URL

---

[2] Doblog(http://www.doblog.com/), provided by NTT Data Corp. and Hotto Link, Inc.

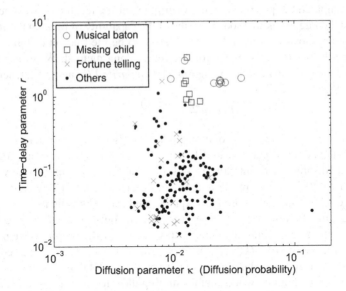

Fig. 2. Results for the Doblog database

propagation paths from the Doblog dataset, where we ignored the URLs that only one blogger mentioned. Out of these, only those that are longer than 10 time steps are chosen for analyses, resulting into 172 sequences. Each sequence data represents a topic, and a topic can be distributed in multiple URLs. The same URL can appear in different sequences. Here note that the time stamp of each blog article is different from each other and thus, the time intervals in the sequence $< t_1, t_2, ..., t_k >$ are not a fixed constant.

5.2 Experimental Results

We ran the experiments for each identified URL and obtained the corresponding parameters κ and r. Figure 2 is a plot of the results for the major URLs. The horizontal axis is the diffusion parameter κ and the vertical axis is the delay parameter r. The latter is normalized such that $r = 1$ corresponds to a delay of one day, meaning $r = 0.1$ corresponds delay of 10 days. We only explain three URLs that exhibit some interesting propagation properties. The circle is a URL that corresponds to the musical baton which is a kind of telephone game on the Internet. It has the following rules. First, a blogger is requested to respond to five questions about music by some other bologger (receive the baton) and the requested blogger replies to the questions and designate the next five bloggers with the same questions (pass the baton). It is shown that this kind of message propagates quickly (less than one day on the average) with a good chance (one out of 25 to 100 persons responds). This is probably because people are interested in this kind of message passing. The square is a URL that corresponds to articles about a missing child. This also propagates quickly with a meaningful probability (one out of 80 persons responds). This is understandable considering the urgency of the message. The cross is a URL that corresponds to articles about fortune telling. Peoples responses are diverse.

Some responds quickly (less than one day) and some late (more than one month after), and they are more or less uniformly distributed. The diffusion probability is also nearly uniformly distributed. This reflects that each individual's interest is different on this topic. The dot is a URL that corresponds to one of the other topics. Interestingly, the one in the bottom right which is isolated from the rest is a post of an invitation to a rock music festival. This one has a very large probability of being propagated but with very large time delay. In general, it can be said that the proposed method can extract characteristic properties of a certain topics reasonably well only from the observation data.

6   Discussion

Being able to handle the time more precisely brings a merit to the analysis of such information diffusion as in a blog data because the time stamp is available in the unit of second. There are subtle cases where it is not self evident to which value to assign the time when the discretization has to be made. We have solved this problem.

There are many pieces of work in which time sequence data is analyzed assuming a certain model behind. Ours also falls in this category. The proposed approach brings in a new perspective in which it allows to use the structure of a complex network as a kind of background knowledge in a more refined way. There are also many pieces of work on topic propagation analyses, but they focus mostly on the analyses of average propagation speed (propagation speed distribution) and average life time. Our method is new and different in that we explicitly address the diffusion phenomena incorporating diffusion probability and time delay as well as the structure of the network.

The proposed method derives the learning algorithm in a principled way. The objective function has a clear meaning of the likelihood by which to obtain the observed data, and the parameter is iteratively updated in such a way to maximize the likelihood, guaranteeing the convergence. Due to the property of continuous time, we excluded the possibility that a node is activated simultaneously by multiple parent nodes. It is also straightforward to formulate the likelihood taking the possibility of the simultaneous activation into account. However, the numerical experiments revealed that the results are not as accurate as the current model. Having to explore millions of paths with very small probability does harm numerical computation. This is, in a sense, similar to the problem of feature selection in building a classifier. It is known that the existence of irrelevant features is harmful even though the classification algorithm can in theory ignore those irrelevant features.

The CTIC model is a continuous-time information diffusion model that extends the discrete-time model by Gruhl et al [15]. We note that their model is based on the popular IC model and they model the time-delay by a geometric distribition. In the CTIC model, we model a time-delay by an exponential distribution. Song et al [16] also modeled time-delays of information flow by exponential distributions in formulating an information flow model by a continuous-time Markov chain (i.e., a random-surfer model). Thus, we can regard the CTIC model as a natural extension to continuous-time information diffusion model based on the IC model, and investigating its characteristics can be an important research issue. As explained in Section 2.2, the CTIC model is rather complicated, and developing a learning algorithm of the CTIC model is challenging. In this paper, we presented an effective method for estimating the parameters of the CTIC model from observed data, and applied it to node-ranking and social behavioral data analysis. To the best of our knowledge, we are the first to formulate a continuous-time information diffusion model based on the IC model and a rigorous learning algorithm to estimate the model parameters from observation. We are not claiming that the model is most accurate. The time-delay distribution for real information diffusion must be more complex, and a power-law distribution and the like might be more suitable. Our future work includes incorpolating various more realistic distributions as the time-delay distribution.

The learning algorithm we proposed is a one-time batch processing. In reality the observation data are keep coming and the environment may change over time. It is not straightforward to convert the algorithm to incremental mode. The simplest way to cope with this is to use a fixed time window to collect data and use the parameters at the previous window as the initial guesses.

We consider that our proposed ranking method presents a novel concept of centrality based on the information diffusion model, i.e., *the CTIC model*. Actually, nodes identified as higher ranked by our method are substantially different from those by each of the conventional methods. This means that our method enables a new type of social network analysis if past information diffusion data are available. Note that this is not to claim to replace them with the proposed method, but simply to propose that it is an addition to them which has a different merit in terms of information diffusion.

We note that the analysis we showed in this paper is the simplest case where κ and r take a single value each for all the links in E. However, the method is very general. In a more realistic setting we can divide E into subsets $E_1, E_2, ..., E_N$ and assign a different value κ_n and r_n for all the links in each E_n. For example, we may divide the nodes into two groups: those that strongly influence others and those not, or we may divide the nodes into another two groups: those that are easily influenced by others and those not. We can further divide the nodes into multiple groups. If there is some background knowledge about the node grouping, our method can make the best use of it.

7   Conclusion

We emphasized the importance of incorporating continuous time delay for the behavioral analysis of information diffusion through a social network, and addressed the problem of estimating the parameters for a continuous time delay independent cascade (CTIC) model from the observed data by rigorously formulating the likelihood of obtaining these data and maximizing the likelihood iteratively with respect to the parameters (time-delay and diffusion). We tested the convergence performance of the proposed method by applying it to the problem of ranking influential nodes using the network structure from two real world web datasets and showed that the parameters converge to the correct values efficiently by the iterative procedure and can predict the high ranked influential nodes much more accurately than the well studied four heuristic methods. We further applied the method to the problem of behavioral analysis of topic propagation using a real world blog data and showed that there are indeed sensible differences in the propagation patterns in terms of delay and diffusion among different topics.

Acknowledgements

This work was partly supported by Asian Office of Aerospace Research and Development, Air Force Office of Scientific Research, U.S. Air Force Research Laboratory under Grant No. AOARD-08-4027, and JSPS Grant-in-Aid for Scientific Research (C) (No. 20500147).

References

1. Newman, M.E.J., Forrest, S., Balthrop, J.: Email networks and the spread of computer viruses. Physical Review E 66, 035101 (2002)
2. Newman, M.E.J.: The structure and function of complex networks. SIAM Review 45, 167–256 (2003)
3. Gruhl, D., Guha, R., Liben-Nowell, D., Tomkins, A.: Information diffusion through blogspace. SIGKDD Explorations 6, 43–52 (2004)
4. Domingos, P.: Mining social networks for viral marketing. IEEE Intelligent Systems 20, 80–82 (2005)
5. Leskovec, J., Adamic, L.A., Huberman, B.A.: The dynamics of viral marketing. In: Proceedings of the 7th ACM Conference on Electronic Commerce (EC 2006), pp. 228–237 (2006)
6. Goldenberg, J., Libai, B., Muller, E.: Talk of the network: A complex systems look at the underlying process of word-of-mouth. Marketing Letters 12, 211–223 (2001)
7. Kempe, D., Kleinberg, J., Tardos, E.: Maximizing the spread of influence through a social network. In: Proceedings of the 9th ACM SIGKDD International Conference on Knowledge Discovery and Data Mining (KDD 2003), pp. 137–146 (2003)
8. Kimura, M., Saito, K., Nakano, R.: Extracting influential nodes for information diffusion on a social network. In: Proceedings of the 22nd AAAI Conference on Artificial Intelligence (AAAI 2007), pp. 1371–1376 (2007)
9. Kimura, M., Saito, K., Motoda, H.: Blocking links to minimize contamination spread in a social network. ACM Transactions on Knowledge Discovery from Data 3, 9:1–9:23 (2009)
10. Saito, K., Kimura, M., Nakano, R., Motoda, H.: Finding influential nodes in a social network from information diffusion data. In: Proceedings of the International Workshop on Social Computing and Behavioral Modeling (SBP 2009), pp. 138–145 (2009)
11. Wasserman, S., Faust, K.: Social network analysis. Cambridge University Press, Cambridge (1994)
12. Brin, S., Page, L.: The anatomy of a large-scale hypertextual web search engine. Computer Networks and ISDN Systems 30, 107–117 (1998)
13. Ng, A.Y., Zheng, A.X., Jordan, M.I.: Link analysis, eigenvectors and stability. In: Proceedings of the 17th International Joint Conference on Artificial Intelligence (IJCAI 2001), pp. 903–910 (2001)
14. Adar, E., Adamic, L.A.: Tracking information epidemics in blogspace. In: Proceedings of the 2005 IEEE/WIC/ACM International Conference on Web Intelligence (WI 2005), pp. 207–214 (2005)
15. Gruhl, D., Guha, R., Liben-Nowell, D., Tomkins, A.: Information diffusion through blogspace. In: Proceedings of the 13th International World Wide Web Conference (WWW 2004), pp. 107–117 (2004)
16. Song, X., Chi, Y., Hino, K., Tseng, B.L.: Information flow modeling based on diffusion rate for prediction and ranking. In: Proceedings of the 16th International World Wide Web Conference (WWW 2007), pp. 191–200 (2007)

Privacy-Preserving Evaluation of Generalization Error and Its Application to Model and Attribute Selection

Jun Sakuma[1] and Rebecca N. Wright[2]

[1] University of Tsukuba, 1-1-1 Tennodai, Tsukuba, Ibaraki, 305-8577, Japan
jun@cs.tsukuba.ac.jp
[2] Rutgers University, 96 Frelinghuysen Road Piscataway, NJ 08854, USA
rebecca.wright@rutgers.edu

Abstract. Privacy-preserving classification is the task of learning or training a classifier on the union of privately distributed datasets without sharing the datasets. The emphasis of existing studies in privacy-preserving classification has primarily been put on the design of privacy-preserving versions of particular data mining algorithms, However, in classification problems, preprocessing and postprocessing— such as model selection or attribute selection—play a prominent role in achieving higher classification accuracy. In this paper, we show generalization error of classifiers in privacy-preserving classification can be securely evaluated without sharing prediction results. Our main technical contribution is a new generalized Hamming distance protocol that is universally applicable to preprocessing and postprocessing of various privacy-preserving classification problems, such as model selection in support vector machine and attribute selection in naive Bayes classification.

1 Introduction

Rapid growth of online services has increased the opportunities to store private or confidential information. Data mining enables to exploit valuable knowledge from data collections while dataminers are forced to treat such private datasets under prudent control in order to prevent the leakage or misuse. Considering such situations, privacy-preserving data mining (PPDM) provides a secure way to compute the output of a particular algorithm applied to the union of privately distributed datasets without sharing them. Privacy-preserving classification is defined as a problem to train a classifier with the union of privately distributed datasets without sharing them. Privacy-preserving classification was pioneered by Lindell and Pinkas, who considered ID3 decision trees [6]. Following this, various privacy-preserving classification methods have been presented, including naive Bayes classifiers [11], k-nearest neighbor [15], and support vector machines [14,5].

These privacy-preserving classifiers have been designed as privacy-preserving versions of particular data mining algorithms. However, the dataminer's task rarely starts and ends with running a particular data mining algorithm. Rather, various preprocessing and postprocessing steps improve the capability of knowledge discovery with data mining. For example, real datasets often include noisy or useless attributes; including them in the classifer learning process reduces classification accuracy. Classification accuracy

Z.-H. Zhou and T. Washio (Eds.): ACML 2009, LNAI 5828, pp. 338–353, 2009.

can be improved by eliminating these unnecessary attributes (i.e., attribute selection). Furthermore, when the target classifier has tunable parameters, adjustment of model parameters (i.e., model selection) can help minimize the generalization error.

The emphasis in privacy-preserving data mining has mostly not been placed on pre- and postprocessing. Earlier work of Yang et al. [12] begins to address privacy-preserving model selection. As we discuss below, our work extends their initial study in many ways.

In a distributed setting, there are many ways that data can be distributed. Two commonly considered cases are vertically partitioned data and horizontally partitioned data. In this paper, we consider data partitioned between two parties. In the vertically partitioned model, each party holds a subset of attributes of the data entries. In the horizontally partitioned model, each party holds a subset of data entries of all attributes [10]. Usually, these subsets are assumed to be disjoint; in the vertically partitioned case, both parties are assumed to know the identity of the complete attribute set.

Most distributed privacy-preserving classification algorithms produce as output an actual classifier known to one or more of the involved parties. In our work, we also consider "extending" the privacy to later stages of the process, sometimes keeping the classifier itself privacy (which we refer to as a *privately shared classifier*) or even the prediction results themselves (which we call *privately shared prediction*). If the classifier and the prediction are not privately shared, we refer to these as regular classifier and regular prediction, respectively. (See Section 2.2 for more detail.) Many combinations of partitioning models and representations are possible in privacy-preserving classification. The most privacy is obtained if both the classifer and the prediction are privately shared; this also enables a modular design of post-processing such as cross validation that requires using multiple generated models for prediction before deciding on the final output model. Table 1 summarizes representations of classifiers and predictions in existing privacy-preserving classification.

Yang et al. [12] presented a privacy-preserving k-fold cross validation that evaluates generalization error of privacy-preserving classifiers. A limitation of their protocol is that it is applicable only to privacy-preserving classifiers that return regular predictions in the vertically partitioned model.

We note that in principle, any private distributed computation can be carried out using secure function evaluation (SFE) [13,4]. It follows that SFE technically allows us to evaluate generalization errors of privacy-preserving classifiers with any partitioning models and representations. However, as we demonstrate in Section 3.3 with experimental results, these computations can be inefficient for practical use, particularly when the input size is large.

Our contribution. We present a new Hamming distance protocol that allows us to evaluate generalization errors of privacy-preserving classifiers using k-fold cross validation (Section 3.2). Our protocol works with privacy-preserving classification with any representation and partitioning model. It is therefore more general and more private than Yang et al.'s protocol [12]. Experimental results show that the computation load of our protocol is much smaller than that of Hamming distance computation implemented with SFE (Section 3.3). To demonstrate our protocol with pre- and postprocessing of privacy-preserving classification, our protocol is applied to model selection in support

Table 1. Data partitioning models and representations of classifiers and predictions in privacy-preserving classification and privacy-preserving cross valiadtion

	Horizontally partitioned model	Vertically partitioned model
Regular classifier Regular prediction	Naive Bayes [11] ID3 [2]	ID3 [6],SVM [14], cross validation [12]
Privately shared classifier Regular prediction	k-NN [15]	cross validation [12] (in restricted cases)
Regular classifier Privately shared prediction		
Privately shared classifier Privately shared prediction	SVM [14]	Naive Bayes [11] SVM [5]

vector machine with polynomial kernels (Section 4) and attribute selection in naive Bayes classification (Section 5). In both applications, the efficiency is examined with experimental analysis.

2   Preliminaries

2.1   Privacy-Preserving Classification

Let a set of training examples be $X_{\mathrm{tr}} = \{(x_i, y_i)\}_{i=1}^n$, where $x_i \in X$ and $y_i \in Y = \{1, ..., m\}$. In classification tasks, classifier $h : X \mapsto Y$ is trained on X_{tr}. Let $\{(x, y)\}$ be test examples, which are unseen in the training phase. The classification problem is to find a classifier $h \in \mathcal{H}$ with a small generalization error $\epsilon = E_{(x,y)}([h(x) \neq y])$ where $E_{(x,y)}$ is the expectation over (x, y). $[z]$ is 1 if predicate z holds and 0 otherwise.

In *privacy-preserving classification*, we assume a setting such that training and test examples are distributed over two or more parties and these examples are desired to be kept private mutually. We assume two parties, denoted as Alice and Bob. Let partitioned sets of training examples of Alice and Bob be X_{tr}^A and X_{tr}^B, respectively.

Typically two types of data partitioning models are considered. In *vertically partitioned model*, a partitioned dataset corresponds to a subset of attributes of all data entries. In *horizontally partitioned model*, a partitioned dataset corresponds to a subset of data entries of all attributes. Test examples are partitioned similarly. If not specifically mentioned, both partitioned models are considered together in this paper.

2.2   Representations of Classifier and Prediction

Privacy-preserving classification essentially includes two problems, privacy-preserving training and privacy-preserving prediction. In the training phase, classifier h is desired to be trained without violating the given partitioning model. Knowledge that Alice and Bob can obtain from the execution of the training phase is determined by the representation of the classifier. Given distributed training examples, if at least one party obtains a trained classifier after the training phase, we say the party learns a *regular classifier*. As long as the training protocol is designed correctly and securely, private inputs will not be exposed to parties explicitly. However, note that the regular classifier itself might leak some private information to the party who receives the classifier. For example, naive Bayes classifier consists of frequency of attributes, which leaks statistics of attributes held by other parties.

The *privately shared classifier* provides enhanced privacy-preservation with the idea of secret sharing by *random shares*. Let $\mathbb{Z}_N = \{0, 1, ..., N - 1\}$. We say Alice and Bob have random shares of secret $y \in \mathbb{Z}_N$ when Alice has $r^A \in \mathbb{Z}_N$ and Bob has $r^B \in \mathbb{Z}_N$ in which r^A and r^B are uniform randomly distributed in $\mathbb{Z}_N$ with satisfying $y = (r^A + r^B) \bmod N$. Assume that classifiers are characterized by a parameter vector $\alpha \in \mathbb{Z}^q$. If two parties obtain random shares of α after the training phase, we say two parties obtain a privately shared classifier. Note that no information associated with the classifier can be obtained from any single share of the classifier. For example, a hyperplane classifier is written as $h(x; \alpha, \beta) = \sum_{i=1}^d \alpha_i x_i + \beta$. Let (α^A, α_i^B) and (β^A, β^B) be random shares of α and β, respectively. Then, (α^A, β^A) and (α^B, β^B) forms a privately shared classifier.

Representations of predictions are stated similarly. Once a party obtains a regular classifier, the party can locally compute predictions of its own test examples. This is referred to as the *regular prediction*. When a privately shared classifier is trained, an additional protocol to obtain predictions is required. If the prediction protocol is designed such that it returns the prediction to at least one party, it is again called regular prediction. If the prediction protocol returns random shares of the prediction, it is referred to as *privately shared prediction*. See Table 1 for the summary of existing privacy-preserving classification with respect to partitioning models and representations.

2.3 Problem Statement

Let $y = (y_1, ..., y_n)$ and $\hat{y} = (\hat{y}_1, ..., \hat{y}_n)$ be a vector of labels of test examples and predicted labels, respectively. Assume binary classification for now, that is, $y, \hat{y} \in \{0, 1\}^n$. Then, the generalization error of binary classification is evaluated as Hamming distance $H(y, \hat{y}) = \sum_{i=1}^n (y_i \oplus \hat{y}_i)$. As shown in Table 1, there are many possible combinations of representations and partitioning models in privacy-preserving classifications, while the protocol for the generalization error evaluation in privacy-preserving classification is desired to be universally applicable to privacy-preserving classifiers with different representations and partitioning models.

First, assume that a prediction protocol which returns regular predictions is given to Alice to Bob. Then, in the vertically partitioning model, predictions can be represented as $\hat{y}^A = (\hat{y}_1, ..., \hat{y}_n)$ and $\hat{y}^B = (0, ..., 0)$. In the horizontally partitioning model, predictions can be represented as $\hat{y}^A = (y_1, ..., y_c, 0, ..., 0)$ and $\hat{y}^B = (0, ..., 0, \hat{y}_{c+1}, ..., \hat{y}_n)$ for some c without loss of generality. Thus, $\hat{y} = \hat{y}^A + \hat{y}^B$ holds for both partitioning models.

Next, assume that a prediction protocol which returns privately shared predictions is given to Alice and Bob. Then, $\hat{y}_i = \hat{y}_i^A + \hat{y}_i^B \bmod N$ holds for all i in both horizontally and vertically partitioning models. Thus, regardless of representations of predictions and partitioning models, private evaluation of generalization errors is reducible to the problem stated as follows:

Statement 1. *Let (y^A, y^B) and $(\hat{y}^A, \hat{y}^B)$ be vectors of random shares of binary vector y and $\hat{y}$, respectively. Assume that Alice and Bob takes $(y^A, \hat{y}^A)$ and $(y^B, \hat{y}^B)$ as inputs,* *respectively. After the execution of the Hamming distance protocol, Alice and Bob obtain random shares s^A and s^B such that $s^A + s^B = H(y, \hat{y}) \bmod N$. Furthermore, neither Alice nor Bob obtains anything else.*

In Section 3.2, we describe our Hamming distance protocol, which is secure and correct in the sense of Statement 1. This protocol provides a way to evaluate classification errors of given privacy-preserving classifiers. Generalization errors of privacy-preserving classifiers can be evaluated by making use of k-fold cross validation with this Hamming distance protocol.

Next, a problem of model/attribute selection in privacy-preserving classification is stated. Let $\mathcal{H}$ be a set of candidate classifiers. Let $\ell : \mathcal{X} \times \mathcal{Y} \times \mathcal{H} \mapsto [0, 1]$ be an estimator of the generalization error. Then, the model selection problem in privacy-preserving classification is stated as follows:

Statement 2. *Let $H = \{h(\cdot; \alpha_i^A, \alpha_i^B) | i = 1, ..., \ell\}$ be a set of candidate privately shared classifiers. Alice takes X_{tr}^A and $(\alpha_1^A, ..., \alpha_\ell^A)$ as private inputs. Bob takes X_{tr}^B and $(\alpha_1^B, ..., \alpha_\ell^B)$ as private inputs. Let X_{ts}^A and X_{ts}^B be test examples of Alice and Bob, respectively. After the execution of model selection, Alice and Bob obtain privately shared classifier h^* such that $h^* = \arg\min_{h \in H} \ell(X_{ts}^A \cup X_{ts}^B, h)$. Furthermore, neither of Alice nor Bob obtains anything else.*

Note that any regular classifiers are special cases of privately shared classifiers. This problem statement therefore includes any possible combinations of different representations listed in Table 1. Furthermore, if we regard $H = \{h(\cdot; \alpha_i^A, \alpha_i^B) | i = 1, ..., \ell\}$ as a set of privately shared classifiers which are trained with candidates of attribute subsets, the statement described above is readily rewritten as the statement of the attribute selection problem in privacy-preserving classification.

We present a protocol to solve problems defined as Statement 2 in Section 3. We then demonstrate the universal applicability of our protocol by taking model selection in privacy-preserving support vector machines (Section 4) and attribute selection in privacy-preserving Naive Bayes classifiers (Section 5) as examples.

3   Privacy-Preserving Evaluation of Generalization Error

This section describes our Hamming distance protocol, which enables evaluation of classification errors in privacy-preserving classification. Then a model selection protocol for privacy-preserving classification is presented by using the Hamming distance protocol, in which the generalization errors are estimated with k-fold cross validation. Before going into the protocol description, we introduce necessary cryptographic tools.

3.1   Cryptographic Tools

Homomorphic Public-key Cryptosystem. Given a corresponding pair of (sk, pk) of private and public keys and a message m, then $c = e_{pk}(m; \ell)$ denotes a (random) encryption of m, and $m = d_{sk}(c)$ denotes decryption. The encrypted value c uniformly distributes over $\mathbb{Z}_N$ if ℓ is taken from $\mathbb{Z}_N$ randomly. An *additive homomorphic cryptosystem* allows addition computations on encrypted values without knowledge of the secret key. Specifically, there is some operation $\cdot$ (not requiring knowledge of sk) such that for any plaintexts m_1 and m_2, $e_{pk}(m_1 + m_2; \ell) = e_{pk}(m_1; \ell_1) \cdot e_{pk}(m_2; \ell_2)$, where ℓ is uniformly random provided that at least one of ℓ_1 and ℓ_2 is. Based on this property, it also follows that given a constant k and the encryption $e_{pk}(m_1; \ell)$, we can compute multiplications by k via repeated application of $\cdot$, denoted as $e_{pk}(km; k\ell) = e_{pk}(m; \ell)^k$. In what follows, we omit the random number ℓ from our encrypt ions for simplicity.

Secure Function Evaluation. In principle, private distributed computations can be carried out using Secure function evaluation (SFE) [13,4] is a general and well studied methodology for evaluating any function privately. Technically, the entire computation of model selection itself can be implemented with SFE; however, these computations can be too inefficient for practical use, particular when the input size is large. Classification often takes large dataset as input. Therefore, we make use of existing SFE solutions for small portions of our computation for a more efficient overall solution. Specifically, we use SFE for a problem of private computation of argmax. Let Alice and Bob take $(s_1^A, ..., s_k^A)$ and $(s_1^B, ..., s_k^B)$ as private inputs. We use SFE for Alice and Bob to learn $i^* = \arg\min_i(s_i^A + s_i^B)$ without sharing their private inputs.

3.2 Hamming Distance Protocol

As discussed in Section 2.3, we can suppose Alice and Bob hold class label vectors (y^A, y^B) and predictions of labels $(\hat{y}^A, \hat{y}^B)$ in the form of random shares regardless of partitioning models representations of classifiers/predictions. The classification error is then evaluated in the form of Hamming distance as

$$H(y, \hat{y}) = \sum_{i=1}^{n} \left((y_i^A + y_i^B) \oplus (\hat{y}_i^A + \hat{y}_i^B) \right) = \sum_{i=1}^{n} \left((y_i^A + y_i^B) - (\hat{y}_i^A + \hat{y}_i^B) \right)^2$$

$$= (y^A - \hat{y}^A)^T \cdot (y^A - \hat{y}^A) + (y^B - \hat{y}^B)^T \cdot (y^B - \hat{y}^B) + 2(y^A - \hat{y}^A) \cdot (y^B - \hat{y}^B), (1)$$

where $\oplus$ is logical exclusive-or and other operations are all arithmetic. The first and the second term of RHS of eq. 1 can be locally and independently evaluated by Alice and Bob, respectively. Therefore, via the private evaluation of $2(y^A - \hat{y}^A) \cdot (y^B - \hat{y}^B)$, random shares of $H(y, \hat{y})$ is privately evaluated . Based on this idea, the protocol to solve the problem of Statement 1 is shown in Fig. 1. In the protocol, operation $\in_r$ means choosing an element of the set uniform randomly. The correctness of the protocol is explained as follows. In Step 2, what Bob computes is rearranged as

$$c \leftarrow e_{pk}\left(-r^B + 2\sum_{i=1}^{n}(y_i^A - \hat{y}_i^A)(y_i^B - \hat{y}_i^B)\right) = e_{pk}\left(-r^B + 2(y^A - \hat{y}^A) \cdot (y^B - \hat{y}^B)\right).$$

Then, In Step 3, Alice obtains

$$s^A = d^A(c) + (y^A - \hat{y}^A)^T \cdot (y^A - \hat{y}^A) = -r^B + 2(y^A - \hat{y}^A) \cdot (y^B - \hat{y}^B) + (y^A - \hat{y}^A)^T \cdot (y^A - \hat{y}^A)$$

and Bob obtains $s^B = r^B + (y^B - \hat{y}^B)^T \cdot (y^B - \hat{y}^B)$. Thus, s^A and s^B are random shares of $H(y, \hat{y})$, which are the desired outputs. The security of this protocol is shown.

- Alice's input: $y^A, \hat{y}^A \in \mathbb{Z}_N^n$, key pair (pk, sk)
- Bob's input: $y^B, \hat{y}^B \in \mathbb{Z}_N^n$, public key pk
- Alice's output: $s^A \in \mathbb{Z}_N$
- Bob's output: $s^B \in \mathbb{Z}_N$, where $s^A + s^B = H(y^A + y^B, \hat{y}^A + \hat{y}^B) \bmod N$

1. For $i = 1, ..., n$, Alice compute $e_{pk}(y_i^A - \hat{y}_i^A)$ and send them to Bob
2. Bob computes $c \leftarrow e_{pk}(-r^B) \cdot \prod_{i=1}^n e_{pk}(y_i^A - \hat{y}_i^A)^{2(y_i^B - \hat{y}_i^B)}$, where $r^B \in_r \mathbb{Z}_N$. Then, Bob sends c to Alice
3. Alice outputs $s^A \leftarrow d^A(c) + \sum_{i=1}^n (y_i^A - \hat{y}_i^A)^2$ and Bob outputs $s^B \leftarrow r^B + \sum_{i=1}^n (y_i^B - \hat{y}_i^B)^2$

Fig. 1. Hamming distance protocol

Lemma 1. *(Security of Hamming distance protocol) If Alice and Bob behave semi-honestly, Hamming distance protocol is secure in the sense of Statement 1.*

Due to space limitations, we give only the intuition behind the security of this protocol. The message Alice receives during the protocol execution is c. Alice can decrypt c but this is randomized by Bob; nothing can be learned by Alice. The messages Bob receives are $e_{pk}(y_i^A - \hat{y}_i^A)(i = 1, ..., n)$. Bob cannot decrypt this and nothing can be learned by Bob, either. Consequently, both learn nothing.

3.3 Preliminary Experiments: Hamming Distance Protocol

The scalability of Hamming distance protocol with randomly generated binary vectors were examined. Programs were implemented in Java 1.6.0. As the cryptosystem, [8] with 1024-bit keys was used. Experiments were carried out under Linux with Core2 2.0GHz (CPU), 2GB (RAM). The results were compared with the computation time of SFE in that the Hamming distance computation with exactly the same input and output is performed. Fairplay [7] was used for the implementation of SFE.

Fig. 2 shows the results of experiments. The results are the average of ten times execution with different binary vectors. A single execution of Hamming distance protocol with n-dimensional vectors includes n times modulo multiplication and n times modulo exponentiation. As expected, the change in the computation time with respect to the dimensionality of input vectors is linear. While the computation time of SFE is polynomially bounded, the results show that the computation time of SFE is more inefficient than that of our Hamming distance protocol.

3.4 Privacy-Preserving Model Selection by Means of k-Fold Cross Validation

In k-fold cross validation, a set X of examples is randomly split into k mutually disjoint subsets $X_1, X_2, ..., X_k$. Classifier h_j is then trained on $X \setminus X_j$. Let ϵ_j be generalization error of h_j. The cross validation estimate of the generalization error is given as $\epsilon = \sum_{j=1}^k \frac{\epsilon_j}{k}$.

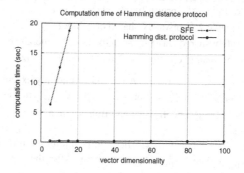

Fig. 2. Computation time of Hamming distance protocol: the dimensionality vs. time (sec)

For model selection, classifiers are trained with candidates of model parameters. For each trained classifier, the generalization error is evaluated with k-fold cross validation. The model that achieves the lowest generalization error is chosen as the output of the model selection.

Suppose that ℓ candidates of model parameters are prepared in the form of privately shared classifiers. Then, the protocol for privacy-preserving model selection by means of k-fold cross validation is described as shown in Fig. 3.

In the protocol, $(X^A_{ts,j}, X^B_{ts,j})$ is the jth fold of training examples. $(\alpha^A_{ij}, \alpha^B_{ij})$ is the privately shared classifier trained with the ith candidate of model parameters and the jth fold of test examples. $(\hat{y}^A_{ij}, \hat{y}^B_{ij})$ is privately shared predictions of jth fold of test examples classified by the ith candidate model. Labels of test examples are evaluated as privately shared prediction in Step 1. Then, the number of misclassifications is evaluated as random shares of Hamming distance by Hamming distance protocol (Step 2). Finally, the model with the lowest generalization error is privately chosen by secure function evaluation (Step 3). The security of this protocol is as follows:

Theorem 1. *Assume Alice and Bob behave semi-honestly. Given a privacy-preserving protocol for prediction, privacy-preserving model selection is secure in the sense of Statement 2.*

A sketch of the proof is shown. In the protocol of Fig. 3, privacy-preserving subprotocols are composed in a way that randomized outputs of a subprotocol are taken as inputs of the next subprotocol. A composition theorem [6] guarantees that if subprotocols invoked at intermediate steps are privacy-preserving, then the resulting composition is privacy-preserving, too. Based on this fact, the proof of this theorem is readily derived via Lemma 1 and the security of SFE [13,4].

Assuming $(\alpha^A_{ij}, \alpha^B_{ij})$ are privately shared classifiers trained from datasets with different attribute sets, the protocol in Fig. 3 is readily available for attribute selection in privacy-preserving classification. Exactly the same security proof is valid for this privacy-preserving attribute selection.

- Alice's input: test example X_{ts}^A, privately shared classifier $\alpha_{ij}^A (i = 1, ..., \ell, j = 1, ..., k)$
- Bob's input: test example X_{ts}^B, privately shared classifier $\alpha_{ij}^B (i = 1, ..., \ell, j = 1, ..., k)$
- Alice's output: privately shared classifier with the lowest generalization error $\alpha_{i^*j}^A$
- Bob's output: privately shared classifier with the lowest generalization error $\alpha_{i^*j}^B$

1. (Evaluation) For $i = 1, ..., \ell, j = 1, ..., k$, Alice and Bob jointly execute privacy-preserving evaluation of shared prediction by taking $(X_{ts,j}^A, \alpha_{ij}^A)$ and $(X_{ts,j}^B, \alpha_{ij}^B)$ as inputs, respectively. After the execution of the protocol, Alice and Bob obtain a vector of prediction shares $\hat{y}_{ij}^A$ and $\hat{y}_{ij}^B$, respectively.

2. (Prediction) For $i = 1, ..., \ell, j = 1, ..., k$, Alice and Bob jointly run Hamming distance protocol by taking $(y_{ts,i}^A, \hat{y}_{ij}^A)$ and $(y_{ts,i}^B, \hat{y}_{ij}^B)$, respectively. After the execution of the protocol, Alice and Bob obtain random shares s_{ij}^A and s_{ij}^B where $H(y_{ts,i}^A + y_{ts,i}^B, \hat{y}_{ij}^A + \hat{y}_{ij}^B) = (s_{ij}^A + s_{ij}^B)$ mod N

3. (Model selection) Alice and Bob jointly run secure function evaluation for

$$i^* \leftarrow \arg\min_i \sum_{j=1}^k (s_{ij}^A + s_{ij}^B \quad \mod N).$$

Alice and Bob respectively obtain shares of classifiers $(\alpha_{i^*j}^A)$ and $(\alpha_{i^*j}^B)$ which achieve the lowest generalization error

Fig. 3. Privacy-preserving model selection

4   Model Selection in Privacy-Preserving Support Vector Machine

4.1   Privacy-Preserving Support Vector Machines

This section demonstrates the application of our model selection protocol to privacy-preserving SVMs with polynomial kernels.

Let $(x_i, y_i)_{1 \le i \le n}$ be a set of training examples where $x_i \in \mathbb{R}^d$ and $y_i \in \{1, -1\}$. Maximal margin classifiers of SVMs are trained in a high dimensional feature space. We substitute $\phi(x_i)$ for each x_i where $\phi : \mathcal{X} \mapsto \mathcal{F}$ and $\mathcal{F}$ is a high-dimensional feature space. When features $\phi(x)$ only occurs in the form of dot products, Mercer kernels k for dot products, $\phi(x_i) \cdot \phi(x) = k(x_i, x)$, can be substituted.

SVMs take advantage of this trick to alleviate computation in the high-dimensional feature space. The linear maximal margin classifier is given as

$$h(x) = \text{sgn}\left(\sum_{i=1}^n \alpha_i \phi(x_i) \cdot \phi(x)\right) = \text{sgn}\left(\sum_{i=1}^n \alpha_i k(x_i, x)\right), \tag{2}$$

which is obtained by solving a convex quadratic programming with respect to α [9].

Laur et al. have proposed a cryptographically private SVM of polynomial kernels for vertically partitioned private datasets [5]. In order to alleviate solving convex quadratic programming with satisfying privacy enforcement, their protocol adopts kernel adatron as the base algorithm.

4.2   Model Selection in Privacy-Preserving Support Vector Machines

SVM with polynomial kernels includes polynomial dimension p and margin parameter c as model parameters. In this section, we demonstrate our model selection protocol by taking Laur et. al.'s privacy-preserving SVM as the base classifier algorithm.

As we addressed in Theorem 1, our protocol is designed assuming that the prediction protocol is privacy preserving—that is, nothing but prediction results are leaked from the execution of the prediction protocol. Laur et. al.'s prediction protocol is designed assuming that $x \cdot x_i \in \mathbb{Z}_N^*$ holds[1] while this does not always necessarily hold. It follows that the security of the prediction protocol might be compromised if $x \cdot x_i \notin \mathbb{Z}_N^*$. In order for our model selection protocol to be secure in the sense of Statement 2, we introduce power-of-sum protocol as a new building block for prediction, which is privacy-preserving for any $x, x_i \in \mathbb{Z}_M$ if $(2M)^p < N$, and then present a new prediction protocol for polynomial kernels using power-of-sum.

Power-of-Sum Protocol.   In the prediction function of SVM, degree-p polynomials need to be privately evaluated. As a building block for prediction, we introduce a protocol, power-of-sum. Let $(2M)^p < N$. Let Alice and Bob have $x \in \mathbb{Z}_M$ and $y \in \mathbb{Z}_M$ as private inputs, respectively. Power-of-sum protocol allows two parties to compute random shares of $r^A + r^B = (x + y)^p \bmod N$ without sharing their private inputs. We assume that degree p is publicly known throughout the paper. Note that the knowledge of p does not violate any private information possessed by parties.

The protocol is shown in Fig. 4. By binomial theorem, $(x+y)^p = \sum_{i=0}^p \binom{p}{i} x^i y^{p-i}$ holds. So, the computation of Step 2 corresponds to

$$c^B \leftarrow e_{\mathsf{pk}}(-r^B) \cdot \prod_{i=0}^p (c^i)^{y^{p-i}} = e_{\mathsf{pk}}\left(-r^B + \sum_{i=0}^p \binom{p}{i} x^i y^{p-i}\right) = e_{\mathsf{pk}}((x+y)^p - r^B). \quad (3)$$

Thus, Alice obtains $s^A = -s^B + (x+y)^p$ and Bob obtains s^B, which are random shares of $(x + y)^p$. Note that both x and y are positive integers s.t. $(x+y)^p \le (2M)^p < N$. Without this, numbers may be wrapped around in the computation of step 2, which does not induce desired outputs.

We show an intuitive explanation of the security. Messages received by Alice are randomized by Bob. Messages received by Bob are encrypted and the private key is possessed only by Alice. Thus, both learn nothing but the final result.

Private Prediction for Degree-p Polynomial Kernels
Our new prediction protocol is shown in Fig. 5. In the protocol, random shares

$$s^A + s^B = \sum_{i=1}^n \alpha_i k(x_i, x) = \sum_{i=1}^n \alpha_i (x_i \cdot x)^p \quad \bmod N \qquad (4)$$

are privately evaluated in Step 3 and then the random shares of the prediction are obtained in Step 4 as $h^A(x) + h^B(x) = \mathrm{sgn}(s^A + s^B) \bmod N$. Eq. 4 is derived as follows.

---

[1] Integers less than N and coprime to N.

Regardless of partitioning models, $x_i = x_i^A + x_i^B$ and $x = x^A + x^B$ hold. So, the scalar product of these is expanded as

$$x \cdot x_i = (x^A + x^B) \cdot (x_i^A + x_i^B) = x^A \cdot x_i^A + x^A \cdot x_i^B + x^B \cdot x_i^A + x^B \cdot x_i^B \quad \text{mod } N. \quad (5)$$

In Step 2-a, random shares of scalar products $x^B \cdot x_i^A$ and $x^A \cdot x_i^B$ are privately computed by scalar product protocol [3]. Note that the other two scalar products are locally computed. Thus, random shares of $x \cdot x_i$ are obtained by Alice and Bob in the form of $x \cdot x_i = (s^{A1} + s^{A2} + x_i^A \cdot x^A) + (s^{B1} + s^{B2} + x_i^B \cdot x^B) \mod N$. By taking these random shares as inputs of power-of-sum in Step 2-b, random shares of $k(x, x_i)$ are obtained (eq. 7). Since eq. 8 of Step 3-b can be rewritten as

$$c \leftarrow \prod_{i=1}^{n} \left(e_{\mathsf{pk}}(\alpha_i^A k^A(x_i, x)) \cdot e_{\mathsf{pk}}(\alpha_i^B k^B(x_i, x)) \cdot e_{\mathsf{pk}}(\alpha_i^A)^{k^B(x_i, x)} \cdot e_{\mathsf{pk}}(k^A(x_i, x))^{\alpha_i^B} \right) \cdot e_{\mathsf{pk}}(-s^B)$$

$$= \prod_{i=1}^{n} e_{\mathsf{pk}} \left(\alpha_i^A k^A(x_i, x) + \alpha_i^B k^B(x_i, x) + \alpha_i^A k^B(x_i, x) + \alpha_i^B k^A(x_i, x) - s^B \right)$$

$$= \prod_{i=1}^{n} e_{\mathsf{pk}} \left(\alpha_i k(x_i, x) - s^B \right) = e_{\mathsf{pk}} \left(\sum_{i=1}^{n} (\alpha_i k(x_i, x)) - s^B \right),$$

Alice obtains $s^A = \sum_{i=1}^{n} \alpha_i k(x_i, x) - s^B$ by decrypting c in Step 3-c. Finally, privately shared prediction with polynomial kernels is obtained by means of secure function evaluation in Step 4. Note that Step 2-a can be skipped in the vertically partitioned model because $x^A \cdot x_i^B = x^B \cdot x_i^A = 0$ holds.

Model Selection. Using private prediction for polynomial kernels (Fig. 5) in the first step of the model selection protocol in Fig. 3, privacy-preserving model selection is readily achieved.

4.3 Experiments

Our privacy-preserving model selection produces the same final output as model selection without privacy preservation. Therefore, objectives of experiments are (1) to demonstrate the usability of our protocol in privacy-preserving classification with relatively large-size datasets and (2) to investigate the computational cost.

Setting. Two datasets (ionosphere and breast-cancer) were taken from the UCI machine learning repository [1].

We tuned two model parameters: polynomial degree $p = 1, 2, 3, 4$ and margin parameter $c = 2^{-8}, 2^{-6}, ..., 2^6$. In total, generalization errors of classifiers trained in 32 settings were evaluated as candidates using 10-fold cross validation. Two kinds of private information models are considered. In "type-I", training examples are not private but test examples are private. In this case, scalar product (Fig. 5, step 2-a) does not have to be performed privately. In "type-II", both training and test examples are private; Step 2-a must be done privately. Given privately shared classifiers, we measured the computational time spent for following five steps included in the model selection phase (the first three items are included in Step 1 of Fig. 3).

- Alice's input: $x \in \mathbb{Z}_M$, a valid key pair $(\mathsf{pk}, \mathsf{sk})$
- Bob's input: $y \in \mathbb{Z}_M$, Alice's public key pk
- Alice's output: random share s^A
- Bobs output: random share s^B where $s^A + s^B = (x + y)^p \bmod N$

1. Alice sends $c_i \leftarrow e_{\mathsf{pk}}\left(\binom{p}{i} x^i\right)$ to Bob for $i = 0, ..., p$.
2. Bob sends $c^B \leftarrow e_{\mathsf{pk}}(-s^B) \cdot \prod_{i=0}^{p}(c^i)^{y^{p-i}}$, where $s^B \in_r \mathbb{Z}_N$. Then, Bob sends this to Alice
3. Alice outputs $s^A \leftarrow d^A(c^B)$. Bob outputs s^B.

Fig. 4. Private sharing of degree-d polynomial kernels

- Alice's input: test example x^A, training examples $(x_1^A, ..., x_n^A)$, share of classifier α^A and a valid key pair $(\mathsf{pk}, \mathsf{sk})$
- Bob's input: test example x^B, training examples $(x_1^B, ..., x_n^B)$, share of classifier α^B and a public key pk
- Alice's output: shared prediction $h^A(x)$
- Bob's output: shared prediction $h^B(x)$ where $h(x) = h^A(x) + h^B(x) \bmod N$

1. For $i = 1, ..., n$, Alice sends $e_{\mathsf{pk}}(\alpha_i^A)$ to Bob. Then Bob computes $e_{\mathsf{pk}}(\alpha_i) \leftarrow e_{\mathsf{pk}}(\alpha_i^A) \cdot e_{\mathsf{pk}}(\alpha_i^B)$.
2. Kernel sharing:
   (a) For $i = 1, ..., n$, Alice and Bob jointly do scalar product protocol to have shares

$$s^{A1} + s^{B1} = x_i^A \cdot x^B \bmod N, \ s^{A2} + s^{B2} = x_i^B \cdot x^A \bmod N \tag{6}$$

   where Alice obtains s^{A1}, s^{A2} and Bob obtains s^{B1}, s^{B2} as outputs
   (b) Alice and Bob jointly do power-of-sum protocol to have shares of kernels

$$k^A(x, x_i) + k^B(x, x_i) = k(x, x_i) = (x \cdot x_i)^p \bmod N \tag{7}$$

   where Alice's input is $s^{A1} + s^{A2} + x_i^A \cdot x^A$ and Bob's input is $s^{B1} + s^{B2} + x_i^B \cdot x^B$.
3. Evaluation:
   (a) For $i = 1, ..., n$, Alice sends $e_{\mathsf{pk}}(k^A(x_i, x))$ and $e_{\mathsf{pk}}(\alpha_i^A k^A(x_i, x))$ to Bob.
   (b) Bob generates $s_B \in_r \mathbb{Z}_N$, computes c as follows and sends c to Alice

$$c \leftarrow \prod_{i=1}^{n}\left(e_{\mathsf{pk}}(\alpha_i^A k^A(x_i, x)) \cdot e_{\mathsf{pk}}(\alpha_i^B k^B(x_i, x)) \cdot e_{\mathsf{pk}}(\alpha_i^A)^{k^B(x_i, x)} \cdot e_{\mathsf{pk}}(k^A(x_i, x))^{\alpha_i^B}\right) \cdot e_{\mathsf{pk}}(-s^B) \tag{8}$$

   (c) Alice: Compute $s^A \leftarrow d^A(c)$
4. Prediction: Alice and Bob jointly compute $h^A(x) + h^B(x) = \mathrm{sgn}(s^A + s^B) \bmod N$ by secure function evaluation

Fig. 5. Private prediction by polynomial kernels

1. kernel sharing: scalar product and power-of-sum protocol (Step 2 of Fig. 5)
2. evaluation of $h(x)$: private prediction by polynomial kernels (Step 3 of Fig. 5)
3. prediction: secure function evaluation of comparison (Step 4 of Fig. 5)
4. evaluation of gen. err.: hamming distance protocol (HDP) or SFE (Step 2 Fig. 3)
5. model selection: secure function evaluation of argmin (Step 3 of Fig. 3)

Results. Fig 6 (left and right) illustrates generalization errors of SVM evaluated by 10-fold cross validation. As shown, the generalization error is significantly improved when model parameters are appropriately chosen. Note that these results are not revealed to the participants by the protocol, but only the parameters which achieves the lowest generalization error is revealed in privacy-preserving settings.

The computation time is summarized in Table 2. The number of steps including cryptographic operations in the first three steps, kernel sharing, evaluation and prediction, are $O(n^2d + np)$, $O(n^2\ell k)$, and $O(n\ell k)$ in type-II. In type-I, the number of steps with cryptographic operations of kernel sharing is decreased to $O(np)$. Private evaluation of $f(x)$ is the most time consuming because it includes large numbers of multiplication and exponentiation of ciphers. The computation complexity of private prediction is not large but this includes SFE. Therefore, the computation time of these steps takes a large portion of the entire computation, too. The complexity of Hamming distance protocol (gen. err.) and argmin by SFE (model sel.) is $O(n\ell k)$ and $O(k)$. The number of iterations of Hamming distance protocol is not small; however this protocol does not include SFE. The model selection step includes SFE; however the number of iteration k is usually not very large (in our case, $k = 32$). Thus, these are not time-consuming.

The computation time of SFE is 20 times more than that of Hamming distance protocol. However, kernel sharing, evaluation, and prediction take a large portion of the entire computation time. Therefore, the improvement in computation time by making use of Hamming distance protocol in this work is no more than 12 % in Type-I and 9 % in Type-II. From these, we can conclude that our Hamming distance protocol actually reduces the computation time of model selection while speeding up of the evaluation of the prediction function is essential to reduce the total amount of the computation time in this privacy-preserving classification.

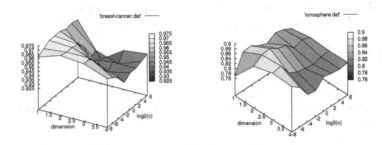

Fig. 6. Generalization error of SVM classifiers with varying polynomial-degree and margin parameter (left: ionosphere, right: breast cancer)

Table 2. Computational time (second) consumed at each step of model selection in privacy-preserving SVM. In the generalization error evaluation step, computation time of Hamming distance protocol (HDP) and SFE are compared.

dataset			ionopshere			
computation			type-I/HDP	type-II/HDP	type-I/SFE	type-II/SFE
kernel sharing	$k(x_i, x_j)$	$d = 2$	2195	13574	2195	13574
		$d = 3$	2217	13596	2217	13596
		$d = 4$	2273	13652	2273	13652
eval.	$f(x_j)$		52158			
pred.	$\hat{y}_j =$ sgn $(f(x_j))$		30576			
gen. err.	$\epsilon = \sum_j [y_j \neq \hat{y}_j]$		89.9		16643	
model sel.	$\ell^* = \min_\ell \epsilon^\ell$		127.6			
total			89996	140686	106189	157239

dataset			breast-cancer			
computation			type-I/HDP	type-II/HDP	type-I/SFE	type-II/SFE
kernel sharing	$k(x_i, x_j)$	$d = 2$	4264	42693	4264	42693
		$d = 3$	4308	47001	4308	47001
		$d = 4$	4417	51418	4417	51418
eval.	$f(x_j)$		195840			
pred.	$\hat{y}_j =$ sgn $(f(x_j))$		59404			
gen. err.	$\epsilon = \sum_j [y_j \neq \hat{y}_j]$		125.1		32473.6	
model sel.	$\ell^* = \min_\ell \epsilon^\ell$		127.6			
total			268485	396878	300833	429227

5 Privacy-Preserving Attribute Selection in Naive Bayes Classification

As another application of our protocol, privacy-preserving attribute selection in naive Bayes classification is demonstrated in this section. Vaidya et. al. have presented privacy-preserving naive Bayes classifiers in both horizontally and vertically partitioned datasets of nominal and numerical attributes [11]. Although our protocol can be combined with any possible variations of Vaidya et. al.'s classifiers, here we restrict our attention to vertically partitioned naive Bayes classifiers of nominal attribute, in which privately shred predictions are returned to parties. See [11] for details of their training protocols and prediction protocol.

Setting. Two datasets (lenses and breast-cancer) were taken from the UCI machine learning repository [1]. The lenses dataset has three classes and the breast-cancer dataset has two classes. Since the lenses dataset is not binary classification, label vectors were transformed into binary vectors using indicator variables. For attribute selection, we enumerated subsets of attributes exhaustively and trained naive Bayes classifiers for each attribute subset. Lenses and breast-cancer dataset have four and nine attribute values, respectively, So we evaluated $\sum_{i=1}^{4} \binom{4}{i} = 15$ and $\sum_{i=1}^{9} \binom{9}{i} = 511$ attribute subsets in

Table 3. Computational time (second) consumed at each step of attribute selection in privacy-preserving naive Bayes. In the generalization error step, computation time of Hamming distance protocol (HD ptcl.) and that of SFE was compared.

dataset	lenses		breast-cancer	
computation	w/ SFE	w/ HD ptcl.	w/ SFE	w/ HD ptcl.
eval. and pred.	179.27		43345	
gen.err.	58.35	3.0324	18329	136.47
att. sel.	5.981		203.76	
total	243.6	188.3	61877	43685

lenses and breast-cancer, respectively. The generalization error of the classifier trained is evaluated by 10-fold cross validation in each setting.

Given privately shared naive Bayes classifiers after the training phase, we measured the computational time spent for (1) evaluation and prediction (Step 1 of Fig. 3), (2) evaluation of generalization error (Hamming distance protocol or SFE, Step 2 of Fig. 3), (3) attribute selection (Step 3 of Fig. 3).

Results. The number of correctly classified examples (accuracy) was 70.83% in lenses dataset and 71.67% in breast-cancer dataset without attribute selection. With attribute selection, the accuracy was improved as 87.50% in lenses dataset (two of five attributes were selected) and 75.17% in breast-cancer dataset (two of nine attributes were selected).

The computation time spent in each step is shown in Table 3. Again, we measured the time of computation in which SFE or Hamming distance protocol is used for the privacy-preserving evaluation of generalized errors. From the results, the computation time is reduced 23% in lenses dataset and 30% in breast-cancer dataset by making use of Hamming distance protocol for the evaluation of generalization error.

6   Conclusion

In this paper, we presented solutions for generalization error evaluation in privacy-preserving classification. We consider both vertically partitioned and horizontally partitioned data. In addition, our solutions can work with both regular prediction and privately shared prediction. Privacy-preserving classification can be designed (and have been designed by other researchers) for various combinations of these partitioning models and representations are possible. In this paper, we introduced a new Hamming distance protocol for generalized error evaluation that works with any of these representations and any data partitioning models.

To show the universal applicability of our protocol, we experimentally demonstrated our protocol with model selection in support vector machine and attribute selection in naive Bayes classification. The result showed that our privacy-preserving model selection and attribute selection could successfully improve the classification accuracy in privacy-preserving naive Bayes and in privacy-preserving SVM. Furthermore, the computation time for Hamming distance computation was reduced to nearly one-fiftieth by making use of our Hamming distance protocol in comparison with that of SFE, while the reduction rate of the total computation time in privacy-preserving classifiers including model/attribute selection with real-world datasets was around 10%–30%. This is because there is a computation bottleneck not only in the evaluation of generalization errors but also in the evaluation of the prediction function. From these observations, we conclude that speeding up of the evaluation of the prediction function is essential to further reducing the total amount of computation time in privacy-preserving classification.

Acknowledgement

This work was carried out partly while the first author was a visitor of the DIMACS Center. The second author is partially supported by the National Science Foundation under grant number CNS–0822269.

References

1. Asuncion, A., Newman, D.J.: UCI machine learning repository (2007)
2. Du, W., Zhan, Z.: Building decision tree classifier on private data. In: Proceedings of the IEEE international conference on Privacy, security and data mining, vol. 14, pp. 1–8. Australian Computer Society (2002)
3. Goethals, B., Laur, S., Lipmaa, H., Mielikainen, T.: On private scalar product computation for privacy-preserving data mining. In: Park, C.-s., Chee, S. (eds.) ICISC 2004. LNCS, vol. 3506, pp. 104–120. Springer, Heidelberg (2005)
4. Goldreich, O.: Foundations of Cryptography, Basic Applications, vol. 2. Cambridge University Press, Cambridge (2004)
5. Laur, S., Lipmaa, H., Mielikäinen, T.: Cryptographically private support vector machines. In: Proceedings of the 12th ACM SIGKDD international conference on Knowledge discovery and data mining, pp. 618–624. ACM Press, New York (2006)
6. Lindell, Y., Pinkas, B.: Privacy Preserving Data Mining. Journal of Cryptology 15(3), 177–206 (2002)
7. Malkhi, D., Nisan, N., Pinkas, B., Sella, Y.: Fairplay: a secure two-party computation system. In: Proceedings of the 13th USENIX Security Symposium, pp. 287–302 (2004)
8. Paillier, P.: Public-Key Cryptosystems Based on Composite Degree Residuosity Classes. In: Stern, J. (ed.) EUROCRYPT 1999. LNCS, vol. 1592, pp. 223–238. Springer, Heidelberg (1999)
9. Schölkopf, B., Smola, A.J.: Learning with kernels. MIT Press, Cambridge (2002)
10. Vaidya, J., Clifton, C., Zhu, M.: Privacy Preserving Data Mining. Series: Advances in Information Security, vol. 19 (2006)
11. Vaidya, J., Kantarcıoğlu, M., Clifton, C.: Privacy-preserving Naïve Bayes classification. The VLDB Journal The International Journal on Very Large Data Bases, 1–20 (2007)
12. Yang, Z., Zhong, S., Wright, R.N.: Towards Privacy-Preserving Model Selection. In: Bonchi, F., Ferrari, E., Malin, B., Saygın, Y. (eds.) PinKDD 2007. LNCS, vol. 4890, pp. 138–152. Springer, Heidelberg (2008)
13. Yao, A.C.-C.: How to generate and exchange secrets. In: Proceedings of the 27th IEEE Symposium on Foundations of Computer Science (FOCS), pp. 162–167 (1986)
14. Yu, H., Jiang, X., Vaidya, J.: Privacy-preserving SVM using nonlinear kernels on horizontally partitioned data. In: Proceedings of the 2006 ACM symposium on Applied computing (SAC), pp. 603–610. ACM Press, New York (2006)
15. Zhan, J., Chang, L.W., Matwin, S.: Privacy Preserving K-nearest Neighbor Classification. International Journal of Network Security 1(1), 46–51 (2005)

Coping with Distribution Change in the Same Domain Using Similarity-Based Instance Weighting

Jeong-Woo Son, Hyun-Je Song, Seong-Bae Park, and Se-Young Park

Department of Computer Engineering
Kyungpook National University
702-701 Daegu, Korea
{jwson,hjsong,sbpark,sypark}@sejong.knu.ac.kr

Abstract. Lexicons are considered as the most crucial features in natural language processing (NLP), and thus often used in machine learning algorithms applied to NLP tasks. However, due to the diversity of lexical space, the machine learning algorithms with lexical features suffer from the difference between distributions of training and test data. In order to overcome the distribution change, this paper proposes support vector machines with example-wise weights. The training distribution coincides with the test distribution by weighting training examples according to their similarity to all test data. The experimental results on text chunking show that the distribution change between training and test data is actually recognized and the proposed method which considers this change in its training phase outperforms ordinary support vector machines.

1 Introduction

Domain adaptation aims to develop well adapted learning algorithms for various domains which can be different with the domain of training data [4]. In domain adaptation, the training data are drawn from a mixture of two distributions which are not identical but related. Then, the goal in learning algorithms is to perform well in one of these distributions, the "target" domain.

Domain adaptation is recently an interesting issue in natural language processing (NLP) since the situation considered in domain adaptation appears frequently in practical NLP tasks. Many studies have been proposed [1,7] in domain adaptation for NLP tasks. In this work, they assume that the general corpus such as Penntree Bank, Wall Street Journal, and so on is sufficiently constructed, while the corpus from the "target" domain which the learning algorithms should be performed is not enough and is a costly enterprise. In such cases, it is general to train learning algorithms with small corpus from the "target" domain and auxiliary information obtained from general corpus. Therefore, the distribution change between corpora should be considered to cope with the performance drop by the domain difference. all of the previous work have focused on this domain change [3,16].

The distribution change occurs even when both training and test data are generated from the same domain due to lexical features. In NLP, lexicons are one of the most crucial aspects to represent information on a sentence, a phrase, and so on. They have a special characteristic that the dimension of lexical space is almost unbounded and is

Z.-H. Zhou and T. Washio (Eds.): ACML 2009, LNAI 5828, pp. 354–366, 2009.

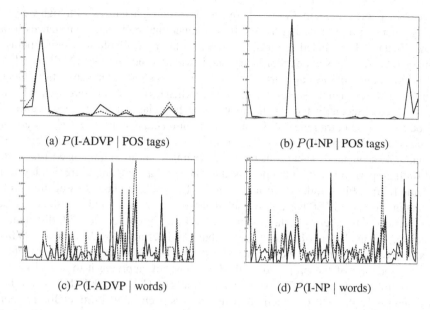

(a) P(I-ADVP | POS tags)   (b) P(I-NP | POS tags)

(c) P(I-ADVP | words)   (d) P(I-NP | words)

Fig. 1. Distribution difference of chunk classes. (a) and (b) are when POS tags are used as features, while (c) and (d) are words are used as features.

still growing. As a result, lexical distribution of training data is generally different from test data even when they are generated from the same domain.

Figure 1 shows $P(y|x)$ of I-ADVP and I-NP classes in the training and test data drawn from a single corpus, Wall Street Journal corpus. In this figure, the solid lines denote the distribution of training data while the dashed lines show that of test data, and y-axis denotes probability values. Figure 1-(a) and 1-(b) are the distributions when x is expressed with POS tags. The number of POS tags are relatively small and fixed. Accordingly, the distributions in the training and test data are quite similar. On the other hand, Figure 1-(c) and 1-(d) show the distributions when words are used for x. The training and test distributions are different in these figures. This phenomenon causes severe problems in many NLP tasks, since most learning algorithms are developed under the basic assumption that both training and test data are independent and identically distributed (i.i.d.). That is, the distribution change between training and test data affects their performance.

As a solution to this distribution change, this paper proposes support vector machines (SVMs) with example-wise weights. It first determines the nongold-standard class labels by the ordinary SVMs, and then the classification result with the ordinary SVMs is used to generate the feature of a test instance including preceding class labels. After that, it computes the weight of each training instance to coincide the training distribution with the test distribution using Kullback-Leibler Importance Estimation Procedure (KLIEP) [20]. That is, the weights of training instances are determined to minimize Kullback-Leibler divergence between the training and the test distributions. After the weights are set, they play a role of constraints which decide the boundary of maximum values of support vectors in SVMs. Therefore, the training instances with high weights have greater effects in learning SVMs. It is assumed that both labeled instances from training data and unlabeled instances from test data are available in training phrase. Note that this setting is not extraordinary, and is widely used in various NLP tasks [2].

The advantages of this paper are three-fold. First, it is shown that distribution change actually occurs even in the same-domain corpora when lexical features are used in learning NLP tasks. Secondly, the dynamic features such as class labels of preceding words can be considered in training SVMs under distribution change. In NLP tasks, every instance is strongly dependent on its neighbor instances since natural language sentences are naturally sequential. That is, the usage of dynamic features in training SVMs is crucial for high performance. In the proposed method, dynamic features are employed in learning KLIEP. Finally, the performance of KLIEP on multi-class classification problems is improved. KLIEP has a drawback that the weight of a training instance is too sensitive to the number of instances in a class, In the proposed method, distribution coincidence is made according to class distributions determined by the ordinary SVMs. As a result, the influence by the number of instances is reduced in computing weights.

The validation of the proposed method is shown by applying it to a well-known NLP task of text chunking. The experimental results on CoNLL-2000 shared task data set demonstrate that the distribution change occurs even in Wall-Street Journal corpus and the proposed method outperforms ordinary support vector machines by using example-wise weighting. The results imply that the proposed method is a plausible way to improve performance of the problems with distribution change by lexical features.

The rest of the paper is organized as follows. Section 2 reviews the related works on domain adaptation and distribution change. Section 3 introduces the idea of support vector machines with example-wise weights. Section 4 explains how to weigh the training instances based on similarity to all test data, and then Section 5 shows the experimental results. Finally, Section 6 draws conclusions.

2   Related Work

Recently learning of natural languages pays attention to distribution change [1]. In [13] and [6], it is revealed that the distribution change results in performance decrease of more than 10%. They performed some experiments with DSO corpus composed of two different corpora of Brown and Wall Street Journal. In word sense disambiguation, various machine learning methods trained with Wall Street Journal corpus showed poor performance when tested with Brown corpus, and vise versa.

This problem can be solved with domain adaptation in which two domains of a target domain and a source domain are assumed. To achieve better performance under domain adaption, many previous studies divide training instances into two categories [1,3,4,7,16]. One category contains the training instances from the target domain, and the other has those from the source domain. After that, target domain instances are applied to train a model while source domain instances are used to obtain auxiliary information for training the model.

For instance, Daumé and Marcu regarded the data from target domain as a mixture of two distributions: a "truly *in-domain*" distribution and a "general domain" distribution

[4]. Similarly, the data from source domain are considered to be drawn from a mixture of a "truly *out-of-domain*" distribution and a "general domain" distribution. They focused in extracting the general domain distribution from source domain data.

The main problem of this kind of work is that it is not valid if labeled target examples are not available. The only method using unlabeled target examples is, at least to our knowledge, the one proposed by Jiang and Zhai [10]. This method divided domain adaption into two kinds of sub-adaptations: labeling adaptation and instance adaptation. Then, the unlabeled target data are used to estimate the parameters for instance adaptation. However, even this work does not consider the distribution change in a single domain.

In the machine learning community, the difference between training and test distributions has been studied under the name of *covariate shift*. The influence of covariate shift can be alleviated by weighting training instances according to their importance. The importance weight of a data point x is defined as

$$w(x) = \frac{P_{Te}(x)}{P_{Tr}(x)},$$

where $P_{Te}(x)$ is the probability of x in the test distribution and $P_{Tr}(x)$ is that in the training distribution [19]. Therefore, the most important task in covariate shift is to estimate the importance weights of training instances.

Estimating both training and test distributions [8] is the most naive way to compute the importance weights. After estimating the distributions, the importance weights are computed as the ratio of the estimated distributions. However, this is ineffective since density estimation is one of the most difficult tasks in machine learning. Therefore, most previous studies in covariate shift focused on the estimation of the importance weights without density estimation.

The Kernel Mean Matching (KMM) method proposed by Huang et al.[9] finds the weights so that the mean values of training and test data are matched in a reproducing kernel Hilbert space. In KMM, finding the weights is expressed as a convex optimization. Thus, quadratic programming is used to solve the optimization problem. However, KMM has problems that it converges extremely slow or diverges under some conditions.

Sugiyama et al. proposed the Kullback-Leibler Importance Estimation Procedure (KLIEP) [20]. KLIEP expresses the weight as a linear model and determines the parameters of the model so that the Kullback-Leibler divergence from test distribution to training distribution is minimized. The optimization of KLIEP is also a convex optimization. Therefore a global solution can be obtained.

3   Support Vector Machines with Example-Wise Weights

The goal of support vector machines (SVMs) is to discover an optimal decision boundary which discriminates positive examples from negative examples. When training data are mapped onto a feature space, there exist a number of decision boundaries, so-called hyperplanes. Among the hyperplanes, SVMs determine the optimal hyperplane which maximizes the distance from the hyperplane to support vectors.

For given training data $\mathcal{D} = \{(x_1, y_1), \ldots, (x_n, y_n)\}$, where x_i is a vector representation for the i-th data point and $y_i \in \mathcal{Y} = \{-1, +1\}$ is its class label, a hyperplane is written as

$$\theta \cdot x_i - b = 0,$$

where θ and b are parameters estimated from $\mathcal{D}$. The parameters θ and b have to satisfy

$$\theta \cdot x_i - b \geq 1 \quad \text{if } y_i = +1,$$
$$\theta \cdot x_i - b \leq 1 \quad \text{if } y_i = -1,$$

for all $(x_i, y_i) \in \mathcal{D}$. Then, the margin, the distance between a hyperplane and support vectors is given as $\frac{2}{\|\theta\|}$. The goal of SVMs is to minimize $\frac{1}{2}\|\theta\|$ satisfying the constraints above. In soft-margin SVMs, by introducing slack variables ξ_i, this goal is changed to minimize

$$\frac{1}{2}\|\theta\|^2 + C \sum_{i=1}^{n} \xi_i,$$

where C is a user-defined constant to control the trade-off between allowing training errors and forcing rigid margin.

The soft-margin SVMs are a typical classifier of which performance drops drastically when training and test data are not drawn from an identical distribution [22]. One possible solution to this problem is to assign weights w_i to the training instances. Then, the goal is changed to minimize

$$\frac{1}{2}\|\theta\|^2 + C \sum_{i=1}^{n} w_i \xi_i$$

subject to

$$\langle \phi(x_i, \hat{y}_i) - \phi(x_i, y_i), \theta \rangle \geq 1 - \frac{\xi_i}{\Delta(\hat{y}_i, y_i)}$$

for all y_i and $\xi_i \geq 0$. Here the function $\phi(x_i, y_i)$ maps a data point into a feature space, $\hat{y}_i$ is the predicted class label of x_i, and $\Delta(\hat{y}_i, y_i)$ denotes a loss function between $\hat{y}_i$ and y_i.

The dual representation of this optimization is given as

$$\min_{\alpha} \frac{1}{2} \sum_{i,j=1}^{n} \alpha_{iy_i} \alpha_{jy_j} k(x_i, y_i, x_j, y_j) - \sum_{i=1; y \in \mathcal{Y}}^{n} \alpha_{iy}$$

$$\text{subject to } \alpha_{iy} \geq 0 \text{ for all } i, y$$

$$\text{and } \sum_{y_i \in \mathcal{Y}} \frac{\alpha_{iy}}{\Delta(\hat{y}_i, y_i)} \leq w_i C,$$

where the kernel function, $k(x_i, y_i, x_j, y_j) = \langle \phi(x_i, y_i), \phi(x_j, y_j) \rangle$, returns the inner product between two data points. Note that the weight w_i restricts the boundary of α_i.

This dual representation problem can be solved by contemporary SVMs with little modification [21]. For detail description on learning SVMs with data-driven weights, refer to [18].

4 Weight Estimation

4.1 Kullback-Leibler Importance Estimation

In order to determine a weight $w(x)$ of a data point x without distribution estimation, the weight is modeled in KLIEP as a linear model. That is,

$$\hat{w}(x) = \sum_{l=1}^{b} \alpha_l \varphi_l(x),$$

where α_l is a parameter to be learned and $\varphi_l(x)$ is a basis function such that

$$\varphi_l(x) \geq 0 \quad \text{for all} \quad x \in \mathcal{D}.$$

Since test distribution can be approximated from $w(x)$ and training distribution [19], the goal is to find an optimal $\hat{w}(x)$'s. The optimal weights $\hat{w}(x)$ are obtained by minimizing Kullback-Leibler divergence between $P_{Te}(x)$ and $\hat{P}_{Te}(x)$, where $\hat{P}_{Te}(x)$ is an empirical distribution of $P_{Te}(x)$ given as

$$\hat{P}_{Te}(x) = \hat{w}(x) P_{Tr}(x). \tag{1}$$

Here, Kullback-Leibler divergence between between $P_{Te}(x)$ and $\hat{P}_{Te}(x)$ is defined as

$$
\begin{aligned}
KL&[P_{Te}(x) \| \hat{P}_{Te}(x)] \\
&= \int_{\mathcal{D}} P_{Te}(x) \log \frac{P_{Te}(x)}{\hat{w}(x) P_{Tr}(x)} dx \\
&= \int_{\mathcal{D}} P_{Te}(x) \log \frac{P_{Te}(x)}{P_{Tr}(x)} dx \\
&\qquad - \int_{\mathcal{D}} P_{Te}(x) \log \hat{w}(x) dx.
\end{aligned}
$$

When x^{Tr} and x^{Te} denote samples in training distribution and test distribution respectively, $\hat{w}(x)$'s which minimize $KL[P_{Te}(x) \| \hat{P}_{Te}(x)]$ can be found by

$$\max_{\{\alpha_l\}_{l=1}^{b}} \left[\sum_{j=1}^{m} \log \left(\sum_{l=1}^{b} \alpha_l \varphi_l(x_j^{Te}) \right) \right]$$

$$\text{subject to} \quad \sum_{i=1}^{n} \sum_{l=1}^{b} \alpha_l \varphi_l(x_i^{Tr})$$

$$\text{and} \quad \alpha_1, \alpha_2, \ldots, \alpha_b \geq 0,$$

where n is the number of training instances and m is that of test instances [20]. Since this is a convex optimization, the global solution can be obtained.

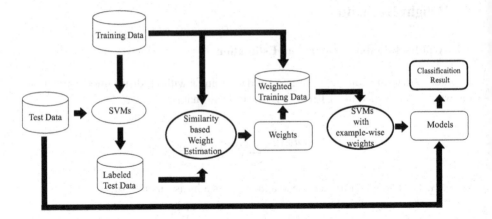

Fig. 2. Overall structure of the support vector machines with example-wise weighting

4.2 Similarity-Based Weight Estimation

The weights for training instances are determined according to how similar the training distribution is to the test distribution. However, test distribution is unknown in most practical problems. Therefore, the similarity is measured for each data point rather than for whole distribution. That is, the weights are determined under the assumption that the density of a training instance which shares high similarity with all test instances will be high in the test distribution, since test instances are sampled from the test distribution.

The similarity between a training instance x and a test instance x_l^{Te} is computed using a kernel function. That is, a kernel $k(x, x_l^{Te})$ is used as a basis function in Equation (1). Then, $w(x)$, the weight of x, is a weighted sum of similarities between x and all test instances. When b is the number of bases,

$$w(x) = \sum_{l=1}^{b} (\alpha_l(x) \times k(x, x_l^{Te})).$$

Note that b is equal to the number of test instances m, since all test instances are used to estimate $w(x)$. The values of $\alpha_l(x)$'s are then learned by KLIEP.

Natural language sentences are naturally sequential, which implies that preceding class labels influence the prediction of current data point. Therefore, the preceding labels are, in general, included in the vector representation of a data point, when a learning method based on i.i.d. sampling is applied to NLP tasks. However, the proposed method first estimates the test distribution of which class labels are unknown, and thus it is difficult to include the preceding labels in the representation of data points.

Figure 2 depicts how this problem is solved in this paper. An ordinary SVMs are first trained only with training data, and then non-gold-standard labels of test data are estimated by the ordinary SVMs. Since SVMs are strong enough to trust its classifications, the class labels determined by the ordinary SVMs are used to generate the feature of a test instance including preceding class labels. Then, the training distribution is coincided with the test distribution. That is, the weight of each training instance is determined according to its similarity to test data by the proposed method. Finally, the SVMs with example-wise weights are trained using the weighted training data.

By adopting the process shown in Figure 2, the proposed method has two advantages. First, the performance is improved by allowing dynamic features such class labels of preceding words. In addition, when KLIEP is applied to multi-class classification, the weight of a training instance drawn from the class with a great number of instances has a tendency to be abnormally high, and thus it results in performance drop. This problem is solved by computing the weights according to class distributions determined by the ordinary SVMs.

5   Experiments

5.1   Task

The proposed method is verified with text chunking, one of the well-known problems in NLP. In order to represent chunks, an IOB model is used, where every word is tagged with a chunk label extended with I (inside a chunk), O (outside a chunk), and B (inside a chunk, but the word is in another chunk from the previous words). That is, each phrase type has two kinds of chunk labels. For instance, NP is extended with B-NP and I-NP. In CoNLL-2000 shared task data set[1], they assume 11 types of phrase so that there exist 23 chunk types including one additional chunk tag, O. Therefore, text chunking can be considered as a multi-class classification.

In order to incorporate contextual information in the feature vector for i-th data point, the information of surrounding words such as POS tags and class labels are included in its vector representation. More formally, x_i, the feature vector of i-th data point is represented as

$$x_i = \{w_i, p_i, w_k, p_k, c_k, w_j, p_j\},$$
$$k = i - 1, i - 2,$$
$$j = i + 1, i + 2,$$

where w_i is the word appeared at i-th position and p_i, c_i are its POS tag and class label respectively. This feature representation is widely used in text chunking [5,12].

5.2   Data Set and Setting

The proposed method is evaluated with the data set used at CoNLL-2000 shared task. This data set is generated from Wall-Street Journal (WSJ) corpus which consists of business and financial news. Since the training data generated from WSJ section 15-18 is composed of 211,727 tokens and the test data from WSJ section 20 consists of 47,377 tokens, both training and test data are drawn from the same domain. Table 1 shows the number of training and test instances with respect to phrase types. The phrase distributions of training and test data are very similar though they are not uniform. In a word, even if $P(y)$ of test data is quite similar to that of training data, $P(y|x)$ can be differentiated according to the feature type of x (see Figure 1).

---

[1] http://lcg-www.uia.ac.be/conll2000/chunking

Table 1. The number of instances in training and test data of CoNLL-2000 shared task data set

Phrase Type	Training Data		Test Data	
	No. of Instances	Ratio (%)	No. of instances	Ratio (%)
ADJP	2,703	1.277	605	1.277
ADVP	4,670	2.206	955	2.015
CONJP	129	0.061	22	0.046
INTJ	40	0.019	2	0.004
LST	10	0.004	7	0.015
NP	118,388	55.915	26,798	56.551
PP	21,572	10.189	4,859	10.254
PRT	558	0.264	106	0.224
SBAR	2,277	1.075	539	1.137
UCP	8	0.004	0	0.000
VP	33,470	15.808	7,314	15.435
O	27,902	13.178	6,180	13.042
All	211,727	100.0	47,387	100.0

SVMs are basically a binary classifier. Even if there are many ways to use SVMs in multi-class classification [17], this paper adopts "one-versus-all" strategy. SVM^{light} [11] is used to implement support vector machines. We experiment four variants of SVMs to evaluate the proposed method under distribution change. The four SVMs are

- SVM1: SVM of which features are words and POS tags,
- SVM2: SVM proposed by Kudoh and Matsumoto [12][2],
- SVMEW1: SVM with example-wise weights of which features are words and POS tags,
- SVMEW2: SVM with example-wise weights in which the weights are determined by the proposed method.

Since SVM1 and SVMEW1 are trained with word and POS tag features, they do not consider any preceding class labels. For all SVMs, a polynomial kernel with degree of two is used.

5.3   Experimental Results

Figure 3 shows the results of comparing SVM1, SVM2, and SVMEW1. This figure proves that the class labels of preceding words are important at least in text chunking. SVM2 which uses preceding class labels in its feature outperforms both SVM1 and SVMEW1 which do not use preceding class labels. SVMEW1 shows higher F-score than SVM1, which implies that it is important to consider distribution change in learning SVMs. However, the more important thing is that the preceding class labels should be included in the feature set due to the characteristics of natural languages.

Table 2 shows the final results of the proposed method, SVMEW2. The F-score of SVMEW2 is 94.19 while that of SVM2 is just 93.21. It is interesting to note that the

---

[2] This SVM showed the best performance in CoNLL-2000 shared task.

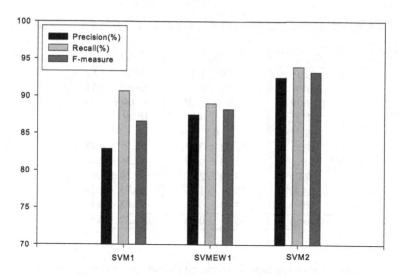

Fig. 3. The effects of preceding class labels in text chunking

Table 2. Performance comparison of SVMEW2 and SVM2

	SVM2			SVMEW2		
	Precision	Recall	$F_{\beta=1}$	Precision	Recall	$F_{\beta=1}$
ADJP	73.35%	71.00%	72.16	73.21%	72.37%	**72.79**
ADVP	80.07%	79.79%	79.93	80.93%	80.37%	**80.65**
CONJP	50.00%	55.56%	**52.63**	41.67%	55.56%	47.62
INTJ	100.00%	50.00%	66.67	100.00%	50.00%	66.67
LST	0.00%	0.00%	0.00	0.00%	0.00%	0.00
NP	92.88%	94.67%	93.76	94.52%	95.64%	**95.07**
PP	95.86%	98.30%	97.06	95.99%	98.92%	**97.43**
PRT	76.77%	71.70%	**74.15**	74.76%	72.64%	73.68
SBAR	91.79%	83.55%	87.48	92.42%	86.54%	**89.38**
VP	92.60%	94.03%	93.31	93.22%	94.98%	**94.09**
Overall	92.53%	93.91%	93.21	93.55%	94.84%	**94.19**

proposed method shows worse F-score than SVM2 for CONJP and PRT, even though it outperforms SVM2 in most phrases. In determining the weights of training instances, a number of training instances are demanded to cover most test distribution. However, these phrases have just less than 600 training instances. SVMEW2 fails in improving SVM2 also for other phrases with small number of training instances like INTJ and LST.

The phrase SBAR is a special case. The F-score of SBAR in SVMEW2 is much improved though SBAR has just 2,277 training instances. The main reason for the errors in SBAR of SVM2 is that SBAR is easy to be misclassified as PP for the words like

Table 3. Five-fold cross validation results. All performances are measured with $F_{\beta=1}$.

Methods	SVM2	SVMEW2
ADJP	**74.38 ± 1.27**	74.29 ± 1.83
ADVP	80.72 ± 1.25	**81.63 ± 2.11**
CONJP	37.54 ± 16.26	**39.66 ± 15.51**
INTJ	14.19 ± 14.45	**18.89 ± 12.17**
LST	0	0
NP	94.22 ± 0.27	**95.59 ± 0.49**
PP	96.90 ± 0.27	**97.36 ± 0.31**
PRT	74.90 ± 2.20	**75.20 ± 1.69**
SBAR	85.01 ± 1.26	**86.73 ± 1.70**
VP	93.18 ± 0.25	**94.03 ± 0.45**
Overall	93.33 ± 0.26	**94.37 ± 0.50**

Table 4. The comparison of experimental results with existing methods

Method	$F_{\beta=1}$
SVMs with example-wise weighting	**94.19**
Ordinary SVMs	93.21
Boosted Maximum Entropy Models	92.82
Regularized Winnow	94.13
Entropy Guided Transformation Learning	92.79

'*as*' which can be used as both conjunction and preposition. Therefore, the improvement of PP leads to the improvement of SBAR. Since the F-score of PP is improved in SVMEW2, that of SBAR is also improved though SBAR has small number of training instances.

Five-fold cross validation is performed to verify evaluations of SVM2 and SVMEW2. Table 3 shows the results. SVMEW2 outperforms SVM2 for all phrases except ADJP. This is also statistically significant result, since p-value from T-test for overall performance is just 0.001 which makes the null hypothesis rejected even with 99% confidence level.

Finally, we compare the proposed method with other methods in text chunking which do not consider the distribution change. The compared methods are (i) ordinary SVMs [12], (ii) boosted maximum entropy models [15], (iii) regularized winnow [23], and (iv) entropy guided transformation learning (ETL) [14]. Table 4 shows their performances. The F-scores of ordinary SVMs and boosted maximum entropy are 93.21 and 92.82 respectively, while those of the regularized winnow and ETL are 94.13 and 92.79. Even though this comparison is not fair, since the compared methods do not use any information from test data, it, at least, implies that the proposed method improves the performance by coinciding both training and test distribution.

6   Conclusion

In this paper we have shown that NLP tasks suffer from distribution change caused by lexical features. As a solution to this problem, we have proposed a novel method to coincide training distribution with test distribution by weighting training instances. The weight of a training instance is determined based on the similarity between it and all test instances. The experiments on CoNLL-2000 shared task data set showed that distribution change occurs even in an identical domain and the proposed method achieves the state-of-the-art performance. This implies that the proposed method is a plausible way to cope with distribution change.

We also discussed that the class labels of preceding words are still important under distribution change. To incorporate the labels into the representation of training instances, in the proposed method, ordinary SVMs first estimate the class labels of test data, and then SVMs with example-wise weighting are trained with the class labels. The experimental results proved that this improves the performance in text chunking.

Acknowledgements

This work was supported in part by MIC & IITA through IT Leading R&D Support Project and by the Korean Ministry of Education under the BK21-IT Program.

References

1. Arnold, A., Nallapati, R., Cohen, W.: Exploiting Feature Hierarchy for Transfer Learning in Naemd Entity Recognition. In: Proceedings of the 46th Annual Meeting of the Association of Computational Linguistics, pp. 245–253 (2008)
2. Chapelle, O., Schölkopf, B., Zien, A.: Semi-Supervised Learning. MIT Press, Cambridge (2006)
3. Chelba, C., Acero, A.: Adaptation of Maximum Entropy Capitalizer: Little Data Can Help a Lot. In: Proceedings of Empirical Methods in Natural Language Processing and Very Large Corpora, pp. 285–292 (2004)
4. Daumé, H., Marcu, D.: Frustratingly Easy Domain Adaptation. In: Proceedings of the 45th Annual Meeting of the Association of Computational Linguistics, pp. 256–263 (2007)
5. Déjean, H.: Learning Rules and Their Exceptions. Journal of Machine Learning Research 2, 669–693 (2002)
6. Escudero, G., Marquez, L., Rigau, G.: An Empirical Study of the Domain Dependence of Supervised Word Sense Disambiguation Systems. In: Proceedings of Empirical Methods in Natural Language Processing and Very Large Corpora, pp. 172–180 (2000)
7. Florian, R., Hassan, H., Ittycheriah, A., Jing, H., Kambhatla, N., Luo, X., Nicolov, N., Roukos, S.: A Statistical Model for Multilingual Entity Detection and Tracking. In: Proceedings of Human Language Technology and North American Chapter of the Association for Computational Linguistics Annual Meeting, pp. 1–8 (2004)
8. Heckman, J.: Sample Selection Bias as a Specification Error. Econometrica 47(1), 153–162 (1979)
9. Huang, J., Smola, A., Gretton, A., Borgwardt, K., Schölkopf, B.: Correcting Sample Selection Bias by Unlabeled Data. In: Advances in Neural Information Processing Systems 19, pp. 601–608. MIT Press, Cambridge (2007)

10. Jiang, J., Zhai, C.: Instance Weighting for Domain Adaptation in NLP. In: Proceedings of the 45th Annual Meeting of the Association of Computational Linguistics, pp. 264–271 (2007)
11. Joachims, T.: Making Large-Scale SVM Learning Practical, LS8, Universitaet Dortmund (1998)
12. Kudoh, T., Matsumoto, T.: Chunking with Support Vector Machines. In: Proceedings of the 2nd Meeting of the North American Chapter of the Association for Computational Linguistics on Language technologies, pp. 1–8 (2001)
13. Martinez, D., Agirre, E.: One Sense per Collocation and Genre/Topic Variations. In: Proceedings of Empirical Methods in Natural Language Processing and Very Large Corpora, pp. 207–215 (2000)
14. Milidiú, R., Santos, C., Duarte, J.: Phrase Chunking Using Entropy Guided Transformation Leanring. In: Proceedings of the 46th Annual Meeting of the Association of Computational Linguistics, pp. 647–655 (2008)
15. Park, S.-B., Zhang, B.-T.: A Boosted Maximum Entropy Models for Learning Text Chunking. In: Proceedings of the 19th International Conference on Machine Learning, pp. 482–489 (2002)
16. Roark, B., Bacchiani, M.: Supervised and Unsupervised PCFG Adaptation to Novel Domains. In: Proceedings of Human Language Technology and North American Chapter of the Association for Computational Linguistics Annual Meeting, pp. 126–133 (2004)
17. Shawe-Taylor, J., Cristianini, N.: Support Vector Machines and Other Kernel-based Learning Methods. Cambridge University Press, Cambridge (2000)
18. Schmidt, M., Gish, H.: Speaker Identification via Support Vector Classifiers. In: Proceedings of IEEE International Conference on Acoustics, Speech, and Signal Processing, pp. 105–108 (1996)
19. Shimodaira, H.: Improving Predictive Inference Under Covariate Shift by Weighting the Log-Likelihood Function. Journal of Statistical Planning and Inference 90(2), 227–244 (2000)
20. Sugiyama, M., Nakajima, S., Kashima, H., Bünau, P., Kawanabe, M.: Direct Importance Estimation with Model Selection and Its Application to Covariate Shift Adaptation. In: Advances in Neural Information Processing Systems 20, pp. 1433–1440. MIT Press, Cambridge (2008)
21. Tsochantaridis, I., Joachims, T., Hofmann, T., Altun, Y.: Large Margin Methods for Structured and Interdependent Output Variables. Journal of Machine Learning Research 6, 1453–1484 (2005)
22. Zadrozny, B.: Learning and Evaluating Classifiers Under Sample Selection Bias. In: Proceedings of the 21st International Conference on Machine Learning, pp. 114–121 (2004)
23. Zhang, T., Damerau, F., Johnson, D.: Text Chunking Using Regularized Winnow. In: Proceedings of the 39th Annual Meeting of the Association of Computational Linguistics, pp. 539–546 (2001)

Monte-Carlo Tree Search in Poker Using Expected Reward Distributions

Guy Van den Broeck, Kurt Driessens, and Jan Ramon

Department of Computer Science, Katholieke Universiteit Leuven, Belgium
{guy.vandenbroeck,kurt.driessens,jan.ramon}@cs.kuleuven.be

Abstract. We investigate the use of Monte-Carlo Tree Search (MCTS) within the field of computer Poker, more specifically No-Limit Texas Hold'em. The hidden information in Poker results in so called MIXIMAX game trees where opponent decision nodes have to be modeled as chance nodes. The probability distribution in these nodes is modeled by an opponent model that predicts the actions of the opponents. We propose a modification of the standard MCTS selection and backpropagation strategies that explicitly model and exploit the uncertainty of sampled expected values. The new strategies are evaluated as a part of a complete Poker bot that is, to the best of our knowledge, the first exploiting no-limit Texas Hold'em bot that can play at a reasonable level in games of more than two players.

1 Introduction

Poker playing computer bots can be divided into two categories. There are the *game-theoretic bots*, that play according to a strategy that gives rise to an approximate Nash equilibrium. These bots are difficult, or impossible to beat, but are also not able to exploit possible non-optimalities in their opponents. Game theoretic strategy development also becomes very complex as the number of opponents increases, so most bots that try to play Nash equilibria, do so in a heads-up setting, i.e., with only one opponent. The other type of bot is the *exploiting bot* that employs game tree search and opponent modeling techniques to discover and exploit possible weaknesses in opponents. In this paper, we focus on the second type of bot.

Research in computer Poker has been mainly dealing with the *limit* variant of Texas Hold'em. In limit Poker, bet sizes are fixed as well as the total number of times a player can raise the size of the pot. Only recently [1], attention has started shifting to the no-limit variant and the increased complexity that it brings. In fact, all exploiting bots developed so far, deal with the heads-up limit version of the game. The limit game tree consists of about 10^{18} nodes and modern bots are capable of traversing the complete tree in search of optimized actions through a full-width depth-first search [2]. Gilpin et al.[1] estimate the game tree in heads-up no-limit Texas Hold'em (with bet-size discretization) to reach a node-count of about 10^{71}.

Z.-H. Zhou and T. Washio (Eds.): ACML 2009, LNAI 5828, pp. 367–381, 2009.

The true tournament version of Texas Hold'em is played with a maximum of 10 players per table, expanding the size of the game tree even more. Traversing the whole game tree is in this case not an option. While existing bots can either play limit or heads-up poker, the bot we describe here can deal with the full complexity of the poker game by employing sampling methods and bet-size discretization.

Monte-Carlo Tree Search (MCTS) is a best-first search technique that estimates game tree node values based on the results from simulated gameplay. Its main goal is to replace exhaustive search through the game tree with well founded sampling methods. The most popular heuristic for the best-first selection in MCTS is UCT [3], which stands for Upper Confidence bounds applied to Trees. UCT is a translation of the upper confidence bound selection algorithms developed for bandit-based problem that puts a limit on the regret caused by exploration moves. In this paper we suggest the use of MCTS methods in no-limit Texas Hold'em Poker to limit the extent of the tree the Poker bot has to search, thereby reducing the computational requirements to reach a reasonable play-level.

As a first contribution, we apply MCTS in the context of MIXIMAX game trees [2] as they are encountered in incomplete information games such as Poker where opponent actions can not simply be predicted as value minimization actions. While adaptations of MCTS have been defined for EXPECTIMAX game trees [4] that deal with non-determinism such as dice rolls, we are not aware of any previous work on MCTS in incomplete information games. As a second contribution, we adapt both the backpropagation and selection steps in MCTS to incorporate explicit sample variance information in the game tree nodes.

The rest of the paper is structured as follows. The following section briefly presents no-limit Texas Hold'em Poker and the game tree it results in. Section 3 discusses MCTS and some standard implementation choices for it. In section 4 we propose extensions of MCTS that deal explicitly with the sampling distribution in each node. The proposed strategies are evaluated in section 5 after which we conclude.

2   The Poker Game Tree

Texas Hold'em Poker is played by two or more people who each hold two hidden cards. The game starts with a betting round where people can invest money in a shared pot by putting in a bet. Alternatively, they can check and let the next player make the first bet. Players must match every bet made by investing an equal amount (calling) or by investing even more (raising). When a player fails to do so and gives up, he folds. The betting round ends when everybody has folded or called. To make sure each game is played, the first two players in the first round are forced to place bets, called small and big blinds. The big blind is double the amount of chips of the small blind.

Next, a set of three community cards (called the Flop) is dealt to the table, followed by a new betting round. These cards are visible to every player and the goal is to combine them with the two hidden cards to form a combination of cards (also called a hand). One card is added to the set of community cards twice more (the first one called the Turn and the last called the River) to make a total of five, each time initiating a new betting round. Finally, the player with the best hand that hasn't folded wins the game and takes the pot.

From a game tree perspective, this gives rise to the following node types:

Leaf nodes: Evaluation of the expected value of the game in these nodes is usually trivial. In Poker, we choose to employ the expected sum of money a player has after the game is resolved. In case the player loses, this is the amount of money he didn't bet. In case the player wins, it is that amount plus the current pot, with a possible distribution of the pot if multiple players have similar poker hands. Leafs can also take probability distributions over currently unknown cards into account.

Decision nodes: These are the nodes where the bot itself is in control of the game. In non-deterministic, fully observable games, these would be the nodes where the child that maximizes the expected outcome is selected. While in Poker, it can be beneficial to not always select the optimal action and remain somewhat unpredictable, they should still be treated as maximization nodes during the game tree search.

Chance nodes: These nodes are comparable to those encountered in EXPECTI-MAX games. In Poker, they occur when new cards are dealt to the table. In these nodes, children are selected according to the probability distribution connected to the stochastic process steering the game at these points.

Opponent nodes: In complete information games, the opponent will try to minimize the value in the nodes where he can make a choice. However, in poker, we do not know the evaluation function of the opponent, which will typically be different because the opponent has different information. Neither does the opponent know our evaluation function. Instead of treating the opponent nodes as min-nodes one can therefore consider these nodes as chance nodes. The big difference is that the probability distribution in opponent nodes is unknown and often not static. The probability distribution over the different options the opponent can select from can be represented (and learned) as an opponent-model.[1]

These nodes make up a MIXIMAX game tree that models the expected value of each game state. Because this tree is too big, it is impossible to construct. We can only approximate the expected values using an incomplete search procedure like MCTS.

The Opponent Model. A full study or overview of possible factorizations of the opponent model in Poker is out of the scope of this paper. However, since the performance of the MCTS bots as described in this paper is greatly dependent on it, we include the employed opponent modeling strategy for completeness reasons.

---

[1] Depending on the expressivity of the opponent-model, this approach can either try to model the hidden information in the game or reduce it to the EXPECTIMAX model.

Assume that the action performed in step i is represented as A_i, that the community cards dealt at step i are represented as C_i and that H represents the hand cards held by all active players. We use an opponent model factored into the two following distributions:

1. $P(A_i|A_0 \ldots A_{i-1}, C_0 \ldots C_i)$: This model predicts the actions of all players, taking previous actions and dealt community cards into account.
2. $P(H|A_0 \ldots A_n, C_0 \ldots C_n)$: This model predicts the hand cards of all active players at showdown (represented as step n).

The first probability can be easily estimated using observations made during the game, as all relevant variables can be observed. This is only possible because we leave hand cards out of the first model. Predicting the exact amount of a bet or a raise is difficult. We decided to deal with this by treating minimal bets and all-in bets (i.e. when a player bets his entire stack of money) as separate cases and discretizing the other values.

In the end we need to know who won the game and therefore, we need to model the hand cards with the second probability. This one can be harder to model as "mucking" (i.e. throwing away cards without showing them to other players) can hide information about the cards of players in a loosing position. To make abstraction of the high number of possible hand-card combinations, we reduce the possible outcomes to the rank of the best hand that the cards of each active player result in [5] and again discretize these ranks into a number of bins. This approach has the advantage that mucked hands still hold some information and can be attributed to bins with ranks lower than that of the current winning hand.

The actual models are learned using the Weka toolkit [6] from a dataset of a million games played in an online casino. Each action in the dataset is an instance to learn the first model from. Each showdown is used to learn the second model. The game state at the time of an action or showdown is represented by a number of attributes. These attributes include information about the current game, such as the round, the pot and the stack size of the player involved. It also includes statistics of the player from previous games, such as the bet and fold frequencies for each round and information about his opponents, such as the average raise frequency of the active players at the table. With these attributes, Weka's M5P algorithm [7] was used to learn a regression tree that predicts the probabilities of each action or hand rank bin at showdown.

The use of features that represent windowing[2] statistics about previous games of a player in a model learned on a non-player-specific data set has the advantage of adapting the opponent model to the current opponent(s). It is important to note that there are usually not sufficient examples from a single individual player to learn something significant. We circumvent this problem by learning what is common for all players from a large database of games. At the same time, to exploit player-specific behavior, we will learn a function from the game situation

---

[2] Statistics are kept over a limited number of past games to adapt to changing game circumstances.

and player-specific statistics to a prediction of his action. A key issue here is to select player-specific statistics which are easy to collect with only a few observed games, while capturing as much as possible about the players' behavior.

3   Monte-Carlo Tree Search

Monte-Carlo Tree Search is a best-first search strategy. It revolutionized research in computer-Go and gave rise to Go-bots such as CRAZYSTONE [8], MOGO [9] and MANGO [10], which represent the current top-ranks in computer Go competitions. The original goal of MCTS was to eliminate the need to search MINIMAX game trees exhaustively and sample from the tree instead. Later, extensions of the MCTS technique were formulated that could deal with EXPECTIMAX trees such as those encountered in backgammon [11]. To the best of our knowledge, MTCS has not yet been adapted to deal with MIXIMAX trees that are encountered in incomplete information games such as Poker.

MCTS incrementally builds a subtree of the entire game tree in memory. For each stored node P, it also stores an estimate $\hat{V}(P)$ of the expected value $V^*(P) = \mathbb{E}[r(P)]$ of the reward $r(P)$ of that node together with a counter $T(P)$ that stores the number of sampled games that gave rise to the estimate. The algorithms starts with only the root of the tree and repeats the following 4 steps until it runs out of computation time:

Selection: Starting from the root, the algorithm selects in each stored node the branch it wants to explore further until it reaches a stored leaf (This is not necessarily a leaf of the game tree). The selection strategy is a parameter of the MCTS approach.

Expansion: One (or more) leafs are added to the stored tree as child(ren) of the leaf reached in the previous step.

Simulation: A sample game starting from the added leaf is played (using a simple and fast game-playing strategy) until conclusion. The value of the reached result (i.e. of the reached game tree leaf) is recorded. MCTS does not require an evaluation heuristic, as each game is simulated to conclusion.

Backpropagation: The estimates of the expected values $V^*(P) = \mathbb{E}[r(P)]$ (and selection counter $T(P)$) of each recorded node P on the explored path is updated according to the recorded result. The backpropagation strategy is also a parameter of the MCTS approach.

After a number of iterations of the previous four steps, an action-selection strategy is responsible for choosing a good action to be executed based on the expected value estimates and the selection counter stored in each of the root's children. It has been suggested to pick the child with the highest expected value or the highest visit count. Our experiments show that the choice of action-selection strategy has no significant impact.

3.1   UCT Sample Selection

Node selection or sample selection as described above is quite similar to the widely studied Multi-Armed Bandit (MAB) problem. In this problem, the goal is to minimize regret[3] in a selection task with K options, where each selection c_i results in a return according to a fixed probability distribution $r(c_i)$. The **Upper Confidence Bound** selection strategy (UCB1) is based on the Chernoff-Hoeffding limit that constrains the difference between the sum of random variables and the expected value. For every option c_i, UCB1 keeps track of the average returned reward $\mathbb{E}[r(c_i)]$ as well as the number of trials $\mathrm{T}(c_i)$. After sampling all options once, it selects the option that maximizes:

$$\hat{V}(c_i) + C\sqrt{\frac{\ln \mathrm{T}(P)}{\mathrm{T}(c_i)}} \tag{1}$$

where $\mathrm{T}(P) = \sum_j \mathrm{T}(c_j)$ is the total number of trials made. In this equation, the average reward term is responsible for the exploitation part of the selection strategy, while the second term, which represents an estimate of the upper bound of the confidence interval on $\mathbb{E}[r(c_j)]$, takes cares of exploration. C is a parameter that allows tuning of this exploration-exploitation trade-off. This selection strategy limits the growth rate of the total regret to be logarithmic in the number of trials [12].

UCB applied to Trees (UCT) [3] extends this selection strategy to Markov decision processes and game trees. It considers each node selection step in MCTS as an individual MAB problem. Often, UCT is only applied after each node was selected for a minimal number of trials. Before this number is reached, a predefined selection probability is used. UCT assumes that all returned results of an option are independently and identically distributed, and thus that all distributions are stationary. For MCTS, this is however not the case, as each sample will alter both $\hat{V}(s)$ and $\mathrm{T}(s)$ somewhere in the tree and thereby also the sampling distribution for following trials. Also, while the goal of a MAB problem is to select the best option as often as possible, the goal of the sample selection strategy in MCTS is to sample the options such that the best option can be selected at the root of the tree in the end. While both goals are similar, they are not identical. Nonetheless, the UCT heuristic performs quite well in practice.

3.2   Sample Selection in CRAZYSTONE

The CRAZYSTONE selection strategy [8] chooses options according to the probability that they will return a better value than the child with the highest expected value. Besides the estimated expected value $\mathbb{E}[r(P)]$ each node P also stores the standard deviation of the estimates of the expected value $\sigma_{\hat{V},P}$.

---

[3] Regret in a selection task is the difference in cumulative returns compared to the return that could be attained when using the optimal strategy.

The probability of selecting an option c_i is set to be proportional to:

$$P(c_i) \sim \exp\left(-2.4\frac{\hat{V}(c_{best}) - \hat{V}(c_i)}{\sqrt{2(\overline{\sigma}(c_{best})^2 + \overline{\sigma}(c_i)^2)}}\right). \tag{2}$$

where c_{best} is the option with the highest expected value. Under the assumption that values follow a Gaussian distribution, this formula approximates the target probability.

The CRAZYSTONE-selection is only applicable for deterministic MINIMAX-trees. Non deterministic nodes cause each node to have a probability distribution over expected rewards that does not converge to a single value. This stops $\overline{\sigma}(n)$ from converging to 0 and causes the number of samples using non-optimal options to remain non-negligible.

3.3 Backpropagation in Mixed Nodes

For both chance and opponent nodes the expected value of child-nodes are given a weight proportional to the amount of samples used:

$$\hat{V}(n) = \sum_j \frac{T(c_j)\hat{V}(c_j)}{T(n)} \tag{3}$$

This is simply the average of all values sampled through the current node.

3.4 Backpropagation in Decision Nodes

In decision nodes, we have a choice how to propagate expected values. The only requirement is that the propagated value converges to the maximal expected value as the number of samples rises. Two simple strategies present themselves:

– Use the average sampling result as it is used in mixed nodes: when using the correct selection strategy, for example UCT, the majority of samples will eventually stem from the child with the highest expected value and this will force equation 3 to converge to the intended value. Before convergence, this strategy will usually underestimate the true expected value, as results from non-optimal children are also included.
– Use the maximum of the estimates of the expected values of all child nodes:

$$\hat{V}(P) = \max_j \hat{V}(c_j). \tag{4}$$

As, in the limit, $\hat{V}(P)$ will converge to the true value $V^*(P)$, the maximum of all estimates will also converge to the correct maximum. This approach will usually overestimate the true expected value as any noise present on the sampling means is also maximized.

4   Sample Selection and Backpropagation Strategy

In this section, we introduce new sample-selection and backpropagation strategies for MCTS that take advantage of information about the probability distribution over the sampling results. Existing techniques do not take fully into account the consequences of having non-deterministic nodes in the search tree. The CRAZYSTONE algorithm does use the variance of the sample in its nodes, but its strategy only works well for deterministic games.

Not only backpropagating an estimate of the value of a node but also a measure of the uncertainty on this estimate has several advantages. First, as one can see in our derivations below, it allows a more accurate estimation of the expected value of nodes higher in the tree based on its children. Second, better sample selection strategies can be used. In particular, instead of the Chernoff-Hoeffding based heuristic from Equation (1), one can use an upper bound of the confidence interval taking the variance into account, and one can select for sampling the child node with highest

$$\hat{V}(c_i) + C.\sigma_{\hat{V},c_i} \tag{5}$$

where C is a constant and $\sigma_{\hat{V},c_i}$ is the standard error on $\hat{V}$. We call this strategy UCT+. Here, the first term is again the exploitation term and the second term an exploration term. In contrast to Equation (1) it takes into account the uncertainty based on the actually observed samples of the child node.

Let P be a node of the search tree. We will denote with $V^*(P)$ the value of P, i.e. the expected value of the gain of the player under perfect play. Unfortunately, we do not know yet these values, nor the optimal strategies that will achieve them. Assume therefore that we have estimates $\hat{V}(P)$ of the $V^*(P)$. It is important to know how accurate our estimates are. Therefore, let

$$\sigma_{\hat{V},P}{}^2 = \mathbb{E}[(\hat{V}(P) - V^*(P))^2].$$

For example, if P is a leaf node, and if we have J samples $x_1 \ldots x_J$ of the value of P, then we can estimate

$$\hat{V}(P) = \frac{1}{J} \sum_{j=1}^{J} x_j$$

The unbiased estimate of the population variance is

$$\frac{1}{J-1} \sum_{j=1}^{J} (\hat{V}(P) - x_j)^2.$$

As $\hat{V}(P)$ is the average of J samples of this population, we can estimate the variance on it (the standard error) with

$$\sigma_{\hat{V},P}{}^2 = \mathbb{E}[(\hat{V}(P) - V^*(P))^2] = \frac{1}{J(J-1)} \sum_{j=1}^{J} (\hat{V}(P) - x_j)^2$$

For large samples, the estimate $\hat{V}(P)$ converges to $V^*(P)$.

Suppose now that P is a node of the search tree and that it has n children $c_1 \ldots c_n$. Given $\hat{V}(c_i)$ and $\sigma_{\hat{V},i}$ for $i = 1..n$, we are interested in finding an estimate $\hat{V}(P)$ for the value $V^*(P)$ of the parent node P and a variance $\sigma_{\hat{V},P}$ on this estimate. In Section 4.1 we will consider this problem for decision nodes, and in Section 4.2 we will discuss opponent nodes and chance nodes.

4.1 Decision Nodes

Let P be a decision node and let c_i ($i = 1 \ldots n$) be its children. We will here assume that when the game would ever get into node P, we will get extra time to decide which of the children c_i, $i = 1 \ldots n$ is really the best choice. Let

$$e_i = \hat{V}(c_i) - V^*(c_i)$$

be the error on the estimate of the value of child c_i. We have

$$\hat{V}(P) = \int_{e_1 \ldots e_n} (\max_{i=1}^{n}(\hat{V}(c_i) - e_i)) \, P(e_1 \ldots e_n) de_1 \ldots de_n$$

If the estimates $\hat{V}(c_i)$ are obtained by independent sampling, then the errors e_i are independent and we can write

$$\hat{V}(P) = \int_{e_1 \ldots e_n} (\max_{i=1}^{n}(\hat{V}(c_i) - e_i)) \prod_{i=1}^{n} P(e_i) de_1 \ldots de_n$$

For the variance of $\hat{V}(P)$ we can write

$$\sigma_{\hat{V},P} = \int_{e_1 \ldots e_n} (\hat{V}(P) - \max_{i=1}^{n}(\hat{V}(c_i) - e_i))^2 \prod_{i=1}^{n} P(e_i) de_1 \ldots de_n$$

If we assume the errors e_i are normally distributed, these integral can be evaluated symbolically, at least if the number of children is not too high. Unfortunately, as the maximum of several normal distributions is not a normal distribution itself, we can not expect such assumption to be correct. However, we can obtain a relatively close and efficiently computable approximation, by ignoring this problem and acting as if all errors would be normally distributed.

4.2 Opponent Nodes and Chance Nodes

As discussed in Section 2 chance nodes and opponent nodes can be treated equivalently, using an opponent model to dictate the probability distribution in opponent nodes. Let P be a chance node. Let c_i, $i = 1 \ldots n$ be the children of P and let their probability be p_i. Then, we have

$$\hat{V}(P) = \sum_{i=1}^{n} p_i \hat{V}(c_i)$$

and

$$\sigma_{\hat{V},P}^2 = \sum_{i=1}^{n} p_i \sigma_{\hat{V},i}^2$$

Both the total expected value and its standard error are weighed by p_i. In chance nodes, p_i describes the stochastic process. In opponent nodes, the probability is provided by the opponent model.

5   Experiments

In this section we will evaluate our new algorithm empirically. In particular, we will investigate the following questions:

Q1 : Is the new algorithm able to exploit weaknesses in rule-based players?
Q2 : How does the proposed backpropagation algorithm compare to existing algorithms?
Q3 : What is the effect of available computing resources (thinking time)?
Q4 : What is the effect of the proposed sample selection algorithm?

5.1   Setup

Ideally, we would let our bots play against strong existing bots. Unfortunately, the few no-limit bots that exist are not freely available. Such a comparson would not really be fair as existing bots only play heads-up poker, whereas our bot is more generally applicable. Because it is the first exploiting no-limit bot, a comparison of exploitative behaviour would also be impossible.

We implemented the following poker playing bots to experiment with:

- STDBOT, the standard MCTS algorithm.
- NEWBOT, our new algorithm as explained in the section 4.
- RULEBOTHANDONLY, a naive rule-based bot. His behaviour only depends on the hand cards he's dealt.
- RULEBOTBESTHAND, a more complex rule-based bot. He estimates the probability that he has the strongest hands and bases his strategy entirely on that information.

STDBOT and NEWBOT both use the opponent model that's described in Section 2. RULEBOTHANDONLY and RULEBOTBESTHAND don't use any opponent model. In our experiments, we each time let the bots (with certain parameters) play a large number of games against each other. Each player starts every game with a stack of 200 small bets (sb). The contant C in Equation (1) is always equal to 50 000. To reduce the variance in the experiments, players are dealt each other's cards in consecutive games. We compute the average gain of a bot per game in small bets.

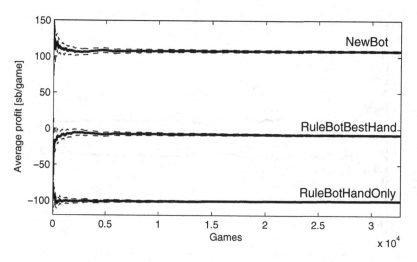

Fig. 1. Average profit of NEWBOT against two rule-based bots

5.2 Exploitative Behaviour (Q1)

We let NEWBOT play against RULEBOTHANDONLY and RULEBOTBESTHAND in a large number of 3-player games. Figure 1 shows the average profit of each bot. The dotted lines are the σ-confidence interval around the estimated average profit.

NEWBOT is clearly superior to both rule-based bots. The more complex RULE-BOTBESTHAND is clearly superior to RULEBOTHANDONLY. This experiment shows that MCTS in combination with an opponent model is capable of strong exploitative behaviour and that exploitative game tree search bots can be developed for no-limit poker with more than 2 players.

5.3 Comparison with Standard MCTS (Q2)

To test the new backpropagation algorithm, we let NEWBOT that uses the maximum distribution backpropagation strategy play against STDBOT that uses the standard backpropagation algorithm, where a sample-weighted average of the expected value of the children is propagated. To avoid overestimation, the maximum distribution only replaces the standard algorithm if a node has been sampled 200 times. Both bots were allowed 1 second computation time for each decision and were otherwise configured with the same settings. Figure 2 shows the average profit of both bots.

The backpropagation algorithm we propose is significantly superior in this experiment. It wins with 3 small bets per game, which would be a very large profit among human players. Using other settings for the UCT selection algorithm in other experiments, the results were not always definitive and sometimes the proposed strategy even performed slightly worse that the standard algorithm.

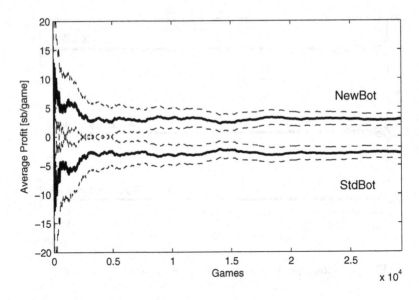

Fig. 2. Average profit of NEWBOT vs. STDBOT

Therefore, it is hard to draw general conclusions for question Q2 based on these experiments. Further parameter tuning for both approaches is necessary before any strong claims are possible either way.

5.4   Influence of Thinking Time (Q3)

In this context, we investigated the influence of available computation time on the effectiveness of the maximum distribution backpropagation algorithm. Figure 3 shows the average profit of NEWBOT in heads-up games against STDBOT for different computation times.

Even though there are some outliers, the balance is cleary in favour of NEW-BOT for longer thinking times. If enough time is available, the advantages of the maximum distribution strategy start to outweigh the disadvantage of being able to sample less.

We expect that, when more elaborate opponent modeling techniques are used that require a proportionally larger amount of time to be evaluated and thereby cause both techniques to sample a comparable number of game outcomes, the scale will also tip further in favor of the new backpropagation algorithm.

5.5   Influence of the Proposed Sample Selection Algorithm (Q4)

To investigate the influence of the proposed selection algorithm, we let a NEW-BOT with the standard UCT algorithm using Equation (1) play against a NEW-BOT using Equation (5) for sampling with C equal to 2. Both bots are allowed

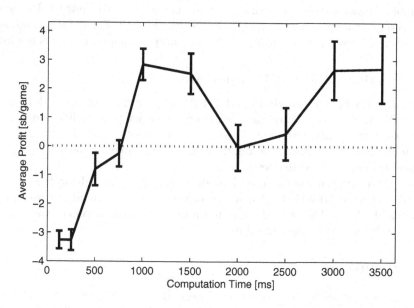

Fig. 3. Average profit of NEWBOT against STDBOT depending on the calculation time

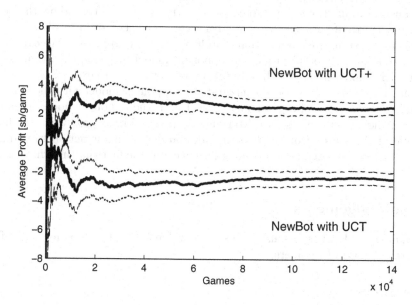

Fig. 4. Average profit of NEWBOT using UCT+ against NEWBOT using standard UCT

to sample the game tree 25 000 times. All else being equal, the new UCT+ sample selection strategy clearly outperforms the standard UCT. As for Q2, not all parameter settings gave conclusive results and more parameter tuning is needed.

5.6 Experiments against Human Players

To estimate the bot's strength we performed tentative experiments with human players. NEWBOT was able to beat a human novice when playing 400 games with a significant profit of 8 sb/game. The same experiment against an experienced human player ended in a draw. From these initial experiments, we conclude that the bot plays at a reasonable level.

Large scale experiments against human players are planned for the future. We will integrate NEWBOT with a free online casino to simultaneously play hundreds of games. This will allow us to optimize the parameters of the search procedure for games against humans.

6   Conclusions

We've introduced MCTS in the context of Texas Hold'em Poker and the MIXIMAX game trees it results in. Using a new backpropagation strategy that explicitly models sample distributions and a new sample selection strategy, we developed the first exploiting multi-player bots for no-limit Texas Hold'em. In a number of experimental evaluations, we studied some of the conditions that allow the new strategies to result in a measurable advantage. An increase in available computation time translates in additional profit for the new approach. We expect a similar trend as the complexity of the opponent model surpasses the complexity of the backpropagation algorithm. In future work, we want to explore several directions. First, we want to better integrate the opponent model with the search. Second, we want to find a good way to deal with the opponent nodes as minimization nodes as well as non-deterministic nodes. Finally, we want to shift the goal of the search from finding the optimal deterministic action to finding the optimal randomized action (taking into account the fact that information is hidden).

Acknowledgements

Jan Ramon and Kurt Driessens are post-doctoral fellows of the Fund for Scientific Research (FWO) of Flanders.

References

1. Gilpin, A., Sandholm, T., Sørensen, T.: A heads-up no-limit Texas Hold'em poker player: discretized betting models and automatically generated equilibrium-finding programs. In: Proceedings of the 7th international joint conference on Autonomous agents and multiagent systems, International Foundation for Autonomous Agents and Multiagent Systems Richland, SC, vol. 2, pp. 911–918 (2008)

2. Billings, D.: Algorithms and assessment in computer poker. PhD thesis, Edmonton, Alta, Canada (2006)
3. Kocsis, L., Szepesvari, C.: Bandit based monte-carlo planning. In: Fürnkranz, J., Scheffer, T., Spiliopoulou, M. (eds.) ECML 2006. LNCS (LNAI), vol. 4212, pp. 282–293. Springer, Heidelberg (2006)
4. Russell, S., Norvig, P.: Artificial intelligence: A modern approach. Prentice Hall, Englewood Cliffs (2003)
5. Suffecool, K.: Cactus kev's poker hand evaluator (July 2007), http://www.suffecool.net/poker/evaluator.html
6. Witten Ian, H., Eibe, F.: Data Mining: Practical machine learning tools and techniques. Morgan Kaufmann, San Francisco (2005)
7. Wang, Y., Witten, I.: Induction of model trees for predicting continuous classes (1996)
8. Coulom, R.: Efficient selectivity and backup operators in Monte-Carlo tree search. In: van den Herik, H.J., Ciancarini, P., Donkers, H.H.L.M(J.) (eds.) CG 2006. LNCS, vol. 4630, pp. 72–83. Springer, Heidelberg (2007)
9. Gelly, S., Wang, Y.: Exploration exploitation in go: UCT for Monte-Carlo go. In: Twentieth Annual Conference on Neural Information Processing Systems, NIPS 2006 (2006)
10. Chaslot, G., Winands, M., Herik, H., Uiterwijk, J., Bouzy, B.: Progressive strategies for monte-carlo tree search. New Mathematics and Natural Computation 4(3), 343 (2008)
11. Van Lishout, F., Chaslot, G., Uiterwijk, J.: Monte-Carlo Tree Search in Backgammon. In: Computer Games Workshop, pp. 175–184 (2007)
12. Auer, P., Cesa-Bianchi, N., Fischer, P.: Finite-time analysis of the multiarmed bandit problem. Machine Learning 47(2-3), 235–256 (2002)

Injecting Structured Data to Generative Topic Model in Enterprise Settings

Han Xiao[1,2,*], Xiaojie Wang[2], and Chao Du[3]

[1] Technische Universität München, D-85748 Garching bei München, Germany
hanxiao@live.de
[2] Beijing University of Posts and Telecommunications, 100876 Beijing, China
xjwang@bupt.edu.cn
[3] Beihang University, 100191 Beijing, China
chaodu.global@gmail.com

Abstract. Enterprises have accumulated both structured and unstructured data steadily as computing resources improve. However, previous research on enterprise data mining often treats these two kinds of data independently and omits mutual benefits. We explore the approach to incorporate a common type of structured data (i.e. organigram) into generative topic model. Our approach, the *Partially Observed Topic model* (POT), not only considers the unstructured words, but also takes into account the structured information in its generation process. By integrating the structured data implicitly, the mixed topics over document are partially observed during the Gibbs sampling procedure. This allows POT to learn topic pertinently and directionally, which makes it easy tuning and suitable for end-use application. We evaluate our proposed new model on a real-world dataset and show the result of improved expressiveness over traditional LDA. In the task of document classification, POT also demonstrates more discriminative power than LDA.

Keywords: probabilistic topic model, structured data, document classification.

1 Introduction

The increased use of computers has augmented the amount of data that can be stored in them. To take advantage of the rich storehouse of knowledge that the data represent, we need to harness the data to extract information and turn it into business intelligence. Any well-structured dataset or database with a large number of observations and variables can be mined for valuable information. Many retail businesses now use the sales record to adjust their marketing strategies. On the other hand, it is equally important to analyze unstructured data like text documents and webpages. A business developer might need to examine the latest news on webpages and manually coded them into categories, and assign them to the related division. However, enterprises with information depositories do not yet use the full resources of their two kinds of data. Although text mining is well-studied in last few decades, it is rarely analyzed with the support of structured data in enterprise. Just a few works [1, 2] discussed extracting interesting patterns from the

---

* This work was done during H.X.'s internship at Hewlett-Packard Laboratories, China.

Z.-H. Zhou and T. Washio (Eds.): ACML 2009, LNAI 5828, pp. 382–395, 2009.
© Springer-Verlag Berlin Heidelberg 2009

structured and unstructured data jointly by Machine Learning approaches. In this paper, we present a straightforward manner to incorporate the structured data of enterprise into generative topic model by Gibbs sampling. The structured data we considered in this work is organigram, which is the most common information in the Enterprise. Our new model, which we refer to henceforth as the *Partially Observed Topic model* (POT), defines a one-to-one correspondence between LDA's latent topics and organizations constrained to organigram. This allows the structured words to guide the evolution of topics in our model.

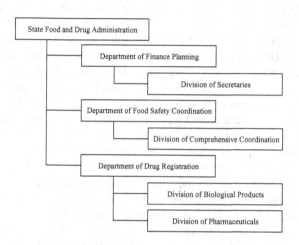

Fig. 1. Example of a three level organigram, consists of organization names and hierarchical structure that shows the hierarchical structure of an organization with its parts

The rest of the paper is organized as follows. In section 2, we discuss past work related to the topic model. In section 3, we describe the POT model in detail including its generative process, and the collapsed Gibbs sampling used for parameter inference. Section 4 presents some qualitative and quantitative evaluation of our proposed model against with traditional LDA. We use the document-specific mixing proportions to feed a downstream SVM classifier, we demonstrate the advantages of POT in document classification task. Section 5 concludes the paper and discusses future work.

2   Related Work

Research in statistical models of co-occurrence has led to the development of a variety of useful topic models. Topic models, such as Latent Dirichlet Allocation(LDA) [3], have been an effective tool for the statistical analysis of document collections and other discrete data. LDA assumes that the words of each document arise from a mixture of topics. The topics are shared by all documents in the collection; the topic proportions are document-specific and randomly drawn from a Dirichlet distribution. LDA allows each document to exhibit multiple topics with different proportions, and it can thus capture the heterogeneity in grouped data which exhibit multiple patterns. In the recent past, several extensions to this model have been proposed such as the Topics Over Time(ToT) [4] model that permits us to analyze the popularity of various topics as a function of time, Hidden Markov Model-LDA [5] that integrates topic modeling with syntax, Author-Persona-Topic models [6] that model words and their authors, etc. In each case, graphical model structures are carefully-designed to capture the relevant structure and co-occurrence dependencies among the data.

However, most of these generative topic models are designed in unsupervised fashion, which made them suffered from non-commutative and unstable of topics. Due to the lack of a priori ordering on the topics, they are unidentifiable between or even within runs of the algorithm. Some topics are stable and will reappear across samples while others topics are idiosyncratic for a particular solution. To avoid this problem, some modifications of LDA to incorporate supervision haven been proposed previously. Supervised LDA [10] posits that a label is generated from each document's empirical topic mixture distribution. DiscLDA [11] associates a single categorical label variable with each document and associates a topic mixture with each label. A third model, presented in [2] showed that the structured and unstructured data can naturally benefit each other in the tasks of entity identification and document categorization.

We follow previous approaches in incorporating structured data into generative topic model, and propose a semi-supervised fashion from a document collection and a back-end organigram database. We motivate how the organigram can reinforce the topic clustering and document categorization. In proposed *Partially Observed Topic model* (POT), we not only learn the topical components, but also map each topic to an organization. Thus each document is represented as mixtures of organizations, which are distributions over the words in the corpus. The POT model interprets the organizations name and other unstructured word in a joint generative process, and automatically learns the posterior distribution of each word conditioned on each organization. We also perform a Gibbs sampling technique to establish of topic hierarchies from the structure embedded in organigram.

3   Hierarchical Organization Topic Model

Our notation used in this paper is summarized in Table 1, and the graphical model of POT models is shown in Figure 2.

Partially Observed Topic (POT) model is a generative model of organization names and unstructured word in the documents. Like standard LDA, it models each document as a mixture of underlying topics and generates each word from one topic. In contrast to LDA, topic discovery in POT is influenced not only by word co-occurrences, but also organigram information, including: organization names and hierarchical structure. This allows POT to predict words distribution for topics with the guide of organigram, which results in organization specific multinomial. The robustness of the topic cluster is greatly enhanced by integrating this structured data. The generative process of POT, which corresponds to the process used in Gibbs sampling for parameter estimation, is given as follows:

1. Draw T multinomial ϕ_z from a Dirichlet prior β, one for each organization z;
2. For each document d, draw a multinomial θ_d from a Dirichlet prior α;
        For each organization name e_{dj} in document d:
3.   Draw an organization name e_{dj} from multinomial θ_d
        For each word w_{di} in document d:
4.   Draw an organization z_{di} from θ_d
5.   Draw a word w_{di} from multinomial $\phi_{z_{di}}$

As shown in the process, each document is represented as mixtures of organizations, which denote by multinomial θ. θ plays the key role in connecting the organization names and the unstructured words in the document. The dependency of θ on both e and z is the only additional dependency we introduce in LDA's graphic model. As the mechanism in LDA, frequently co-occurrence of some words often indicated that

Table 1. Notations used in POT model

Symbol	Description
T	number of organizations
D	number of documents
V	number of unique words
θ_d	the multinomial distribution of organizations specific to document d
ϕ_z	the multinomial distribution of words specific to organization z
z_{di}	the organization associated with the ith token in the document d
z^{sup}	the parent organization of z
z^{sub}	the child organization of z
w_{di}	the ith token (unstructured word) in the document d
e_{di}	the ith organization name mentioned in the document d
α	Dirichlet prior of θ
β	Dirichlet prior of ϕ
γ	Smooth for unseen organization

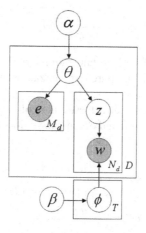

Fig. 2. Graphical model of POT

they belong to the same topic or organization. On the other hand, if we observe an organization's name in a document, it will give higher probability on corresponding parameter of θ_d. This will result in the unstructured words in this document more likely related to this organization. Although unstructured words do not directly mention an organization, they may provide significant indirect evidence for inferring it. Especially in a document without any organization names, the unstructured words will imply the latent organizations of document as they work in traditional Multinomial Naive Bayes classifier [12].

We use Gibbs sampling to conduct parameter inference. The Gibbs sampler provides a clean method of exploring the parameter space. Note that we adopt conjugate prior(Dirichlet) for the multinomial distributions, and thus we can easily integrate out θ and ϕ, analytically capturing the uncertainty associated with them. Since we need not sample θ and ϕ at all, the sampling procedure becomes facility. Starting from the joint distribution $P(w, z, e|\alpha, \beta)$, we calculate the conditional probabilities $P(z_{di}|z_{-di}, w, e, \alpha, \beta)$ conveniently using Bayes Rule, where z_{-di} denotes the organization assignment for all words token except w_{di}. In the Gibbs sampling procedure, we draw the organization assignment z_{di} iteratively for all unstructured word token w_{di} according to the following conditional probability distribution:

$$P(z_{di}|z_{-di}, w, e, \alpha, \beta) \propto (n_{d,e_{dj}} + r - 1)(n_{d,z_{di}} + \alpha - 1) \times$$
$$\frac{n_{z_{di},w_{di}} - m_{z^{sup},w_{di}} + \beta_{w_{di}} - 1}{\sum_{v=1}^{V}(n_{d_i,v} + \beta_v) - 1} \tag{1}$$

where $n_{d,e}$ represent how many times the name of organization e mentioned in document d, we interpolate r here to avoid probability from zero, since some documents may not mention any organization names. $n_{d,z}$ is the number of tokens in document d that assigned to organization z, $n_{z,v}$ is the number of tokens of word v are assigned to the organization z. $m_{z^{sup},v}$ describes how many time the tokens of word v are assigned to all parent organizations of z. According to the hierarchical structure of organigram, it can work out level-by-level recursively as follow:

$$m_{z,v} = \frac{n_{z,v}}{n_{z^{sub}}} + m_{z^{sup},v} \tag{2}$$

where $n_{z^{sub}}$ is the number of child organizations of z. Intuitively, using the counts that eliminate words co-occurs with parent organizations will refine the distribution for current organization.

The sampling algorithm gives direct estimates of z for every word. θ and ϕ of the word-organization distributions and organization-document distributions can be obtained from the count matrices as follows:

$$\phi_v^z = \frac{n_{z,v} + \beta_v}{\sum_{v=1}^{V}(n_{z,v} + \beta_v)} \tag{3}$$

$$\theta_d^z = \frac{n_{d,z} + \alpha_z}{\sum_{z=1}^{Z}(n_{d,z} + \alpha_z)} \tag{4}$$

In state-of-the-art hierarchical Bayesian models, it is important to get the right values for the hyper-parameters. For many topic models, the final distribution is sensitive to hyper-parameters. One could use Gibbs EM algorithm [7] to infer the value of hyper-parameters. In the paper, for simplicity, we use fixed symmetric Dirichlet distributions and empirical value of r smooth in our experiments.

As we highlighted in Section 1, we seek an approach that can automatically learn word-organization correspondences. Towards this objective, we separated the structured word from unstructured words in our generative process to refine the topic evolution. We also note that there are some variants of our purposed model. For example, one could consider an arbitrary restrictive model [14] that restrict θ_d to be defined only over the topics that correspond to its organization mentions, while keep zero value on others. Thus, all topic assignments over words are limited to the organizations mentioned in current document. However, this restriction ignores the effect of common and background topics. The background topics may crops up throughout in corpus, nevertheless without apparent mentions. For instance, as we illustrated in Figure 1, "Department of Drug Registration (DDR)" and "Department of Food Safety Coordination (DFSC)" are sub-organizations that attached to "State Food and Drug Administration (SFDA)". In our news story about food and drug security, "DDR" and "DFSC" own different representative words respectively, they may also share some words such as "health", "legal", "safety", "permission" that actually generated by "SFDA". In case the "SFDA" itself is not directly mentioned in document, the restrictive version of model will fail to recover the word distribution over "SFDA" accurately. This is a reasonable assumption because the organigram itself is embedded with hierarchy. From the view of sub-organizations, their parents could be seen as background topics. However, the parent-organizations may not necessarily be mentioned in every document, and therefore become latent topics. In contrast to standard LDA and the restrictive model, we model the topics in document as neither fully latent nor clearly observed. When structured word is seen in document, the related topics will be assigned with high probability convincingly, meanwhile the unstructured words in document are free to be sampled from any of the T organizations, swing the probability in favor of the corresponding topics, and thus makes a full-scale and supervised exploration in latent topic space.

4 Experiments

4.1 Dataset

In this section, we present experiments with POT model on HPWEB dataset, which consists of a database and a webpage collection. The database includes the information of all organizations and departments in Hewlett-Packard Company, and the webpage collection has 50 thousand webpages crawled from Hewlett-Packard external web. On the structured side, we extract the name, parent and child for each organization from the database in HPWEB. This form a small backend database (i.e. organigram) used in our experiment as in Table 2 showed.

Table 2. Example organigram based on the HPWEB database

ID	Name	Parent	Child
0	Personal System Group	NULL	1, 2
1	Marketing	0	3, 6
2	Supply Chain	0	4, 5
3	Notebook Sales	1	NULL

This table describes a hierarchical tree structure with a set of linked nodes, where each node represents an organization. Like the definition of the tree structure, each organization has a set of zero or more child organizations, and at most one parent organization. The organigram we build for the experiment is two levels, and has 68 organizations in all (10 parent organizations and 58 sub organizations).

To create our unstructured document collection, we parsed all webpages and tokenized the words, then omitted stop-words and words that occurred fewer than 100 times. This resulted in a total of 34,036 unique words in vocabulary. For each time of experiment, we randomly draw 10,000 documents from HPWEB dataset, and omit pages length shorter than 5 tokens.

4.2 Preprocessing

Each document was processed to obtain its list of unstructured words and the list of organization mentions. To locate the organization names in the document, we need to annotate the proper entities that may indicate organizations. However, it needs to capture all different variations, abbreviations, acronym and spelling mistakes. For example, for the organization 'HP Finance Service', the variations need to be considered includes 'Hewlett Packard Finance Service', 'Finance Service Group', 'HP Finance Group' and 'HPFS'. To address this problem, we first use LEX algorithm [8] to locate the name entities in document. LEX is based on the observation that complex names tend to be multi-word units(MWUs), that is, sequences of words whose meaning cannot be determined by examining their constituent parts. This lexical statistics-based approach is simple, fast and with high recall. Next we compute the candidate matches for each entity that annotated by LEX against the organizations names in our organigram database. We use an open-source Java toolkit SecondString[1] for matching names and records. After experimenting with several similarity metrics, we finally choose the *JaroWinkler-TFIDF* [9] metrics which is hybrid of cosine similarity and the Jaro-Winkler method. By computing the similarity scores for these pairs, and filtering out those scores below 0.8, we get the organization name list that mentioned in document. The list of unstructured words is found simply by removing the organizations' names from the document.

4.3 Topic Interpretation

We first examine the topic interpretation of POT, and how it compares with standard LDA. We fit 68-topic POT and standard unsupervised LDA on HPWEB dataset(6499

---

[1] http://secondstring.sourceforge.net/

POT   LDA

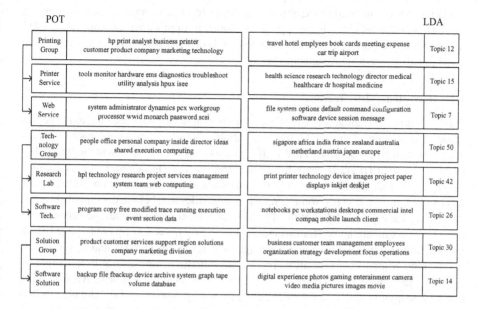

Printing Group	hp print analyst business printer customer product company marketing technology	travel hotel emplyees book cards meeting expense car trip airport	Topic 12
Printer Service	tools monitor hardware ems diagnostics troubleshoot utility analysis hpux isee	health science research technology director medical healthcare dr hospital medicine	Topic 15
Web Service	system administrator dynamics pcx workgroup processor wwid monarch password scsi	file system options default command configuration software device session message	Topic 7
Tech-nology Group	people office personal company inside director ideas shared execution computing	sigapore africa india france zealand australia netherland austria japan europe	Topic 50
Research Lab	hpl technology research project services management system team web computing	print printer technology device images project paper displays inkjet deskjet	Topic 42
Software Tech.	program copy free modified trace running execution event section data	notebooks pc workstations desktops commercial intel compaq mobile launch client	Topic 26
Solution Group	product customer services support region solutions company marketing division	business customer team management employees organization strategy development focus operations	Topic 30
Software Solution	backup file fbackup device archive system graph tape volume database	digital experience photos gaming enterainment camera video media pictures images movie	Topic 14

Fig. 3. Topic clustering result of 8 topics learned by POT (left) and traditional LDA (right). The descriptors are selected by the 10 highest probability words under organization. POT's topic are named by their associated organization name and linked by their hierarchy in organigram. We also selected the 8 most expressive topics in LDA for fairly comparison.

documents, 8,191,292 words in total). Both models ran for 4500 Gibbs sampling iterations, with symmetric priors $\alpha = 0.2$, $\beta = 0.02$, the smooth factor is $\gamma = 2$ in POT.

In POT, topics are directly mapped to appropriate organizations name. The model has nicely captured the words using an auxiliary organigram. Examine the result of POT, we found that words under each organizations are quite representative. For example, the parent organizations often related with some generic words (such as "product", "service", "customer" and "company"). Although these words are too general to interpret, they also imply the differences on business area between two organizations. "product" and "service" are ranked high in both "Printing Group (PG)" and "Solution Group (SG)", on the other hand, they do not appear in the top word lists of "Technology Group (TG)". This indicates that PG and SG have more similarity on business than they compare to TG. At the next level, topic descriptors specify more clearly domain of sub-organizations. The cluster of "Research Lab (RL)" provides an extremely salient summary of research area, while "Software Technology (ST)" assembles many common words used in network administration. POT model separated the words pertaining to different organizations. On the other hand, sub-organizations also inherited some words from their parents, such as "computing" and "execution", which appears in RL and ST respectively are also listed in TG's descriptors.

On the side of standard LDA, although its topic descriptors are also quite expressive, it offers no obvious clues for mapping the topics to appropriate organizations. LDA seems to discover topics that model the fine details of documents with no regard to their discrimination power on organizations.

4.4   Document Visualization

We then study the evaluation of POT performance on document classification. The first group of experiments are qualitative evaluations. We use some popular visualization method to intuitively demonstrate the effectiveness of POT on document modeling. It is worth noting that HPWEB dataset does not lend itself well to document classification analysis. The lack of class-labels on webpages is a major frustration on evaluation. In this experiment, we first use some simple heuristic rules to label the document class and then verify each page's label manually. The rules used here include: the organization mentioned in domain name (for instance, "HR" in http://www.*hr*.hp.com), page title or html meta data. On the other hand, webpages are full of noisy and contain some highly overlapping content: because all webpages are inherited from several base templates, the pages that share same template will naturally overlap a great deal. After filtering out the low-quality and duplicate webpages, we finally categorize 1192 documents into 10 class, that is the 10 top-level organizations. illustrate that the POT model is also feasible for document classification by Figure 4. It shows the 2D embedding of the expected topic proportions of POT and standard LDA by using the t-SNE stochastic neighborhood embedding [13], where each dot represents a document and color-shape pairs represent class labels, each pair is marked with an organization acronym in the corner legend. Obviously, the POT produces a better grouping and separation of the documents in different categories. In contrast, standard LDA does not produce a well separated embedding, and documents in different categories tend to mix together.

It is also interesting to examine the discovered topics and their association with organization mentions in document. In Figure 5, we randomly sample 100 documents

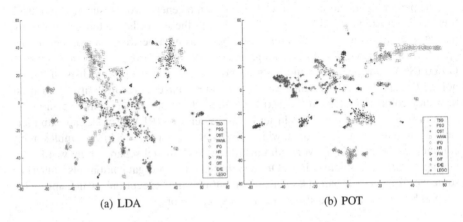

(a) LDA   (b) POT

Fig. 4. t-SNE embedding of topic representation by: tranditional unsupervised LDA (left) and POT model (right)

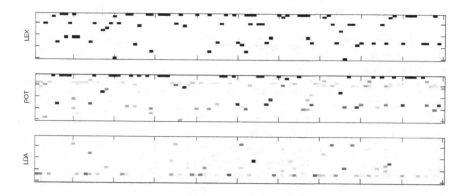

Fig. 5. Document-organization co-occurrence matrix. Each row represents a document, and each column corresponds to an organization. The upper part of graph shows the LEX result, that is, if LEX finds an organizations name in the document, the co-occur value is set to 1 corresponding-ly. The middle and bottom parts of the figure describe the recovered document-organization co-occurrence matrix by POT and traditional LDA respectively. Each row corresponds to a multi-nomial distribution of organizations specific to document. Grayscale indicates the value of θ_d^z, with black being the highest probability and white being zero.

and 20 organizations (10 top-organizations and 10 sub-organizations) from our dataset and visualize their co-occur matrix. Each row represents a document, and each column corresponds to an organization. It can be clearly seen that, POT's interpretation of documents used the LEX result on organization mentions as guidance, nevertheless, with less arbitrary. It treats the structured data as important evidences, and used them to supervise the topic evolution implicitly. Unlike LEX, by inspection the unstructured word, POT can infer the organization labels on documents, when organization name is not mentioned in document. We can also see that by ignoring supervised responses, standard LDA's co-occur matrix seems hard to interpret, which makes it not appropri-ate for document classification. POT trade off between the arbitrary LEX labeling and totally unsupervised LDA. Since each document is soft-assigned to every organization with a certain probability and described by a multinomial distribution, POT gives a more precise description of the document across the organizations. It should be appar-ent to the reader that POT can be adapted as an effective multi-label classifier for docu-ments. Straight-forward manners such as inferring the document's most likely labels by suitably thresholding its posterior probability over topics, or developing a downstream classifier are natural extensions for further investigation.

Another notable feature of POT model is that it discovers the latent relationship be-tween different organizations. Table 3 listed three documents with organizations found by LEX and top four organizations assigned by POT. By comparing the result with LEX and POT, some hidden relations between organizations can be uncovered. For example, the "Linux" department is strongly related with "Business PC" in the context of doc-ument 36. Document 148 introduces a high-speed video and graph transmission tech-nology developed by "Research Lab", which is relevant with "Digital Entertainment"

Table 3. The interpretation of document on organizations by POT

Doc. ID	Annotated organizations	Top four organizations assigned by POT		POT	LDA
		Names	Prob.		
36	Linux	Business PC	0.28		
		Linux	0.15		
		Human Resources	0.09		
		Service Group (America)	0.06		
148	Research Lab	Digital Entertainment	0.49		
		IT Management	0.10		
		Worldwide Marketing	0.09		
		Business Service	0.02		
691	Large-scale Computing	Software	0.37		
		Large-scale Computing	0.19		
		Business Service	0.05		
		IT Management	0.05		

— a high performance video game console. Finally, the "Large-scale Computing" is involved in some area of "Software" in document 691. These results suggest that POT can be an effective tool in mining business relationship of organizations under certain context. On the right side of Table 3, we also give the document-topics distribution recovered by POT and standard LDA respectively. POT yields yields sharper and sparser distribution over topics, which results in a sharper low dimensional representation.

4.5   Document Classification

To study the practical benefit of injecting the structured data into topic models, we also provide quantitative evaluation of POT on document classification. We perform a multi-class classification on our HPWEB dataset as we previously used. However, many topic models based on LDA are not suit for classification. LDA and POT can not be directly used as classifier. To obtain a baseline, we fit our data to standard LDA model with different number of topics, and then use the latent topic representation of the training documents as features to build a multi-class SVM classifier. For POT, we first build a straight-forward classifier by selecting the topic (organization) with highest posterior probability. The second classifier based on the same strategy on feature selection and training as LDA+SVM, except the topic numbers are fixed to 68. We denote these three classifier as LDA+SVM, POT and POT+SVM, and evaluate their average precision of 5 times experiments. We use SimpleSVM [2] to build a 1-vs-all SVM classifiers. which is popular and used by many previous works. The parameter C is chosen from 5 fold cross-validation on 1192 documents. The precision of different models w.r.t topic numbers are shown in Figure 6.

---

[2] http://gaelle.loosli.fr/research/tools/simplesvm.html

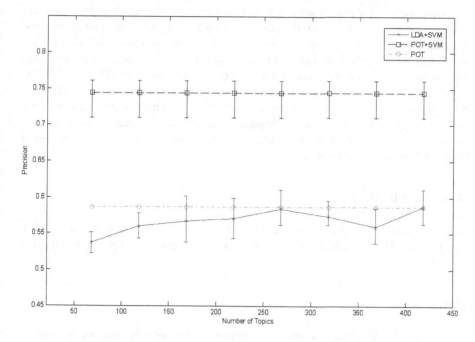

Fig. 6. Precision of multi-class document classification on selected HPWEB dataset(1192 documents, 10 class in all). We first fit POT and LDA models on our dataset, and then use the learned doc-topic distribution as input features to feed the a downstream SVM classifier.

We can clearly see that both purely POT and POT+SVM outperform traditional unsupervised LDA significantly. The average precision of POT+SVM is 74%, POT yields 58%, while standard LDA only holds 55% on average. By injecting the structured word into parameter estimation as shown in Eq. 1, the latent topic space of POT yields a better discrimination power, and thus more suitable for document classification. Although LDA gains slightly better performance as the number of topics arise, it still suffers from low precision even with 5~6 times number of topics than POT.

5   Conclusion

In this paper, we have explored the way to formally incorporate the enterprise structured data into generative topic model. We have presented the Partially Observed Topic model that generates both organization mentions and unstructured words. In contrast to traditional LDA, the topics mixtures are partially observed in parameter estimation. The topics evolution are guided by structured word in document, which results in one-to-one correspondences between latent topics and organizations. Results on a real-world dataset have shown that POT can use back-end organigram to supervise topic evolution implicitly. In contrast to traditional unsupervised LDA, POT discovers more expressive topics that appropriate for document classification.

Enterprise data mining on structured and unstructured data are often processed separately. Our work opens a new direction of research in injecting structured data into generative topic model. In ongoing work, we will explore to benefit more from structured data in enterprise database(e.g. employee information) and discover the meaningful pattern that redound to enterprise. Moreover, we also notice that POT can be adapted to more general field other than enterprise. Given a training set consisting of documents with human-annotated labels or tags in context, POT can derive supervision from these annotations and automatically learn the posterior distribution of each unstructured word on label set. This mechanism can be utilized and extended to automatic summarize of contextual information about labels.

Acknowledgement

We thank Yuhong Xiong and Ping Luo for helpful comments and constructive criticism on a previous draft. We also thank Hewlett-Packard Laboratories for providing HPWEB dataset.

References

[1] Lu, Y., Zhai, C.: Opinion integration through semi-supervised topic modeling. In: Proceedings of WWW International World Wide Web Conference, pp. 121–130 (2008)

[2] Bhattacharya, I., Godbole, S., Joshi, S.: Structured entity identification and document categorization: two tasks with one joint model. In: Proceedings of 14th ACM SIGKDD International Conference on Knowledge Discovery and Data Mining (2008)

[3] Blei, D.M., Ng, A.Y., Jordan, M.I.: Latent Dirichlet allocation. Journal of Machine Learning Research 3, 993–1022 (2003)

[4] Wang, X., McCallum, A.: Topics over Time: A Non-Markov Continuous-Time Model of Topical Trends. In: Proceedings of 12th ACM SIGKDD International Conference on Knowledge Discovery and Data Mining (2006)

[5] Griffiths, T., Steyvers, M., Blei, D., Tenenbaum, J.: Integrating topics and syntax. In: Advances in Neural Information Processing Systems 17, pp. 537–544. MIT Press, Cambridge (2005)

[6] Mimno, D., McCallum, A.: Expertise modeling for matching papers with reviewers. In: Proceedings of the 13th ACM SIGKDD International Conference on Knowledge Discovery and Data Mining (2007)

[7] Andrieu, C., de Freitas, N., Doucet, A., Jordan, M.: An introduction to MCMC for machine learning. Machine Learning 50, 5–43 (2003)

[8] Downey, D., Broadhead, M., Etzioni, O.: Locating complex named entities in web text. In: Proceedings of IJCAI International Joint Conferences on Artificial Intelligence (2007)

[9] Cohen, W.W., Ravikumar, P., Fienberg, S.: A Comparison of String Metrics for Matching Names and Records. In: Proceedings of 9th ACM SIGKDD International Conference on Knowledge Discovery and Data Mining Workshop on Data Cleaning and Object Consolidation (2003)

[10] Blei, D.M., McAuliffe, J.D.: Supervised Topic Models. In: Proceedings of NIPS Neural Information Processing Systems (2007)

[11] Lacoste-Julien, S., Sha, F., Jordan, M.I.: DiscLDA: Discriminative learning for dimensionality reduction and classification. In: Blei, D.M., McAuliffe, J.D. (eds.) Proceedings of NIPS Neural Information Processing Systems (2008)

[12] McCallum, A., Nigam, K.: A comparison of event models for naive Bayes text classification. In: AAAI 1998 Workshop on Learning for Text Categorization. Tech. rep. WS-98-05. AAAI Press, Stanford (1998), http://www.cs.cmu.edu/~mccallum

[13] van der Maaten, L.J.P., Hinton, G.E.: Visualizing High-Dimensional Data Using t-SNE. Journal of Machine Learning Research 9, 2579–2605 (2008)

[14] Ramage, D., Hall, D., Nallapati, R., Manning, C.D.: Labeled LDA: A supervised topic model for credit attribution in multi-labeled corpora. In: Proceedings of the 2009 Conference on Empirical Methods in Natural Language Processing, Singapore, August 6-7, pp. 248–256 (2009)

Weighted Nonnegative Matrix Co-Tri-Factorization for Collaborative Prediction

Jiho Yoo and Seungjin Choi

Department of Computer Science
Pohang University of Science and Technology
San 31 Hyoja-dong, Nam-gu, Pohang 790-784, Korea
{zentasis,seungjin}@postech.ac.kr

Abstract. Collaborative prediction refers to the task of predicting user preferences on the basis of ratings by other users. Collaborative prediction suffers from the *cold start problem* where predictions of ratings for new items or predictions of new users' preferences are required. Various methods have been developed to overcome this limitation, exploiting side information such as content information and demographic user data. In this paper we present a matrix factorization method for incorporating side information into collaborative prediction. We develop Weighted Nonnegative Matrix Co-Tri-Factorization (WNMCTF) where we jointly minimize weighted residuals, each of which involves a nonnegative 3-factor decomposition of target or side information matrix. Numerical experiments on MovieLens data confirm the useful behavior of WNMCTF when operating from a cold start.

1 Introduction

Weighted nonnegative matrix factorization (NMF) is a method for low-rank approximation of nonnegative data, where the target matrix is approximated by a product of two nonnegative factor matrices (2-factor decomposition) [1,2]. Various extensions and modifications of NMF have been studied in machine learning, data mining, and pattern recognition communities. Orthogonality constraints on factor matrices were additionally considered to improve the clustering performance of NMF, leading to orthogonal NMF [3,4,5]. A variety of divergences were employed as an error function to yield different characteristics of learning machines, including α-divergence [6], Bregman divergence [7], and generalized divergences [8]. The 3-factor decomposition was incorporated into NMF, where a nonnegative data matrix is decomposed into a product of 3 nonnegative factor matrices, referred to as *nonnegative matrix tri-factorization* [3]. Probabilistic matrix tri-factorization [9] was recently developed, which is closely related to nonnegative matrix tri-factorization. Nonnegative Tucker decomposition (NTD) [10,11] is a multiway generalization of NMF, where nonnegativity constraints on mode matrices and a core tensor are incorporated into Tucker decomposition. In fact, nonnegative matrix tri-factorization or nonnegative 3-factor decomposition

Z.-H. Zhou and T. Washio (Eds.): ACML 2009, LNAI 5828, pp. 396–411, 2009.

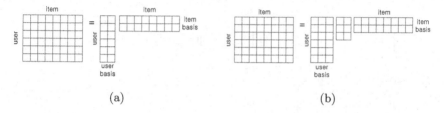

(a)   (b)

Fig. 1. A user-item matrix is factored into a product of user-related and item-related factor matrices: (a) 2-factor decomposition; (b) 3-factor decomposition. In the 2-factor decomposition, the target matrix X is decomposed into a product of two factor matrices U and V such that $X = UV^\top$, where U and V are referred to as left and right factor matrices, respectively. In 2-factor decomposition, the number of columns in both U and V should be equal so that the target matrix is represented by a sum of outer products between corresponding columns in U and V which are related by one-to-one mapping. On the other hand, the 3-factor decomposition assumes $X = USV^\top$ where a non-square factor matrix is introduced in the center, allowing different numbers of columns in U and V so that columns in U and V are interacted to represent X. In this sense, the 3-factor decomposition captures more complex relations, compared to the 2-factor decomposition. In addition, in the 3-factor decomposition, the factor matrix in the center absorbs scaling effect when the other two factor matrices are normalized.

is a special case of NTD, where only 2-way is considered. Pictorial illustration for 2-factor and 3-factor decomposition is shown in Fig. 1.

A large body of past work on NMF has focused on the case of complete data matrix where all entries are observed without missing values. In practice, however, the data matrix is often incomplete with some of entries are missing or unobserved. For instance, most of the entries in user-rating matrix are zeros (unobserved), so that matrix completion is necessary to predict unobserved ratings, which becomes a popular approach to *collaborative prediction* [12,13,14]. Weighted nonnegative matrix factorization (WNMF) seeks a nonnegative 2-factor decomposition of the target matrix by minimizing weighted residuals [15,14,16]. With zero/one weights, only observed entries in the target matrix are considered in the decomposition and unobserved entries are estimated by learned factor matrices. While WNMF was successfully applied to collaborative prediction [14,16], its performance is degraded when the number of rated items is very small and it even fails when operating from a *cold start* where corresponding items are rated by none of users in the data.

Side information such as demographic user data and content information helps bridge the gap between existing items (or users) and new (cold start) items (or users). Various methods for blending pure collaborative filtering with content-based filtering have been developed. For example, these include content-boosted collaborative filtering [17], probabilistic models [18,19], pairwise kernel methods [20], and filterbots-based method [21] to name a few. A method based on the predictive discrete latent factor models was proposed [22], which makes use of the additional information in the form of pre-specified covariates. Recent advances in

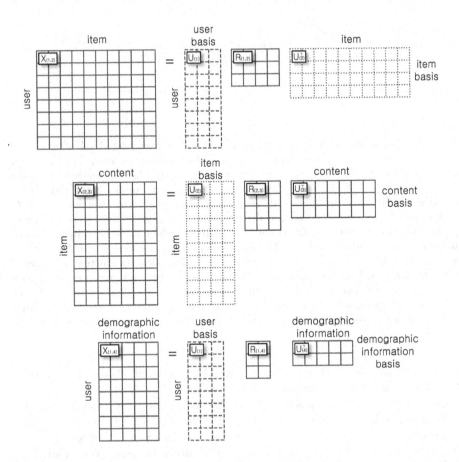

Fig. 2. Co-tri-factorizations of three matrices (user-item matrix, item-genre matrix, and user-demography matrix) are illustrated. User-related factor matrices are shared in the 1st and 3rd decompositions and item-related factor matrices are shared in the 1st and 2nd decompositions.

matrix factorization methods suggest *collective matrix factorization* or *matrix co-factorization* to incorporate side information, where several matrices (target and side information matrices) are simultaneously decomposed, sharing some factor matrices. Matrix co-factorization methods have been developed to incorporate label information [23], link information [24], and inter-subject variations [25]. Collective matrix factorization for relational learning was recently studied [26] Most of the existing methods are based on collective 2-factor decompositions. Some limitations of the 2-factor decomposition are described in Fig. 1 and they carry over to collective 2-factor decompositions.

Matrix tri-factorization or nonnegative 3-factor decomposition has been recently studied for co-clustering [3,27,9]. In this paper, we present a method for weighted nonnegative matrix co-tri-factorization (WNMCTF) where several matrices are simultaneously factored by 3-factor decompositions through minimizing weighted residuals, sharing some factor matrices. Pictorial illustration for matrix co-tri-factorization is shown in Fig. 2. Some useful aspects of 3-factor decomposition (as illustrated in Fig. 1) carry over to WNMCTF. Moreover, we highlight why WNMCTF is preferred over existing collective matrix factorizations or weighted nonnegative matrix co-factorization (WNMCF).

- WNMCTF includes existing nonnegative matrix co-factorization methods as special cases.
- As in the 3-factor decompositions where many-to-many interactions between column vectors in factor matrices are allowed to approximate the target matrix, more complex relations are captured by WNMCTF, compared to WNMCF where only one-to-one interactions between column vectors in factor matrices are allowed.
- The non-square factor matrix in the middle allows different numbers of bases for the left and right factor matrices. It absorbs scaling effect so that the normalization for factor matrices in WNMCTF is much easier, compared to existing collective matrix factorizations.
- WNMCTF can easily handle multi-relational data. For example, it can handle cold-start users and cold-start items at the same time, without further careful manipulation. This is mainly due to the easiness of normalization for factor matrices which are shared in the decompositions.

The rest of this paper is organized as follows. We begin with weighted nonnegative matrix co-factorization which is collective 2-factor decomposition in Section 2. Section 3 presents WNMCTF and corresponding multiplicative updating algorithms. Consistency for shared factor matrices is considered in developing the algorithm for WNMCTF. Numerical experiments for cold-start users and cold-start users/items are provided in Section 4, confirming the useful behavior of WNMCTF over WNMCF. Conclusions are drawn in Section 5.

2   Weighted Nonnegative Matrix Co-Factorization

The classical *learning from data* have been focusing on a feature-based representation of data, which is usually given by a data-by-feature matrix. However, in many practical situations, the data is given by a set of matrices representing the pairwise relations between multiple objects in the domain. The natural way to model this kind of *relational data* is using the entity-relationship model, which represents the data as the relationships between the entities. For example, in the collaborative prediction of movie ratings, some examples of entity types are user, movie, genre, actor, user age and occupation. The exemplary relationships between the entity types are user's ratings of movies, actor's roles in movies, and user's employments in occupations. The set of all n number of entity types in consideration is denoted by $\mathcal{E} = \{e_1, e_2, ..., e_n\}$, where each element e_i represents the i-th entity type. We assume that the binary relationship between two entity types e_i and e_j is given by the form of a nonnegative matrix $\boldsymbol{X}_{(i,j)} \in \mathcal{X}$, where

$\mathcal{X}$ is the set of all given input matrices. Then, the set of all pairwise relations can be defined as $\mathcal{R} = \{(i,j)|\text{if } \boldsymbol{X}_{(i,j)} \text{ exists in } \mathcal{X}\}$.

Basically, WNMCF factors each nonnegative relation matrix $\boldsymbol{X}_{(i,j)}$ with the product of the nonnegative basis matrices $\boldsymbol{U}_{(i)}$ of entity type e_i and $\boldsymbol{U}_{(j)}$ of e_j. That is, we find the matrices $\boldsymbol{U}_{(i)}$ and $\boldsymbol{U}_{(j)}$ such that

$$\boldsymbol{X}_{(i,j)} = \boldsymbol{U}_{(i)}\boldsymbol{U}_{(j)}^{\top}, \text{ for all } (i,j) \in \mathcal{R}. \tag{1}$$

We share the same basis matrix $\boldsymbol{U}_{(i)}$ for all the pairwise relationships $(i,\cdot) \in \mathcal{R}$, that the corresponding entity type is involved in. If we denote the number of instances in the entity type e_i as $|e_i|$, the input matrix $\boldsymbol{X}_{(i,j)} \in \mathbb{R}_+^{|e_i| \times |e_j|}$. The dimensions of factor matrices are determined by the pre-specified number of basis $|b|$, such that $\boldsymbol{U}_{(i)} \in \mathbb{R}_+^{|e_i| \times |b|}$.

The probabilistic interpretation of the factorization (1) is based on the following parametrization of the joint probability $p(e_i, e_j)$ between the entity types,

$$p(e_i, e_j) = \sum_b p(e_i|b)p(e_j|b)p(b).$$

By applying appropriate normalization for the matrices in (1) [28], the matrix $\boldsymbol{X}_{(i,j)}$ corresponds to the joint probability $p(e_i, e_j)$ of two entity types, and each basis matrix $\boldsymbol{U}_{(i)}$ can be interpreted as the conditional probability $p(e_i|b)$.

The input matrix $\boldsymbol{X}_{(i,j)}$ might have missing (unobserved) values in it. To jointly factorize all such $\boldsymbol{X}_{(i,j)}$ into the form of (1), we use the objective function based on the weighted divergence, which can be formulated as

$$\mathcal{L} = \frac{1}{2} \sum_{(i,j)\in\mathcal{R}} \alpha_{(i,j)} \left\| \boldsymbol{W}_{(i,j)} \odot \left(\boldsymbol{X}_{(i,j)} - \boldsymbol{U}_{(i)}\boldsymbol{U}_{(j)}^{\top} \right) \right\|_F^2 \tag{2}$$

where $\|\cdot\|_F$ denotes the Frobenius norm, $\odot$ represents the Hadamard product (element-wise multiplication). $\boldsymbol{W}_{(i,j)}$ is the weight matrix consists of the value one for the observed input values in $\boldsymbol{X}_{(i,j)}$, and the value zero for the unobserved ones. $\alpha_{(i,j)}$'s are the scalar parameters adjusting different scales between the data matrices. The above objective function represents the difference between the observed input data and reconstructed data.

The derivative of the objective function with respect to the basis matrix $\boldsymbol{U}_{(i)}$ can be calculated as

$$\nabla_{U_{(i)}} \mathcal{L} = - \sum_{(i,j)\in\mathcal{R}} \left(\widehat{\boldsymbol{W}}_{(i,j)} \odot \boldsymbol{X}_{(i,j)} \right) \boldsymbol{U}_{(j)}$$

$$+ \sum_{(i,j)\in\mathcal{R}} \left(\widehat{\boldsymbol{W}}_{(i,j)} \odot \left(\boldsymbol{U}_{(i)}\boldsymbol{U}_{(j)}^{\top} \right) \right) \boldsymbol{U}_{(j)}, \tag{3}$$

where $\widehat{\boldsymbol{W}}_{(i,j)} = \alpha_{(i,j)}\boldsymbol{W}_{(i,j)}$. This gradient of the objective function can be decomposed into the form

$$\nabla\mathcal{L} = \left[\nabla_{U_{(i)}}\mathcal{L}\right]^+ - \left[\nabla_{U_{(i)}}\mathcal{L}\right]^-, \tag{4}$$

where $\left[\nabla_{U_{(i)}}\mathcal{L}\right]^{+} > 0$ and $\left[\nabla_{U_{(i)}}\mathcal{L}\right]^{-} > 0$. From this decomposition, we can construct the multiplicative update for $U_{(i)}$ as

$$U_{(i)} \leftarrow U_{(i)} \odot \frac{\left[\nabla_{U_{(i)}}\mathcal{L}\right]^{-}}{\left[\nabla_{U_{(i)}}\mathcal{L}\right]^{+}}. \tag{5}$$

It can be easily seen that the above multiplicative update preserves the nonnegativity of $U_{(i)}$, while $\nabla_{U_{(i)}}\mathcal{L} = 0$ when the convergence is achieved. Therefore, the multiplicative update rules for the basis matrix $U_{(i)}$ can be derived as follows,

$$U_{(i)} \leftarrow U_{(i)} \odot \frac{\sum_{(i,j)\in\mathcal{R}} \left(\widehat{W}_{(i,j)} \odot X_{(i,j)}\right) U_{(j)}}{\sum_{(i,j)\in\mathcal{R}} \left(\widehat{W}_{(i,j)} \odot U_{(i)}U_{(j)}^{\top}\right) U_{(j)}}. \tag{6}$$

Update rule for $U_{(j)}$ can be derived in the similar way, or we can transpose the relation into $X_{(j,i)} = U_{(j)}U_{(i)}^{\top}$ and apply the above update rule.

3   Weighted Nonnegative Matrix Co-Tri-Factorization

Now we will introduce our method, weighted nonnegative matrix co-tri-factorization model and the iterative update algorithm with consistency constraints. As in the case of WNMCF, WNMCTF jointly factorizes given input matrices, but other than the two-factor decomposition (1), it applies three-factor decomposition model for each input matrix $X_{(i,j)}$,

$$X_{(i,j)} = U_{(i)}S_{(i,j)}U_{(j)}^{\top}, \text{ for all } (i,j) \in \mathcal{R}, \tag{7}$$

where the full matrix $S_{(i,j)}$ is put between the two-factor decomposition model. The new matrix $S_{(i,j)}$ represents the many-to-many relationships between the bases in the matrices $U_{(i)}$ and $U_{(j)}$, that is, a basis in the matrix $U_{(i)}$ can be multiplied to any basis in $U_{(j)}$ through the matrix $S_{(i,j)}$. In the two-factor decomposition model, an i-th basis in $U_{(i)}$ can be multiplied only to the corresponding i-th basis in $U_{(j)}$, so only one-to-one relationship between the basis can be modeled. Moreover, the matrix $S_{(i,j)}$ is not restricted to be a square matrix, therefore, we can set the number of basis for entity types differently. If we denote the number of basis for the entity type e_i as $|b_i|$, then $U_{(i)} \in \mathbb{R}_{+}^{|e_i|\times|b_i|}$, $U_{(j)} \in \mathbb{R}_{+}^{|e_j|\times|b_j|}$ and $S_{(i,j)} \in \mathbb{R}_{+}^{|b_i|\times|b_j|}$. Note that the two-factor model (1) introduced in the previous section can be considered as a special case of this model, where the number of bases $|b_i|$ are restricted to be the same ($|b|$) for all entities, and all matrices $S_{(i,j)}$ are restricted to be square and diagonal.

The probabilistic interpretation of the tri-factorization (7) can be followed by the probability decomposition

$$p(e_i, e_j) = \sum_{b_i, b_j} p(e_i|b_i)p(e_j|b_j)p(b_i, b_j). \tag{8}$$

If appropriately normalized, the joint probability $p(e_i, e_j)$ corresponds to the matrix $\boldsymbol{X}_{(i,j)}$, and the conditional probabilities $p(e_i|b_i)$ and $p(e_j|b_j)$ to $\boldsymbol{U}_{(i)}$ and $\boldsymbol{U}_{(j)}$, respectively. The joint probability between the bases, $p(b_i, b_j)$ corresponds to $\boldsymbol{S}_{(i,j)}$.

To jointly factorize all $\boldsymbol{X}_{(i,j)}$ into the form of (7), we introduce the following objective function,

$$\mathcal{L} = \frac{1}{2} \sum_{(i,j)\in\mathcal{R}} \alpha_{(i,j)} \left\| \boldsymbol{W}_{(i,j)} \odot \left(\boldsymbol{X}_{(i,j)} - \boldsymbol{U}_{(i)}\boldsymbol{S}_{(i,j)}\boldsymbol{U}_{(j)}^{\top} \right) \right\|_F^2$$
$$+ \frac{1}{2} \sum_{i} \sum_{\substack{\{i,j\}\in\mathcal{R} \\ \{i,k\}\in\mathcal{R}\setminus\{i,j\}}} \gamma \left\| \frac{\boldsymbol{S}_{(i,j)}\boldsymbol{1}_{(j)}}{C_{(i,j)}} - \frac{\boldsymbol{S}_{(i,k)}\boldsymbol{1}_{(k)}}{C_{(i,k)}} \right\|^2, \tag{9}$$

where $\boldsymbol{1}_{(i)}$ represents an $|e_i|$-dimensional vector of one's, $\|\cdot\|$ represents the vector 2-norm, and $C_{(i,j)} = \boldsymbol{1}_{(i)}^{\top}\boldsymbol{S}_{(i,j)}\boldsymbol{1}_{(j)}$. We used the braced pair $\{i, j\}$ to represent an un-ordered pair of i and j, that is, (i, j) or (j, i) for the notational brevity. The first term is defined in the same way as WNMCF objective function (2), only with different decomposition model. The second term is introduced to impose the consistency between the interpretation of the resulting marginal probability of the basis, $p(b_i)$. The probability $p(b_i)$ can be obtained implicitly from the resulting matrix $\boldsymbol{S}_{(i,j)}$ by applying appropriate normalization and summing with respect to b_j, that is,

$$\boldsymbol{p}(b_i) = \frac{\boldsymbol{S}_{(i,j)}\boldsymbol{1}_{(j)}}{\boldsymbol{1}_{(i)}^{\top}\boldsymbol{S}_{(i,j)}\boldsymbol{1}_{(j)}}, \tag{10}$$

in the vector form $\boldsymbol{p}(b_i)$ of $p(b_i)$. The marginal probability $p(b_i)$ can be calculated from several $\boldsymbol{S}_{(i,j)}$'s involved with the entity type e_i, but these probabilities are not guaranteed to be consistent. By minimizing the third term, we can reduce the difference between the marginal probabilities computed from different $\boldsymbol{S}_{(i,j)}$'s involved. The parameter γ determines the amount of this consistency constraint. In the objective function, we treat the denominator $\boldsymbol{1}_{(i)}^{\top}\boldsymbol{S}_{(i,j)}\boldsymbol{1}_{(j)}$ as a constant $C_{(i,j)}$, because it only depends on the scale of the input matrix $\boldsymbol{X}_{(i,j)}$ if we apply the normalization of the factor matrices which will be explained in the next section.

The derivative of the objective function with respect to the basis matrix $\boldsymbol{U}_{(i)}$ is

$$\nabla_{U_{(i)}}\mathcal{L} = - \sum_{\{i,j\}\in\mathcal{R}} \left(\widehat{\boldsymbol{W}}_{(i,j)} \odot \boldsymbol{X}_{(i,j)} \right) \boldsymbol{U}_{(j)}\boldsymbol{S}_{(i,j)}^{\top}$$
$$+ \sum_{\{i,j\}\in\mathcal{R}} \left(\widehat{\boldsymbol{W}}_{(i,j)} \odot \left(\boldsymbol{U}_{(i)}\boldsymbol{S}_{(i,j)}\boldsymbol{U}_{(j)}^{\top} \right) \right) \boldsymbol{U}_{(j)}\boldsymbol{S}_{(i,j)}^{\top}, \tag{11}$$

where $\widehat{\boldsymbol{W}}_{(i,j)} = \alpha_{(i,j)}\boldsymbol{W}_{(i,j)}$. The derivative with respect to $\boldsymbol{U}_{(j)}$ can be computed in the similar way.

The derivative for $S_{(i,j)}$ can be computed as follows,

$$
\nabla_{S_{(i,j)}} \mathcal{L} = -U_{(i)}^{\top} \left(\widehat{W}_{(i,j)} \odot X_{(i,j)} \right) U_{(j)}
$$
$$
+U_{(i)}^{\top} \left(\widehat{W}_{(i,j)} \odot \left(U_{(i)} S_{(i,j)} U_{(j)}^{\top} \right) \right) U_{(j)}
$$
$$
+\gamma'(m_{(i|j)} - \overline{m}_{(i)}) + \gamma'(m_{(j|i)} - \overline{m}_{(j)}) \tag{12}
$$

where the parameter $\gamma' = N\gamma$, $m_{(i|j)} = \left(S_{(i,j)} 1_{(j)} 1_{(j)}^{\top} \right) / C_{(i,j)}$ and $\overline{m}_{(i)} = \frac{1}{N} \sum_{(i,k) \in \mathcal{R} \backslash (i,j)} \left((S_{(i,k)} 1_{(k)} 1_{(j)}^{\top}) / C_{(i,k)} \right)$ with number of relations N in which e_i is involved.

The above derivatives can be decomposed as in (4), therefore we can build the multiplicative update rule for the matrices $U_{(i)}$ and $S_{(i,j)}$ by using the similar way to (5). Then the rules become,

$$
U_{(i)} \leftarrow U_{(i)} \odot \frac{\sum_{(i,j) \in \mathcal{R}} \left(\widehat{W}_{(i,j)} \odot X_{(i,j)} \right) U_{(j)} S_{(i,j)}^{\top}}{\sum_{(i,j) \in \mathcal{R}} \left(\widehat{W}_{(i,j)} \odot \left(U_{(i)} S_{(i,j)} U_{(j)}^{\top} \right) \right) U_{(j)} S_{(i,j)}^{\top}} \tag{13}
$$

$$
S_{(i,j)} \leftarrow S_{(i,j)} \odot \frac{U_{(i)}^{\top} \left(\widehat{W}_{(i,j)} \odot X_{(i,j)} \right) U_{(j)} + \gamma \left(\overline{m}_{(i)} + \overline{m}_{(j)} \right)}{U_{(i)}^{\top} \left(\widehat{W}_{(i,j)} \odot \left(U_{(i)} S_{(i,j)} U_{(j)}^{\top} \right) \right) U_{(j)} + \gamma \left(m_{(i|j)} + m_{(j|i)} \right)}, \tag{14}
$$

where we denote the parameter γ' simply as γ for the notational simplicity.

We can prove the convergence of the above algorithm without consistency term by adapting the auxiliary function used in the convergence proof of NMF [2]. The critical part of the auxiliary function for WNMCTF which should be positive semi-definite can be formed as a sum of the positive semi-definite parts of the auxiliary function of NMF, hence becomes positive semi-definite. By vectorizing 3-factor decomposition model, the update rule for S can be formed in a 2-factor decomposition model, so the same auxiliary function method can be applied if we do not concern about the additional consistency term. Although the convergence of the algorithm with consistency term in (14) is not proved in this way, but at least the update rule (13) does not increase the objective function, and the main part of the update rule (14) also does.

3.1   Normalization of Factor Matrices

The two-factor decomposition model (1) aims to learn the basis vectors for the data and encodings for the vectors, from the input matrix consisting of data and corresponding feature values. In this case, we only have to normalize the basis matrix to have certain scale, and the encoding matrix is inversely scaled to maintain the scale of the input matrix. We assume that the input matrix is proportional to the joint probability $p(e_i, e_j)$ by the factor of $C_{(i,j)}$, that is, $X_{(i,j)} = C_{(i,j)} \overline{X}_{(i,j)}$, where $\overline{X}_{(i,j)}$ consists of the probability $p(e_i, e_j)$. The scale

$C_{(i,j)}$ goes into the encoding matrix if we normalize the basis matrix to have the probabilistic scale. In other words, in the two-factor decomposition model, we cannot normalize basis and encoding matrices simultaneously in the same probabilistic scale.

The three-factor decomposition model (7) arose to learn from the dyadic data, that is, the input matrix represents the relations between two sets of objects (entities). Each factor matrix has the same semantic in the domain, and therefore should be normalized in the uniform scale. Uniform normalization of the factor matrices brings clear interpretation of the factor matrices, and is also useful to impose the additional constraints such as orthogonality on the factor matrices. If we apply the three-factor decomposition of the input matrix scaled by $C_{(i,j)}$ from its probabilistic scale, we can get the following normalized results,

$$X_{(i,j)} = U_{(i)} S_{(i,j)} U_{(j)}^{\top} \tag{15}$$

$$= \overline{U}_{(i)} \left(C_{(i,j)} \overline{S}_{(i,j)} \right) \overline{U}_{(j)}^{\top}, \tag{16}$$

where $\overline{U}_{(i)}$ is the normalized matrix from $U_{(i)}$ having the values of $p(e_i|b_i)$, and $\overline{S}_{(i,j)}$ is from $S_{(i,j)}$ having $p(b_i, b_j)$, based on the following scaled probabilistic decomposition,

$$C_{(i,j)} p(e_i, e_j) = \sum_{b_i, b_j} C_{(i,j)} p(e_i|b_i) p(b_i, b_j) p(e_j|b_j) \tag{17}$$

$$= \sum_{b_i, b_j} p(e_i|b_i) \left(C_{(i,j)} p(b_i, b_j) \right) p(e_j|b_j). \tag{18}$$

Therefore, if we normalize the factor matrices to have probabilistic scale, the scale of input matrix can be absorbed in the center matrix $S_{(i,j)}$ as $S(i,j) = C_{(i,j)} \overline{S}_{(i,j)}$. The matrix $S_{(i,j)}$ is involved only in the input matrix $X_{(i,j)}$, so this absorbtion of the scales in the matrix $S_{(i,j)}$ does not conflict with factorizations of other input matrices.

Algorithm outline: WNMCTF

---

1. Initialize $U_{(i)}$'s and $S_{(i,j)}$'s with random positive values.
2. Iterate until convergence
   For each factor matrix $U_{(i)}$,
   (a) Update $U_{(i)}$ using the related matrices (13)
   (b) Normalize the matrix $U_{(i)}$ to the probabilistic scale

   $$U_{(i)} \leftarrow U_{(i)} \left(\text{diag} \left(U_{(i)} \mathbf{1}_{(i)} \right) \right)^{-1}$$

   (c) Update all $S_{\{i,j\}}$'s related to $U_{(i)}$ (14)

---

4 Numerical Experiments

We tested WNMCTF algorithm for the collaborative prediction problems, especially with item and user cold-start cases. We used two different datasets, MovieLens-100K which consists of the movie ratings of 943 users for 1682 movies, and MovieLens-1M which consists of the ratings of 6040 users for 3952 movies. MovieLens dataset is one of the most suitable dataset to test the algorithms because the dataset is packed with additional user and movie information. The preference of the user for a specific movie is rated by the integer score from 1 to 5, and 0 value indicates that the movie is not rated by the user.

User information consists of the demographic information about the user, including age, gender, and occupation. For the MovieLens-100K dataset, we partitioned the age into five groups, under 20, 21 to 31, 31 to 40, 41 to 50, and over 51, and mark the value of 1 for the age group of the user in the 5-dimensional age group vector. The age information of MovieLens-1M dataset is coded in the similar way, but uses seven age groups which have more smaller ranges. The gender is represented in the two dimensional vector. There are 21 occupation categories in both datasets, so we used 21-dimensional indicator vector to represent the occupations. As a result, the demographic information of MovieLens-100K dataset is written in a 943-by-28 dimensional user-demographic information matrix, and MovieLens-1M dataset in a 6040-by-30 dimensional matrix. Movie genre information matrix is built in a similar way. There are 19 genre categories in the MovieLens-100K dataset and 18 categories in the MovieLens-1M dataset, so we construct 1682-by-19 movie-genre matrix for MovieLens-100K dataset and 3952-by-18 matrix for MovieLens-1M dataset.

We have four entity types (user, movie, demographic information, and genre information) represented in the three relationship matrices (user-movie matrix, user-demographic matrix, and movie-genre matrix). We want to predict the rating values for the un-rated values in the user-movie rating matrix, using all of these information. We can reconstruct the input rating matrix as (1) or (7), and the missing target ratings are predicted with the reconstructed values. To evaluate the performance of the algorithm, we used the Mean Absolute Error (MAE) defined as follows,

$$ \text{MAE} = \frac{1}{N} \sum_{i=1}^{N} |r_i - \bar{r}_i|, \qquad (19) $$

where N is the number of held-out (test) data points, r_i is the predicted ranking value for the i-th test point, and $\bar{r}_i$ is the true value for the point.

We perform the test on the three different settings. First one compares the WNMCTF algorithm with or without the additional consistency constraints to show the necessity of the constraints in the algorithm. Second one compares the WNMCTF algorithm with several algorithms, on the different number of given user ratings in the test data set. Zero given data case corresponds to the user cold-start problem, which is no ranking information is given by the user.

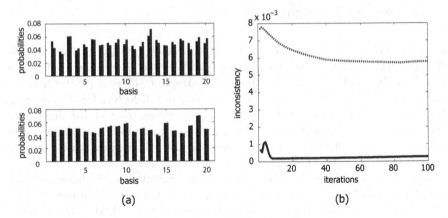

(a)   (b)

Fig. 3. The result of consistency test. (a) Comparison of two marginal probabilities computed from different relationship matrices for 20 basis components. Upper graph compares the results when the consistency constraints are not given by setting the parameter γ value to be zero, and lower graph compares the results when the consistency constraints with the parameter γ value of 10^{-5} are applied. We can obtain fairly consistent result by using the consistency constraints on WNMCTF. (b) The change of mean difference between the marginal probability elements, dotted line shows the case without consistency constraints, solid line shows the case with consistency constraints. After a few iterations, the WNMCTF algorithm with consistency constraints finds fairly consistent results.

In addition to this, the third one eliminates all given rankings for some items, to simulate the situation of item cold-start.

4.1   Consistency of the Results

First we tested the consistency of the results from WNMCTF. For the MovieLens-100K, we used the user-rating matrix and demographic information matrix in addition to the user-movie rating matrix. We set the entity types e_1 to represent user, e_2 to movie, and e_3 to demographic information. We obtained the relationships between the user bases and movie bases $S_{(1,2)}$, and between user bases and demographic information bases $S_{(1,3)}$. We check the difference between the marginal probabilities of the user bases $p(b_1)$ obtained from both matrices $S_{(1,2)}$ and $S_{(1,3)}$. By setting the parameter value of γ in (14) to be zero, we can simulate the case when the consistency constraints are not given. Fig. 3 compares the results of WNMCTF with consistency constraints and without consistency constraints. The average differences between the marginal probability values of each basis are used as a consistency measure. Without consistency constraints, the marginal probabilities appears to be much inconsistent than that with the consistency constraints.

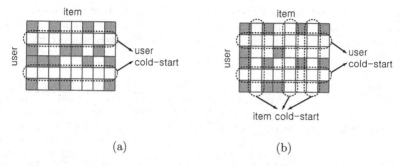

Fig. 4. Pictorial illustration for: (a) cold-start users; (b) cold-start user/items

4.2   User Cold-Start Problem

We tested six algorithms, pure collaborative filtering (PCF), pure content-based prediction (PCB), content-boosted collaborative filtering (CBCF), WNMF, WN-MCF, and WNMCTF for the collaborative prediction problem. PCF is the user neighborhood-based collaborative filtering whose neighborhood size is set to 30, PCB is the naive Bayes classifier for each user trained by the movie genre information, and CBCF is a collaborative filtering method based on the predicted ratings from PCB. The details of these algorithms can be found in [17]. WNMF is implemented as a special case of WNMCF, which uses a single input user-movie rating matrix only. We randomly picked 200 users as a test set, and set some of their ratings to be zero. We tested the algorithms for different number of remaining (given) ratings for each user in the test set. In the extremal setting, no ratings were given from users in the test dataset, which is called the user cold-start case (Fig. 4 (a)). In applying WNMCF and WNMCTF, we use the user demographic information with the typical user-movie rating data to deal with this case.

We measured the MAE for the different numbers of given ratings in test set (Table 1). For each case, we performed the test on the 30 different randomly chosen testsets. As a baseline of the performance, we measured the MAE value when we set the unknown rating values as the mean of the known rating values, and call it MEAN. For the NMF-based algorithms, we used the number of bases as 20, which follows the typical settings [14,29]. To determine the suitable parameter $\alpha_{(1)}$ for the user-movie rating matrix and $\alpha_{(2)}$ for the user-demographic information matrix, we tested the algorithm with several different ratios of α, which were $(\alpha_{(1)}, \alpha_{(2)}) = (1,0), (0.9,0.1), \cdots, (0,1)$. The settings $(0.8,0.2)$ and $(0.9,0.1)$ showed better performance than using only a single input matrix, and we used $(0.9,0.1)$ in our experiments. γ value was also tested on the several scales, and we set γ to be 10^{-5} based on the performances.

WNMCTF showed significantly better performance for all the cases. The complex relationship captured by WNMCTF model was helpful in collaborative prediction problems. PCF, PCB, CBCF and WNMF are failed to generate appropriate predictions in the cold-start cases. PCF requires the mean of the

Table 1. Average and standard deviations of MAE measured for the different number of given ratings per user in the test dataset. The symbol * is used to indicate the cases significantly worse than the best result, based on the Wilcoxon signed-rank test with p-value 0.01.

Dataset	# Given	MEAN	PCF	PCB	CBCF	WNMF	WNMCF	WNMCTF
MovieLens-100K	0	0.9382*	-	-	-	-	0.8472*	**0.8100**
	5	0.9396*	1.0116*	1.3729*	1.3709*	0.8355*	0.8329*	**0.7929**
	10	0.9417*	0.9766*	1.2724*	1.2744*	0.7941*	0.7944*	**0.7661**
	15	0.9447*	0.9614*	1.1608*	1.1748*	0.7685*	0.7682*	**0.7543**
	20	0.9504*	0.9548*	1.0653*	1.0958*	0.7560*	0.7569*	**0.7502**
MovieLens-1M	0	0.9427*	-	-	-	-	0.8099*	**0.7800**
	5	0.9349*	0.9330*	1.4284*	1.4258*	0.8101*	0.8097*	**0.7714**
	10	0.9348*	0.8872*	1.3409*	1.3406*	0.7724*	0.7713*	**0.7452**
	15	0.9348*	0.8697*	1.2601*	1.2649*	0.7556*	0.7544*	**0.7354**
	20	0.9309*	0.8618*	1.1993*	1.2117*	0.7457*	0.7457*	**0.7306**

given ratings, which is impossible to compute when the number of given ratings is zero. PCB fails to build a classifier for the users who have no ratings, and as a result, CBCF cannot predict well for the case. WNMF completely failed to predict the ratings when the given number of ratings is 0, and produces zero values for all the cold-start users. In other cases, PCB brings worse results because the algorithm only uses very coarse content data, which is 18(or 19)-dimensional genre indicator vectors. CBCF performance are degraded because of the poor PCB results. In our experimental settings, the content information itself does not have much information to predict appropriate ratings, but we can achieve good performance by using this information in WNMCTF.

4.3   User and Item Cold-Start Problem

We tested the algorithms on more realistic case, when some movies has no ratings on them, as well as some users (Fig. 4 (b)). To deal with this situation, we jointly used movie genre information with user-movie ratings and user demographic information in WNMCF and WNMCTF.

The MAE values of the results are shown in Table 2. MEAN is the rating scheme that uses the average rating value to predict all unknown ratings. PCF and WNMF used the user-item rating matrix only, PCB and CBCF exploited the genre information to build the content-based predictions, and the WNMCF and WNMCTF co-factorized additional matrices about users and movies. We randomly chose 200 movies in the user-movie rating matrix, and set the ratings for them to be zero. To determine the $\alpha_{(3)}$ value for the movie genre matrix, we performed the similar test with different values for the user-movie rating matrix and movie genre matrix. Based on the performance, we used $(\alpha_{(1)}, \alpha_{(2)}, \alpha_{(3)}) = (0.8, 0.1, 0.1)$ in the experiments. γ is set to be 10^{-5} as in the previous section.

Table 2. Average and standard deviations of MAE measured for the different number of given ratings for test users. We eliminate all ratings for 200 randomly chosen items to simulate item cold-start cases. The symbol * is used to indicate the cases significantly worse than the best result, based on the Wilcoxon signed-rank test with p-value 0.01.

Dataset	# Given	MEAN	PCF	PCB	CBCF	WNMF	WNMCF	WNMCTF
MovieLens-100K	0	0.9442*	-	-	-	-	0.8630*	**0.8273**
	5	0.9401*	0.9902*	1.3790*	1.3766*	0.9024*	0.8460*	**0.8080**
	10	0.9477*	0.9530*	1.2768*	1.2784*	0.8361*	0.8049*	**0.7802**
	15	0.9441*	0.9471*	1.1659*	1.1780*	0.7948*	0.7733*	**0.7607**
	20	0.9431*	0.9425*	1.0784*	1.1088*	0.7708*	0.7579*	**0.7548**
MovieLens-1M	0	0.9348*	-	-	-	-	0.8214*	**0.7916**
	5	0.9348*	0.9257*	1.4419*	1.4390*	0.8527*	0.8144*	**0.7796**
	10	0.9347*	0.8842*	1.3495*	1.3477*	0.8026*	0.7785*	**0.7508**
	15	0.9347*	0.8707*	1.2678*	1.2711*	0.7761*	0.7588*	**0.7392**
	20	0.9309*	0.8589*	1.1878*	1.2003*	0.7616*	0.7483*	**0.7341**

The usefulness of co-factorization in the cold-start cases of collaborative prediction was more clearly shown by this experiment. WNMCF and WNMCTF showed better performance than WNMF in all the cases, and WNMCTF was even better than WNMCF. If more than 15 ratings are given, performances of WNMCF and WNMCTF was not degraded much from the performance showed in the previous section, but WNMF performance was seriously degraded for all the cases. Also in these experiments, PCF, PCB and CBCF cannot produce a good predictions.

5   Conclusions

We have presented WNMCTF algorithm for learning from multiple data matrices, which jointly factorizes each data matrix into the three matrices. WNMCTF can be used as a general framework for the problem having several entities and complex relationships between them. The numerical experiments on the collaborative prediction problem, which uses additional information about users and items, shows the superior performance of WNMCTF, especially on the cold-starting cases.

Acknowledgments. This work was supported by Korea Research Foundation (Grant KRF-2008-313-D00939), Korea Ministry of Knowledge Economy under the ITRC support program supervised by the IITA (IITA-2009-C1090-0902-0045), KOSEF WCU Program (Project No. R31-2008-000-10100-0), and Microsoft Research Asia.

References

1. Lee, D.D., Seung, H.S.: Learning the parts of objects by non-negative matrix factorization. Nature 401, 788–791 (1999)
2. Lee, D.D., Seung, II.S.: Algorithms for non-negative matrix factorization. In: Advances in Neural Information Processing Systems (NIPS). vol. 13. MIT Press, Cambridge (2001)
3. Ding, C., Li, T., Peng, W., Park, H.: Orthogonal nonnegative matrix tri-factorizations for clustering. In: Proceedings of the ACM SIGKDD Conference on Knowledge Discovery and Data Mining (KDD), Philadelphia, PA (2006)
4. Choi, S.: Algorithms for orthogonal nonnegative matrix factorization. In: Proceedings of the International Joint Conference on Neural Networks (IJCNN), Hong Kong (2008)
5. Yoo, J., Choi, S.: Orthogonal nonnegative matrix factorization: Multiplicative updates on Stiefel manifolds. In: Fyfe, C., Kim, D., Lee, S.-Y., Yin, H. (eds.) IDEAL 2008. LNCS, vol. 5326, pp. 140–147. Springer, Heidelberg (2008)
6. Cichocki, A., Lee, H., Kim, Y.D., Choi, S.: Nonnegative matrix factorization with α-divergence. Pattern Recognition Letters 29(9), 1433–1440 (2008)
7. Dhillon, I.S., Sra, S.: Generalized nonnegative matrix approximations with Bregman divergences. In: Advances in Neural Information Processing Systems (NIPS), vol. 18. MIT Press, Cambridge (2006)
8. Kompass, R.: A generalized divergence measure for nonnegative matrix factorization. Neural Computation 19, 780–791 (2007)
9. Yoo, J., Choi, S.: Probabilistic matrix tri-factorization. In: Proceedings of the IEEE International Conference on Acoustics, Speech, and Signal Processing (ICASSP), Taipei, Taiwan (2009)
10. Kim, Y.D., Choi, S.: Nonnegative Tucker decomposition. In: Proceedings of the IEEE CVPR 2007 Workshop on Component Analysis Methods, Minneapolis, Minnesota (2007)
11. Kim, Y.D., Cichocki, A., Choi, S.: Nonnegative Tucker decomposition with alpha-divergence. In: Proceedings of the IEEE International Conference on Acoustics, Speech, and Signal Processing (ICASSP), Las Vegas, Nevada (2008)
12. Srebro, N., Jaakkola, T.: Weighted low-rank approximation. In: Proceedings of the International Conference on Machine Learning (ICML), Washington DC (2003)
13. Rennie, J.D.M., Srebro, N.: Fast maximum margin matrix factorization for collaborative prediction. In: Proceedings of the International Conference on Machine Learning (ICML), Bonn, Germany (2005)
14. Zhang, S., Wang, W., Ford, J., Makedon, F.: Learning from incomplete ratings using non-negative matrix factorization. In: Proceedings of the SIAM International Conference on Data Mining, SDM (2006)
15. Mao, Y., Saul, L.K.: Modeling distances in large-scale networks by matrix factorization. In: Proceedings of the ACM Internet Measurement Conference, Taormina, Sicily, Italy (2004)
16. Kim, Y.D., Choi, S.: Weighted nonnegative matrix factorization. In: Proceedings of the IEEE International Conference on Acoustics, Speech, and Signal Processing (ICASSP), Taipei, Taiwan (2009)
17. Melville, P., Mooney, R.J., Nagarajan, R.: Content-boosted collaborative filtering for improved recommendations. In: Proceedings of the AAAI National Conference on Artificial Intelligence (AAAI), Edmonton, Canada, pp. 187–192 (2002)

18. Popescul, A., Ungar, L.H., Pennock, D.M., Lawrence, S.: Probabilistic models for unified collaborative and content-based recommendation in sparse-data environment. In: Proceedings of the International Conference on Uncertainty in Artificial Intelligence, UAI (2001)
19. Schein, A.I., Popescul, A., Pennock, D.M.: Generative models for cold-start recommendations. In: Proceedings of the SIGIR Workshop on Recommender Systems, New Orleans, LA (2001)
20. Basilico, J., Hofmann, T.: Unifying collaborative and content-based filtering. In: Proceedings of the International Conference on Machine Learning (ICML), Banff, Canada, pp. 65–72 (2004)
21. Park, S.T., Pennock, D., Madani, O., Good, N., DeCoste, D.: Naïve filterbots for robust cold-start recommendations. In: Proceedings of the ACM SIGKDD Conference on Knowledge Discovery and Data Mining (KDD), Philadelphia, PA (2006)
22. Agarwal, D., Merugu, S.: Predictive discrete latent factor models for large scale dyadic data. In: Proceedings of the ACM SIGKDD Conference on Knowledge Discovery and Data Mining (KDD), San Jose, California, USA (2007)
23. Yu, K., Yu, S., Tresp, V.: Multi-label informed latent semantic indexing. In: Proceedings of the ACM SIGIR Conference on Research and Development in Information Retrieval (SIGIR), Salvador, Brazil (2005)
24. Zhu, S., Yu, K., Chi, Y., Gong, Y.: Combining content and link for classification using matrix factorization. In: Proceedings of the ACM SIGIR Conference on Research and Development in Information Retrieval (SIGIR), Amsterdam, The Netherlands (2007)
25. Lee, H., Choi, S.: Group nonnegative matrix factorization for EEG classification. In: Proceedings of the International Conference on Artificial Intelligence and Statistics (AISTATS), Clearwater Beach, Florida (2009)
26. Singh, A.P., Gordon, G.J.: Relational learning via collective matrix factorization. In: Proceedings of the ACM SIGKDD Conference on Knowledge Discovery and Data Mining (KDD), Las Vegas, Nevada (2008)
27. Long, B., Zhang, Z., Yu, P.S.: Co-clustering by block value decomposition. In: Proceedings of the ACM SIGKDD Conference on Knowledge Discovery and Data Mining (KDD), Chicago, IL (2005)
28. Li, T., Ding, C.: The relationships among various nonnegative matrix factorization methods for clustering. In: Proceedings of the IEEE International Conference on Data Mining (ICDM), Hong Kong (2006)
29. Chen, G., Wang, F., Zhang, C.: Collaborative filtering using orthogonal nonnegative tri-factorization. In: Proceedings of the IEEE International Conference on Data Mining (ICDM), Omaha, NE (2007)

Author Index